INERT GASES

HALOGENS

NONMETALS, METALLOIDS

			IIIA	IVA	VA	VIA	VIIA	0
								2 **He** Helium 4.003
			5 **B** Boron 10.811	6 **C** Carbon 12.011	7 **N** Nitrogen 14.007	8 **O** Oxygen 15.999	9 **F** Fluorine 18.998	10 **Ne** Neon 20.183
	IB	IIB	13 **Al** Aluminum 26.982	14 **Si** Silicon 28.086	15 **P** Phosphorus 30.974	16 **S** Sulfur 32.064	17 **Cl** Chlorine 35.453	18 **Ar** Argon 39.948
28 **Ni** Nickel 58.71	29 **Cu** Copper 63.546	30 **Zn** Zinc 65.37	31 **Ga** Gallium 69.72	32 **Ge** Germanium 72.59	33 **As** Arsenic 74.922	34 **Se** Selenium 78.96	35 **Br** Bromine 79.904	36 **Kr** Krypton 83.80
46 **Pd** Palladium 105.4	47 **Ag** Silver 107.868	48 **Cd** Cadmium 112.40	49 **In** Indium 114.82	50 **Sn** Tin 118.69	51 **Sb** Antimony 121.75	52 **Te** Tellurium 127.60	53 **I** Iodine 126.904	54 **Xe** Xenon 131.30
78 **Pt** Platinum 195.09	79 **Au** Gold 196.967	80 **Hg** Mercury 200.59	81 **Tl** Thallium 204.37	82 **Pb** Lead 207.19	83 **Bi** Bismuth 208.98	84 **Po** Polonium 210	85 **At** Astatine 210	86 **Rn** Radon 222

RARE EARTH METALS

63 **Eu** Europium 151.96	64 **Gd** Gadolinium 157.25	65 **Tb** Terbium 158.92	66 **Dy** Dysprosium 162.5	67 **Ho** Holmium 164.93	68 **Er** Erbium 167.26	69 **Tm** Thulium 168.93	70 **Yb** Ytterbium 173.04	71 **Lu** Lutetium 174.97
95 **Am** Americium 243	96 **Cm** Curium 245	97 **Bk** Berkelium 245	98 **Cf** Californium 248	99 **Es** Einsteinium 253	100 **Fm** Fermium 254	101 **Md** Men-delevium 256	102 **No** Nobelium 253	103 **Lw** Lawrencium 257

BASIC CHEMISTRY for the HEALTH SCIENCES

Basic Chemistry for the Health Sciences

Ralph J. Fessenden   **Joan S. Fessenden**
University of Montana

ALLYN AND BACON, INC.
Boston   London   Sydney   Toronto

Library of Congress Cataloging in Publication Data

Fessenden, Ralph J., 1932–
   Basic chemistry for the health sciences.

   Includes index.
   1. Chemistry. I. Fessenden, Joan S. II. Title.
QD31.2.F486 1984   540   83-15373
ISBN 0-205-08016-2

Printed in the United States of America.
10 9 8 7 6 5 4 3 2 1   88 87 86 85 84

Contents

part 2 Organic Chemistry   219

part 3 Important Biological Compounds   325

part 4 Metabolism of Biological Compounds   407

part 5 Special Topics in Biochemistry   463

Preface

Over the past dozen years or so, the standard texts for one- or two-semester general, organic, and biological chemistry courses for students in the allied health sciences have undergone many metamorphoses. Many of the early books either were too rigorous or were not rigorous enough. Today's texts are taking a middle line that is more appropriate for the needs of the students. Most of today's texts are aimed at presenting sufficient chemistry, but not too much, and at explaining how chemical principles apply to the health sciences. We have tried to go one step further—by treating chemistry as a *logical* science upon which many health science applications are based. To achieve this goal, we have occasionally found it necessary to delete classical chemical subject material that bears little relation to the applications presented later in the text. For example, although Boyle's and Charles' laws are presented in optional sections, the *combined* gas laws have not been included.

Organization.   For the most part, the organization for this chemistry course is dictated by necessity: general chemistry is presented first, followed by organic chemistry and biochemistry. Within the general chemistry portion, the general order of subject material is atoms, then bonding, followed by gases, ions, and acid-base chemistry. We place oxidation–reduction in Chapter 11, following ionic chemistry, to build upon the minimal amount of redox information presented with bonding (Chapter 4) and to introduce some simple covalent chemistry before the formal presentation of organic chemistry. Nuclear chemistry (Chapter 12) is placed at the end of the general chemistry portion of the book for two reasons. First, by this time, the student has learned about chemical reactions, chemical equations, and the chemical differences between elements and compounds. Second, many instructors do not attempt to cover nuclear chemistry in this course.

We have attempted to keep the organic chemistry part of the book short. Its main purposes, of course, are to familiarize students with organic structures and to present reactions that will be encountered in biochemistry.

Optional Topics.   It is not easy to teach the large amount of chemistry required for this course in the time allotted, especially in a one-semester course. When time is short, some topics must be deleted from the course presentation. For this reason, we have compartmentalized and marked with a colored box sections that are optional. Study problems pertaining to these sections are starred. Examples of optional topics are the brief introduction to orbitals, the gas laws, and petroleum. Some health-related examples that are only extensions of the chemistry presented are also marked as optional. (The role of *cis,trans* isomerism in vision, in Chapter 15, is an example of this type.) Many of these supplemental topics could be assigned as outside reading. The various options are discussed in greater detail in the instructor's guide.

Health-Related Topics.   Learning chemistry is more enjoyable if applications are used to exemplify theory. To this end, we have tried to use as many health-related and biochemical examples as possible. For example, the body's O_2-CO_2 exchange system is introduced under the topic of partial pressures of gases (Chapter 6), and again in acid-base chemistry (Chapter 10). Digestion is mentioned when the solubilities of organic compounds are discussed (Chapter 13). Phenylketonuria (PKU) is discussed in the section on aromatic substitution in Chapter 15.

Also, we have included many topics that are more important to the applied health sciences than to chemistry itself—for example, the apothecaries' system of units (Chapter 1) and milliequivalents (Chapter 10).

Learning and Teaching Aids.   To help the student organize his or her time and study effectively, we have included brief learning objectives at the beginning of each chapter, a summary and list of key terms at the end of each chapter, and an extensive number of worked-out examples and problems. Important definitions are generally either set off from the text or exemplified by formulas or equations set off from the text. To point out important parts of formulas or equations, we make liberal use of annotations. A glossary of some frequently encountered and important terms has been included.

Problems.   The book includes worked-out examples, in-chapter problems with answers at the end of the book, and end-of-chapter study problems. We have attempted to cover each important point with problems, except where material is purely descriptive. The worked-out examples demonstrate the mathematics to be used and stress the factor-label (dimensional analysis) technique of problem solving.

A review of simple algebraic manipulations (percents, exponents, and the like) is presented in the appendix as a review or reference for the students. The number-line system of metric conversions is also described in the appendix for those instructors who prefer this technique.

Acknowledgments

We are indebted to the reviewers of the manuscript of this book, whose many well-thought-out suggestions were very useful to us in the final development of the manuscript: Professor David L. Anderson of Westbrook College, Professor James W. Barnes of Pearl River Junior College, Professor Sharon L. Coleman of Southeast Missouri State University, Professor Catherine M. Daley of Laboure Junior College, Professor William H. Day of Fresno City College, Professor Ronald E. DiStefano of Northampton County Area Community College, Professor James E. Hardcastle of Texas Woman's University, Professor Richard A. Hendry of Westminster College, Professor Garry McGlaun of Gainesville Junior College, Professor Margaret N. Oglesby of St. Petersburg Junior College, Professor Kenneth O. Pohlman of Lakeland College, Professor Patricia Ann Redden of St. Peter's College, Professor David C. Shroeder of Emporia State University, Professor Charles D. Warner of Hastings College, Professor Joseph G. Watlington of Volunteer State Community College, Professor Beverly Wince of Central Ohio Technical College, and Professor Peter A. Wong of Andrews University.

We also are grateful to our typists: Mary Jeanne Doyle, Paulette Floyd, and Laurie Palmer, all of whom did an excellent job of typing difficult material.

Ralph J. Fessenden
Joan S. Fessenden
University of Montana
Missoula, Montana

PART 1 General Chemistry

1 Science, Chemistry, and Measurements

The Alchemist. (*Courtesy of Dr. Alfred Bader, The Aldrich Chemical Company.*) *Chemistry is the study of matter and its changes. In the Middle Ages, virtually the only people studying matter were the alchemists, who searched for a magical substance that would cure all disease and turn lead into gold. Because the alchemists often recorded their findings in mystical codes, the chemists who followed them had no prior scientific reports to guide their research. Even though some of the important principles of chemistry were discovered in the 1700s, chemistry did not become a modern science until late in the nineteenth century.*

Objectives   Define chemistry and its subdivisions. ☐ List the important features of the scientific method. ☐ Differentiate hypothesis, theory, and scientific law. ☐ Define the prefixes and basic units in the metric system. ☐ Define mass and weight. ☐ List some common units in the apothecaries' system. ☐ Convert units from one system to another. ☐ Define density and specific gravity. ☐ Convert temperatures between °C and °F.

During the twentieth century, tremendous strides have been made in our understanding of the universe, the earth on which we live, and ourselves. We have been to the moon. We can cure some kinds of cancer and control, to an extent, heart disease. We can only speculate as to what new knowledge the closing years of the century will bring.

Many of the giant strides in scientific progress are a direct result of research in **chemistry**: the study of the composition of substances and changes in their composition. Because of the great diversity of things and creatures in our universe and because problems can be approached by different routes, the area of study called chemistry has been subdivided into areas of study and research.

Inorganic chemistry is the study of substances from materials that have never been alive: minerals, metals, and so forth. Historically, *organic chemistry* was the study of substances that came from living things—substances such as petroleum and extracts from plants. However, organic chemistry today embraces a host of drugs, plastics, and other substances that have never been part of a living organism but are synthesized in the laboratory. A better definition of organic chemistry might be the study of substances that contain the element *carbon* in combination with itself and other elements. Similarly, *inorganic chemistry* might be defined as the study of substances that do *not* contain carbon. We find that these definitions also have limitations in that limestone and some other inorganic materials contain carbon.

Biochemistry is the study of substances found in living things and how these substances interact within the living organism. Studies of the composition of viruses or the chemistry of DNA (which carries our genetic code) are examples of biochemical research.

Physical chemistry is the application of the principles of physics to chemistry. What happens to substances at extremely low temperatures? What are the effects of expansion on a gas? How much energy can be obtained from a gram of uranium in a nuclear reactor? These are some topics of physical chemistry.

The boundaries between the subdivisions of chemistry are not sharp. For example, the ingestion of table salt (sodium chloride, an *inorganic* compound) can affect how the body cells function (a *biochemical* topic). As you progress through your study of chemistry, you will notice that the classical subdivisions of chemistry seem to be intertwined, as indeed they are.

In Part 1, we will discuss general chemistry, focusing primarily on the principles of physical and inorganic chemistry. Part 2, on organic chemistry, will provide a necessary base for Part 3, which deals with biochemistry.

In this introductory chapter, we discuss the role of experimentation and observation in scientific progress, and we introduce the *metric system*, which is the system of measurement used in science throughout the world. *Density*, *specific gravity*, and *temperature* will also be described.

1.1   THE SCIENTIFIC METHOD

One reason for scientific advances has been the gradual change over the centuries in the approach to scientific problems. People in some ancient cultures expended a great amount of thought and contemplation on the workings of the universe, the world, and themselves. Today's technological culture combines *observations* and *experimentation* with contemplation. Instead of viewing disease as an inevitable part of living or as some "divine retribution," modern researchers try to determine what causes a disease (a microorganism? a vitamin deficiency? something else?) and what can be done to cure the disease or, better, prevent it. Approaching problems in a systematic manner, using observations and experiments, has been so successful in furthering our knowledge that this approach has been named **the scientific method**.

Of course, not all advances in science have been made by carefully organized research programs. The antibiotic activity of penicillin, for example, was discovered quite by accident in 1928. The Englishman Alexander Fleming was studying variants of *Staphylococcus* bacteria when he observed that a contaminant in his cultures (a variety of *Penicillium* mold) inhibited the growth of his cultures. Unfortunately, it was another twelve years before this observation was put to practical, life-saving use. Albert Einstein formulated his equations of mass, energy, space, and time relationships by brilliant intuition and not by routine data gathering. (However, later observations are still being used to test Einstein's ideas.) Therefore, imagination, observation, and intuition might well be added to disciplined thinking as techniques to use when approaching a scientific problem.

Hypotheses and Theories

Research leads to ideas, or hypotheses. *Hypothesis* is a Greek word from *hypo-*, under, and *tithenai*, to place. A **hypothesis** is a supposition, or the groundwork, used to explain certain facts and to guide further investigation. When a hypothesis has been tested repeatedly and judged to be sound, it may graduate to being called a theory. (We will shortly encounter the atomic theory.) Whereas a hypothesis is an idea, a **theory** is an explanation accepted by most scientists because it is supported by considerable evidence.

In any scientific text, you will also find statements called scientific laws. A

scientific law is a generalization that describes the actual behavior of some aspect of the physical universe. Newton's *law of gravity* is an example.

1.2   SYSTEMS OF WEIGHTS AND MEASURES

In any science-related field, such as chemistry or pharmacy, we need to be able to measure things. In the United States, the system of measurement commonly used is the **English system**. The units in this system are foot, mile, pound, quart, and so forth. The principal disadvantage of the English system is that conversions between units are cumbersome. How long would it take you to determine the number of inches in $5\frac{1}{4}$ miles? On the other hand, the monetary system used in the U.S. is based on the decimal system. In a decimal system, conversions are easy. You would not need much time to calculate the number of pennies in $5.25.

The system of measurement used in most scientific work is the **metric system** or a modern version of the metric system called the **International System of Units** (abbreviated *SI*, from its French name *Le Système International*). Like the dollar system, the metric system is based upon decimals; therefore, conversions between units are easily performed.

In medicine and pharmacy, another system of weights and measures, the **apothecaries' system**, is still in use. We will briefly discuss this system later in this section.

The Metric System

Prefixes Used in the Metric System.   In the metric system, the root names of the basic units of measure are the *gram* (weight), *meter* (length), and *liter* (volume). Prefixes are used with these basic units to designate the other units of measure: for example, kilogram, milligram, and deciliter.

The common prefixes used in the metric system are listed in Table 1.1. These prefixes are interchangeable. For example, *milli-* means "one thousandth." One millimeter is 1/1000 of a meter, while one milliliter is 1/1000 of a liter. In Table 1.1, we have expressed the meanings of the prefixes as fractions,

Table 1.1.   *Some Common Prefixes Used in the Metric System*

Prefix	Fraction	Decimal	Exponent
deci- (d)	1/10	0.1	1×10^{-1}
centi- (c)	1/100	0.01	1×10^{-2}
milli- (m)	1/1000	0.001	1×10^{-3}
micro- (μ)	1/1,000,000	0.000001	1×10^{-6}
nano- (n)	1/1,000,000,000	0.000000001	1×10^{-9}
kilo- (k)	1000	1000	1×10^{3}

as decimals, and as exponents to the base 10. (A detailed explanation of exponents is found in the appendix.)

Mass.   In everyday usage, the terms "weight" and "mass" are used interchangeably; however, there is a difference between the two. The **mass** of an object is defined as the amount of matter in that object. (Theoretically, mass is a measure of the *inertia* of the object—its resistance to being moved or its resistance to a change in motion once it is moving.) **Weight** is the measure of an object's mass under the influence of gravity. The mass of an astronaut who goes to the moon does not change. However, the astronaut's weight on the moon is only about one-fifth what it would be on earth.

The scientist, who deals with measured quantities of matter, is interested, therefore, in the mass of that matter, rather than its weight, which is a variable measure. However, since most of our chemistry is carried out on earth, if a scientist says "weight" while actually meaning "mass," the error is forgiveable.

The basic unit of mass in the metric system is the *gram* (abbreviated g). One thousand grams is a *kilogram* (kg). There are 454 grams in 1 pound (lb); therefore, 1 kilogram contains 2.2 pounds. If scientists are dealing in small quantities of a substance, they may report the mass in *milligrams* (mg). One milligram is one-thousandth of a gram. Table 1.2 lists some common units of mass in the metric and English systems.

Table 1.2.   *Some Common Units of Mass in the Metric System*

Unit	Metric definition	English-system equivalent
gram (g)	—	0.03527 ounce (oz)
milligram (mg)	0.001 g	—
microgram (μg)	0.000001 g	—
kilogram (kg)	1000 g	2.205 pound (lb)

While most drugs are dispensed by pharmacists or doctors in milligram quantities, some are dispensed in *grains* (0.0648 gram). Grain is an apothecaries'-system unit; it is not used by most chemists. However, there is a possibility of confusion between gram (g) and grain (gr) by those involved in the medical sciences.

Length.   The basic unit of length in the metric system is the *meter*, abbreviated m. A meter is just a little longer than a yard. One of the track events in the Olympic games is the 100-meter dash. U.S. athletes use the 110-yard dash to train for this event. One-hundredth of a meter is a *centimeter*, abbreviated cm. There are 2.54 centimeters to an inch. One-tenth of a centimeter (one-thousandth of a meter) is called a *millimeter*, abbreviated mm. If we need to use a unit larger than the meter, we use the *kilometer* (km), which is 1000

Table 1.3.   *Some Common Units of Length in the Metric System*

Unit	Metric definition	English-system equivalent
meter (m)	1.0 m	39.37 inches (in.)
centimeter (cm)	0.01 m	0.3937 in.
millimeter (mm)	0.001 m	0.03937 in.
micrometer (μm), or micron (μ)	0.000001 m	—
kilometer (km)	1000 m	0.6214 mile (mi.)

meters. The common units of length are listed in Table 1.3, and some comparisons between the metric and English systems are shown in Figure 1.1.

Volume.   Volume is not truly a fundamental measure because it is based on length. The volume of a cube that has sides 1 cm long is 1 cubic centimeter. The *cubic centimeter* (abbreviated cc, cu cm, or cm^3) is a basic unit of volume in the metric system. In liquid measurements, the cubic centimeter is also referred to as a *milliliter* (mL):

$$1 \text{ cubic centimeter (cc)} = 1 \text{ milliliter (mL)}$$

One milliliter is one-thousandth of a *liter* (L or l). A liter is just slightly larger than a quart (qt). Table 1.4 summarizes the metric units for volume; some comparisons between the metric and English units are shown in Figure 1.2.

The Apothecaries' System

The English system of weights and measures commonly used in the United States, called the *avoirdupois system* (avdp.), is based on 16 avdp. oz to 1 avdp.

Figure 1.1   *Comparisons between the metric- and English-system measures of length.*

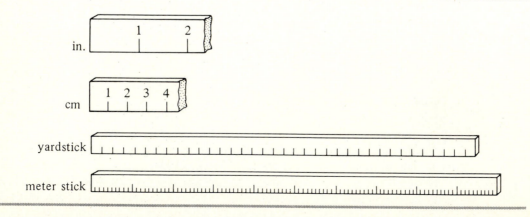

Table 1.4.   *Some Common Units of Volume in the Metric System*

Unit	Metric definition	English-system equivalent
liter (L or l)	1000 cc, or 1000 mL	1.057 quarts (qt)
milliliter (mL or ml)	1 cc, or 0.001 L	0.0338 fluid ounce (fl oz)
microliter (μL or μl)	1×10^{-6} L	—

lb. The *apothecaries' system* (ap.) is different; it is based on 12 ap. oz (slightly larger than avdp. oz) to 1 ap. lb. The apothecaries' system also uses some different units from those in the other systems. Table 1.5 lists some of the

Table 1.5.   *Relationships Within the Apothecaries' System*

weight:
    60 grains (gr) = 1 dram (dr or ℨ)
    8 drams   = 1 ap. ounce (ap. oz)[a]
    12 ap. ounces = 1 ap. pound (ap. lb)[a]

liquid measure:
    60 minims (♏) = 1 fluid dram (fl dr or f ℨ)
    8 fluid drams = 1 fluid ounce (fl oz)[a]

[a] The *fluid ounce* is the same volume in the apothecaries' and the avoirdupois systems, but the *ounce as a unit of mass* differs in the two systems (as, therefore, does the pound). (See Table 1.6.)

Figure 1.2.   *Comparisons between the metric- and English-system measures of volume.*

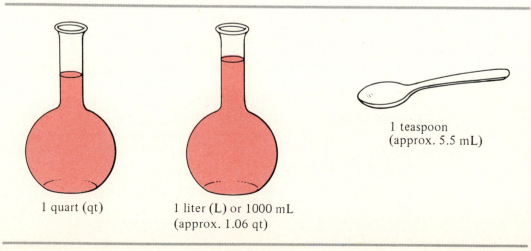

1 quart (qt)

1 liter (L) or 1000 mL
(approx. 1.06 qt)

1 teaspoon
(approx. 5.5 mL)

common units and their symbols, while Table 1.6 shows some relationships between the apothecaries' system and the other systems.

Table 1.6.   *Selected Values for Conversions Between the Apothecaries' System and the Metric or English System*

Apothecaries' unit	Metric equivalent	English (avdp.) equivalent
weight:		
1 grain (gr)	0.0648 g	0.00229 avdp. oz
1 dram (dr or ℨ)	3.887 g	0.137 avdp. oz
1 ap. ounce (ap. oz)	31.10 g	1.097 avdp. oz
volume:		
1 minim (♏)	0.0616 mL	0.00208 fl oz
1 fluid dram (fl dr or *f*ℨ)	3.697 mL	0.125 fl oz
1 fluid ounce (fl oz)	29.57 mL	1 fl oz

1.3   CONVERTING ONE UNIT TO ANOTHER

Because all the systems of weights and measures that we have presented are encountered in medical, pharmaceutical, and related sciences, we need to be able to change milligrams to grains and to perform other conversions. Conversions between metric units or between metric and English units are easily accomplished if we set up equations and *include the units*. If an equation is set up correctly, all unwanted units will cancel, resulting in the correct answer with the correct units. This method of converting units and solving problems is called the **factor-label technique**.

Values to be used in conversion problems can be found in Tables 1.2–1.6. Any equality from these tables can be used as a *conversion factor*. For example, if 1.0 mL = 0.0338 fl oz, then either of the following conversion factors could be used to solve a problem, depending on the information given and the information desired.

$$\frac{1.0 \text{ mL}}{0.0338 \text{ fl oz}} \quad \text{or} \quad \frac{0.0338 \text{ fl oz}}{1.0 \text{ mL}}$$

To make a conversion, we designate the desired information as x. Then, we multiply the given information by the appropriate conversion factor. The conversion factor is chosen so that all unwanted units on the right-hand side of the equation (as we have shown it) cancel.

desired number x = given number × conversion factor

Example   Convert 200 mL to fl oz.

Solution   Setting up the equation,

desired   *given*   *conversion factor*

$$x \text{ fl oz} = 200 \text{ mL} \times \left(\frac{0.0338 \text{ fl oz}}{1.0 \text{ mL}} \right)$$

Canceling units,

$$x \text{ fl oz} = 200 \times 0.0338 \text{ fl oz}$$

$$= 6.76 \text{ fl oz}$$

Example   Convert 25.0 mg to grams.

Solution   From Table 1.2, 1 mg = 0.001 g; therefore,

$$x \text{ g} = 25.0 \text{ mg} \times \frac{0.001 \text{ g}}{1 \text{ mg}}$$

$$= 0.025 \text{ g}$$

An alternative method for conversions within the metric system is described in the appendix.

Problem 1.1.   A standard aspirin tablet contains 5.0 grains of aspirin. How many milligrams of aspirin are in an aspirin tablet? (Refer to Table 1.6.)

Problem 1.2.   If directions specify that 1.0 oz (avdp.) of a powder is to be dissolved in 2.0 cups of water, calculate how many grams of the powder and how many milliliters of water should be used.

1.4   DENSITY AND SPECIFIC GRAVITY

Density is defined as *mass per unit volume*, such as g/cc. For example, a cubic centimeter of lead weighs 11.3 g; therefore, the density of lead is 11.3 g/cc. The densities of a few solid materials are found in Table 1.7.

$$\text{density} = \frac{\text{mass in g}}{\text{volume in cc}}$$

Closely related to density is specific gravity, which is often used for liquids. The **specific gravity** of a liquid is the mass of that liquid divided by the mass of an equal volume of water.

$$\text{specific gravity} = \frac{\text{mass of a substance in g}}{\text{mass of the same volume of water in g}}$$

Table 1.7.   *Densities of Some Common Solids*

Substance	Density in g/cc at 20°C
cork	0.22–0.26
aluminum	2.70
iron	7.86
silver	10.5
lead	11.3
sodium chloride[a]	2.16[b]
sucrose[c]	1.58[b]

[a] Common table salt.
[b] Density of a solid crystal containing no air.
[c] Table sugar.

Because the mass of 1.0 mL of water is 1.0 g, the numerical value of the specific gravity of a substance is the same as its density. Specific gravity is a unitless quantity, however, because the units cancel in the division. The values of both density and specific gravity vary with temperature because the volumes of liquids and solids change somewhat with temperature (although their masses do not change). Therefore, the densities or specific gravities of different substances should be compared at the same temperature. Table 1.8 lists some typical specific gravities.

Table 1.8.   *Specific   Gravities   of   Some Common Liquids*

Substance	Specific gravity at 20°C
gasoline	0.66–0.69
ethanol	0.79
olive oil	0.91
pure water	1.00
sea water	1.025
urine	1.003–1.030[a]
blood	1.054–1.060[a]
mercury metal	13.6

[a] Variable, depending on the quantity of dissolved substances.

Example   The density of a sample of urine at 20°C (room temperature) is 1.015 g/mL. What is its specific gravity?

Solution   If

$$\text{mass of 1.0 mL urine} = 1.015 \text{ g}$$

and

$$\text{mass of 1.0 mL water} = 1.000 \text{ g}$$

then

$$\text{specific gravity} = \frac{1.015 \ \cancel{\text{g/mL}}}{1.000 \ \cancel{\text{g/mL}}}$$

$$= 1.015$$

Specific gravities are conveniently measured with a **hydrometer**. There are many types of hydrometers available. Service station mechanics use one type to test the strength of battery acid and another to check the antifreeze in the radiator. Hydrometers called *urinometers* are used to check the specific gravities of urine specimens. An abnormal specific gravity may be an indication of some pathological condition. For example, a higher than normal specific gravity may be an indication of sugar in the urine, a condition that can arise from *diabetes mellitus*. A lower than normal specific gravity may indicate high fluid intake (diluting the urine more than usual) or *diabetes insipidus*, a rare condition in which the kidneys produce large quantities of very dilute urine.

The principle behind a hydrometer is that it sinks just enough to displace liquid equal to its mass. The less dense the liquid, the farther the hydrometer will sink. (See Figure 1.3.)

Figure 1.3.   *A hydrometer.*

scale on stem

float to keep
the tube
upright

lead weight
to provide the
proper mass

in a less dense
liquid

in a more dense
liquid

1.5   MEASURING TEMPERATURES

The **temperature** of an object is a measure of its heat content. In the metric system, the most commonly used temperature scale is the *Celsius*, or *centigrade*, *scale* (abbreviated °C). In the English system, the *Fahrenheit scale* (°F) is used. A third temperature scale, called the *Kelvin scale* (K), will be discussed in Chapter 6.

In the Celsius scale, the freezing point of water is set at 0°C and the boiling point of water at 100°C. Therefore, one degree on the Celsius scale is $\frac{1}{100}$ of the temperature change between these two reference points.

The Fahrenheit scale was designed by the German physicist G. D. Fahrenheit about 1700. At that time, there were no accurate measuring devices. The two reference points he chose were body temperature, which he set at 100°F, and the temperature of an ice-salt bath, which he set at 0°F. Unfortunately, he missed the true temperatures slightly at both ends. The Fahrenheit scale has 180 divisions between the freezing point of water (32°F) and the boiling point of water (212°F). (See Figure 1.4.)

Temperature Conversions

When conversions between the Celsius and Fahrenheit scales are necessary, the following equations are useful:

$$°F = 1.8(°C) + 32°   °C = 0.56(°F - 32°)$$

In the first equation, the number 1.8 relates the size of a °C to that of a °F. (The °C is $\frac{180}{100}$, or 1.8, larger.) The term $+32°$ is the difference that the two scales are offset from each other. The second equation is derived similarly.

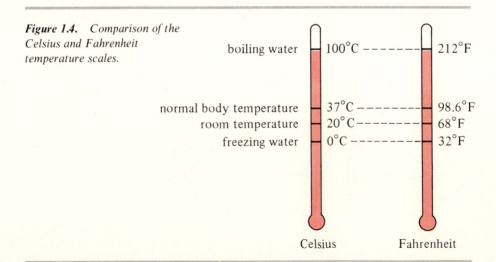

Figure 1.4.   *Comparison of the Celsius and Fahrenheit temperature scales.*

boiling water — 100°C – – – – – – – 212°F

normal body temperature — 37°C – – – – – – – 98.6°F
room temperature — 20°C – – – – – – – 68°F
freezing water — 0°C – – – – – – – 32°F

Celsius   Fahrenheit

The following examples show how the two equations may be used to convert a temperature from one scale to the other.

Example   Convert 30°C to °F.

Solution   We substitute in the first equation:

$$°F = 1.8(30°) + 32°$$
$$= 54° + 32°$$
$$= 86°$$

Example   Convert 65°F to °C.

Solution   We substitute in the second equation:

$$°C = 0.56(65° - 32°)$$
$$= 0.56(33°)$$
$$= 18°$$

Problem 1.3.   Make the following temperature conversions:
(a) 50°F to °C   (b) 50°C to °F

Problem 1.4.   If a patient's temperature is found to be 104.3°F, what is the patient's temperature in °C?

SUMMARY

Chemistry, the study of the composition of substances and changes in their composition, is conveniently subdivided into inorganic, organic, bio-, and physical chemistry.

The *scientific method* of approaching a problem is a systematic, experimental approach of testing *hypotheses* (suppositions) and leads to *theories* (tested explanations).

Scientists use the *metric system* of measurement, which is a decimal system. Some prefixes used in these systems are shown in Table 1.1. Units of length are based on the *meter*, units of mass on the *gram*, and units of volume on the *liter*. The *apothecaries' system* (ap.) is similar to the English *avoirdupois system* (avdp.), but is based upon 12 ap. oz to the ap. lb. (See Tables 1.5 and 1.6.) Conversions between units can be accomplished by the *factor-label technique*.

The *density* of a substance is its mass per volume (g/cc), while the *specific gravity* is the mass per volume divided by the mass of the same volume of water. The numerical values of density and specific gravity are the same. A *hydrometer* is a device used to measure specific gravity.

The temperature scale in the metric system is the *Celsius scale.* The following equations are used to convert °C and °F:

$$°F = 1.8(°C) + 32° \qquad °C = 0.56(°F - 32°)$$

KEY TERMS

chemistry	metric system	density
scientific method	apothecaries' system	specific gravity
hypothesis	gram	Celsius scale
theory	meter	
English system	liter	

STUDY PROBLEMS

1.5.   Identify each of the following statements as a hypothesis, a theory, or a law:

(a) The amount of matter and energy in the universe is fixed, although there may be some interconversion.

(b) In all probability, there are other planets besides Earth in the universe that are capable of supporting life.

1.6.   Make the following conversions within the metric system:

(a) 352 mg to grams

(b) 5.0 mg to micrograms

(c) 581 g to kilograms

(d) 2.3 kg to grams

1.7.   Make the following metric-system conversions:

(a) 5.0 cm to meters

(b) 5.0 cm to millimeters

(c) 5.0 cc to milliliters

(d) 0.25 L to milliliters

(e) 380 mL to liters

(f) 53 μL to milliliters

1.8.   Referring to the tables in this chapter, make the following conversions:

(a) 3.0 in. to centimeters

(b) 250 g to pounds

(c) 5.0 lb to kilograms

(d) 1.0 cup ($\frac{1}{4}$ qt) to liters

(e) 250 mg to grains

(f) 15.0 mL to fluid drams

1.9.   Calculate your own weight and height in kilograms and centimeters, respectively.

1.10.   A liquid medication is to be administered orally as 0.10 mL per kilogram of body weight. Approximately how many teaspoons of the medication should be given to a patient who weighs 120 lb?

1.11.   Using Table 1.8, determine the approximate mass of 2.10 L of (a) sea water, (b) ethanol, (c) pure water, and (d) mercury.

1.12.   If 250 mL of urine weighs 262 g at 20°C, what is the specific gravity of the urine?

1.13.   A metal reputed to be pure silver displaces 3.00 mL of water and weighs 33.9 g. Is the metal pure silver?

1.14.   A home brewer of beer finds that two tubs of malt–sugar–water mixture have different specific gravities of 1.03 and 1.07. Which of the two tubs will likely yield the beer of higher alcohol content?

1.15.   Listed below are the boiling points of a few solvents. Express these boiling points in °F or °C, whichever is not given.

(a) water, 100°C   (b) ethyl ether, 95°F

(c) ethanol, 174°F   (d) gasoline, 39°–204°C

1.16.   A patient's temperature is reported as 38.5°C. What is the patient's temperature in °F?

2 Matter, Energy, and Physical Change

A laboratory distillation apparatus. We distill a liquid by boiling it and collecting the condensed vapors, called the distillate. Distillation is used to purify a liquid or to separate a mixture of liquids with different boiling points. For example, distillation is used to separate ethanol (beverage alcohol) from fermentation mixtures. Moonshiners—people who make illegal alcoholic beverages—sometimes use old automobile radiators as condensers to cool and liquefy the ethanol vapors. This practice can cause lead poisoning in the consumers of the distilled beverage.

Objectives   Define and list examples of matter and energy. □ State the law of conservation of matter and energy. □ Define molecules and atoms. □ Define element, compound, and mixture. □ State the law of definite proportions. □ Define the three states of matter and describe how they differ in their molecular attractions. □ Differentiate between physical and chemical changes and between physical and chemical properties. □ Define heat, calorie, heat capacity, specific heat, heat of fusion, and heat of vaporization; calculate the number of calories needed for changes in temperature and state.

The world around us and our own bodies are composed of physical substances whose properties are controlled, in part, by energy. For example, the radiant energy of the sun is responsible for light and warmth on the earth and thus is responsible for all life. In this chapter, we will discuss matter and its composition, some different types of matter (elements, compounds, mixtures), and physical changes of matter.

2.1   MATTER AND ENERGY

Chemistry encompasses the study of matter and energy and their relationships. **Matter** is something that has mass and occupies space. We can see and feel a stone; it is matter. Water is also matter, and so is the air around us, even though we cannot see it.

Matter can exist in three different forms or states. These **three states of matter**, which we will discuss in Section 2.3, are (1) *solid*, (2) *liquid*, and (3) *gas*. When matter changes from one state to another, we refer to the event as a *change in the state of matter*. When water freezes, it has changed its state from liquid to solid. Matter may be changed from one state to another or even changed into another substance, but only in unusual circumstances (as in the sun or in a nuclear reactor, where mass is converted to energy) is mass actually lost.

Although energy is not quite so easy to perceive as matter, it is real nonetheless. **Energy** is defined as the *capacity to do work or to cause change*. Electricity, heat, and light are familiar forms of energy. Some important general forms of energy are *potential energy* (such as the energy possessed by water behind a dam) and *kinetic energy* (the energy of motion).

Energy, like matter, can be changed from one form to another. Water behind a dam has potential energy. When the water flows through the dam, the potential energy is changed to kinetic energy. Some of this kinetic energy can be used to turn a turbine and generate electricity. The electricity is transported to our homes, where it can be changed to heat and light. In all these processes, energy is neither gained nor lost; the energy is merely changed from one form to another.

Although energy is rarely actually lost, it is often "lost" in a practical sense because of inefficient energy conversions. For example, a light bulb converts electrical energy to light. However, some of the electrical energy used by the light bulb is "lost" as heat. Similarly, the human body can convert the potential chemical energy in nutrients to the energy required for motion, thought, growth, and reproduction; however, during these energy conversions, some of the original energy is lost from the body as heat.

Matter and energy can change in form but cannot be created or lost. This fact has been formulated into a scientific law:

Law of Conservation of Matter and Energy. The total quantity of matter and energy in the universe is fixed, even though there may be some interconversion.

2.2   COMPOSITION OF MATTER

We have defined matter as something that has mass and occupies space. But what is matter made of? Picture a drop of water enlarging until it is the size of the earth. Look very closely at the earth-sized drop. You will see that it is made up of identical particles, each one about the size of a hamburger. These particles are the **molecules** of water. Although a molecule is extremely small, it has mass and occupies space. It is matter. If we summed the masses of all the water molecules in a water drop, the value would equal the mass of the drop.

If we could examine a single molecule of water, we would find that it is composed of three smaller particles, called **atoms**. Two of these atoms are the same and are called hydrogen (H) atoms, while the other atom is called an oxygen (O) atom. The three atoms are bonded together to form the molecule of water. We represent the water molecule as H—O—H, or H_2O.

According to modern atomic theory, *all matter is composed of individual atoms, of groups of atoms bonded together as molecules, or of electrically charged atomic particles called ions.* We observe differences between various types of matter because they are composed of different types of atoms bonded together in different ways. Table salt and water, for example, differ both in the kinds of atoms they contain and in how these atoms are bonded together. We will discuss atoms in more detail in Chapter 3 and how atoms are bonded together as molecules or ions in Chapter 4.

Elements and Compounds

A substance composed of just one kind of atom is called an **element**. In some elements, the smallest individual particles are simply the atoms themselves. Neon gas (Ne), for example, is composed of only neon atoms. Other elements are composed of molecules that are made from only one type of atom. Oxygen gas (O_2), nitrogen gas (N_2), and sulfur (S_8) are examples of this group; a

Figure 2.1. *Elements and compounds.*

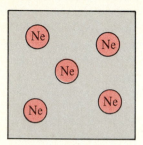

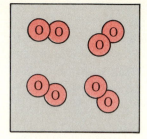

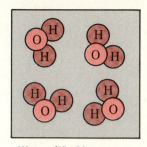

Neon (Ne) is an
element composed
of neon atoms.

Oxygen (O_2) is an
element composed
of oxygen molecules.

Water (H_2O) is a
compound composed
of water molecules.

molecule of oxygen is made from two oxygen atoms, a molecule of nitrogen is
made from two nitrogen atoms, and a molecule of sulfur is made from eight
sulfur atoms.

A **compound**, such as water, is a substance made up of more than one
element; its molecules contain more than one kind of atom, but all its molecules are identical. Figure 2.1 depicts the difference between elements and a
compound.

Compounds and Mixtures

A compound such as water cannot be changed to another compound or
broken into its elements by ordinary physical means. We cannot make hydrogen gas (H_2) and oxygen gas (O_2) by pulling water apart with tweezers or even
by boiling it. (We will find later that a compound can be changed by *chemical*
means, however.)

If we mix pure iron filings and pure sugar, we no longer have a pure
substance, but a **mixture** of two substances. The two components keep their
identities and can be separated by physical means. We could use a pair of
tweezers or a magnet to separate the sugar-iron mixture.

A mixture of iron filings and sugar is an example of a *heterogeneous
mixture*, a mixture that does not have a constant composition. Sugar (sucrose)
can dissolve in water to yield a *homogeneous mixture*, a mixture that has the
same composition throughout, even when viewed through an ordinary microscope (which cannot resolve particles as small as most molecules). Even
though the solution of sugar and water is homogeneous, if we could enlarge a
drop of this solution sufficiently, we would find two types of molecules—sugar
molecules and water molecules (see Figure 2.2). Each kind of molecule has
retained its identity. Again, we have a mixture and could separate the two by
physical means. For example, the sugar could be recovered by boiling away

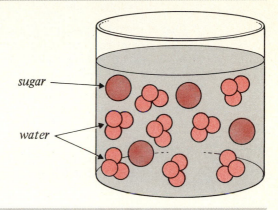

Figure 2.2. *Sugar dissolved in water is a mixture of two types of molecules.*

sugar

water

the water, which is a physical process. Table 2.1 lists a few physical processes used to separate ordinary mixtures.

We can formalize the definitions of compounds and mixtures:

A **compound** cannot be broken down into its elements by physical means.
A **mixture** can be separated into its pure components by physical means.

Table 2.1. *Some Physical Means of Separating Mixtures*

Physical process	Description	Example
extraction	removing a substance from a mixture with a liquid	Brewing coffee is the extraction of caffeine and other compounds from the ground coffee with water.
filtration	removing a solid substance from a liquid by passing the mixture through filter paper	Brewed coffee can be separated from the coffee grounds by being poured through filter paper.
distillation	purifying a liquid by boiling it, then collecting the condensed vapors	Water containing dissolved solid impurities can be purified by distillation.

Problem 2.1. Classify each of the following items as an element, a pure compound, or a mixture:
(a) salt water   (b) iron filings (Fe)
(c) rust (Fe_2O_3)   (d) carbon monoxide (CO)
(e) air

Law of Definite Proportions

One of the distinguishing features of a compound is that it always contains the same weight ratio of its elements. Pure water *always* contains 8.0 g of oxygen for every 1.0 g of hydrogen. The *weight ratio* (not the number of atoms) in water is thus 8.0 parts O to 1.0 part H, or 8 : 1. Pure sodium chloride (table salt, NaCl) always contains 1.54 parts Cl by weight to 1.00 part Na, or 1.54 : 1. In mixtures, by contrast, the weight ratios of the elements need not be constant. Different mixtures of the same components can contain different weight ratios of the elements. For example, we can dissolve 1 g of salt (NaCl) in a pan of water or we can dissolve 10 g. These two mixtures would contain different weight ratios of the elements sodium, chlorine, hydrogen, and oxygen.

Law of Definite Proportions. Any compound has its elements combined in a definite proportion by weight.

Why are the elements in a compound found in definite proportions by weight? The reason is that a pure compound is composed of identical molecules (for example, H_2O molecules in water). Since an oxygen atom has a mass 16 times that of a hydrogen atom, the weight ratio of oxygen to hydrogen in H_2O must be 16 : 2, or 8 : 1. Atomic masses will be discussed in Chapter 3.

2.3   THE THREE STATES OF MATTER

The physical state (solid, liquid, or gas) of a particular bit of matter depends on how strongly its atoms or molecules attract one another and to what extent the atoms or molecules can move around.

A **solid**, such as ice or iron metal, is a substance that maintains its volume and shape without requiring a container because the atoms or molecules are packed closely together, attract one another strongly, and are not free to move about.
A **liquid**, such as liquid water, maintains its volume but can assume the shape of its container because the atoms or molecules are attracted to one another but are free to move about, sliding past one another.
A **gas**, such as water vapor or oxygen in the air, has no definite shape or volume. A gas can expand to fill any sized container because its atoms or molecules are far apart, move about, and have little, if any, attraction for one another.

Figure 2.3 depicts the differences in the particles of a solid, a liquid, and a gas.

Physical Change versus Chemical Change

Water can freeze and become ice. Ice, however, is still a form of water. If we could enlarge ice, we would find that it is composed of the same molecules as

Figure 2.3.   *The differences among the atoms or molecules of solids, liquids, and gases.*

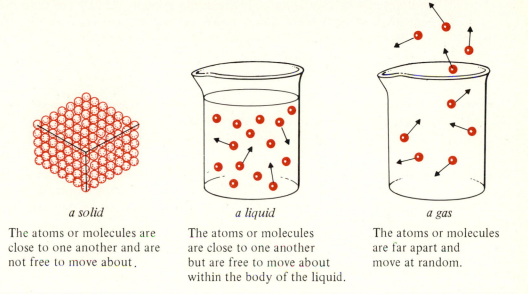

a solid

The atoms or molecules are close to one another and are not free to move about.

a liquid

The atoms or molecules are close to one another but are free to move about within the body of the liquid.

a gas

The atoms or molecules are far apart and move at random.

water. Water can evaporate and exist as a gas that we call water vapor. The gaseous form still contains water molecules. When water is frozen or evaporates, it has changed its state from liquid to solid or liquid to gas. A change in the state of matter is called a **physical change**.

If water (H_2O) were changed to its elements, hydrogen gas (H_2) and oxygen gas (O_2), we would have a **chemical change**. A chemical change involves a change in the structure of a molecule and, therefore, changes in the properties of the substance or substances.

Table 2.2.   *Physical Properties versus Chemical Properties*

Some physical properties	Some chemical properties
melting point	combustibility (ability to burn)
boiling point	
color	ability to react with H_2 to yield a new compound
odor	
ability to conduct an electric current	ability to react with an acid to yield a new compound
specific heat	
density	ability to corrode metals
crystal structure	
solubility in water	

A **physical property** of a substance is any feature we can measure or observe that does not convert the substance to another substance. The temperatures at which water melts and boils are examples of physical properties. A chemical property is the way in which a substance interreacts with another substance to undergo a chemical change. In a **chemical reaction**, substances are transformed into new substances. Table 2.2 lists a few common physical and chemical properties.

2.4   THE EFFECTS OF HEAT ON MATTER

Measuring Heat

Heat is energy supplied to a substance to raise its temperature or cause it to melt or boil. The metric unit commonly used for measuring a quantity of heat is the **calorie** (cal), the amount of heat required to raise the temperature of one gram of water from 14.5°C to 15.5°C. (For practical purposes, one calorie raises the temperature of water one degree at any temperature.)

Probably everyone has heard of Calories in connection with diet. For example, one piece of cheese pizza contains about 185 Calories. But these Calories are spelled with a capital C; they are truly kilocalories, or 1000 true calories. In this book, we use only the true calorie (lower-case c). If we need a larger unit, we will use kilocalories (1000 calories), abbreviated as kcal.

Example   How many calories would be required to raise the temperature of 350 g water from 14°C to 15°C?

Solution   Set up and solve the equation:

$$x \text{ cal} = 350 \text{ g} \times \left(\frac{1.0 \text{ cal}}{1.0 \text{ g}} \right)$$

$$= 350 \times 1.0 \text{ cal}$$

$$= 350 \text{ cal}$$

Heat Capacity and Specific Heat

The **heat capacity** of a substance is defined as the *quantity of heat required to raise the temperature of the substance 1°C*. The greater the amount of mass present, the greater is the heat capacity. A lake has a greater heat capacity than a pond. To compare the heat capacities of different substances, we must specify a given mass—for example, 1.0 g. The quantity of heat required to raise the temperature of 1.0 g of a substance 1°C is referred to as the **specific heat** of

Table 2.3.   *Some Specific Heats*

Substance	Specific heat (cal/g-deg) at 25°C
water (at 14°C)	1.00
iron metal (Fe)	0.106
silver metal (Ag)	0.057
copper metal (Cu)	0.092

that substance. By definition, the specific heat of water is 1.0 calorie/gram-degree (cal/g-deg). Other substances have different specific heat values. (See Table 2.3.) Water has one of the highest specific heats, while silver and copper are at the low end of the scale. This means that more heat must be applied to 1.0 g of water to raise its temperature 1°C than would be needed for the same amount of silver or copper.

Heat capacity is the number of calories required to raise the temperature of the total mass of a substance by 1°C.

Specific heat is the number of calories required to raise the temperature of 1.0 g of a substance by 1°C.

Problem 2.2.   (a) Calculate the number of calories required to heat 250 g of water (approximately one cup) from 20°C (room temperature) just to its boiling point (100°C). (b) What is the heat capacity of 250 g of water? (c) What is the heat capacity of 250 g of silver?

Heating and Changes in State

When a solid is heated, its temperature increases. The absorbed heat is converted to kinetic energy (energy of motion) of the molecules, and the molecules oscillate, or vibrate, more vigorously. Solids expand slightly as they are heated. When the temperature of a solid reaches the point where the molecules actually break out of their positions and move about, the solid melts and becomes a liquid. This temperature is called the **melting point**.

When a solid reaches its melting point, energy is needed to convert all of the solid to a liquid. As heat is applied, *the temperature of the melting solid does not increase until all the solid has melted.* The amount of heat needed for the transition from solid to liquid is called the **heat of fusion** (ΔH_f). In this term, the Greek letter delta (Δ) means "change in," while H means "heat content" at the melting point.

$$\Delta H_f = \text{number of calories required to melt 1.0 g of}$$
$$\text{solid at its melting point}$$

Different substances have different melting points and different heats of fusion. Table 2.4 lists a few substances along with their melting points and heats of fusion.

Table 2.4. *Some Melting Points and Heats of Fusion*

Substance	Mp (°C)	ΔH_f (cal/g)
water (H_2O)	0	79.7
iron (Fe)	1535	63.7
silver (Ag)	961	25.0
copper (Cu)	1083	49.0
sodium chloride (NaCl)	801	123.5
tristearin (common fat)	55	18.6

A few substances can go directly from the solid state to the gaseous state without ever becoming a liquid. This process is called **sublimation**. Dry ice (solid carbon dioxide) and old-fashioned mothballs (naphthalene) are examples of solids that sublime.

Example　A person stranded in a snowstorm may be tempted to eat snow to alleviate thirst. However, eating snow can lower the body's internal temperature (a condition called **hypothermia**) sufficiently to cause death. Calculate the number of calories necessary (a) to melt enough snow at 0°C to fill a 250-mL glass (approximately one cup of water) and (b) to raise the temperature of the water to 37°C (body temperature). (Note how the units are used in setting up the following equation.)

Solution　(a) Water weighs 1.0 g/mL. The heat of fusion of H_2O is 79.7 cal/g (from Table 2.4). Therefore, to melt the snow would require:

$$x \text{ cal} = 250 \text{ mL} \times \left(\frac{1.0 \text{ g}}{\text{mL}}\right) \times \left(\frac{79.7 \text{ cal}}{\text{g}}\right)$$

$$= 19{,}900 \text{ cal}$$

(b) The specific heat of water is 1.0 cal/g-deg (from Table 2.3). To raise the temperature of the melted snow to 37° would require:

$$x = 37° \times 250 \text{ g} \times \frac{1.0 \text{ cal}}{\text{g-deg}}$$

$$= 9250 \text{ cal}$$

Note that *more than twice* the energy is required to melt the snow than to bring the water from 0°C to body temperature!

Once a solid has completely melted, adding heat raises the temperature of the liquid until it starts to bubble vigorously and to become gaseous. The temperature at which this occurs is called the **boiling point**. At the boiling point, applying more heat does not increase the temperature of the boiling liquid; additional heat simply causes the liquid to boil more vigorously. The amount of heat required to change a substance from a liquid to a gas at its boiling point is called the **heat of vaporization** (ΔH_v).

$$\Delta H_v = \text{number of calories required}$$
$$\text{to vaporize 1.0 g of liquid at its boiling point}$$

Energy is needed to vaporize a liquid, even when the liquid is not at its boiling point; this is why the evaporation of rubbing alcohol (boiling point 82°C) cools the skin. Ethyl chloride (boiling point 12°C) is more volatile than rubbing alcohol. At room temperature, it will not remain liquid unless it is kept in a pressurized container. When ethyl chloride is sprayed on the skin, its vaporization actually freezes and numbs the skin. Because of this property, ethyl chloride is sometimes used as a local anesthetic.

Table 2.5 lists the "normal" boiling points (boiling points at atmospheric pressure) and heats of vaporization of a few substances. Note the high heat of vaporization of water compared to those of the other substances listed. This high heat of vaporization indicates a strong attraction among H_2O molecules, a topic that will be discussed in Chapter 4.

Table 2.5. *Some Boiling Points and Heats of Vaporization at the Normal Boiling Point*

Substance	Bp (°C)	ΔH_v (cal/g)
water (H_2O)	100.0	540
ethanol (CH_3CH_2OH)	78.5	204
hydrogen sulfide (H_2S)	−60.7	132
acetic acid (CH_3CO_2H)	117.9	97
mercury (Hg)	356.6	71

Once a material has become completely gaseous, additional heat will increase the temperature of the gas. A gas such as water vapor can be heated above the boiling point of water. Medical instruments are sometimes sterilized in this "superheated steam."

Problem 2.3.   Mercury ("quicksilver") is a silvery liquid metal with toxic vapors that can cause nerve damage. [It has been postulated that the fictional Mad Hatter in *Alice in Wonderland* suffered from mercury poisoning from compounds used by hatters of that era.] (a) Calculate the number of calories required to vaporize 5.0 g of mercury at its boiling point. (b) Compare the value calculated in (a) with the number of calories needed to vaporize an equivalent amount of water at its boiling point. (c) Suggest a reason for the difference between the number of calories required in (a) and in (b).

SUMMARY

Matter is something that has mass and occupies space. The three physical states of matter are *solid*, *liquid*, and *gas*. In ordinary chemical reactions, mass is neither lost nor gained, although it may be changed in form. *Energy* is the capacity to do work or cause change. Energy may change in form (from potential to kinetic, for example), but it is not ordinarily gained or lost.

Matter is made up of small particles composed of *atoms* and *molecules*. *Elements* are substances composed of just one type of atom; these atoms may or may not be combined into molecules. *Compounds* are composed of molecules containing more than one type of atom and cannot be broken down by physical means. *Mixtures* are composed of more than one type of substance (elements or compounds) and can be separated by physical means. In compounds, elements are combined in *definite proportions by weight* (law of definite proportions).

A *physical change* is a change in the state of a substance, while a *chemical change* is a change in the molecular structure of the substance. A *physical property*, such as color or melting point, is a property that we can observe or measure without the substance undergoing a chemical change. A *chemical property* is the ability (or lack of it) to undergo a particular chemical change under the prescribed conditions.

The *heat capacity* of a substance is the number of *calories* required to raise the temperature of a substance 1°C. The number of calories required to raise the temperature of 1 g of a substance 1°C is the *specific heat* of that substance. The specific heat of water is 1 cal/g-deg.

The physical state of a substance can be changed by heating or cooling. When a solid is heated, its temperature rises to the *melting point*; then a quantity of heat (*heat of fusion*, ΔH_f) must be added to melt the solid. The liquid can then be heated to its boiling point, at which temperature added heat (*heat of vaporization*, ΔH_v) is needed to vaporize the substance.

KEY TERMS

matter	liquid	heat capacity
energy	gas	specific heat
atom	physical change	melting point
molecule	physical property	heat of fusion
element	chemical change	sublimation
compound	chemical property	boiling point
mixture	calorie	heat of vaporization
solid		

STUDY PROBLEMS

2.4. Identify the type of energy contained in or exhibited by the following:
(a) a boulder resting at the top of a cliff
(b) an avalanche

2.5. When a campfire burns, the _____ energy of the wood is converted to _____ and _____ .

2.6. From its chemical formula, identify each of the following substances as an element or a compound:
(a) nitrogen gas, N_2
(b) carbon dioxide, CO_2
(c) isopropyl alcohol, C_3H_8O
(d) potassium metal, K

2.7. Match each item on the left with its description on the right:

(a) $N_2 + O_2$ (1) a pure compound
(b) $C_6H_{12}O_6$ (2) a mixture of
                                compounds
(c) a solution of   (3) an element
    ethanol in water
(d) a chocolate chip   (4) a mixture of
    cookie   elements

2.8. State whether each of the following statements refers to a solid, a liquid, or a gas:
(a) Molecules are not attracted to one another and move at random.
(b) Molecules are fairly strongly attracted to one another but can move around one another.
(c) Molecules are held in place by strong attractions.

2.9. Which of the following processes involves a physical change and which involves a chemical change?
(a) iron metal melting
(b) food spoiling
(c) brandy flaming on plum pudding
(d) rock salt melting ice
(e) dew forming on grass

2.10. Tell whether each of the following statements represents a physical property or a chemical property:

(a) An electric current can be passed along copper wire.
(b) An electric current can be passed through some water solutions, resulting in the formation of hydrogen and oxygen gases.
(c) Gasoline can evaporate.
(d) Gasoline can burn.

2.11. Referring to Table 2.3, calculate how many calories are required to increase the temperature of 100 g of each of the following substances from 22°C to 100°C:
(a) water   (b) silver

2.12. (a) If 25 g of a substance requires 750 cal to increase its temperature by 10°C, what is its heat capacity in that temperature range?
(b) What is its specific heat?

2.13. How many calories must be removed from 1.0 L of water to cool it from its boiling point to its freezing point?

2.14. The body contains approximately 5.7 L of blood. Assuming blood and water have the same specific heats, how many kcal are required to raise the temperature of this amount of blood 1.0°C?

2.15. (a) A solid is heated to its melting point; as the solid is melting, the temperature (does/does not) increase.
(b) After the solid becomes a liquid, added heat (does/does not) cause an increase in temperature.
(c) When the liquid reaches its boiling point, added heat (does/does not) raise the temperature as the liquid boils.
(d) As heat is added to a gas, the temperature (does/does not) increase.

2.16. Referring to Tables 2.3 and 2.4, determine the following:
(a) How many calories are required to raise 1.0 kg of copper from 22°C to its melting point?
(b) How many calories are required to melt 1.0 kg of solid copper at its melting point?
(c) How many calories are required to melt 1.0 kg of copper starting from room temperature?

3 Atomic Theory and the Periodicity of the Elements

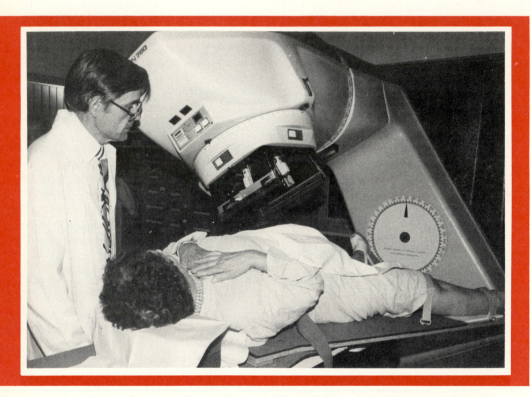

A cancer patient undergoing radiation therapy with a cobalt machine. (Courtesy of Massachusetts General Hospital.) Some atoms contain unstable nuclei that decay, or decompose, and emit powerful radiation similar to x rays. This radiation can penetrate living tissue and kill cells. A cobalt machine, which contains a radioactive form of cobalt, is used in cancer therapy. A narrow beam of radiation from the cobalt source is aimed toward a tumor in a patient's body to kill the cancer cells. The radiation source is rotated around the patient to minimize damage to intervening healthy tissue.

Objectives   List the atomic symbols for a few common elements. □ List the properties of protons, neutrons, and electrons. □ Define atomic number, mass number, isotope, and atomic mass. □ Show how the first two electron shells fill with electrons with increasing atomic number. □ Define atomic orbital; show how the $1s$, $2s$, and three $2p$ orbitals fill with electrons (optional). □ Define valence electrons; show how the number of valence electrons of many elements can be determined from the periodic table.

The Greek philosopher Democritus (about 460–370 B.C.) is credited with the concept that matter is composed of atoms, a concept that lay dormant through the Middle Ages. The idea of atoms was brought to life again about 1800 by John Dalton, an English chemist and physicist. From Dalton's time to the present, the atomic theory has undergone many modifications as new facts have emerged and improved our understanding of nature.

We have already used some concepts of atomic theory in discussing the physical changes in matter. We will find that understanding the atom is even more important in studying chemical change. Indeed, without the atomic theory, chemistry would be not much further along than it was in medieval times.

In this chapter, we will first discuss the symbols that are used to represent atoms and molecules, then turn our attention to the structures of atoms and their masses. We will conclude this chapter with an introduction to the periodic table of the elements, a very useful tool for chemists and students of chemistry.

3.1   ATOMIC SYMBOLS

Each of the over one hundred elements known today has a name as well as an alphabetic symbol, called the **atomic symbol**, which can be used to represent either the element or one of its atoms. When elements are combined in compounds, the shorthand representation, or **formula**, for the compound is the combination of the atomic symbols for the elements found in that compound.

H   O

atomic symbol for hydrogen   *atomic symbol for oxygen*

H—O—H   or H_2O *subscript 2 means two*
formula for water   *hydrogen atoms*

Some elements are represented by just one capital letter of the alphabet, and others are represented by two alphabetic characters. In the latter case, the first letter is always capitalized, and the second letter is lower case. For example, Co is the atomic symbol for cobalt, but CO is the formula for carbon monoxide, a compound containing carbon and oxygen.

Table 3.1 lists some common elements and their atomic symbols. The majority of the symbols (such as Al and C) are abbreviations of the English name for the element. Several of the atomic symbols are derived from their Latin names. For example, the symbol for iron, Fe, is taken from the Latin *ferrum*.

Table 3.1. *Some Common Elements and Their Atomic Symbols*

Name	Symbol	Name	Symbol
calcium	Ca	magnesium	Mg
carbon	C	mercury	Hg
chlorine	Cl	neon	Ne
fluorine	F	nitrogen	N
helium	He	oxygen	O
hydrogen	H	phosphorus	P
iodine	I	potassium	K
iron	Fe	sodium	Na
lithium	Li	sulfur	S

3.2 STRUCTURE OF THE ATOM

In Dalton's time, it was believed that atoms were the smallest particles of matter. We now know that an atom is composed of two principal parts—a heavy inner kernel called the **nucleus** and **electrons** that surround the nucleus (see Figure 3.1).

Although the nucleus is heavy, it is very small in comparison to the overall volume of space occupied by the atom. It has been calculated that the nucleus is only 1/10,000 to 1/100,000 the total size of the atom, which itself is only about 10^{-8} cm in diameter. To get an idea of these size relationships, consider an atom whose nucleus has been enlarged to the size of a beehive and

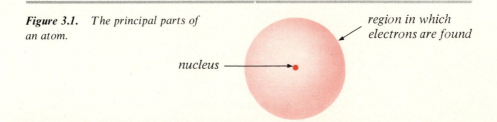

Figure 3.1.   *The principal parts of an atom.*

nucleus

region in which electrons are found

whose electrons are the size of bees. In our enlarged example, the electron bees would range two to twenty miles away from the nuclear hive. In both our enlarged atom and a real atom, most of the volume occupied by the atom is empty space.

The Nucleus

The nucleus of an atom is composed of a number of different particles. In this book we will consider only the two most important nuclear particles: the **neutron** and the **proton**.

A neutron and a proton have essentially equal masses. Their masses are so small that it would be impractical to discuss these masses in terms of grams. (For example, the mass of a proton is 1.6×10^{-24} g.) Instead, we say that each proton and each neutron has a mass of one *atomic mass unit* (amu). The difference between a proton and a neutron is that a proton (p^+) carries a positive charge while a neutron (n^0) is electrically neutral.

Nuclei of atoms of different elements contain varying numbers of protons and neutrons. Therefore, nuclei vary in mass and in the amount of positive charge they carry. For example, a helium nucleus contains two protons and two neutrons; it has a mass of 4 amu and carries an electrostatic charge of $+2$. A carbon nucleus usually contains six protons and six neutrons; it has a mass of 12 amu and carries a charge of $+6$.

Electrons

An electron (e^-) carries a negative charge that is equal to the positive charge of a proton; however, the mass of an electron is only 1/1836 amu. Electrons are held in their locations around their respective nuclei because opposite charges attract each other.

The charges of a single proton and a single electron are equal, but opposite. An atom has the same number of electrons as protons; thus, an atom has no net electrical charge. An atom of helium (He) is neutral because it has two protons in the nucleus and two electrons outside the nucleus. A neutral atom of carbon (C), with six protons in the nucleus, must have six electrons around the nucleus. (In the following diagrams, we show only a portion of the electronic "cloud" surrounding the nucleus, as an arc rather than as a full circle or sphere.)

A neutral helium atom has two electrons outside the nucleus.

A neutral carbon atom has six electrons outside the nucleus.

Table 3.2 summarizes the properties of the principal atomic particles.

Table 3.2.   *Properties of the Principal Atomic Particles*

Name	Mass, in amu	Charge	Where found
proton (p^+)	1.0	$+1$	in nucleus
neutron (n^0)	1.0	0	in nucleus
electron (e^-)	1/1836	-1	surrounding nucleus

3.3   ATOMIC NUMBERS AND ATOMIC MASSES

Atomic Number

Atoms of various elements differ from one another in the number of electrons, protons, and neutrons they contain. The simplest atom is that of the element hydrogen (H), which usually has one proton in the nucleus and one electron outside the nucleus. The next lightest element is helium (He), which has two protons, two neutrons, and two electrons.

It is the *number of protons* in the nucleus that determines to which element the atom belongs. An atom with 79 protons in the nucleus is gold (Au), regardless of the number of neutrons the nucleus contains. If an atom has 78 protons, it is platinum (Pt). If it has 80 protons, it is mercury (Hg). The number of protons in one atom of an element is called the **atomic number** of that element.

$$\text{atomic number} = \text{number of } p^+ \text{ in nucleus}$$

Also, for a neutral atom:

$$\text{atomic number} = \text{number of } e^- \text{ surrounding nucleus}$$

Mass Number

The **mass number** of an atom is the sum of the masses of the protons and neutrons in the nucleus. The mass number of helium is usually 4 amu (the sum of the masses of two protons and two neutrons). The mass number of carbon is usually 12 amu (the sum of the masses of six protons and six neutrons).

$$\text{mass number, in amu} = \text{number of } p^+ + \text{number of } n^0$$

Because electrons are so lightweight compared to protons and neutrons, the mass number of an atom is essentially the mass in amu of the atom.

Example   Nitrogen (N) has an atomic number of 7 and a mass number of 14. Determine the numbers of protons, neutrons, and electrons in a neutral atom of N.

Solution   We have

$$\text{number of } p^+ = \text{atomic number} = 7$$

$$\text{number of } e^- = \text{number of } p^+ = 7$$

Because mass number = number of p^+ + number of n^0,

$$\text{number of } n^0 = \text{mass number} - \text{number of } p^+$$

$$= 14 - 7$$

$$= 7$$

Problem 3.1.   Oxygen (O) has an atomic number of 8 and a mass number of 16. Determine the numbers of protons, neutrons, and electrons in a neutral atom of O.

Isotopes

All atoms that contain only one proton have an atomic number of 1 and are said to be hydrogen. In nature we find that the majority of hydrogen atoms have only one proton (and no neutrons) in their nuclei. However, a small percentage of hydrogen atoms contain one proton and one neutron in their nuclei. These atoms are still hydrogen atoms because they contain only one proton, but they are twice as heavy as most hydrogen atoms. We call these two types of hydrogen isotopes of each other.

Isotopes are forms of an element in which the atoms have the same number of protons but different numbers of neutrons.

The common isotope of hydrogen is called hydrogen-1, or simply *hydrogen*. The isotope that has an additional neutron is called hydrogen-2, or *deuterium*. The number following the name (the "2" in hydrogen-2) refers to the mass number (1 p^+ + 1 n^0 for hydrogen-2). There is also a third isotope of hydrogen, called hydrogen-3, or *tritium*, that contains one proton and two neutrons. Tritium is unstable and undergoes a slow and spontaneous disintegration in which the tritium nuclei break up and give off particles and energy called *radiation*. Tritium is said to be *radioactive*.

isotopes of hydrogen, each with one p^+:

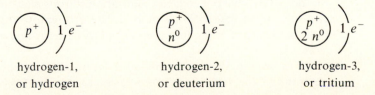

| hydrogen-1, | hydrogen-2, | hydrogen-3, |
| or hydrogen | or deuterium | or tritium |

Problem 3.2. Most carbon atoms are carbon-12 ($6p^+ + 6n^0$), but a small proportion of naturally occurring carbon atoms are carbon-13. Draw a diagram similar to the preceding diagrams for a neutral carbon-13 atom.

Atomic Mass

The **atomic mass**, or **atomic weight**, of an element is the average of the mass numbers of the naturally occurring isotopes, weighted to account for the relative abundance of each isotope.

atomic mass = weighted average of mass numbers of
naturally occurring isotopes

For example, the natural abundance of hydrogen (elemental or in compounds) is 98.985% hydrogen-1 and 0.015% hydrogen-2. The atomic mass of hydrogen is usually listed as 1.008 amu, a value that is slightly higher than 1 because of the presence of a small amount of hydrogen-2.

3.4   ELECTRONIC STRUCTURE OF THE ATOM

When we discuss mass and mass relationships among atoms, we are discussing their nuclei, since the mass of an atom is due primarily to the mass of the nucleus. However, a complete atom contains a nucleus surrounded by electrons. Even though electrons contribute very little to the mass of an atom, these electrons control the chemical properties of the elements. Consequently, we will consider electrons in some detail.

Electron Shells

The electrons of an atom are not scattered at random around the nucleus but are restricted to certain specified regions of space, called *shells*, around the nucleus. The different shells in which the electrons are found are referred to as the first shell (which is closest to the nucleus), the second shell (which is slightly farther away), and so on up to the seventh shell (which is the farthest from the nucleus). (See Figure 3.2.)

Attractions between oppositely charged particles are similar to gravitational attractions in one respect. These attractions are stronger when the particles are closer together, and they decrease as the particles become farther apart. Therefore, electrons in the closest shell are more strongly attracted by the positively charged nucleus than are electrons that are farther away. The electrons in the first shell are more difficult to remove from the atom. We say that the electrons in the first shell have *lower energy* and are at the *first energy level*.

The second shell is farther away from the nucleus than the first. The electrons in the second shell are easier to remove from the atom than are

Figure 3.2. *Two-dimensional diagram of the first three shells in which electrons are found around a nucleus. In three dimensions, the shells might be compared to balloons within balloons.*

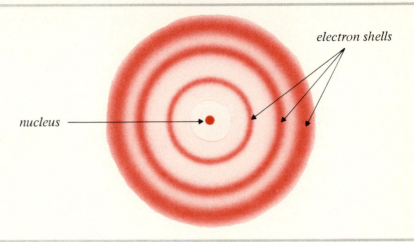

electron shells

nucleus

electrons in the first shell. The electrons in the second shell are said to be at the *second energy level* and are of slightly higher energy than electrons in the first shell. The third shell is farther out than the second and is of higher energy than the second. Each succeeding shell through the seventh is of successively higher energy.

Each shell can accommodate a limited number of electrons. The first shell can hold no more than two electrons. When there are three electrons around a nucleus, two of the electrons are in the first shell and the third is in the second shell. The second shell can have no more than 8 electrons; the third, 18; and the fourth, 32. (See Table 3.3.)

Table 3.3. Maximum Number of Electrons in the First Four Shells

Shell	Number of electrons
1	2
2	8
3	18[a]
4	32

[a] The third and fourth shells overlap; therefore, the fourth shell begins to fill when the third shell is only partially filled with 8 electrons instead of 18.

Filling the Shells

Because the positively charged nucleus attracts the negatively charged electrons as closely as possible, electrons seek the closest shell and the lowest energy level possible. As the number of protons in the nucleus is increased, the closest shells fill with electrons first, then the more distant shells fill in succession. Table 3.4 shows the electron structures of some of the elements. You can see how the shells are filled when the nucleus contains an increasing number of protons.

Table 3.4. *Electronic Structures of Some Elements*

		Number of electrons in each shell		
Element	Atomic number	1st shell	2nd shell	3rd shell
hydrogen (H)	1	1	—	—
helium (He)	2	2	—	—
lithium (Li)	3	2	1	—
beryllium (Be)	4	2	2	—
boron (B)	5	2	3	—
carbon (C)	6	2	4	—
nitrogen (N)	7	2	5	—
oxygen (O)	8	2	6	—
fluorine (F)	9	2	7	—
neon (Ne)	10	2	8	—
sodium (Na)	11	2	8	1
magnesium (Mg)	12	2	8	2

Example   Determine the electron structure of a neutral atom of aluminum (Al), atomic number 13.

Solution   First, determine the number of electrons in a neutral atom.

$$\text{number of } e^- = \text{number of } p^+ = \text{atomic number}$$

Therefore, for Al, the number of e^- is 13.

Then, beginning with the first shell, fill each one until all 13 electrons have been distributed. (The outer shell may or may not be filled.)

	Shell:	1	2	3
maximum number of e^-		2	8	18
for Al (13 e^-)		2	8	3

Problem 3.3.   Determine the electron structures of
(a) phosphorus (P), atomic number 15
(b) sulfur (S), atomic number 16

■ 3.5   ORBITAL MODEL OF THE ATOM (optional)

The concept of electrons being distributed in shells was originally developed by the Danish chemist-physicist Niels Bohr in 1912. However, the chemistry of the elements cannot be explained by electron shells alone. To account for the chemical properties of the elements, Bohr proposed that each shell consists of *subshells* with slightly different energies from one another. The term subshell is something of a misnomer, however, because it implies that the electrons in these subshells are situated in spheres around the nucleus. You will see shortly that this is not necessarily true.

Modern ideas concerning atomic structure are based on the theory that each electron shell is composed of compartments of space around the nucleus called atomic orbitals. An **orbital** is defined as the *region in space around the nucleus in which a particular electron or pair of electrons is likely to be found.*

The first shell contains only one orbital, called the 1s orbital. The number 1 refers to the first shell, and the letter *s* refers to a spherical orbital with the nucleus in the center. The hydrogen atom has only one electron, which is found somewhere in the 1s orbital. Usually, we do not think of that one electron as a discrete particle but as a fuzzy cloud of electronic charge, as shown in Figure 3.3.

The second shell also contains an *s* orbital, called the 2s orbital. Like the 1s orbital, the 2s orbital can be thought of as a fuzzy sphere. Because the 2s orbital is farther from the nucleus than is the 1s orbital, it surrounds the 1s orbital. (See Figure 3.3.)

The second shell also contains three *p* orbitals, called the 2p orbitals, that are of slightly higher energy than the 2s orbital. Thus, the second shell contains a total of *four* orbitals: 2s, 2p, 2p, 2p. A *p* orbital (Figure 3.4) is not spherical. Rather, it is shaped like a dumbbell. A *p* orbital has two lobes with

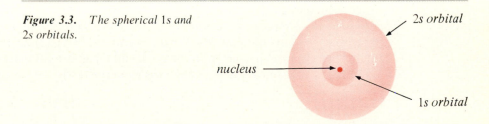

Figure 3.3.   *The spherical 1s and 2s orbitals.*

nucleus ⟶

2s orbital

1s orbital

Note: In this text, the colored boxes indicate optional sections. The key terms and end-of-chapter study problems pertaining to these sections are starred.

Figure 3.4. *The 2p orbitals.*

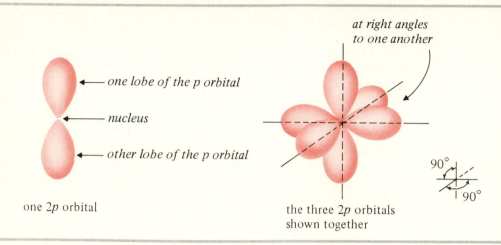

one lobe of the p orbital

nucleus

other lobe of the p orbital

one 2*p* orbital

at right angles to one another

the three 2*p* orbitals shown together

the nucleus in the center. The most likely place to find a 2*p* electron is in this dumbbell-shaped region in space.

The third shell has one 3*s* orbital, three 3*p* orbitals, and five 3*d* orbitals, for a total of nine orbitals. The fourth shell has sixteen orbitals: one 4*s*, three 4*p*, five 4*d*, and seven 4*f* orbitals. Larger shells contain even greater numbers of orbitals. Most of the chemistry that we will be discussing is concerned with the first two shells only; therefore, we will have very little to say about orbitals other than *s* and *p*.

Pairing of Electrons in Orbitals

One orbital can hold zero, one, or two electrons, but no more than two. Two electrons in one orbital are said to be *paired*. Because electrons are paired in orbitals, the first shell, with only a 1*s* orbital, can hold up to two electrons. The second shell, with one 2*s* orbital and three 2*p* orbitals, can hold up to eight electrons. (These maximum numbers of electrons are the same as those listed earlier in Table 3.3.)

To represent two electrons paired in an orbital, a pair of arrows ($\uparrow\downarrow$) are often used. If we represent an orbital by a box, we can then represent (1) an orbital that contains no electrons (no arrows in the box), (2) an orbital that contains one electron (one arrow in the box), and (3) an orbital that is filled (an up-pointing arrow and a down-pointing arrow in the box).

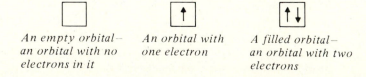

An empty orbital—an orbital with no electrons in it

An orbital with one electron

A filled orbital—an orbital with two electrons

Electron Configurations

In previous sections, we showed how the electrons of an atom of an element are distributed in electron shells. Let us take this concept one step further and show how electrons are distributed not only in shells but also in the orbitals within these shells. As in filling the shells, electrons fill orbitals by occupying the lowest-energy atomic orbital available. Keep in mind that each orbital can hold a maximum of two electrons.

Hydrogen (atomic number 1) has its one electron in the first shell in the $1s$ orbital. Helium (atomic number 2) has its two electrons in the first shell in the $1s$ orbital. Now the $1s$ orbital (as well as the first shell) is filled. Lithium (atomic number 3) has two of its three electrons in the $1s$ orbital and its third electron in the next lowest-energy orbital, the $2s$ orbital. Table 3.5 lists the electron configurations of the elements through magnesium. You can see that this table is simply an extension of Table 3.4.

Table 3.5.   *Electron Configurations of Some Elements*

		Shell								
		1	2				3			
Element	Atomic number	s	s	p	p	p	s	p	p	p + five d
hydrogen (H)	1	↑	—	—	—	—	—			
helium (He)	2	↑↓	—	—	—	—	—			
lithium (Li)	3	↑↓	↑	—	—	—	—			
beryllium (Be)	4	↑↓	↑↓	—	—	—	—			
boron (B)	5	↑↓	↑↓	↑	—	—	—			
carbon (C)	6	↑↓	↑↓	↑	↑[a]	—	—			
nitrogen (N)	7	↑↓	↑↓	↑	↑	↑	—			
oxygen (O)	8	↑↓	↑↓	↑↓	↑	↑	—			
fluorine (F)	9	↑↓	↑↓	↑↓	↑↓	↑	—			
neon (Ne)	10	↑↓	↑↓	↑↓	↑↓	↑↓	—			
sodium (Na)	11	↑↓	↑↓	↑↓	↑↓	↑↓	↑			
magnesium (Mg)	12	↑↓	↑↓	↑↓	↑↓	↑↓	↑↓			

[a] The three p orbitals (equal energies) receive one electron each before pairing occurs.

In general, you need not be concerned with the orbitals in the third and fourth shells. However, because potassium (K), atomic number 19, and calcium (Ca), atomic number 20, are important elements found in the human body, let us consider their electron configurations. The electron shells farther from the nucleus become progressively closer together and overlap. The third and fourth shells overlap slightly, so that the $4s$ orbital is of lower energy than the $3d$ orbital. For this reason, when the $3s$ and $3p$ orbitals are filled, the next electrons go into a $4s$ orbital instead of a $3d$ orbital. Potassium and calcium thus have the following electron configurations:

	1s	2s 2p 2p 2p	3s 3p 3p 3p	4s
K (atomic no. 19)	↑↓	↑↓ ↑↓ ↑↓ ↑↓	↑↓ ↑↓ ↑↓ ↑↓	↑
Ca (atomic no. 20)	↑↓	↑↓ ↑↓ ↑↓ ↑↓	↑↓ ↑↓ ↑↓ ↑↓	↑↓

Problem 3.4. The element following magnesium in the periodic table is aluminum (Al), atomic number 13. Write out its electron configuration as we have just done for potassium and calcium.   ■

3.6   VALENCE ELECTRONS AND THE PERIODIC TABLE

Valence Electrons

The number of electrons in the outermost unfilled shell of each element determines the chemical properties of the element. These outer electrons are often called the **bonding electrons**, or **valence electrons**. Referring to Table 3.4, you will see that hydrogen, lithium, and sodium have *one* valence electron each. We also have just shown that potassium has one valence electron. Similarly, beryllium, magnesium, and calcium have *two* valence electrons each.

Each of the following elements has one valence electron:

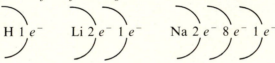

H $1e^-$ Li $2e^-$ $1e^-$ Na $2e^-$ $8e^-$ $1e^-$

Problem 3.5. How many valence electrons do the following elements contain?
(a) carbon   (b) nitrogen   (c) oxygen

The Periodic Table of the Elements

Early chemists were struck by the similarities in chemical behavior among certain elements. There were many attempts to group the elements in some pictorial way to emphasize these similarities. These efforts culminated in what we now call the **periodic table of the elements** (Figure 3.5). Credit for the periodic table is given to both Dimitri Mendeleev, a Russian, and Lothar Meyer, a German. Mendeleev and Meyer jointly received the Royal Society's Davy medal in England in 1882 for their work on the periodic table. (It is an interesting historical sidelight that the periodic table is the epitaph on Mendeleev's tomb.)

The atomic symbol of each element is shown in the periodic table along with the element's atomic number and atomic mass. Note that the elements are listed in order of atomic number starting at the top left (H, atomic number

Figure 3.5.   *The periodic table of the elements.*

Group IA — *alkali metals* — *alkaline earth metals* — *inert gases* — *halogens* — *nonmetals, metalloids* — *transition metals* — *rare earth metals*

IA	IIA	IIIB	IVB	VB	VIB	VIIB	VIII	VIII	VIII	IB	IIB	IIIA	IVA	VA	VIA	VIIA	0
1 H 1.008																	2 He 4.003
3 Li 6.939	4 Be 9.0122											5 B 10.811	6 C 12.011	7 N 14.007	8 O 15.999	9 F 18.998	10 Ne 20.183
11 Na 22.990	12 Mg 24.312											13 Al 26.982	14 Si 28.086	15 P 30.974	16 S 32.064	17 Cl 35.453	18 Ar 39.948
19 K 39.102	20 Ca 40.08	21 Sc 44.956	22 Ti 47.90	23 V 50.942	24 Cr 51.996	25 Mn 54.938	26 Fe 55.847	27 Co 58.933	28 Ni 58.71	29 Cu 63.546	30 Zn 65.37	31 Ga 69.72	32 Ge 72.59	33 As 74.922	34 Se 78.96	35 Br 79.904	36 Kr 83.80
37 Rb 85.47	38 Sr 87.62	39 Y 88.905	40 Zr 91.22	41 Nb 92.906	42 Mo 95.94	43 Tc 99	44 Ru 101.07	45 Rh 102.905	46 Pd 106.4	47 Ag 107.868	48 Cd 112.40	49 In 114.82	50 Sn 118.69	51 Sb 121.75	52 Te 127.60	53 I 126.904	54 Xe 131.30
55 Cs 132.905	56 Ba 137.34	57 La 138.91	72 Hf 178.49	73 Ta 180.948	74 W 183.85	75 Re 186.2	76 Os 190.2	77 Ir 192.2	78 Pt 195.09	79 Au 196.967	80 Hg 200.59	81 Tl 204.37	82 Pb 207.19	83 Bi 208.98	84 Po 210	85 At 210	86 Rn 222
87 Fr 223	88 Ra 226.05	89 Ac 227	104 Ku 261	105 Ha 260	106												

58 Ce 140.12	59 Pr 140.91	60 Nd 144.24	61 Pm 145	62 Sm 150.35	63 Eu 151.96	64 Gd 157.25	65 Tb 158.92	66 Dy 162.5	67 Ho 164.93	68 Er 167.26	69 Tm 168.93	70 Yb 173.04	71 Lu 174.97
90 Th 232.03	91 Pa 231	92 U 238.03	93 Np 237	94 Pu 242	95 Am 243	96 Cm 247	97 Bk 247	98 Cf 251	99 Es 254	100 Fm 254	101 Md 256	102 No 253	103 Lw 257

1) and moving left to right (He, 2), then starting again at the left of the second row (Li, 3). The second row follows:

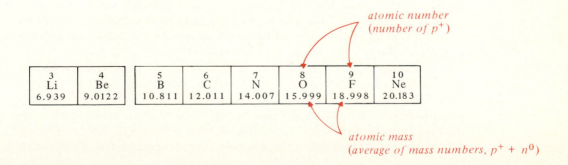

Of the known elements, only the first 92, through uranium, are found to any measurable extent in nature. Elements from 93 up, called *transuranium elements*, are unstable elements that have been synthesized by nuclear chemists.

You can see from the periodic table that most of the elements are metals—elements that are solid at room temperature (except for mercury, which is a liquid), are generally shiny, and can conduct an electric current. Note the heavy line in the periodic table in Figure 3.5. This line separates the metals (all elements to the left of the line plus the rare earth metals below) from the nonmetals (the elements to the right).

Although approximately three-fourths of the elements are metals, the ancient Greeks were familiar with only seven of them: gold, silver, iron, copper, tin, lead, and mercury. These metals either are found in the elemental state in nature or are easily obtained from their naturally occurring compounds—by heating, for example. (Note that the atomic symbols for these metals are all derived from their Latin names; for example, the symbol Au for gold is from *aureum*.) Today we use almost all of the metals in the periodic table in commerce and industry.

Groups in the Periodic Table

The elements are grouped in the periodic table according to similarities in chemical properties. One chemical property is the **ionization potential**: the amount of electrical force needed to remove a single valence electron from each of 6×10^{23} neutral atoms of an element.

Figure 3.6 is a graph showing atomic number versus ionization potential. Note that helium (He), neon (Ne), and argon (Ar) have high ionization potentials—that is, a relatively large amount of energy is required to remove an electron. On the other hand, lithium (Li), sodium (Na), and potassium (K) have low ionization potentials—very little energy is needed to remove an electron. Also, note that the ionization potential rises and falls in this group in a periodic fashion, according to atomic number.

Based on this type of observation, He, Ne, and Ar (with high ionization potentials) are placed in one group of elements (Group 0, the column at the far right of the periodic table). The elements in this group are called the **inert gases**, or **noble gases**, because they are, for the most part, chemically unreactive. The elements Li, Na, and K (with low ionization potentials) are placed in another group (Group IA) called the **alkali metals**. The alkali metals each can lose an electron readily and are quite reactive.

The elements in Group IIA are referred to as the **alkaline earth metals**. The next group is referred to as the **transition metals** and is further divided into ten subgroups (Groups IIIB through IIB). Following these are four groups containing other metals, nonmetals, and *metalloids* (substances intermediate between metals and nonmetals in their properties). The next-to-last group (Group VIIA) is called the **halogens**. The class called **rare earth metals**

Figure 3.6.   *Ionization potentials of the first 19 elements, showing that helium (He), neon (Ne), and argon (Ar) have high ionization potentials, while lithium (Li), sodium (Na), and potassium (K) have low ionization potentials.*

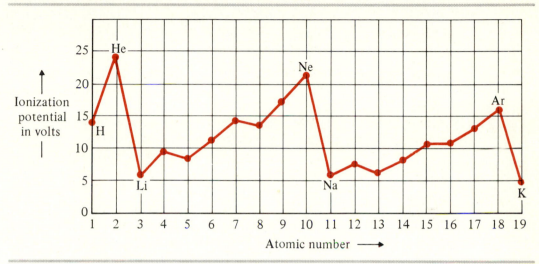

arises from overlapping electronic energy levels. To keep the periodic table a manageable shape and size, chemists usually show the rare earth metals below the rest of the table.

The horizontal rows in the periodic table are called *periods,* or simply *rows.* The elements in a row do not show striking chemical similarities as do the elements in a group.

Maybe you have guessed why the Group IA elements exhibit similar chemical properties. The elements in Group IA (H, Li, Na, K, etc.) have *one* valence electron each. The elements in Group IIA have *two* valence electrons each. Skipping over the transition metals (the number of whose valence electrons is sometimes unpredictable because of overlapping energy levels), we find that the Group IIIA elements have *three* valence electrons each, and so forth. In general, the group number is the same as the number of valence electrons.

	IA	IIA	...	IIIA	IVA	VA	VIA	VIIA	0
no. of outer e^-	1	2		3	4	5	6	7	8

Problem 3.6.   Suggest a reason why helium (He) and neon (Ne) are in a group called Group 0 (zero).

SUMMARY

The elements are often represented by *atomic symbols*, such as H for hydrogen. An atom is composed of a positively charged nucleus containing *protons* (p^+) and *neutrons* (n^0), surrounded by negatively charged *electrons* (e^-). The following terms are used in discussions of elements:

atomic number = number of p^+

mass number = number of p^+ + number of n^0

atomic mass = weighted average of mass numbers of naturally occurring isotopes of an element

Isotopes are atoms of the same element (atoms having the same atomic number) that contain differing numbers of neutrons in their nuclei (have different mass numbers).

Electrons are found in a series of *shells* surrounding a nucleus. Each shell is subdivided into *atomic orbitals*. Because each orbital can hold a maximum of two electrons, each shell is limited in the number of electrons it can contain (first shell, $2 \, e^-$; second shell, $8 \, e^-$). As the atomic number increases in a series of elements, electrons are added to the lowest-energy orbitals available.

Electrons in an outer unfilled shell are called *valence electrons*. The *periodic table of the elements* groups elements according to the number of valence electrons.

KEY TERMS

atomic symbol

chemical formula

nucleus

neutron

proton

electron

atomic number

mass number

isotope

atomic mass

electron shell

*atomic orbital

valence electron

periodic table

ionization potential

noble gas

alkali metal

alkaline earth metal

transition metal

halogen

STUDY PROBLEMS

3.7. Identify each of the following as an atomic symbol or as a formula:
(a) C_6H_6 (b) H_2 (c) Ne   (d) CO

3.8. Write the name for each of the following elements:
(a) C   (b) H   (c) O   (d) N
(e) P   (f) He   (g) S

3.9. Label each of the following phrases as applicable to a neutron, a proton, both of these, or neither of these:
(a) is found in the nucleus of an atom
(b) has a negative electrical charge
(c) has no electrical charge
(d) has a mass of 1 amu

3.10. Referring to the periodic table, give (1) the number of protons in the nucleus, (2) the atomic number, and (3) the atomic mass of each of the following elements:

(a) nitrogen (N)   (b) aluminum (Al)

(c) carbon (C)   (d) copper (Cu)

3.11. Which of the following phrases is true for neutral atoms of a pair of isotopes of an element? More than one answer may be correct.

(a) have the same mass number

(b) have the same number of electrons

(c) have the same number of protons

3.12. Without referring to the text, fill in the numbers of electrons in the following table.

Element	Atomic number	Electron shell		
		1	2	3
carbon (C)	6	__	__	__
fluorine (F)	9	__	__	__
sodium (Na)	11	__	__	__

3.13. Referring to Problem 3.12, state the number of valence electrons in a neutral atom of

(a) carbon

(b) fluorine

(c) sodium

3.14. Referring to the periodic table, state the number of valence electrons of

(a) rubidium (Rb), atomic no. 37

(b) silicon (Si), atomic no. 14

(c) sulfur (S), atomic no. 16

4 Chemical Bonds

Synthesized gem diamonds. (Courtesy of General Electric Research and Development Center.) Carbon exists in several elemental states including soot, coal, graphite, and diamonds. The photograph shows two of these elemental states—graphite (as a powder and as pencil lead) and diamonds. These particular diamonds were synthesized in the laboratory from graphite. The properties of a diamond are different from those of graphite because the carbon atoms in these two substances are bonded together differently. Graphite contains sheets of molecules in which the carbon atoms are joined together in flat six-membered rings; a graphite molecule resembles a hexagonal-tiled floor. A diamond is a single giant molecule in which the carbon atoms are bonded together in a three-dimensional network.

Objectives   Define the octet rule and give examples of ion-formation using this rule. ☐ Define oxidation number and relate it to an element's position in the periodic table. ☐ Define and illustrate oxidation, reduction, oxidizing agent, reducing agent, cation, and anion. ☐ Describe the structure of an ionic compound in the solid state and in solution. ☐ Write formulas and names for ionic compounds. ☐ List the numbers of covalent bonds that some common elements form. ☐ Write formulas for compounds containing single, double, and triple bonds. ☐ Show how atoms share electrons by overlapping atomic orbitals (optional). ☐ Define electronegativity, polar bonds, and hydrogen bonds. ☐ Name simple covalent compounds and polyatomic ions. ☐ Draw Lewis, structural, and molecular formulas. ☐ Calculate oxidation numbers of atoms in simple covalent compounds (optional).

In Chapter 3, we discussed the structures of atoms. However, most matter is not composed of simple atoms but of *molecules,* or combinations of atoms. In this chapter, we will consider the ways in which atoms can be bonded together and the types of elements and compounds that can result. You will see how the number of valence electrons determines if an element is likely to form *ions* or *molecules* and whether an element is likely to be *oxidized* or *reduced,* and you will learn how to name simple chemical compounds.

4.1 TYPES OF CHEMICAL BONDS

Chemical bonds, which hold atoms together as molecules, arise from the interactions of the valence electrons of atoms. One type of chemical bond arises from the *transfer of electrons* from one atom to another. The result is two or more electrically charged particles called **ions**. Because opposite charges attract, oppositely charged ions in solid substances are held together by attractions called **ionic bonds**.

For example, a sodium atom (Na) can lose an electron (with one negative charge) to a chlorine atom (Cl). In the process, the sodium atom becomes a positively charged *sodium ion* (Na^+) while the chlorine atom becomes a negatively charged *chloride ion* (Cl^-). In the following chemical equation, only the valence electrons are shown as dots. Electrons in filled inner shells do not participate in bonding and, therefore, are not shown. A formula in which the bonding electrons are shown as dots is called a **Lewis formula**, after the American chemist G. N. Lewis, who developed these formulas early in the twentieth century. In this particular example, the result of the electron transfer is *sodium chloride,* or common table salt.

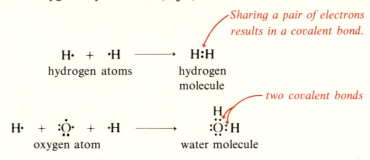

valence e⁻ (inner e⁻ not shown)

An ionic bond arises from the attraction between + and −.

$$Na\cdot \; + \; \cdot\ddot{\underset{\cdot\cdot}{Cl}}\colon \longrightarrow Na^+ \; \colon\ddot{\underset{\cdot\cdot}{Cl}}\colon^-$$

This equation reads: A sodium atom (Na) and a chlorine atom (Cl) undergo a chemical reaction to yield a sodium ion (Na⁺) and a chloride ion (Cl⁻).

A second type of chemical bond, called a **covalent bond**, arises from the *sharing of electrons.* Two hydrogen atoms (H) can share one pair of electrons to yield a hydrogen molecule (H_2). Similarly, hydrogen can form covalent bonds with oxygen to yield water (H_2O).

Sharing a pair of electrons results in a covalent bond.

$$H\cdot \; + \; \cdot H \longrightarrow H\colon H$$
hydrogen atoms   hydrogen molecule

two covalent bonds

$$H\cdot \; + \; \colon\ddot{O}\cdot \; + \; \cdot H \longrightarrow \; \colon\ddot{O}\colon H$$
          oxygen atom   water molecule

A third type of chemical bond, which we will not discuss further in this book, is the **metallic bond**. In a metal, all the atoms contribute their outer electrons to a general electron "pool." A metal is held together by the attractions between the positively charged ions and the negative electrons. In a metal, the positive and negative charges are equal in number and cancel. Therefore, we write the formula of a metal as its atomic symbol, such as Fe or Na.

Na^+	e^-	Na^+	e^-	Na^+
e^-	Na^+	e^-	Na^+	e^-
Na^+	e^-	Na^+	e^-	Na^+

A metallic bond arises from the attraction between + and −.

The importance of metals to society is due to such properties as moldability and ability to conduct an electric current, both of which are a direct result of the metallic bond. For example, metals can conduct electricity, which is a flow of electrons, because of the mobile, "loose" electrons.

In the following sections, first we will discuss the transfer of electrons and the formation of ionic compounds; then we will discuss covalent bonding.

4.2   OCTET RULE

Why do atoms gain, lose, and share electrons? How can we predict what will happen when atoms get together? To answer these questions, let us first consider one group of elements that hardly ever forms chemical bonds.

Group 0 in the periodic table of the elements is composed of the chemically inert *noble gases*. Although individual atoms of some elements are found in the near-vacuum of outer space, the noble gases are among the few elements that normally exist as individual atoms here on earth. These gases are said to be nonreactive, or inert, because, with very few exceptions, they do not combine with other elements to yield compounds.

The noble gases are inert because an electronic structure in which the outer shell is filled with electrons is exceptionally stable. The outer shell of helium is filled with 2 electrons (all that the first shell can hold), while the outer shell of neon is filled with 8 electrons. Although the third shell can contain a maximum of 18 electrons, this shell often behaves as a filled outer shell when it contains only 8 electrons. Therefore, argon behaves as if its outer shell were also filled to capacity with electrons.

	Shell: 1	2	3
helium (He)	$2\,e^-$			*filled outer shell (no valence electrons)*
neon (Ne)	$2\,e^-$	$8\,e^-$	
argon (Ar)	$2\,e^-$	$8\,e^-$	$8\,e^-$	*behaves as a filled outer shell*

About 1920, the physical chemists G. N. Lewis, I. Langmuir, and W. Kössel formulated the following concept:

Atoms gain, lose, or share electrons to attain electronic configurations of the noble gases.

Because most (but not all) elements of interest gain, lose, or share electrons to attain an outer shell of *eight* electrons (the configuration of neon or argon), this statement is often referred to as the **octet rule**. The formation of sodium fluoride (NaF), an ionic compound used to fluoridate water to prevent tooth decay, from sodium (Na) and fluorine (F) atoms exemplifies the octet rule:

$$\text{Na } 2\,e^-\ 8\,e^-\ \left(1\,e^-\right) \ + \ \text{F } 2\,e^-\ 7\,e^- \longrightarrow$$

$$\left[\text{Na } 2\,e^-\ 8\,e^-\right]^+ \ + \ \left[\text{F } 2\,e^-\ 8\,e^-\right]^-$$

sodium ion (Na$^+$)   fluoride ion (F$^-$)

Each of these ions has the electron configuration of neon (filled outer shell).

Note that sodium ions, fluoride ions, and neon atoms all have the same electron configuration; however, these species are not identical because the nuclei of the ions or atoms contain different numbers of protons. It is the atomic number, not the number of electrons, that determines the identity of an element, regardless of whether the element is in the form of atoms or is bound up in a compound.

Lithium and beryllium are examples of metals that do not form an octet in their outer second shells, but rather lose electrons to attain a helium configuration of two electrons in the first shell. Note that beryllium must lose *two* electrons to attain a noble-gas configuration.

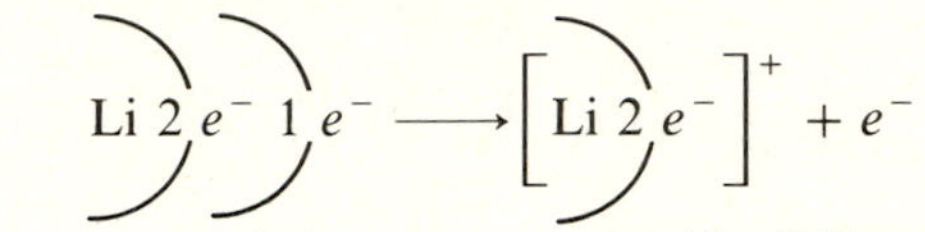

$$\text{Li } 2\,e^-\ 1\,e^- \longrightarrow \left[\text{Li } 2\,e^-\right]^+ + e^-$$

A lithium atom (Li) *loses one* e^- *to yield a lithium ion* (Li$^+$).

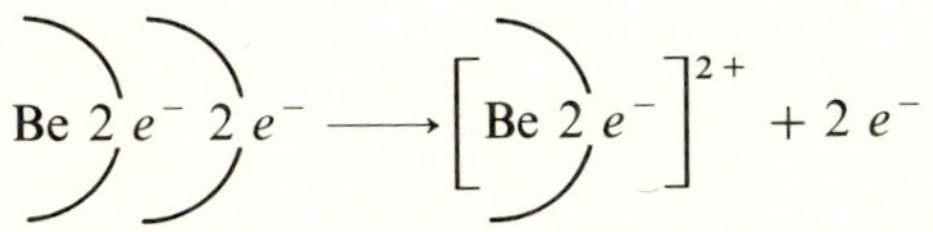

$$\text{Be } 2\,e^-\ 2\,e^- \longrightarrow \left[\text{Be } 2\,e^-\right]^{2+} + 2\,e^-$$

A beryllium atom (Be) *loses two* e^- *to yield a beryllium ion* (Be^{2+}).

Problem 4.1.   Predict the number of electrons that an atom of each of the following elements would gain or lose in the formation of an ion:
(a) potassium (K)   (b) calcium (Ca)   (c) aluminum (Al)

The Octet Rule and the Periodic Table

Now you can understand one reason why the various elements in each group of the periodic table exhibit similar behavior. Each atom of all the elements in Group IA contains one valence electron. With the exception of the hydrogen atom, which tends to share electrons, atoms of all the elements in Group IA tend to lose this single valence electron in chemical reactions.

	Shell: 1	2	3	4	
lithium (Li)	$2\,e^-$	$1\,e^-$			*one valence electron*
sodium (Na)	$2\,e^-$	$8\,e^-$	$1\,e^-$		
potassium (K)	$2\,e^-$	$8\,e^-$	$8\,e^-$	$1\,e^-$	

All the elements in Group IIA have *two* valence electrons, which they tend to lose in chemical reactions.

	Shell: 1	2	3	4	
beryllium (Be)	$2\,e^-$	$2\,e^-$			two valence electrons
magnesium (Mg)	$2\,e^-$	$8\,e^-$	$2\,e^-$		
calcium (Ca)	$2\,e^-$	$8\,e^-$	$8\,e^-$	$2\,e^-$	

Except in the case of certain metals such as iron (Fe), atoms tend not to lose three or more electrons. The loss of this many electrons would place a large amount of positive charge on the ion (from $+3$ to $+7$), an unstable situation that would require a large amount of energy. Therefore, the elements in Groups IIIA, IVA, and VA tend to share electrons instead of undergoing electron transfer. The elements in Groups VIA and VIIA can share electrons but are also capable of *gaining* electrons to attain a noble-gas configuration.

	Shell: 1	2	3	
oxygen (O)	$2\,e^-$	$6\,e^-$		six valence electrons;
sulfur (S)	$2\,e^-$	$8\,e^-$	$6\,e^-$	tend to gain two e^-

	Shell: 1	2	3	4	
fluorine (F)	$2\,e^-$	$7\,e^-$			seven valence electrons;
chlorine (Cl)	$2\,e^-$	$8\,e^-$	$7\,e^-$		tend to gain one e^-
bromine (Br)	$2\,e^-$	$8\,e^-$	$8\,e^-$	$7\,e^-$	

Problem 4.2. Write the symbol for the ion that oxygen forms (the oxide ion). Represent the outer, valence electrons by dots and show the electrostatic charge of the ion.

4.3 OXIDATION AND REDUCTION

In ionic compounds, the amount of positive or negative charge of a simple ion is called its **oxidation number** or sometimes its **valence**. Table 4.1 shows some ions with their oxidation numbers. Note that iron can form ionic compounds by losing either two electrons or three electrons. For example, in rust (iron oxide, Fe_2O_3), iron is found as Fe^{3+} with an oxidation number of $+3$; in hemoglobin, the oxygen carrier in red blood cells, iron is found as Fe^{2+} with an oxidation number of $+2$. The ability of some elements to form ions with different oxidation numbers generally arises from overlapping energy levels and unfilled inner levels, to and from which electrons can migrate. When an element is in its elemental state (such as iron metal, Fe), its oxidation number is zero.

Table 4.1.   *Oxidation Numbers of the Usual Ions of Some Elements*

Element	Ion	Oxidation number
Group IA:		
lithium (Li)	lithium ion (Li^+)	$+1$
sodium (Na)	sodium ion (Na^+)	$+1$
potassium (K)	potassium ion (K^+)	$+1$
Group IIA:		
magnesium (Mg)	magnesium ion (Mg^{2+})	$+2$
calcium (Ca)	calcium ion (Ca^{2+})	$+2$
transition metals:		
iron (Fe)	iron(II) ion (Fe^{2+})	$+2$ or $+3$
	iron(III) ion (Fe^{3+})	
silver (Ag)	silver ion (Ag^+)	$+1$
Group IIIA:		
aluminum (Al)	aluminum ion (Al^{3+})	$+3$
Group VIA:		
oxygen (O)	oxide ion (O^{2-})	-2
sulfur (S)	sulfide ion (S^{2-})	-2
Group VIIA:		
fluorine (F)	fluoride ion (F^-)	-1
chlorine (Cl)	chloride ion (Cl^-)	-1
bromine (Br)	bromide ion (Br^-)	-1
iodine (I)	iodide ion (I^-)	-1

When a neutral atom forms an ion, its oxidation number changes. If an atom loses electrons, its oxidation number increases. An increase in the oxidation number of an atom or ion is referred to as **oxidation**.

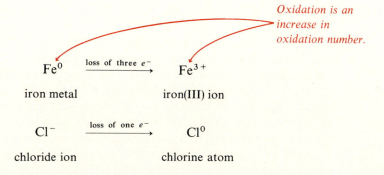

Oxidation is an increase in oxidation number.

Conversely, if an atom gains electrons, its oxidation number decreases. A decrease in oxidation number is called **reduction**.

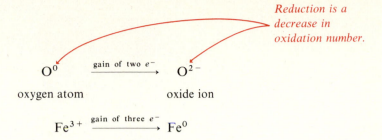

Fe^{3+} $\xrightarrow{\text{gain of three } e^-}$ Fe^0

Because electrons cannot be truly lost or gained but only transferred, any oxidation (loss of electrons) must be accompanied by the reduction (gain of electrons) of something else. In the case of iron metal being oxidized by rusting, oxygen gas (O_2) is reduced. In this reaction, oxygen is acting as an *oxidizing agent* and iron is acting as a *reducing agent*.

oxidized   *reduced*

2 Fe   +   3 O$_2$ ⟶   Fe$_2$O$_3$

iron   oxygen   iron(III) oxide

the reducing agent   *the oxidizing agent*   (rust)

An **oxidizing agent** causes oxidation; an oxidizing agent is reduced in an oxidation-reduction reaction.

A **reducing agent** causes reduction; a reducing agent is oxidized in an oxidation-reduction reaction.

Problem 4.3. Each of the following chemical equations represents a chemical reaction that involves an oxidation and a reduction. In each case, identify (1) what is oxidized, (2) what is reduced, (3) the oxidizing agent, and (4) the reducing agent.

(a) Na + Cl$_2$ $\longrightarrow$ 2 NaCl

(b) SnCl$_2$ + Cl$_2$ $\longrightarrow$ SnCl$_4$

4.4 IONIC COMPOUNDS

Electrostatic attractions are the attractions between + and − charges. An *ionic compound* is composed of oppositely charged ions that are held together by electrostatic attractions in a three-dimensional structure called a *crystal lattice*. Figure 4.1 is a representation of a portion of a sodium chloride crystal. In an ionic crystal, the positive ions, called **cations**, are found in alternate positions with the negative ions, called **anions**. Each ion is thus surrounded by and attracted to oppositely charged ions. In an ionic compound, the total number of + charges equals the number of − charges. Under normal circum-

Figure 4.1.   *An ionic compound in*
the solid state.

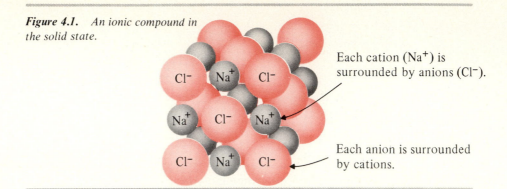

Each cation (Na^+) is
surrounded by anions (Cl^-).

Each anion is surrounded
by cations.

stances, it is not possible to separate anions from cations. To do so would
require enormous amounts of energy because like charges repel each other.

Chemically and physically, ionic compounds behave completely differently from the elements of which they are composed. Sodium metal (Na) is a
reactive metal that explodes on contact with water. Elemental chlorine (Cl_2) is
a toxic, greenish-yellow gas that is used as a bactericide in drinking water and
swimming pools. Yet the ions of these elements, Na^+ and Cl^-, are quite stable
and are necessary components of body fluids. The phrase "low-sodium diet"
refers to sodium ions, not elemental sodium.

Because of strong attractions between cations and anions, ionic com-

Figure 4.2.   *An ionic compound dissolved in water.*

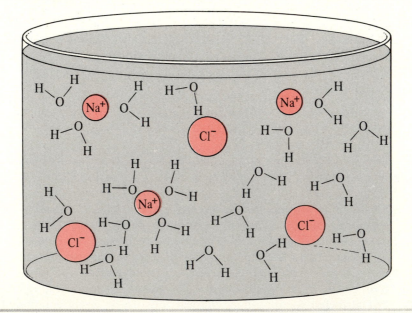

pounds tend to have very high melting points. For example, sodium chloride melts at 804°C.

When an ionic compound is dissolved in water, the cations (Na^+) and the anions (Cl^-) separate from one another and become surrounded by water molecules. (See Figure 4.2.) Because ions become separated in solution, dissolved *mixtures* of ionic compounds do not contain the original ionic compounds as such, but simply a mixture of the different ions. For example, a water solution of KBr and NaCl contains a mixture of K^+, Na^+, Br^-, and Cl^- ions. The identity of the original KBr and NaCl is lost when these compounds are dissolved in water and their ions are separated. Body fluids, such as blood or urine, contain dissolved ions. Even though body fluids contain Na^+ and Cl^-, we cannot say that these fluids actually contain sodium chloride because the fluids contain other cations and anions as well.

4.5   FORMULAS FOR IONIC COMPOUNDS

Because the $+$ and $-$ electrical charges in an ionic compound must be equal, the elements in an ionic compound are found in definite proportions. For example, solid sodium chloride is composed of an equal number of sodium ions and chloride ions. Thus, Na^+ and Cl^- ions are present in the solid in a ratio of $1:1$. We write the formula for sodium chloride as Na^+Cl^-, or simply NaCl. By convention, the symbol for the cation is written first, followed by the symbol for the anion. The formula NaCl implies a $1:1$ ratio of Na^+ and Cl^-.

In an ionic compound containing ions with charges of different magnitudes (such as $+2$ and -1), *the charges must be balanced*; that is, the sum of the oxidation numbers must be zero. In calcium chloride, composed of Ca^{2+} and Cl^-, we must have *twice* as many chloride ions as calcium ions, or $Ca^{2+} + 2\,Cl^-$. When we write a formula for this compound, we use a subscript after the Cl symbol to represent $2\,Cl^-$.

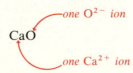

When calcium ions are combined with oxide ions, the electrical charges are balanced when equal numbers of Ca^{2+} and O^{2-} are present.

$$\underset{\text{one } Ca^{2+}\text{ ion}}{\overset{\text{one } O^{2-}\text{ ion}}{CaO}}$$

Problem 4.4. Write formulas for the ionic compounds that would be formed from the following pairs of ions:

(a) Ca^{2+} and S^{2-} (b) Na^+ and O^{2-} (c) Al^{3+} and O^{2-}

Problem 4.5. Formulas for three ionic compounds of copper (Cu) follow. From the formulas and from Table 4.1, determine the oxidation number of copper in each case.
(a) CuI   (b) CuI$_2$ (c) Cu$_2$S

4.6 NAMING IONIC COMPOUNDS

In the name of an ionic compound, the cation's name is given first, followed by the anion's name as a separate word. Generally, the cation is a metal ion, which goes by its atomic name. The name of the anion is derived from the atomic name, but the ending is changed to tell us that it is an anion. For example, chlorine forms a chlor*ide* ion (Cl$^-$); sulfur forms a sulf*ide* ion (S^{2-}); oxygen forms an ox*ide* ion (O^{2-}). The ending *-ide* is generally used when the anion contains only one element. (Exceptions are the hydroxide ion, OH$^-$, and the cyanide ion, CN$^-$.) Table 4.1 (Section 4.3) lists a few simple anions and their names.

LiBr   CaO

lithium bromide   calcium oxide

Certain metals can form ions that have different oxidation numbers. Iron, for example, is found in some compounds as Fe^{2+} and in other compounds as Fe^{3+}. When naming a compound of iron, we need to differentiate between the two possibilities. Iron chloride is an incorrect name because we cannot tell whether FeCl$_2$ or FeCl$_3$ is the formula for the compound. There are two common ways of differentiating the oxidation numbers in a name. One is by using a Roman numeral. Using this method, Fe^{2+} is called the iron(II) ion, and Fe^{3+} is called the iron(III) ion. The other method is to use different suffixes in the names. For example, the name ferr*ous* is used for the Fe^{2+} ion, and ferr*ic* is used for the Fe^{3+} ion.

FeCl$_2$ FeCl$_3$

iron(II) chloride,   iron(III) chloride,
or ferrous chloride   or ferric chloride

(Iron has an oxidation   *(Iron has an oxidation*
number of +2.)   *number of +3.)*

Problem 4.6. Referring to Table 4.1 and Problem 4.5, name the following ionic compounds:
(a) KBr   (b) Ag$_2$O   (c) CuI   (d) CuI$_2$

4.7 COVALENT COMPOUNDS

A **covalent bond** is the *sharing of two electrons by two atoms*. A **covalent compound**, also called a **molecular compound**, is a compound that contains only covalent bonds. A covalent compound thus consists of discrete molecules, not ions. These molecules have weak attractions among themselves compared to the attractions between ions. For this reason, covalent compounds generally have lower melting points and lower boiling points than do ionic compounds. If a covalently bonded compound is dissolved in water, its atoms do not generally become separated as do ions. For example, when ethanol (grain alcohol, CH_3CH_2OH) dissolves in water, its molecules retain their identity. There are exceptions to this generalization, however. Some molecules that are covalent when pure, such as HCl, break into ions when dissolved in water. This topic will be discussed in Chapter 9.

Carbon (C), nitrogen (N), and phosphorus (P) are biologically indispensable elements that are almost never found as simple ions. The energy required for carbon, nitrogen, or phosphorus to gain or lose a sufficient number of electrons to attain a noble-gas configuration is too high. Carbon, nitrogen, and phosphorus almost always form covalent bonds.

We mentioned earlier that hydrogen (H) rarely forms actual cations (H^+). Hydrogen is placed above Group IA in the periodic table because hydrogen atoms, like the atoms of elements in Group IA, contain one valence electron each. Hydrogen does not behave like an alkali metal because only a small positively charged proton (unshielded by electrons) would remain if an electron were removed from a hydrogen atom. Because a naked proton is a high-energy particle, hydrogen tends to share electrons instead of forming true ions. The other nonmetals (such as oxygen or chlorine) can form either ionic or covalent bonds.

Table 4.2 lists some important elements commonly found in covalent compounds along with the number of covalent bonds they usually form. The number of covalent bonds an element forms, like the oxidation number of an ion in ionic compounds, depends on the number of valence electrons in the element's neutral atoms.

Table 4.2.   *The Number of Covalent Bonds Formed by Some Elements*

Element	Number of valence electrons	Usual number of covalent bonds	Example
hydrogen (H)	1	1	HCl, H_2O
carbon (C)	4	4	CH_4, CCl_4
nitrogen (N)	5	3	NH_3, N_2
oxygen (O)	6	2	H_2O, CO_2
halogens (F, Cl, Br, I)	7	1	HCl, CCl_4

4.8   COVALENT BONDING AND THE NOBLE-GAS CONFIGURATION

Hydrogen, the simplest of atoms, forms covalent bonds by sharing its single electron with another atom. The hydrogen molecule (H_2) consists of two hydrogen atoms sharing their combined two electrons. By sharing two electrons, each hydrogen atom gains a filled first shell, the electronic configuration of the noble gas helium. We can represent the sharing of electrons between two H atoms in H_2 in several ways.

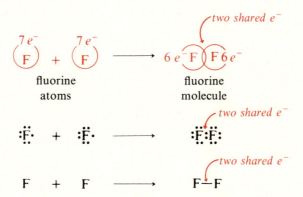

Fluorine is also capable of forming covalent bonds. A fluorine atom has seven electrons in the second shell, but eight are needed to fill this outer shell. A molecule of fluorine (F_2) is formed from two atoms of fluorine sharing one pair of electrons.

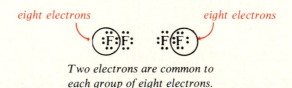

In the fluorine molecule, each atom has given up one of its seven electrons to form the covalent bond. Each fluorine atom still has six unshared electrons in the outer shell but now has two shared electrons also, for a total of eight electrons.

eight electrons   *eight electrons*

:F:F: :F:F:

Two electrons are common to
each group of eight electrons.

Example   Using dots to represent outer electrons, draw the formulas for the covalent compounds (a) HCl and (b) H_2O.

Solution

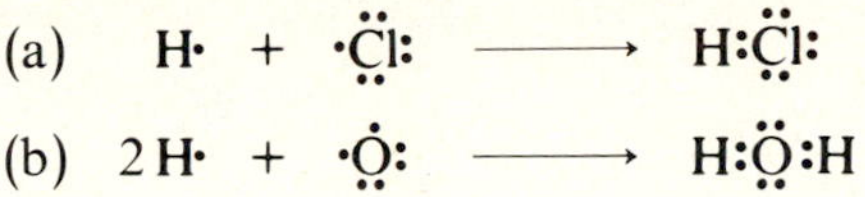

(a)   H· + ·C̈l: ⟶ H:C̈l:

(b)   2 H· + ·Ö: ⟶ H:Ö:H

Problem 4.7. Write Lewis (dot) formulas for the following compounds; circle the eight electrons around each C and N atom.
(a) ammonia, NH_3 (b) methane, CH_4
(c) chloroform, $CHCl_3$

Except for the noble gases, virtually none of the nonmetallic elements on the earth are found as free neutral atoms, such as H· or :C̈l·. Instead, they are found as covalently bonded molecules, such as H_2, Cl_2, I_2, O_2, or N_2. Many of the nonmetallic elements are too reactive to be found in nature even as molecules; instead, they are synthesized by chemical manufacturers for use in research and industry. Chlorine gas (Cl_2) is an example of a highly reactive gas that is manufactured from NaCl. Oxygen (O_2) and nitrogen (N_2), which form the bulk of the earth's atmosphere, of course, are relatively stable, nonmetallic, gaseous elements on the earth.

Single, Double, and Triple Bonds

When atoms share one pair of electrons between them, the resultant covalent bond is called a **single bond**.

single bonds

H:H   :F̈:F̈:   H:Ö:H

Sometimes a pair of atoms shares *two* pairs of electrons, and two covalent bonds join the two atoms. We say that the atoms are joined by a **double bond**. In each of the following cases, C and O are joined by a double bond, represented as C::O or C=O.

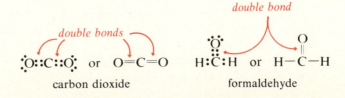

double bonds   *double bond*

:Ö::C::Ö: or O=C=O   H:C̈:H or H—C—H

carbon dioxide   formaldehyde

Atoms can also share *three* pairs of electrons, resulting in a **triple bond**.

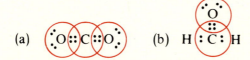

triple bond

H:C::C:H   or   H—C≡C—H   H:C::N: or   H—C≡N

acetylene   hydrogen cyanide

In each of these examples, note that hydrogen always forms one covalent bond (never more than one), carbon forms four bonds, nitrogen forms three bonds, and oxygen forms two bonds. These are the same number of bonds as shown in Table 4.2. For example,

Each O has two bonds.

$$O=C=O \qquad H-\overset{\overset{\displaystyle O}{\|}}{C}-H$$

Each C has four bonds.

Example   Write Lewis formulas for (a) carbon dioxide and (b) formaldehyde, and circle the eight electrons around each carbon and oxygen atom.

Solution

(a)   :O::C::O: (b)   H:C:H   with O above

Problem 4.8.   Write Lewis formulas and structural (line-bond) formulas for (a) nitrogen gas, N_2, and (b) ethylene, C_2H_4. Circle the eight electrons around each C and N atom.

■ 4.9   ORBITAL THEORY OF COVALENT BONDING (optional)

Let us now briefly consider the **molecular orbital theory of covalent bonding**. Electrons around the nucleus of an atom are found in orbitals. When atoms share two electrons, they overlap their orbitals; the shared electrons become paired and occupy the same region in space.

We call an orbital of a nonbonded atom an *atomic orbital*. When two atoms overlap their atomic orbitals, these atomic orbitals merge together to form a new orbital that is shared jointly by both atoms. This new orbital is called a **molecular orbital**. Like the atomic orbital, the molecular orbital can hold up to two electrons. Thus, *any covalent bond is formed by one pair of electrons in a molecular orbital.*

Figure 4.3.   *The bonding between the 1s orbitals of two hydrogen atoms in the hydrogen molecule (H_2).*

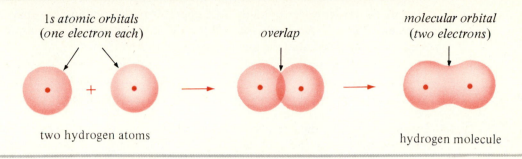

1s *atomic orbitals*
(one electron each)   *overlap*   *molecular orbital*
                                                              (two electrons)

two hydrogen atoms   hydrogen molecule

Figure 4.3 shows how the 1s atomic orbitals of two hydrogen atoms overlap to form an H_2 molecule with one molecular orbital containing two electrons. Figure 4.4 shows a somewhat different case: that of two hydrogen atoms and a sulfur atom forming hydrogen sulfide (H_2S), a toxic gas with the odor of rotten eggs.

two unfilled p orbitals

$$H\cdot \ + \ \cdot\ddot{\underset{\cdot\cdot}{S}}: \ + \ H\cdot \ \longrightarrow \ H:\overset{\overset{\textstyle H}{\cdot\cdot}}{\underset{\cdot\cdot}{S}}:$$

hydrogen sulfide (H_2S)

Figure 4.4.   *The bonding of two 3p orbitals of a sulfur atom with two 1s orbitals of hydrogen atoms to yield hydrogen sulfide (H_2S).*

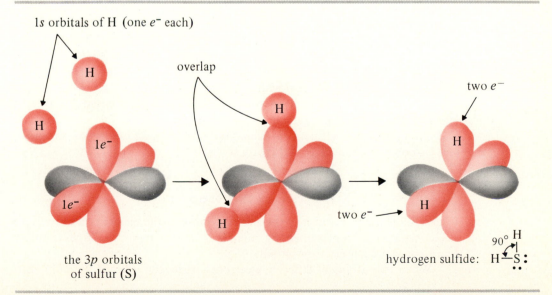

1s orbitals of H (one e^- each)

overlap

two e^-

$1e^-$

$1e^-$

two e^-

the 3p orbitals
of sulfur (S)

hydrogen sulfide: $H\overset{90°\ H}{\underset{\cdot\cdot}{-}S}:$

In H_2S, the H atoms' orbitals overlap with p orbitals of the S atom. Because the p orbitals of sulfur (or any other atom) are at right angles to one another, the H—S bonds are also at right angles to each other. Thus, the types of orbitals used in covalent bonding are important in determining the shapes of molecules. Later in this book you will learn that the shapes of proteins, for example, are instrumental in the functioning of all living organisms. ■

4.10   ELECTRONEGATIVITY AND POLAR COVALENT BONDS

Electrons are held by their respective nuclei because of the attraction between the positive protons in the nucleus and the negative charges of the electrons. The nuclei of some elements have a greater attraction for their outer, bonding electrons than do those of other elements.

Electronegativity is the magnitude of the attraction of an atom for its outer, valence electrons, especially in covalent compounds.

In a molecule, an atom that has a greater attraction for bonding electrons is said to be *more electronegative* than the atom to which it is bonded. For example, in hydrogen chloride (HCl), the chlorine atom has a greater attraction for the bonding electrons than does hydrogen: chlorine thus has a greater electronegativity than does hydrogen.

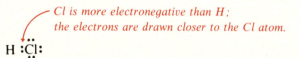

*Cl is more electronegative than H ;
the electrons are drawn closer to the Cl atom.*

The reason for electronegativity differences is twofold. First, atoms vary in size. Because attractions between + and − decrease with increasing distance, a smaller atom has a greater electronegativity than does a comparable larger one. Second, a greater number of protons in the nucleus means an increasing attraction for electrons and thus a greater electronegativity. The principal elements we will deal with in this text are those in the first full row of the periodic table, where electronegativity increases as we proceed left to right because of an increasing number of protons. Hydrogen (H) is a special case and a very common element in compounds. It is located between boron (B) and carbon (C) in the electronegativity series.

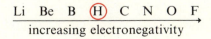

increasing electronegativity

Figure 4.5 shows a portion of the periodic table of the elements with relative electronegativities represented by numbers. In this figure, a larger number represents a greater electronegativity.

When two atoms of differing electronegativities are joined by a covalent

Figure 4.5. *Relative electronegativities of some elements related to their positions in the periodic table. A larger number (and darker shading) indicates a greater electronegativity.*

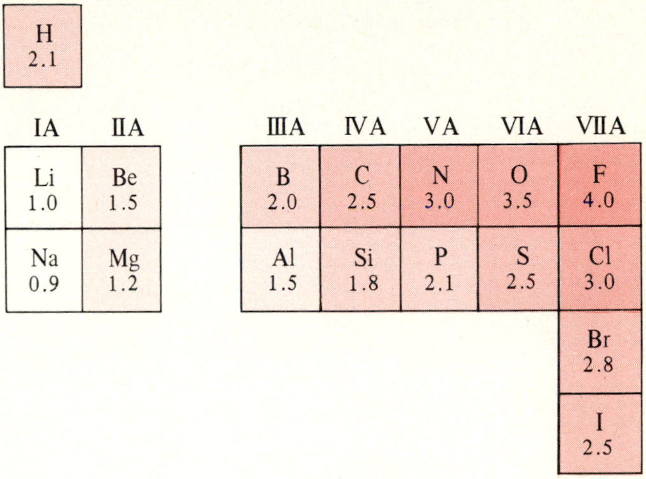

H 2.1

IA	IIA
Li 1.0	Be 1.5
Na 0.9	Mg 1.2

IIIA	IVA	VA	VIA	VIIA
B 2.0	C 2.5	N 3.0	O 3.5	F 4.0
Al 1.5	Si 1.8	P 2.1	S 2.5	Cl 3.0
				Br 2.8
				I 2.5

bond, the bonding electrons are drawn toward the more electronegative atom. A covalent bond in which the electronic charge is not shared equally by the two atoms is said to be a **polar bond**. In the case of HCl, the bonding electrons lie closer to the Cl than to the H. Thus, hydrogen may be said to have a *partial positive charge*, and chlorine may be said to have a *partial negative charge*. We symbolize a partial positive charge by the Greek letter delta followed by a plus sign: $\delta+$. A partial negative charge is shown as $\delta-$.

partial positive charge ──→ $\underset{\text{H}}{\overset{\delta+}{}}$ ── $\underset{\text{Cl}}{\overset{\delta-}{}}$ ←── *partial negative charge*

If the electronegativity difference between two atoms is great enough, the bonding electrons may be completely transferred to the more electronegative atom to yield ions. This is the reason that sodium chloride, as one example, is *not* covalent. Sodium (Na), at the extreme left of the periodic table, is much less electronegative than chlorine (Cl), at the extreme right.

When two atoms of the same element are bonded together, they both have an *equal* pull on shared electrons. Since both atoms have the same electronegativity, there is an equal distribution of electrons between the two atoms. We call such a bond a **nonpolar bond**. Because C and H have very similar electronegativity values, a C—H bond is relatively nonpolar.

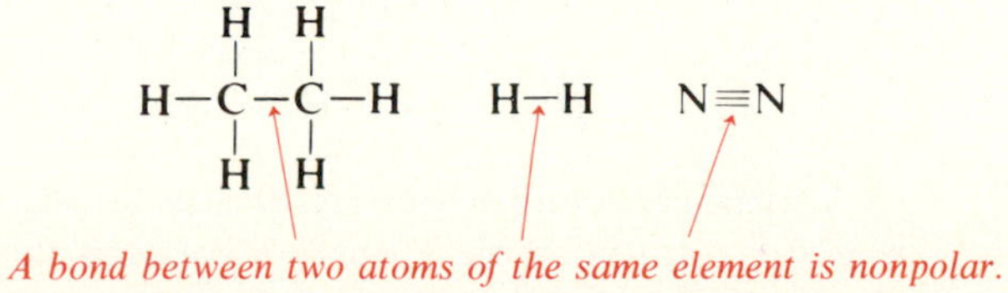

A bond between two atoms of the same element is nonpolar.

We may summarize what we have said about nonpolar, polar, and ionic bonds as follows:

nonpolar bonds   polar bonds   ionic bonds

such as H—H   such as $\overset{\delta+}{H}$—$\overset{\delta-}{Cl}$ such as Na^+Cl^-

increasing electronegativity difference between two atoms

Problem 4.9. Use the symbols $\delta+$ and $\delta-$ to show the polarity of the bonds in

(a) 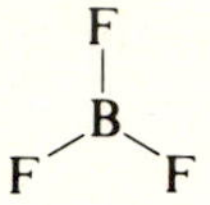
$$F - \underset{F}{\overset{|}{B}} - F$$

(a)   F—B with F and F   (b)   O=C=O   (c) 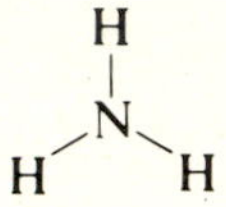   H—N with H and H

The Importance of Water's Polarity

Life as we know it could not exist without water. Therefore, let us examine the structure of a water molecule. Although we write the formula for water as H—O—H or simply H_2O, the angle between the two OH bonds is not 180°, but 104.5°—slightly larger than a right angle. In a molecule of water, the more electronegative oxygen draws the bonding electrons toward itself and away from the hydrogen atoms. In addition to being "bent" and polar, a water molecule contains two unshared pairs of valence electrons on the oxygen atom.

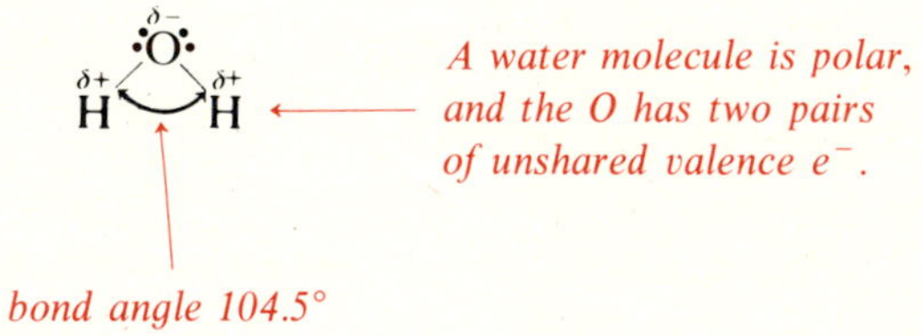

A water molecule is polar, and the O has two pairs of unshared valence e^-.

bond angle 104.5°

Because of the polarity, water molecules attract each other—the partially positive H of one H_2O molecule is attracted to the partially negative O of another H_2O molecule. This attraction is strong relative to most intermolecular attractions because the H can become associated with the unshared electrons on the O. This association of the H with an O of another water molecule is called a **hydrogen bond**.

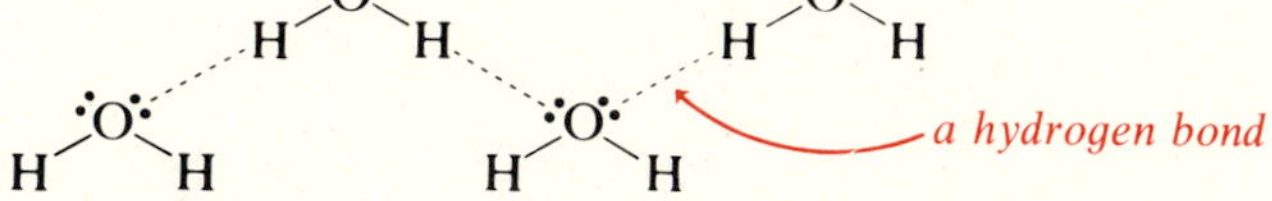

a hydrogen bond

A hydrogen bond is only 5–10% as strong as a covalent bond but still represents a considerable attraction between the molecules. In fact, it is the hydrogen bonds holding H_2O molecules together that cause water to be a liquid instead of a gas at room temperature; most other compounds formed from small molecules, such as O_2 or CO_2, are gases at room temperature.

The polarity of water molecules and the unshared electrons on the oxygen also allow water molecules to attract ions and other polar molecules. For this reason, water is a good solvent for many ionic compounds and small polar organic molecules. For example, blood is a water solution of many simple ions (such as Na^+, Mg^{2+}, Ca^{2+}, and Cl^-), many complex ions (such as HCO_3^- and protein anions), and polar organic molecules (such as the sugar *glucose*).

4.11   FORMULAS AND NAMES FOR SIMPLE COVALENT COMPOUNDS

We have been using a number of different types of formulas for covalent compounds. Any one of these is correct, depending on the application. We use *Lewis formulas* for counting electrons. We use line-bond formulas, or *structural formulas*, for showing the *structure*, or order of attachment of atoms, in a molecule. In many cases, we can condense a structural formula so that the structure of a molecule is still evident, even though all the bonds are not shown. We use *molecular formulas*, showing only the types of atoms and their numbers, for convenience when the order of attachment of the atoms is well known or of no importance. You will encounter other types of formulas later in this book. Table 4.3 shows examples of some types of formulas for a few simple covalent compounds.

Except for very common covalent compounds (such as water, H_2O), covalent compounds containing only two elements are named similarly to ionic compounds. The principal difference is that prefixes denoting the number of each type of atom are included in the name. These prefixes are necessary

Table 4.3.   *Some Types of Formulas for Covalent Compounds*

Name	Lewis formula	Structural formula	Condensed structural formula	Molecular formula
water	H:Ö:H	H—O—H	H_2O	H_2O
ammonia	H:N̈:H with H above	H—N—H with H above	NH_3	NH_3
methanol	H:C̈:Ö:H with H below	H—C—O—H with H above and below	CH_3OH	CH_4O
formaldehyde	Ö double H:C:H	O double H—C—H	$H_2C{=}O$	CH_2O
acetylene	H:C⋮⋮C:H	H—C≡C—H	CH≡CH	C_2H_2

Table 4.4.   *Prefixes Denoting Number*

Prefix	Number	Prefix	Number
mono-	1	penta-	5
di-	2	hexa-	6
tri-	3	hepta-	7
tetra-	4	octa-	8

because there may be more than one covalent compound that the elements can form. For example, carbon forms two different compounds with oxygen: carbon monoxide (CO) and carbon dioxide (CO_2). Table 4.4 lists the prefixes for the first eight numbers. (These prefixes are taken from the Greek names for the numbers.)

In the name of a covalent compound that contains two elements, the first element named is the one to the *left* in the periodic table. If only one atom of this element is present, the prefix *mono-* (meaning one atom) is usually omitted. The second element, the one to the right in the periodic table, is then included in the name as the second word with the ending *-ide*, just as for an ionic compound. Generally, the prefix mono- is not deleted from the name of the second word.

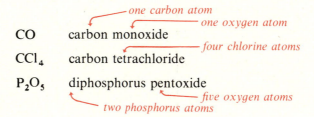

CO   carbon monoxide

CCl_4 carbon tetrachloride

P_2O_5 diphosphorus pentoxide

Problem 4.10.   Name the following compounds:
(a) NO_2 (b) N_2O_3 (c) CS_2

Problem 4.11.   Write molecular formulas for (a) sulfur trioxide, (b) diphosphorus trioxide, and (c) phosphorus pentachloride.

4.12   POLYATOMIC IONS

Polyatomic ions, also called **complex ions** or **radicals**, are ions that contain more than one atom. The atoms in a polyatomic ion are bonded together by covalent bonds, but one or more atoms in the group carry a positive or negative ionic charge. Therefore, the group as a whole has an ionic charge.

$$^-:\overset{..}{\underset{..}{O}}:H \quad \text{or} \quad ^-OH \quad \text{or} \quad OH^-$$

hydroxide ion
an anion

$$\left[H:\overset{\overset{H}{..}}{\underset{\underset{H}{..}}{N}}:H \right]^+ \quad \text{or} \quad \overset{+}{N}H_4 \quad \text{or} \quad NH_4^+$$

ammonium ion
a cation

When a compound containing polyatomic ions dissolves in water, the ions separate from one another, but the polyatomic ion usually retains its identity. For example, a water solution of sodium hydroxide (NaOH) contains Na^+ and ^-OH ions.

Oxygen is an especially important element found in many polyatomic ions. As may be seen in the list of some common polyatomic ions in Table 4.5, the names of most anions containing oxygen end in *-ate*, as opposed to names for monatomic (one atom) ions, which usually end in *-ide*. As we have mentioned, the hydroxide ion is an exception.

$$KMnO_4 \qquad\qquad Na_2SO_4$$

potassium permanganate · · · · · · · · · · · · · · · · · · sodium sulfate

The ending -ate usually signifies oxygen in the anionic portion of a compound.

In a formula for an ionic compound containing more than one polyatomic ion, parentheses are used to enclose the ion. The subscript denoting the number of complex ions follows the second parenthesis.

two ^-OH ions

$$Ca^{2+} + 2\ ^-OH \quad \text{or} \quad Ca(OH)_2$$

$$2\ NH_4^+ + SO_4^{2-} \quad \text{or} \quad (NH_4)_2SO_4$$

two NH_4^+ ions

Table 4.5. *Some Common Polyatomic Ions*

Name of ion	Formula	Ionic charge	Example of compound
hydroxide	OH^-	-1	$NaOH$
carbonate	CO_3^{2-}	-2	Na_2CO_3
bicarbonate	HCO_3^-	-1	$NaHCO_3$
sulfate	SO_4^{2-}	-2	Na_2SO_4
nitrate	NO_3^-	-1	$NaNO_3$
phosphate	PO_4^{3-}	-3	Na_3PO_4
permanganate	MnO_4^-	-1	$KMnO_4$
ammonium	NH_4^+	$+1$	NH_4Cl

The names of ionic compounds containing polyatomic ions are like those of simple ionic compounds: the name of the cation followed by the name of the anion. The names of the preceding two compounds are calcium hydroxide and ammonium sulfate.

Problem 4.12.   Name the following ionic compounds:
(a) $Mg(OH)_2$ (b) $(NH_4)_2CO_3$
(c) $Ca(HCO_3)_2$ (d) K_2SO_4

Problem 4.13.   Write formulas for and name the compounds formed from the following ions. (Be sure to determine the correct number of each ion before writing the formula.)
(a) $Mg^{2+} + CO_3^{2-}$ (b) $Fe^{3+} + SO_4^{2-}$ (c) $NH_4^+ + PO_4^{3-}$

■ 4.13   OXIDATION NUMBERS OF COVALENTLY BONDED ATOMS (optional)

In Section 4.3, we described the assigning of oxidation numbers to ions in simple ionic compounds. The oxidation number of an atom in a covalently bonded molecule or polyatomic ion can be calculated using the following assumptions:

1. An atom of any *element* (such as H_2 or O_2) has an oxidation number of zero.
2. In compounds, oxygen (O) has an oxidation number of -2.
3. In compounds, hydrogen (H) and Group IA elements (Li, Na, K) have oxidation numbers of $+1$.
4. In the formula for a compound, the sum of the oxidation numbers equals zero.
5. In the formula for a polyatomic ion, the sum of the oxidation numbers equals the charge of the ion.

Example   Calculate the oxidation number of N in nitric acid, HNO_3.

Solution   H = $+1$. Each O = -2; therefore, three O's = -6:

$$\text{partial sum} = (+1) + (-6) = -5$$

Because HNO_3 is a neutral compound, the oxidation number of N must be $+5$.

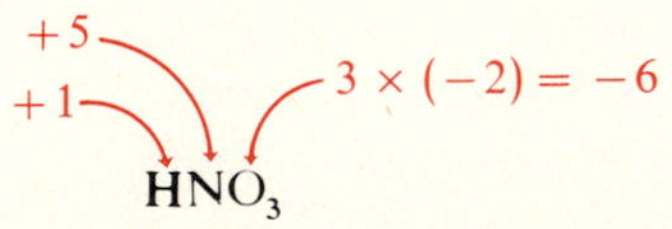

$$\overset{+1 \quad +5}{H}\,\overset{}{N}\overset{3 \times (-2) = -6}{O_3}$$

$$\text{sum} = +1 + 5 + (-6) = 0$$

Problem 4.14.   Calculate the following oxidation numbers:
(a) Fe in FeO   (b) Al in Al_2O_3
(c) N in NO_3^- (d) Mn in MnO_4^-
(e) S in K_2SO_4

Just as for simple atoms and ions, an increase in the oxidation number of an atom in a covalent compound is called an oxidation of that atom, while a decrease is called a reduction. For example, nitric oxide (NO) is an oxide of nitrogen formed when nitrogen compounds are burned. This compound reacts spontaneously with atmospheric oxygen to yield nitrogen dioxide (NO_2). Because the oxidation number of N increases from $+2$ to $+4$, the nitrogen atom undergoes oxidation in this reaction.

$$\overset{(+2)(-2)}{2\ NO}\ +\ \overset{(0)}{O_2}\ \longrightarrow\ \overset{(+4)}{2\ NO_2}\qquad\textit{Each O is } -2.$$

N *is oxidized from* $+2$ *to* $+4$ *state.*
O *in* O_2 *is reduced from 0 to* -2 *state.* ■

SUMMARY

The *octet rule* states that atoms gain, lose, or share electrons in order to attain electronic configurations of the *noble gases*, usually an outermost shell containing eight (or, in some cases, two) electrons.

Ionic compounds, which are composed of positively charged *cations* and negatively charged *anions*, arise from electron transfer. The *oxidation number* of a simple ion is the electrostatic, or ionic, charge on the ion. An increase in oxidation number of an atom or ion is called *oxidation*, while a decrease in oxidation number is called *reduction*.

The chemical formula for an ionic compound shows the ratio of ions by number, while the name is composed of the cation's name plus the anion's name. The anion ending *-ide* usually means a simple monatomic anion, while *-ate* implies a polyatomic anion containing oxygen.

Covalent compounds are compounds in which the atoms are combined into molecules by *covalent bonds*, which arise from the sharing of pairs of electrons. Electrons are shared by the formation of *molecular orbitals*. Some atoms may be joined by single, double, or triple bonds in molecules. Some types of formulas for covalent compounds are shown in Table 4.3. Covalent compounds containing two elements are named like ionic compounds, but with prefixes denoting the numbers of atoms in each molecule.

Electronegativity is a measure of an atom's attraction for its outer (bonding) electrons. Electronegativity differences give rise to *polar covalent bonds*, in which a greater amount of electronic charge is associated with one of the two atoms involved in the bond.

Polyatomic ions are ions containing groups of atoms covalently bonded together. The rules for determining the oxidation number of covalently bonded atoms are listed in Section 4.13.

KEY TERMS

ion	oxidizing agent	*molecular orbital
ionic bond	reducing agent	electronegativity
Lewis formula	cation	polar bond
octet rule	anion	hydrogen bond
noble gas	covalent bond	structural formula
oxidation number	single bond	molecular formula
oxidation	double bond	polyatomic ion
reduction	triple bond	

STUDY PROBLEMS

4.15. Referring to the periodic table of the elements inside the cover of this book, write equations that illustrate the number of electrons the following elements would gain or lose in ion-formation:
(a) F   (b) Mg   (c) S

4.16. Write the equation for the formation of lithium bromide (LiBr) from lithium atoms and bromine atoms. In your equation, indicate the electrons in the outer shell of each species with dots.

4.17. Which of the following elements would form cations fairly readily? Explain.
(a) magnesium (Mg)   (b) nickel (Ni)
(c) oxygen (O)   (d) iodine (I)

4.18. Give the oxidation number of each of the following ions:
(a) Sr^{2+} (b) Cr^{3+} (c) S^{2-} (d) I^-

4.19. Determine the oxidation number for each of the ions in the following compounds:
(a) $ZnBr_2$ (b) FeS
(c) K_2O (d) Hg_2Cl_2

4.20. Identify each of the following conversions as oxidation or reduction:
(a) $Cu^{2+} + 2\ e^- \rightarrow Cu$
(b) $Cl_2 + 2\ e^- \rightarrow 2\ Cl^-$
(c) $Fe \rightarrow Fe^{2+} + 2\ e^-$

4.21. In the following equations, identify the oxidizing agent, the reducing agent, the cation, and the anion:
(a) $2\ Cu + O_2 \rightarrow 2\ Cu^{2+} + 2\ O^{2-}$
(b) $2\ Fe + 3\ Cl_2 \rightarrow 2\ FeCl_3$

4.22. Which of the following phrases describe the ionic compound potassium chloride (KCl)?
(a) is a high-melting solid
(b) is a greenish-yellow gas
(c) is a reactive metal
(d) breaks apart into K^+ and Cl^- when dissolved in water

4.23. Write formulas for the following compounds:
(a) magnesium iodide   (b) potassium sulfide
(c) aluminum oxide

4.24. Name the following ionic compounds:
(a) CaO   (b) AgCl   (c) Fe_2S_3

4.25. Referring to the periodic table, tell how many covalent bonds each of the following elements usually forms:
(a) O   (b) C   (c) N

4.26. Write a chemical equation for the formation of a chlorine molecule from two chlorine atoms. Indicate the number of electrons in the outer shell of each chlorine atom before and after their combination.

4.27. Write Lewis (dot) formulas and structural (line-bond) for the following covalent compounds. [Compounds (d) and (e) contain one or more double bonds.]

(a) methane, CH_4

(b) carbon tetrachloride, CCl_4

(c) water, H_2O

(d) carbon disulfide, CS_2

(e) phosgene, Cl_2CO

4.28. Rewrite the following formulas, showing *each* bond as a line:

(a) $CH_3CH_2CH_3$ (b) $CH_3\overset{\displaystyle O}{\overset{\|}{C}}OH$

4.29. In each pair, which is the more electronegative element? (Refer to the periodic table inside the cover or to Figure 4.5, Section 4.10.)

(a) H or F   (b) Na or Cl

(c) C or O   (d) C or N

4.30. Place partial charges ($\delta+$ and $\delta-$) where appropriate in the following structures:

(a) $H-Br$ (b) $Cl-Cl$ (c) $H\overset{\displaystyle S}{\diagup\diagdown}H$

4.31. (a) Draw two molecules of water, showing a hydrogen bond between them. (b) Ammonia (NH_3) can also form hydrogen bonds with water. Draw formulas to show *two different types* of hydrogen bonds between NH_3 and H_2O.

4.32. Which of the following properties of water would you attribute to hydrogen bonding?

(a) is a good solvent for ionic compounds

(b) is a liquid at room temperature

(c) has an unusually high heat of vaporization

4.33. Name the following covalent compounds:

(a) SO_2 (b) SO_3 (c) N_2O_5

4.34. Write molecular formulas for the following covalent compounds:

(a) disulfur trioxide   (b) sulfur dichloride

(c) dinitrogen monoxide

4.35. Name the following compounds:

(a) $NH_4^+ Cl^-$ (b) $K^+ {}^-OH$

(c) $Na^+ HCO_3^-$

4.36. Write formulas for

(a) ammonium nitrate

(b) sodium sulfate

(c) calcium carbonate

4.37. When the compounds in Problem 4.35 are dissolved in water, what species are found in the water?

*4.38. Determine the oxidation number of each atom in the following equations and tell which atoms are oxidized and which are reduced.

(a) $2\,SO_2 + 3\,O_2 \rightarrow 2\,SO_3$

(b) $N_2 + 3\,H_2 \rightarrow 2\,NH_3$

5 Chemical Reactions

Demolition of the Madison Hotel in Boston, Massachusetts (photograph courtesy of the Boston Globe*). The 19-story Madison Hotel was built over a two-year period in 1928–1930. On May 15, 1983, it was demolished in 9.7 seconds. Explosive charges were set on the lower floors to shatter support columns and to cause the rest of the building to implode, or cave in. This demolition illustrates the power of an exergonic (energy releasing) chemical reaction to cause physical change.*

Objectives   List four general types of chemical reactions. ☐ Define reaction mechanism, and state the premises upon which it is based. ☐ Draw an energy diagram and define its parts. ☐ Differentiate exothermic and endothermic reactions. ☐ Describe the action of a catalyst. ☐ List ways of increasing a reaction rate. ☐ Define side reaction. ☐ Define reversible reaction and equilibrium; describe the conditions required for these to occur. ☐ Define Le Châtelier's principle and describe, with an equation, its effect on a chemical reaction. ☐ Balance chemical equations. ☐ Calculate formula weight and the weight of 1.0 mole; use moles to calculate weight relationships in chemical reactions.

The digestion of food, the action of aspirin as an analgesic (pain reliever), human movement or thought—in fact, all aspects of life—depend on **chemical reactions**, changes in the composition of matter. When a chemical reaction occurs, atoms (or ions) and electrons are exchanged or transferred to yield new substances.

A chemical reaction is represented on paper by a **chemical equation**, a shorthand way of showing what is happening in a test tube, in our bodies, or in a blast furnace. A typical chemical equation is shown below:

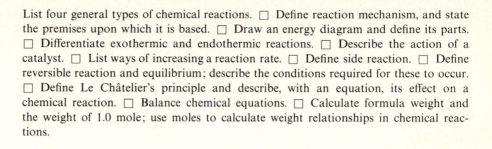

a chemical equation:

$$2\,H_2 + O_2 \longrightarrow 2\,H_2O$$

reactants   *product*

In this reaction, we start with two gases, hydrogen (H_2) and oxygen (O_2). We call these the *reactants*, and they appear on the left-hand side of the equation. On the other side of the arrow we find the *products*—in this case, water. The arrow means *yield* or *yields* and, in a chemical equation, is used to separate the reactants from the products. We read the equation as *hydrogen and oxygen yield water*, or *hydrogen undergoes reaction with oxygen to yield water*.

This chapter deals with chemical reactions and chemical equations. We will discuss some types of chemical reactions, the role of energy in chemical reactions, what affects the rate of a reaction, and the weight relationships of reactants and products.

5.1   TYPES OF CHEMICAL REACTIONS

From the great variety of substances we see around us, we might guess that there is an almost infinite range of types of chemical reactions. Surprisingly, this is not true. The number is actually rather limited. Table 5.1 lists a few general types of reactions, with A, B, C, and D representing atoms, ions, or molecules.

Table 5.1.   *Some General Types of Chemical Reactions*

Type of reaction	General equation	Example
combination	$A + B \rightarrow AB$	$C + O_2 \rightarrow CO_2$
decomposition	$AB \rightarrow A + B$	$H_2CO_3 \rightarrow H_2O + CO_2$
simple, or single, displacement	$A + BC \rightarrow AC + B$	$Cu + 2\,AgNO_3 \rightarrow$ $Cu(NO_3)_2 + 2\,Ag$
double displacement	$AB + CD \rightarrow AD + CB$	$NaOH + HCl \rightarrow NaCl + H_2O$

The names of the reaction types shown in Table 5.1 are reasonably self-explanatory. A *combination reaction* is a combination of two particles (atoms, ions, or molecules) to yield one molecule. A *decomposition reaction* is a reaction in which one larger molecule is converted to two or more smaller particles. A *simple displacement* is a reaction in which one atom, molecule, or ion replaces another. A *double displacement* is a reaction in which atoms or ions are exchanged. A simple example of each reaction type is shown in the table. Other examples will be found throughout the rest of the text.

Problem 5.1.   Categorize the following reactions according to the classifications in Table 5.1:

(a) $AgNO_3 + NaCl \longrightarrow AgCl + NaNO_3$

(b) $CH_3I + NaOH \longrightarrow CH_3OH + NaI$

(c) $CH_3I + NH_3 \longrightarrow CH_3\overset{+}{N}H_3 + I^-$

(d) $H_2O + CO_2 \longrightarrow H_2CO_3$

5.2   HOW DOES A REACTION OCCUR?

A chemical equation is a convenient device for representing a chemical reaction. However, the chemical equation does not tell us the whole story about a reaction. The equation does not tell us if two chemicals will explode when they are mixed or if they will not undergo reaction until they are heated. The equation does not tell us if *other* reactions might occur between the reactants. Furthermore, the equation does not tell us *how* a particular reaction occurs—that is, how reactant molecules get together and form products. The explanation of how a reaction occurs is called a **reaction mechanism** and is based on the following premises:

1. Molecules must collide with one another in order for a reaction to occur.
2. Colliding molecules need a certain minimum amount of energy for the collision to result in reaction.

Collision Between Molecules

The first of the preceding two premises is fairly obvious. If two molecules are five miles apart, they cannot undergo reaction with each other. Molecules must collide with each other in order to react.

Molecules in the gaseous state are continuously moving and collide frequently with one another. Molecules in a liquid also move around and collide. Molecules in a solid are restricted in their motion and are less likely to collide with other molecules. We find that reactions occur more readily in the gaseous and liquid states than in the solid state. Reactions of a solid generally occur only on the surface of the solid. For example, an iron bar rusts on the surface of the metal where the metal is in contact with oxygen and moisture. The rusting does not begin at the center of the bar.

In the gaseous and liquid states, collisions between molecules are commonplace. Yet, not all these collisions result in a chemical reaction. Most collisions between molecules result only in a change in direction of the movement of the molecules. This brings us to our second premise—for a reaction to occur, colliding molecules require a certain minimum amount of *energy*.

Energy of Colliding Molecules

Molecules have both *kinetic energy*, or energy of motion, and *internal energy*, which we may think of as the potential energy of vibrations of atoms and bonds within the molecules. When heat is added to a substance, the heat energy is converted to kinetic energy; on the average, the molecules move faster, collide more frequently, and collide with greater impact. When the molecules collide with greater impact, some of the kinetic energy is converted to internal energy; the atoms within a molecule vibrate more vigorously and bonds within the molecule are more likely to be broken. The result of bond-breaking is a chemical reaction.

An example of a chemical reaction is the reaction of methane and oxygen:

$$CH_4 \ + \ 2\,O_2 \ \xrightarrow{\text{spark}} \ CO_2 + 2\,H_2O + \text{heat}$$

methane   oxygen   carbon   water
                                    dioxide

Methane is the principal ingredient of natural gas. We know that it will burn readily in the presence of oxygen and will form carbon dioxide and water. However, if we just mix methane and oxygen in a bottle, the methane does not burn. Energy in the form of a spark is needed to start the reaction. In the preceding equation, we added the word "spark" over the arrow to show that it is a necessary reaction condition. Showing extra information about a reaction in this way is common in chemistry.

When a spark is introduced to a methane-oxygen mixture, the energy is transferred to a few of the molecules. These few molecules gain enough internal energy that, when they collide, they undergo reaction. We say they have gained enough energy to cross the **energy barrier** and go to products.

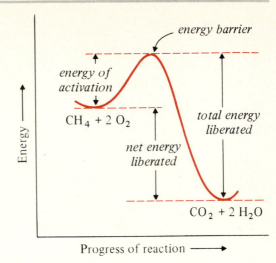

Figure 5.1. *Energy diagram for the reaction of methane and oxygen.*

As methane and oxygen molecules cross the energy barrier to yield carbon dioxide and water, energy is released. Some of this energy is transferred to other molecules of methane and oxygen in the mixture; then, when they collide, they too can cross the energy barrier and give up still more energy. This process continues until the entire mixture of methane and oxygen has undergone reaction. We can see and feel the excess energy released from the overall process as light and heat.

Energy Diagrams

The energy requirements of a reaction can be represented by an **energy diagram**, a graph of the progress of reactants going to products versus the energy of the intermediate species. Figure 5.1 shows an energy diagram for the burning of methane. Starting at the left-hand side of the graph, we see that the reactants $CH_4 + 2 O_2$ have a certain average amount of internal energy. As we move to the right on the graph, we see that the average energy of the reactants must be increased to bring them up to the energy barrier. Once they are over the energy barrier, the reacting species *lose* energy as they become products. The process is similar to driving a car over a mountain pass. Energy must be supplied to drive to the top of the pass, but rolling downhill is easy.

The energy required for the reactants to reach the top of the energy barrier is called the **energy of activation**. The energy lost by the reacting species as they proceed from the top of the energy barrier to the average energy of the products is the *total energy liberated*. The difference between the average energy of reactants and that of products is the *net energy liberated* (or, in some cases, gained).

Not all substances have the same amount of potential energy. From the

Figure 5.2. *Energy diagrams showing the differences between an exothermic reaction and an endothermic reaction.*

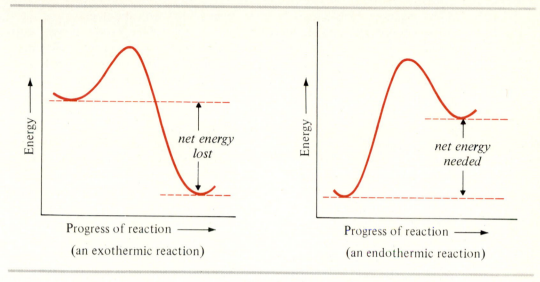

(an exothermic reaction)   (an endothermic reaction)

energy diagram, we can see that carbon dioxide and water contain less energy than methane and oxygen. When the products of a reaction have less energy than do the reactants, there is a net release of energy during the course of the reaction. When methane burns, heat and light are given off. A reaction that gives off heat as it proceeds is called an **exothermic reaction**. (*Exo-* means *out*, as in the word exit. *Therm* means *heat*, as in thermal underwear or thermometer.)

If the products of a reaction contain more energy than do the reactants, the reaction is called an **endothermic reaction**.* (*Endo-* comes from the Greek word *endon*, meaning *within*.) We must supply heat continuously to an endothermic reaction to keep it going. The heat helps the molecules over the energy barrier and makes up the energy difference between the reactants and the products. (See Figure 5.2.)

Problem 5.2.   Refer to the methane-oxygen energy diagram in Figure 5.1.
(a) Would you expect the following reaction to be exothermic or endothermic?

$$CO_2 + 2\ H_2O \longrightarrow CH_4 + 2\ O_2$$

(b) Do the products of the reaction in (a) contain more energy or less energy than do the reactants?
(c) Would this reaction give off energy once initiated?

---

* Because energy exists in forms other than heat, sometimes the more general terms *exergonic* (energy-releasing) and *endergonic* (energy-absorbing) are used to describe reactions. These words are from the Greek word *ergon*, which means work or energy.

5.3   CATALYSTS

Some reactions have energy barriers so high that we cannot make the reactions proceed by heating; however, the same reactions may proceed easily when the appropriate catalyst is present. A **catalyst** is a substance that does not itself undergo reaction under the reaction conditions (or is regenerated during the course of the reaction), but lowers the energy barrier for the reaction of other substances. Since it does not undergo a net reaction itself, only a small amount of a catalyst is needed to promote the reaction of comparatively large amounts of reactants. Different catalysts function in different ways. In some cases, a catalyst interacts with a reactant molecule to weaken its bonds so the reaction can proceed.

Living organisms could not function without catalysts, which in living systems are called **enzymes**. An enzyme permits a reaction to take place quickly at body temperature under mild conditions, whereas the same reaction may require hours of heating in the laboratory. For example, *hydrolysis* (cleavage with water) of protein molecules, a process that is essential in the digestion of meats, eggs, etc., requires boiling with strong acid in the laboratory. In the human digestive tract, this same reaction occurs at 37°C (body temperature) and under less acidic conditions because the reaction is catalyzed by enzymes such as pepsin and trypsin.

a large protein molecule + H_2O $\xrightarrow{\text{enzymes}}$ many small molecules of amino acids

Figure 5.3 shows two typical energy curves superimposed—one for a

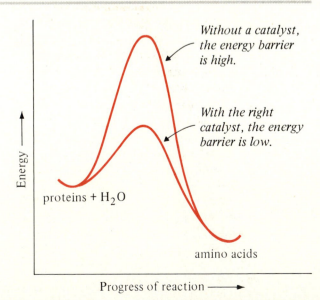

Figure 5.3.   *A catalyst lowers the energy barrier for a reaction.*

Without a catalyst, the energy barrier is high.

With the right catalyst, the energy barrier is low.

proteins + H_2O

amino acids

Energy ⟶

Progress of reaction ⟶

reaction with a catalyst and one for the same reaction without a catalyst. It can be seen that the energies of the reactants and products are not changed by the addition of the catalyst; only the height of the energy barrier is changed.

5.4   RATE OF A REACTION

The speed at which a reaction proceeds is called the **rate of reaction**. The rate of any reaction depends on the number of molecules of reactants passing over the energy barrier in a given period of time.

Experimentally, we find that certain *types* of reactions proceed at different rates. Reactions of ionic compounds in solution usually proceed very quickly because of the mobility of the ions and the low energy of activation needed for combination. In reactions of covalently bonded compounds, bonds must be broken and reformed. In general, reactions of covalent compounds have much higher energies of activation and proceed more slowly than ionic reactions.

The rate of a reaction can be increased in a number of ways. An *increase in temperature* usually increases the rate of a reaction because the added heat increases the average kinetic and internal energies of the reactant molecules. The increased energy allows molecules to cross the energy barrier at a greater rate for two reasons. First, the number of molecular collisions is increased. Second, each collision is more likely to cause a reaction. Roughly, a temperature increase of 10°C (a difference of about 18°F) doubles the rate of reaction. When a person is running a fever, the person's biochemical reactions proceed at a faster rate than normal, as evidenced by increased respiration and pulse rate. For each 1°C rise in body temperature, the body tissues require about 13% more oxygen.

Conversely, a *decrease in temperature* usually causes a decrease in the rate of a reaction. In a human, a drop in body temperature, called *hypothermia*, of only 1–2°C (2–3°F) causes violent shivering (the body's attempt to rewarm itself). A further drop in temperature can cause stupor, unconsciousness, and finally death, which usually occurs when the internal body temperature reaches 25°C (78°F).

Besides an increase in temperature, another factor that increases the rate of reaction is an *increase in the concentration* of a reactant. The increased concentration means simply that there are more molecules of a reactant in a given volume. Therefore, the number of collisions between reactant molecules is greater, and more molecules can cross the energy barrier in a given period of time. Air, for example, contains about 20% oxygen. A person with respiratory problems (such as pneumonia or emphysema) may be given air enriched with oxygen, because increasing the concentration of oxygen increases the efficiency of respiration.

A reaction rate can also be increased by the *addition of a catalyst*. As we have mentioned, a catalyst lowers the energy barrier and allows molecules to cross more easily.

5.5 SIDE REACTIONS

In medicine, drugs are prescribed to relieve pain, to inhibit bacterial growth, and for other purposes. Many drugs also cause *side effects*, generally undesirable, in some or all users. For example, a side effect of a birth control pill is water retention and weight gain.

In chemistry, a reaction that occurs simultaneously with the principal reaction is called a **side reaction**. When methane burns, we expect to get carbon dioxide and water, but carbon monoxide (CO) may also be formed. Because carbon monoxide is poisonous, most states in the United States have laws that require proper venting of gas furnaces so that any carbon monoxide formed will pass outside.

the expected reaction:

$$CH_4 + 2\,O_2 \longrightarrow \underset{\text{carbon dioxide}}{CO_2} + 2\,H_2O$$

a side reaction:

$$2\,CH_4 + 3\,O_2 \longrightarrow \underset{\text{carbon monoxide}}{2\,CO} + 4\,H_2O$$

5.6 REVERSIBLE REACTIONS AND CHEMICAL EQUILIBRIA

The tart taste of a carbonated beverage is due in part to the presence of carbonic acid (H_2CO_3). When a beverage is carbonated, carbon dioxide gas is forced under pressure into solution. The dissolved carbon dioxide undergoes combination with the water to form carbonic acid.

$$CO_2 + H_2O \longrightarrow \underset{\text{carbonic acid}}{H_2CO_3}$$

When a bottle of soda is opened, it fizzes. When the pressure is released, carbon dioxide is formed from the dissociation of carbonic acid in the soda.

$$H_2CO_3 \longrightarrow CO_2 + H_2O$$

These two reactions are just the reverse of each other. Carbon dioxide and water form carbonic acid, while carbonic acid forms carbon dioxide and water. The two reactions can proceed either backward or forward. Such a reaction is called a **reversible reaction**. We use two arrows ($\rightleftharpoons$), instead of just one, to show the reversibility of a reaction.

$$CO_2 + H_2O \rightleftharpoons H_2CO_3$$

The double arrow means a reversible reaction.

Figure 5.4.   *Energy diagrams for a reversible reaction and an irreversible reaction.*

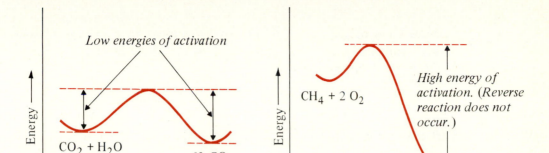

This particular reversible reaction plays an important role in the functioning of the human body. The body's cells convert glucose, obtained from carbohydrates (sugars and starches), to H_2O, CO_2, and energy. The CO_2 dissolves in the blood to yield H_2CO_3. When the blood reaches the lungs, some of the H_2CO_3 decomposes to H_2O and gaseous CO_2, which is then exhaled.

For a reaction to be reversible, the energies of activation of both the forward reaction and the reverse reaction must be fairly low. This means that the energy of the reactants and that of the products must be fairly close to each other. An energy diagram for the reversible reaction of carbon dioxide and water is shown in Figure 5.4.

If the energy of the products of a reaction is much lower than the energy of the reactants, there is a high energy of activation for the reverse reaction. In this case, the reverse reaction does not proceed. For example, the burning of methane in oxygen produces carbon dioxide and water. The energy that would be required to form methane and oxygen from carbon dioxide and water is prohibitive. No reverse reaction occurs. Therefore, we say that such a reaction is *irreversible*.

If an aqueous solution of H_2CO_3 is enclosed in a stoppered bottle, some of the H_2CO_3 decomposes to H_2O and CO_2. However, some of the CO_2 gas above the solution collides with the surface of the solution and recombines with water to yield H_2CO_3. Eventually, the amount of CO_2 gas above the liquid becomes constant, even though molecules are going back and forth from H_2CO_3 to CO_2, because the number of CO_2 molecules returning to the solution is equal to the number leaving.

A state of balance for a reversible reaction, in which the rates of the forward and reverse reactions are equal, is called a **chemical equilibrium**.

Because molecules are actively moving back and forth, a chemical equilibrium is also called a **dynamic equilibrium.** (See Figure 5.5.)

Problem 5.3.   An acid is a compound that can lose H^+ to water. For example, vinegar contains about 5% acetic acid, which undergoes the following reversible reaction with water:

$$CH_3CO_2H + H_2O \rightleftharpoons CH_3CO_2^- + H_3O^+$$

acetic acid

Carbonic acid also can lose H^+ to water. Write an equation for this reaction.

Problem 5.4.   When a beaker of distilled water is allowed to sit open to the air, the water becomes slightly acidic (contains H_3O^+). Explain.

Le Châtelier's Principle

If the stopper is removed from a jar of aqueous H_2CO_3 or if a can of carbonated beverage is opened, CO_2 gas can escape. In this case, the system is no longer at equilibrium, and more H_2CO_3 decomposes than is formed. In 1888,

Figure 5.5.   A dynamic equilibrium is a state of balance in which the rate of the forward reaction and the rate of the reverse reaction are equal.

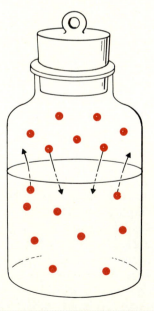

equal numbers of CO_2 molecules leaving and returning

the French chemist Henry Louis Le Châtelier described what happens when one of the factors keeping a system in equilibrium is changed.

Le Châtelier's Principle.   If one of the factors affecting an equilibrium is changed, the equilibrium will shift in the direction toward restoring the equilibrium to its original state.

Let us use the H_2CO_3-CO_2 equilibrium to illustrate Le Châtelier's principle. If CO_2 is allowed to escape, the equilibrium shifts to the right (as we have written the equation). This shift relieves the stress on the system due to the lack of CO_2 over the solution by allowing more CO_2 to form.

$$H_2CO_3 \rightleftharpoons H_2O + CO_2$$

*In a closed system, the concentrations
of H_2CO_3 and CO_2 remain constant.*

$$H_2CO_3 \rightleftharpoons H_2O + CO_2\uparrow$$

*If CO_2 is allowed to bubble away,
more H_2CO_3 decomposes.*

In any reversible reaction in equilibrium, adding a reactant to the left side of the equation (or removing a product from the right side) drives the reaction to the right.

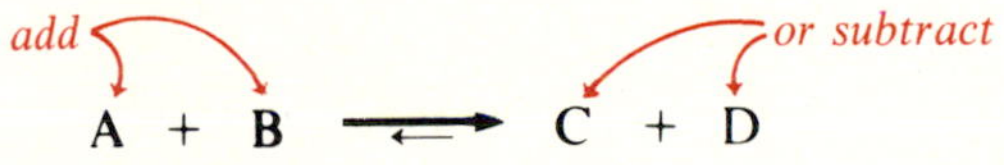

add ⟋   *or subtract*
$$A \ + \ B \ \rightleftharpoons \ C \ + \ D$$

Conversely, adding one of the substances on the right (or removing one on the left) drives the reaction to the left.

subtract ⟋   *or add*
$$A \ + \ B \ \rightleftharpoons \ C \ + \ D$$

Problem 5.5.   When ammonia gas (NH_3) is dissolved in water, small amounts of ammonium ions ($NH_4{}^+$) and hydroxide ions (OH^-) are formed in the following reversible reaction:

$$NH_3 + H_2O \rightleftharpoons NH_4{}^+ + OH^-$$

(a) What would be the effect on this reaction of dissolving ammonium chloride ($NH_4{}^+ + Cl^-$) in the ammonia solution?
(b) What would be the effect of increasing the amount of NH_3 in the solution?

5.7   BALANCING CHEMICAL EQUATIONS

We defined chemical equations in the introduction to this chapter. Let us take a more detailed look at chemical equations and see how we set them up and use them.

Mass is neither gained nor lost in a chemical reaction; what goes in has to come out. Both sides of a chemical equation must have the same number of each type of atom. If there are four hydrogen atoms on the reactant side of an equation, there must also be four hydrogen atoms on the product side.

$$2\,H_2 + O_2 \longrightarrow 2\,H_2O$$

reactants   *product*

Each side of the equation shows four H's and two O's.

A subscript number in a formula cannot be changed to balance an equation, because changing a subscript amounts to changing a formula. Instead, we use a **coefficient**, a number placed before a formula, to indicate the proper numbers of atoms in an equation. In the preceding equation, the coefficients tell us that *two* molecules of H_2 undergo reaction with *one* molecule of oxygen to yield *two* molecules of water. (Note that the coefficient 1 is not written in equations.) An equation that contains the proper coefficients to equalize the numbers of atoms on each side of an equation is said to be a **balanced equation**.

Let us consider the reaction of sodium metal with chlorine gas. Because it has been investigated in the laboratory, we know that sodium and chlorine undergo reaction to yield sodium chloride.

not balanced:

$$Na \;+\; Cl_2 \longrightarrow NaCl$$

sodium   chlorine   sodium chloride

The preceding equation is not balanced. We do not have the same number of chlorine atoms on each side. We can put a 2 in front of NaCl. This gives us two Cl atoms on each side.

not balanced:

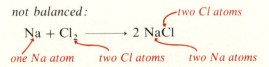

$$Na + Cl_2 \longrightarrow 2\,NaCl$$

one Na atom   two Cl atoms   two Na atoms   *two Cl atoms*

Now the sodium is not balanced. We must put a 2 in front of Na on the left-hand side of the equation. Now all the atoms are balanced.

balanced:

$$2\,Na + Cl_2 \longrightarrow 2\,NaCl$$

Most chemical equations can be balanced by just such a hit-or-miss method. We call this method of balancing equations *balancing by inspection*.

Problem 5.6.   Balance the following equations:
(a) $Mg(OH)_2 + HCl \rightarrow MgCl_2 + H_2O$
(b) $CaCO_3 + HCl \rightarrow CaCl_2 + H_2O + CO_2$
(c) $Fe + O_2 \rightarrow Fe_2O_3$

5.8   WEIGHT RELATIONSHIPS FROM CHEMICAL EQUATIONS

Formula Weight

In Chapter 3, you learned that atomic mass is the average mass in atomic mass units (amu) of one atom of an element. For example, the atomic mass of carbon is 12.0 amu. The **formula weight**, or **molecular weight**, of an ionic or covalent compound is the *sum of the atomic masses in the compound's formula*.

formula weight = sum of atomic masses in formula

Example   Calculate the formula weights to the nearest 0.1 amu of (a) NaCl, (b) $CaCl_2$, and (c) H_2O.

Solution
(a)   atomic mass of Na   = 23.0 amu
      atomic mass of Cl   = 35.5 amu
      ─────────────────────────────────
      formula weight of NaCl = 58.5 amu
      (sum of atomic masses)

(b)   atomic mass of Ca   = 40.1 amu
      atomic mass of 2 Cl   = 2×35.5 amu = 71.0 amu
      ─────────────────────────────────
      formula weight of $CaCl_2$ = 111.1 amu

(c)   atomic mass of 2 H   = 2×1.0 amu = 2.0 amu
      atomic mass of O   = 16.0 amu
      ─────────────────────────────────
      formula weight of H_2O = 18.0 amu

Problem 5.7.   Using the atomic masses in the periodic table (to one decimal place), calculate the formula weights of the following compounds:
(a) KCl   (b) $MgCl_2$ (c) Al_2O_3 (d) $CH_2{=}CH_2$

Problem 5.8.   Using the atomic mass of sodium and appropriate formula weights, calculate the *weight percentage* of sodium ions in (a) sodium chloride (table salt, NaCl) and (b) sodium bicarbonate (baking soda, $NaHCO_3$). (See the appendix for calculations of percentages.)

The Mole

It is extremely important in research, in industry, and in clinical laboratories to know exactly how much of one substance will undergo complete reaction with a certain weight of another substance. Otherwise, a reactant present in excess may just be wasted. Because atoms, ions, and molecules have different masses, we cannot simply mix equal weights of two substances. To circumvent this problem, the concept of the mole was developed.

The **weight of 1.0 mole** of any substance is its formula weight expressed in grams instead of in atomic mass units.

Example   Calculate the weight of 1.0 mole of H_2SO_4.

Solution   (a) Calculate the formula weight of H_2SO_4 in atomic mass units, using the periodic table to find the atomic masses.

formula weight of H_2SO_4 = (2 × atomic mass of H)

$$+ (1 \times \text{atomic mass of S}) + (4 \times \text{atomic mass of O})$$

formula weight of H_2SO_4 = (2 × 1.0 amu) + (1 × 32.1 amu) + (4 × 16.0 amu)

$$= 2.0 \text{ amu} + 32.1 \text{ amu} + 64.0 \text{ amu}$$

$$= 98.1 \text{ amu}$$

(b) Change the units of the formula weight to grams:

$$\text{weight of 1.0 mole} = 98.1 \text{ g}$$

Example   Calculate the weight of 1.0 mole of H_2O.

Solution   Find the formula weight of H_2O and change the units to grams:

$$\text{formula weight of } H_2O = 2(1.0 \text{ amu}) + 16.0 \text{ amu}$$

$$= 18.0 \text{ amu}$$

$$\text{weight of 1.0 mole of } H_2O = 18.0 \text{ g}$$

Problem 5.9.   Calculate the weight of 1.0 mole of the following substances. [*Hint:* In (c), the formula weight is the same as the atomic mass.]
(a) CO_2 (b) Cl_2 (c) Na

The most important ramification of the concept of moles is that *1.0 mole of any substance contains the same number of formula units (molecules, atoms, ion pairs, etc.) as 1.0 mole of any other substance.* The number of formula units in 1.0 mole of any substance is 6.02×10^{23} or, rounded, 6×10^{23} units. The number 6×10^{23} is called **Avogadro's number** in honor of the Italian physicist Amedeo Avogadro (1776–1856), an early worker in the study of atomic theory.

Let us look at some examples to illustrate what we mean by "formula units."

formula unit circled

1.0 mole H_2O contains 6×10^{23} molecules H_2O

1.0 mole $\text{Na}^+ \text{Cl}^-$ contains 6×10^{23} ion pairs of $Na^+ \; Cl^-$

1.0 mole OH^- means 6×10^{23} OH^- ions

A mole of one substance contains the same number of units as a mole of another because the weight of a mole is proportional to formula weight. For example, it has been determined that one atom of carbon (12.0 amu) is 12 times heavier than one atom of hydrogen (1.0 amu). To have equal numbers of atoms of C and H, we need a weight of carbon 12 times greater than the weight of hydrogen. Indeed, 1.0 mole of C has a weight 12 times that of 1.0 mole of H.

Each of the following examples is 1.0 mole and contains 6×10^{23} units:

18.0 g H_2O

58.5 g NaCl

17.0 g OH^- (if we could weigh out negative ions alone)

Moles and Chemical Equations

Let us consider the formation of water from hydrogen and oxygen. The balanced equation tells us that two molecules of H_2 undergo reaction with one molecule of O_2 to yield two molecules of H_2O. If this is the case, then 12×10^{23} molecules (2.0 moles) of H_2 react with 6×10^{23} molecules (1.0 mole) of O_2 to yield 12×10^{23} molecules (2.0 moles) of H_2O.

$$2\, H_2 + O_2 \longrightarrow 2\, H_2O$$

$$2 \text{ moles } H_2 + 1 \text{ mole } O_2 \longrightarrow 2 \text{ moles } H_2O$$

$$4.0 \text{ g } H_2 + 32.0 \text{ g } O_2 \longrightarrow 36.0 \text{ g } H_2O$$

We cannot work with individual atoms and molecules in the laboratory, but we can work with weights of substances; and the concept of moles allows us to do just that. Using a balanced chemical equation and moles, we can calculate the *actual weights of reactants needed for complete reaction and the weights of products to expect.*

The weights of reactants and products in a reaction need not be just 1.0 mole—they can be varied. The important point is that the *molar ratio* is followed. For example,

$$2\ H_2\quad +\quad O_2\quad \longrightarrow\quad 2\ H_2O$$

$$2\ moles\ +\quad 1\ mole\quad \longrightarrow\quad 2\ moles$$

$$or\quad 10\ moles\ +\quad 5\ moles\quad \longrightarrow\quad 10\ moles$$

$$or\quad 0.5\ mole\ +\ 0.25\ mole\quad \longrightarrow\quad 0.5\ mole$$

Example   Iron metal undergoes reaction with oxygen to form ferric oxide (Fe_2O_3), or rust. How many moles of oxygen would be used for the complete rusting of 1.0 mole of Fe?

Solution   The balanced equation is

$$4\ Fe\quad +\quad 3\ O_2\quad \longrightarrow\quad 2\ Fe_2O_3$$

1.0 mole   *x* moles

given

Using the factor-label method,

conversion factor:
molar ratio from equation

given

$$x\ \text{moles}\ O_2 = 1.0\ \text{mole Fe} \times \left(\frac{3\ \text{moles}\ O_2}{4\ \text{moles Fe}}\right)$$

Cancel units and solve:

$$x\ \text{moles}\ O_2 = 1.0 \times \tfrac{3}{4}\ \text{mole}\ O_2$$

$$= 0.75\ \text{mole}\ O_2$$

Problem 5.10.   Balance the following equation and calculate the number of moles of water vapor liberated when 0.50 mole of ethanol (CH_3CH_2OH) is burned.

$$CH_3CH_2OH + O_2 \xrightarrow{\text{spark}} CO_2 + H_2O$$

Conversions Between Moles and Grams

Converting Moles to Grams. To calculate the weights of reactants needed or products expected in a chemical reaction, regardless of the size of the reaction, a chemist needs to know the weight of a fraction of a mole or of several moles. To convert a certain number of moles to grams, we use the following equation.

$$x \text{ grams} = \text{number of moles} \times \frac{\text{grams}}{1.0 \text{ mole}}$$

Example   How many grams of HCl are present in 0.33 mole?

Solution   (1) Find the formula weight of HCl:

$$1.0 \text{ amu} + 35.5 \text{ amu} = 36.5 \text{ amu}$$

(2) Set up the equation:

$$x \text{ grams} = \text{number of moles} \times \frac{\text{grams}}{1.0 \text{ mole}}$$

(3) Substitute, cancel units, and solve:

$$x \text{ grams} = 0.33 \text{ mole} \times \frac{36.5 \text{ g}}{1.0 \text{ mole}}$$

$$= 12 \text{ g of HCl}$$

Problem 5.11. Convert the following molar quantities to grams:
(a) 0.25 mole CH_3CH_2OH (b) 12.0 moles H_2O (c) 0.65 mole NaCl

Converting Grams to Moles. To convert a certain number of grams of a substance to moles, we use the following equation. (This is the same as the equation used in the preceding section; it is simply rearranged for convenience.)

$$\text{number of moles} = \frac{\text{weight of sample in grams}}{\text{grams}/1.0 \text{ mole}}$$

Example   How many moles of NaOH are in 10.0 g?

Solution   (1) Find the formula weight of NaOH:

$$23.0 \text{ amu} + 16.0 \text{ amu} + 1.0 \text{ amu} = 40.0 \text{ amu}$$

(2) Set up the equation:

$$\text{number of moles} = \frac{\text{number of grams}}{\text{grams}/1.0 \text{ mole}}$$

(3) Substitute and solve:

$$\text{number of moles in 10.0 g of NaOH} = \frac{10.0 \cancel{g}}{40.0 \cancel{g}/1.0 \text{ mole}}$$

$$= 0.25 \text{ mole}$$

Problem 5.12.   Convert each of the following sample weights to moles:
(a) 5.0 g NaCl   (b) 50.0 g H_2O (c) 0.90 g CH_3CH_2OH

Use of Moles and Grams in an Equation

Now let us consider chemical reactions on a weight basis. Given the weight of any one substance in a balanced equation, whether it is a reactant or a product, we can calculate all the other weights. However, *we still use moles because only the moles give us the right proportions of reactants and products.* This is the procedure that we follow:

1. Balance the equation.
2. Convert grams to moles.
3. Calculate the desired number of moles from the proportions in the balanced equation.
4. Finally, convert the desired number of moles to grams.

Example   Starches and sugars are converted to glucose in the digestive system. The glucose is then absorbed into the bloodstream and carried to the cells of the body, where some of it is oxidized to carbon dioxide and water. (The series of reactions in this conversion gives off energy that can be used by the body.) How many grams of carbon dioxide would be produced from 5.0 g of glucose?

$$C_6H_{12}O_6 + 6 \ O_2 \xrightarrow{\text{many steps}} 6 \ CO_2 + 6 \ H_2O$$
$$\text{glucose}$$

Solution   (1) The equation is balanced.
(2) Since the formula weight of glucose = 180 amu and the weight of 1.0 mole of glucose = 180 g,

$$\text{number of moles of glucose} = \frac{5.0 \cancel{g}}{180 \cancel{g}/\text{mole}} = 0.028 \text{ mole}$$

(3) From the balanced equation,

$$\text{number of moles of } CO_2 \text{ produced} = 6 \times \text{moles of glucose}$$
$$= 6 \times 0.028 \text{ mole}$$
$$= 0.17 \text{ mole}$$

(4) Since the formula weight of $CO_2 = 44.0$ amu and the weight of 1.0 mole of $CO_2 = 44.0$ g,

$$\text{weight of } CO_2 \text{ in } 0.17 \text{ mole} = 0.17 \; \cancel{\text{mole}} \times 44.0 \text{ g/}\cancel{\text{mole}}$$
$$= 7.5 \text{ g}$$

Problem 5.13. The burning of hydrocarbons (compounds containing only C and H) such as gasoline, kerosene, natural gas, and propane supplies most of the energy that society uses. There is evidence that this form of energy generation is resulting in a net increase in atmospheric CO_2. Calculate how many grams of CO_2 are produced when 100 grams of propane is burned completely. (*Unbalanced equation:* $C_3H_8 + O_2 \rightarrow CO_2 + H_2O$.)

SUMMARY

A chemical reaction is represented by a *balanced chemical equation*, which shows the formulas and proportions of *reactants* and *products*. Some types of reactions are *combination*, *decomposition*, *simple displacement*, and *double displacement*.

To undergo reaction, molecules must collide and have sufficient internal energy (*energy of activation*) to cross the energy barrier. An *energy diagram* is a graph depicting the energy requirements of a reaction.

In an *exothermic reaction*, which releases more energy than it takes in, the products contain less potential energy than do the reactants. In an *endothermic reaction*, which absorbs more energy than it releases, the products contain more energy than do the reactants.

A *catalyst* can lower the energy of activation of a reaction, causing the reaction to proceed more readily, but a catalyst does not affect the relative energies of reactants and products. The *rate of a reaction* is affected by temperature, concentration of reactants, and an appropriate catalyst.

A *side reaction* is a second reaction that a set of reactants can undergo. A *reversible reaction* is one that can proceed forward or backward. A *chemical equilibrium* is a state of balance in which the rates of forward and reverse reactions of a reversible reaction are equal. If equilibrium conditions are changed, the equilibrium will shift so as to restore balance (*Le Châtelier's principle*).

One *mole* of a substance has a weight equal to the *formula weight* in grams and contains 6.02×10^{23} units, such as molecules or atoms (*Avogadro's number*). The coefficients in a balanced chemical equation show the molar ratios of reactants and products. The following equations can be used to convert grams to moles and moles to grams.

$$x \text{ g} = \text{number of moles} \times (\text{grams}/1.0 \text{ mole})$$

$$\text{number of moles} = \frac{\text{weight in g}}{(\text{grams}/1.0 \text{ mole})}$$

To carry out weight calculations for a reaction, one always starts with a balanced equation, converts grams to moles, carries out the desired calculation, then converts moles to grams.

KEY TERMS

chemical reaction	double displacement reaction	reversible reaction
chemical equation	reaction mechanism	chemical equilibrium
reactant	energy of activation	Le Châtelier's principle
product	exothermic reaction	formula weight
combination reaction	endothermic reaction	mole
decomposition reaction	catalyst	Avogadro's number
displacement reaction	side reaction	

STUDY PROBLEMS

5.14. Express the following equation in words:

$$2 \text{ Na} + 2 \text{ H}_2\text{O} \longrightarrow 2 \text{ NaOH} + \text{H}_2$$

5.15. Without referring to the text, label parts (a)–(e) of the following energy diagram:

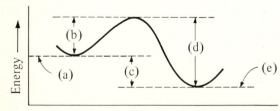

$$2 \text{ H}_2 + \text{O}_2 \longrightarrow 2 \text{ H}_2\text{O}$$

5.16. Explain why a very small quantity of catalyst can speed the reaction of a large quantity of reactants.

5.17. Explain briefly how a reversible reaction can be recognized by its energy diagram.

5.18. The following equation shows how hemoglobin (represented as HHb) reacts with oxygen in the capillaries in the lungs before it is carried to body's tissues by the blood.

$$\text{HHb} + \text{O}_2 \rightleftharpoons \text{H}^+ + \text{HbO}_2^{\,-}$$

(a) What would be the effect on this equilibrium if the blood contained excess H^+ ions? What would be the effect on the body's tissues?

(b) What would be the effect on this equilibrium if a person were to breathe pure oxygen instead of air (21% oxygen)?

5.19. Balance the following equations:
(a) $\text{K} + \text{H}_2\text{O} \rightarrow \text{KOH} + \text{H}_2$
(b) $\text{H}_2 + \text{Br}_2 \rightarrow \text{HBr}$
(c) $\text{CH}_3\text{CH}_3 + \text{O}_2 \rightarrow \text{CO}_2 + \text{H}_2\text{O}$
(d) $\text{Zn} + \text{HCl} \rightarrow \text{ZnCl}_2 + \text{H}_2$

5.20. Calculate the formula weights (to the nearest 0.1 amu) of (a) HCl and (b) CH_4.

5.21. Given the following equation, what would be the weight ratio of the reactants?

$$H_2SO_4 + 2 \; NaOH \longrightarrow Na_2SO_4 + 2 \; H_2O$$

5.22. How many grams are present in 0.125 mole of (a) CH_4 and (b) $CHCl_3$?

5.23. How many moles are present in 25.0 grams of (a) H_2O and (b) NaCl?

5.24. Fill in the blanks:

$$H_2SO_4 \quad + \quad 2 \; NaOH \longrightarrow$$

10 moles   (a) _____ moles

$$Na_2SO_4 \quad + \quad 2 \; H_2O$$

(b) _____ moles   (c) _____ moles

5.25. Chlorine gas (Cl_2) can be produced from molten sodium chloride (NaCl) by an electric current. How many grams of sodium chloride would be required to produce 142 grams of chlorine gas?

5.26. In the presence of a catalyst, methane (CH_4) can react with chlorine gas (Cl_2) to produce carbon tetrachloride (CCl_4) and hydrogen chloride gas (HCl). If 4.00 grams of methane were converted to CCl_4 by this procedure, how many grams of HCl would be produced?

6 Gases and Their Properties

A hot-air balloon. (© Jack Iddon, 1983.) When an unconfined gas is heated, it expands to occupy a larger volume. For this reason, hot air is less dense than cool air, and a balloon filled with hot air rises in cool air. Balloonists use a portable propane heater to regulate the temperature of the air in the balloon and thus to control the rate of ascent or descent.

Objectives   Describe the composition of the atmosphere, the nitrogen cycle, and the oxygen cycle; describe the origins and importance of ozone (optional). ☐ Describe the uses of some gases in medicine (optional). ☐ List the assumptions in the kinetic molecular theory of gases; relate this theory to the properties of gases. ☐ Define atmosphere of pressure and mmHg. ☐ Calculate a change in pressure or volume of a gas by Boyle's law, and calculate a change in temperature or volume of a gas by Charles' law (optional). ☐ Relate the partial pressures of gases in a mixture to the total pressure; describe how partial pressures aid the body's oxygen–carbon dioxide exchange. ☐ Relate the solubility of a gas to temperature and pressure. ☐ Define vapor pressure and describe its relation to evaporation and boiling of a liquid.

Our earth is bathed in a mixture of gases that we call the atmosphere. Although we cannot see, smell, or taste the common atmospheric gases, life depends on them. This chapter will describe the roles of oxygen, nitrogen, and ozone in the atmosphere; how oxygen and nitrogen are used by living systems; and some medical uses of gases. We will discuss some physical properties of gases, how these properties can be explained by the "kinetic molecular theory," and how some of these properties control, in part, the oxygen–carbon dioxide balance in the body. Finally, the solubilities of gases in liquids and liquid-gas conversions will be considered.

■ 6.1 OXYGEN AND NITROGEN IN THE ATMOSPHERE (optional)

The composition of the atmosphere is shown in Table 6.1. Most of the air that surrounds us is relatively inert nitrogen gas (N_2). About one-fifth of the air is

Table 6.1.   *Composition of the Atmosphere*

Name of gas	Formula	Percent of atmosphere	Comments
nitrogen	N_2	78	relatively inert, indirectly used by plants
oxygen	O_2	21	supports respiration, combustion, and aerobic decay
argon	Ar	0.93	inert
carbon dioxide	CO_2	0.03	used by plants to synthesize carbohydrates
others[a]	—	0.04	—

[a] Water vapor is also present in varying amounts. Relative humidity of 20% is approximately equivalent to air containing 5% water vapor.

Figure 6.1.   *The oxygen and nitrogen cycles.*

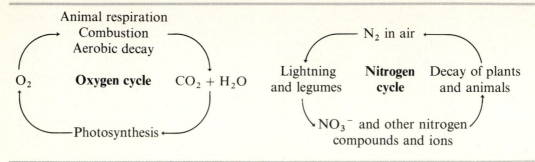

oxygen (O_2), while other gases are found in only trace amounts. The composition of the atmosphere is relatively constant, but it is in a continuous state of renewal. Let us consider how oxygen and nitrogen are used and renewed.

The Oxygen Cycle

In animals, oxygen is used to oxidize foodstuffs and produce the energy that supports life. Carbon dioxide is a product of these oxidation reactions. When wood, gasoline, natural gas, or any other carbonaceous fuel is burned, oxygen is used, and carbon dioxide is again produced. When plant or animal material decays, the complex compounds of the once-living systems are oxidized; again, oxygen is used and carbon dioxide is produced. (This type of decay is called *aerobic decay*, aerobic referring to a process using air or oxygen. Decay in the absence of air, such as in a cesspool or septic tank, is called *anaerobic decay*.) Therefore, we have three major types of oxygen consumption—respiration, combustion, and aerobic decay—each one a producer of carbon dioxide.

From Table 6.1, you can see that the atmosphere is composed of less than 1% carbon dioxide. The carbon dioxide produced by respiration, decay, and combustion is removed from the atmosphere by plant life on land and in the oceans. Green plants use carbon dioxide and water in a lengthy series of enzyme-catalyzed reactions called *photosynthesis* to produce carbohydrates and oxygen. We call the consumption and subsequent regeneration of oxygen the **oxygen cycle**. (See Figure 6.1.) It is this cycle that keeps the percentages of oxygen and carbon dioxide in the atmosphere surprisingly constant.

The Nitrogen Cycle

Plants obtain carbon, hydrogen, and oxygen from air and water. Another element important to plants is nitrogen, which is an indispensable part of protein molecules and nucleic acids, as well as of chlorophyll, the green substance in plants that traps energy from sunlight for photosynthesis. Plants cannot use atmospheric nitrogen directly; it is too inert. The conversion of atmospheric nitrogen to nitrogen compounds that can be used by plants is called **nitrogen fixation**. One way nitrogen is fixed is by lightning during thunderstorms. The electrical discharge supplies the energy for nitrogen to react with oxygen in the air. In the following flow equation, O_2 (one reactant) is shown over the arrow. This is simply a shorthand technique for representing a series of reactions.

$$N_2 \xrightarrow[\text{lightning}]{O_2} \quad 2\,NO \xrightarrow[\text{spontaneous}]{O_2} \quad 2\,NO_2$$

nitric oxide   nitrogen dioxide

Rain washes the nitrogen dioxide from the air to the ground, where the nitrogen dioxide is converted to the nitrate ion ($NO_3{}^-$), which can be readily used by plants.

Another process of nitrogen fixation is carried out by legumes—plants such as clover, beans, and alfalfa. These plants have nodules on their roots that harbor nitrogen-fixing bacteria. Legumes, therefore, indirectly increase the amount of nitrogenous material in the soil.

When plants and animals decay, the nitrogen in their systems returns to the soil and air in the form of nitrate ions, ammonia gas, and nitrogen gas, and the **nitrogen cycle** is completed. (See Figure 6.1.)

Ozone

The action of sunlight on NO_2 can cause it to cleave and yield oxygen atoms. These oxygen atoms can undergo reaction with oxygen gas to form ozone (O_3).

$$NO_2 \xrightarrow{\text{sunlight}} NO + O \xrightarrow{O_2} O_3$$

ozone

Ozone is also produced by the action of electrical discharge on oxygen gas. We can often detect the sharp, irritating odor of ozone near an electrical device or after a lightning storm.

A human breathing 2 parts ozone per million parts of air experiences respiratory symptoms and loss of coordination. (The usual level of ozone encountered in smog is under 0.3 ppm.) However, under proper conditions, ozone is useful; for example, it is used as a bactericide in municipal water systems and industrially as a bleach.

Ozone is also found in the upper atmosphere, where ultraviolet (uv) rays from the sun provide the energy to cause an exothermic reaction in which oxygen is converted to ozone. This ozone can go back to oxygen in another exothermic reaction. Thus, the formation and decomposition of ozone protects living organisms on the surface of the earth by removing some of the ultraviolet radiation from the sun's rays and helping to keep the earth warm.

in the ozone layer:

$$3\ O_2 \xrightarrow{\text{uv radiation}} 2\ O_3 + \text{heat}$$
$$\longrightarrow 3\ O_2 + \text{heat}$$

■

■ 6.2   USE OF GASES IN MEDICINE (optional)

Anesthetics

Many gases are indispensable to those dealing in health sciences. For example, nitrous oxide (N_2O), sometimes called "laughing gas," has been used for over 100 years as a general anesthetic (an agent that induces general anesthesia, or unconsciousness) in medicine and dentistry. This gas is often used in conjunction with other anesthetics, such as halothane or injected Sodium Pentothal, to induce a deeper state of anesthesia.

Other gases or low-boiling liquids that have been used as inhalation anesthetics include diethyl ether (sometimes simply called ether), divinyl ether (vinethene), and cyclopropane. Halothane is nonflammable, but cyclopropane and the ethers can explode if ignited in air. (You may want to review the condensed structural formulas shown in Table 4.3, Section 4.11, and relate them to the following formulas.)

some inhalation anesthetics:

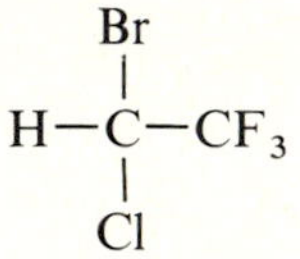

$$\begin{array}{c} Br \\ | \\ H-C-CF_3 \\ | \\ Cl \end{array} \qquad CH_3CH_2-O-CH_2CH_3$$

halothane   diethyl ether

$$CH_2{=}CH-O-CH{=}CH_2 \qquad \begin{array}{c} CH_2 \\ \diagup \quad \diagdown \\ CH_2-CH_2 \end{array}$$

divinyl ether   cyclopropane

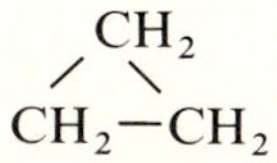

Oxygen

Oxygen-enriched air is frequently used for patients with respiratory problems such as pneumonia (a bacterial or viral infection that causes inflammation of lung tissue and fluid in the lungs), emphysema (a condition in which the air

Photo 6.1.   *The young woman is inhaling oxygen gas. Oxygen is administered when a patient's lungs cannot absorb sufficient oxygen from the air because of illness, injury, carbon monoxide poisoning, or some other abnormal condition. (Photograph by Martin M. Rotker/Taurus Photos.)*

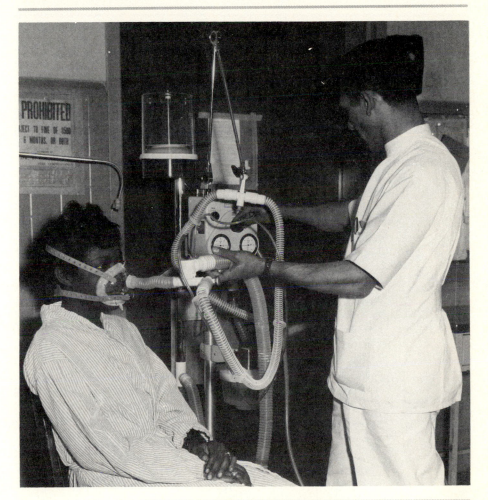

sacs in the lungs lose their elasticity and become distended), smoke inhalation, and carbon monoxide poisoning. Oxygen is also administered along with inhalation anesthetics to prevent suffocation. Compounds that burn in air (only 21% O_2) can burn with incredible speed in an oxygen-enriched atmosphere. Therefore, one must take great care to avoid sparks when oxygen or oxygen-enriched air is in use.

Commercially, oxygen is prepared by liquefying air, then allowing the components to boil off in order of increasing boiling point and collecting each component. This process, called *distillation*, is commonly used to separate or purify liquids. In the case of liquid air, nitrogen (boiling point, −196°C) boils before oxygen (boiling point, −183°C). ■

6.3   SOME PROPERTIES OF GASES

Gases are the least dense form of matter. Although 1.0 mL of liquid water weighs 1.0 gram, 1.0 mL of water vapor (at 0°C at sea level) weighs less than 0.001 gram. Nitrogen and oxygen are slightly more dense than water vapor, while hydrogen and helium are the least dense gases. The low density of helium explains why a balloon filled with this gas rises in air, which is composed mainly of nitrogen and oxygen.

Even though most gases, such as nitrogen and oxygen, are invisible to us, they have physical properties. For example, a gas can exert pressure. It is the gas in a champagne bottle exerting pressure that can cause the cork to fly across the room if the bottle is opened carelessly. A gas can be heated and cooled. It can expand, or occupy a larger volume. Unlike a liquid or a solid, a gas can be compressed, or forced to occupy a smaller volume. Gases also can undergo *diffusion* (spreading out) to occupy any size or shape container or to mix with other gases present.

The physical properties of a gas can be explained by the **kinetic molecular theory of gases**. The following assumptions are made in this theory:

1. Gases are composed of atoms and molecules that are far apart from one another. (This is why a gas can be compressed.)
2. The atoms or molecules are in continuous rapid motion (almost 1000 miles per hour at room temperature). Between collisions, they move in straight lines in a completely random fashion. (The motion of gas molecules explains why a gas undergoes diffusion.)
3. Moving atoms or molecules can collide with one another or with their container and rebound without loss of energy, similar to billiard balls rebounding around a billiard table. (The pressure of a gas is due to collisions with the walls of the container.)
4. The average kinetic energy of molecules of all gases is the same at a given temperature and increases with an increase in temperature.
5. There are no attractions or repulsions among the molecules of a gas.

Problem 6.1.   Using the assumptions in the kinetic molecular theory of gases, suggest a reason why an aerosol can that contains a compressed gas may explode if it is heated.

The kinetic molecular theory and some of the scientific laws governing the behavior of gases are true only for an *ideal gas*, a hypothetical gas with molecules that have no attractions for one another and cannot be liquefied.

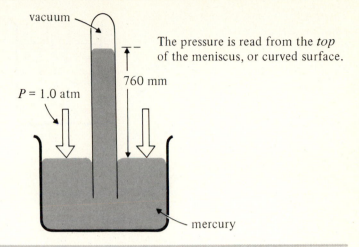

Figure 6.2.   *Diagram of a mercury barometer.*

vacuum

The pressure is read from the *top* of the meniscus, or curved surface.

760 mm

P = 1.0 atm

mercury

Most gases behave similarly to ideal gases at normal temperatures and pressures, but all real gases can be liquefied if subjected to sufficiently low temperatures and high pressures.

Units of Pressure

The atmosphere surrounding us exerts a pressure (*P*) that, at sea level, averages 14.7 pounds per square inch. The average atmospheric pressure at 0°C at sea level is called **1.0 atmosphere of pressure (atm)**, or **standard pressure**.

A barometer is a device used to measure atmospheric pressure. One common type of barometer is simply an inverted tube of liquid mercury metal (Hg). The tube is calibrated so that pressure can be read directly at the top of the column of mercury (see Figure 6.2). One atmosphere of pressure supports a column of mercury that is 760 mm (76.0 cm, or about 30 inches) tall. Gas pressures are often reported as mmHg. Blood pressure is also reported in this unit.

Other liquids could be used in a barometer, but most liquids are too lightweight to be practical. The density of mercury is 13.5 g/mL, while the density of water is 1.0 g/mL. Standard atmospheric pressure can support a column of water 34 feet high.

■ Pressure-Volume Relationship: Boyle's Law (optional)

When a gas is compressed to a smaller volume, the molecules have less volume in which to move, and they strike the surface of the container more frequently. Therefore, a decrease in the volume gives rise to an increase in pressure. In 1662, the English physicist and chemist Robert Boyle quantified this relationship into what we now call *Boyle's law.*

Figure 6.3. As the volume of a given quantity of gas is decreased, the pressure of the gas increases.

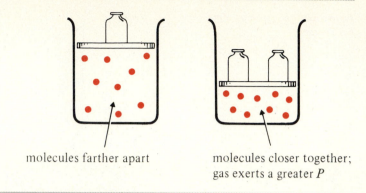

molecules farther apart
molecules closer together; gas exerts a greater P

Boyle's Law. At a constant temperature, the pressure of a given mass of a gas is inversely proportional to the volume.

If we let P represent the pressure of a given mass of gas, V its volume, and C a constant that is characteristic of the system at hand, we can state Boyle's law as

$$P = \frac{C}{V} \quad \text{or} \quad PV = C \qquad \text{at constant } T$$

Figure 6.3 depicts the relationship between pressure and volume.

Example If we have 5.0 L of a gas at 1.0 atm, what will be the pressure if the gas is compressed to 2.0 L?

Solution (1) Calculate the constant.

$$PV = C$$

$$(1.0 \text{ atm})(5.0 \text{ L}) = 5.0 \text{ L-atm}$$

(2) Using the constant, calculate the unknown pressure at the new volume.

$$PV = C$$

$$(P)(2.0 \text{ L}) = 5.0 \text{ L-atm}$$

$$P = \frac{5.0 \cancel{\text{L}}\text{-atm}}{2.0 \cancel{\text{L}}}$$

$$= 2.5 \text{ atm}$$

Problem 6.2. If 1.0 L of oxygen gas at 10 atm is allowed to expand to 5.0 L, what is the new pressure?

Boyle's law has many practical applications. A *manometer* is a device to measure pressure. A *sphygmomanometer* (from the Greek *sphygmos*, "the pulse") is the familiar device used to measure blood pressure. The pressure in the cuff that is wrapped around the arm is increased by squeezing a rubber bulb, which decreases the volume of air in the bulb and increases the pressure of air in the bulb and cuff. A simple medicine dropper works by the same principle. When the bulb of a medicine dropper is squeezed, the increased air pressure above the liquid forces the liquid out.

Breathing is an example of Boyle's law. We inhale by contracting the diaphragm, which increases the size of the lungs and decreases the air pressure. Because a gas flows from an area of higher pressure to one of lower pressure, air enters the lungs. When the diaphragm expands, the lung volume is diminished, the air pressure within the lungs increases, and the air is exhaled.   ■

■ **Temperature-Volume Relationship: Charles' Law (optional)**

As long as the pressure of a given amount of gas is kept the same, heating the gas increases the volume it occupies. Conversely, cooling a gas decreases the volume.

The temperature scale used in work with gases is called the *Kelvin scale*, or the *absolute temperature scale*. This scale is closely related to the Celsius scale. The only difference between the two scales is that zero is set at a different place. In the Kelvin scale, 0K (zero K) is placed at the lowest temperature possible, called *absolute zero*, which is approximately −273°C. (Note that no degree symbol is used in the Kelvin scale.) To convert °C to K, we simply add 273. Thus, 20°C (room temperature) is 293K.

$$K = °C + 273$$

The Kelvin temperature scale is used in temperature-volume relationships because the temperature at which an ideal gas would have a volume of zero is 0K. (A real gas would liquefy before this temperature was reached.)

In 1787, J. A. C. Charles, a French physicist, formulated what is now known as *Charles' law*, which relates the volume of a gas to the absolute temperature.

Charles' Law. For a given mass of gas at constant pressure, the volume is directly proportional to the absolute temperature (temperature in K).

If we let V be the volume of a gas, T its absolute temperature, and C a constant for the system, then Charles' law can be stated:

$$V = CT \quad \text{or} \quad \frac{V}{T} = C \qquad \text{at constant } P$$

Problem 6.3.   If 1.0 L of oxygen gas is warmed from 20°C (room temperature) to 37°C (body temperature) at the same pressure, what is its new volume?

A *temperature inversion*, an example of gases behaving in accordance with Charles' law, is a phenomenon in which denser cool air is trapped in a valley by an overlying layer of warm air. Air pollutants from automobiles, home heating, and industry can accumulate in this trapped air until weather conditions change and the inversion layer is dissipated. Warm air is less dense than cool air because it occupies a greater volume. When cooler air is trapped in a valley, it remains there, much as does the cold air in open chest-type freezers common in grocery stores. The lighter warm air simply rests on top.   ■

The Law of Partial Pressures

Molecules of a gas behave independently of one another. If two gases are mixed, each behaves as if the other were not there. Suppose we take a 1-L container of oxygen at 1 atm and then pump in 2 L of nitrogen, also at 1 atm. Now the combined pressure of the gases in the jug will be 3 atm (1 atm from O_2 and 2 atm from N_2).

This is an example of the law observed by the Englishman John Dalton in 1801.

Law of Partial Pressures. The total pressure of a mixture of gases is equal to the sum of the partial pressures of all the individual gases in the mixture:

$$P_{total} = P_a + P_b + P_c + \cdots$$

Example   At 1.0 atm pressure, a mixture of 1.0 L helium, 1.0 L nitrogen, and 1.0 L oxygen is pumped into a 1.0-L container. What is the total pressure of the mixture of gases in the container?

Solution   Write the equation, substitute, and solve.

$$P_{total} = P_{He} + P_{N_2} + P_{O_2}$$

$$= 1.0 \text{ atm} + 1.0 \text{ atm} + 1.0 \text{ atm}$$

$$= 3.0 \text{ atm}$$

Example   If air at 760 mmHg contains 5.0% by volume water vapor, what is the partial pressure of the water vapor?

Solution   The partial pressure of water vapor is 5.0% of the total pressure, or

$$0.05 \times 760 \text{ mmHg} = 38 \text{ mmHg}$$

Problem 6.4.   At 760 mmHg total air pressure, what is the partial pressure in mmHg of nitrogen in the atmosphere? (Refer to Table 6.1.)

6.4   PARTIAL PRESSURES AND RESPIRATION

How our bodies take in oxygen and release carbon dioxide is partly a result of the law of partial pressures and the tendency of a gas to diffuse from a phase rich in that gas to a phase not so rich. The atmosphere is 21% oxygen and has a pressure of 760 mmHg. The partial pressure of O_2, or P_{O_2}, in the air is thus 0.21×760 mmHg, or 160 mmHg. However, P_{O_2} in the alveoli (air sacs in the lungs) is only 108 mmHg.* When we inhale, O_2 molecules move from the oxygen-rich air into the alveoli, which are less rich in O_2.

The P_{O_2} in venous blood is about 50 mmHg. This lower partial pressure helps the venous blood absorb O_2, which is then taken up by the hemoglobin in the red blood cells. Blood containing oxygenated hemoglobin (called

Figure 6.4.   *Oxygen is taken in and used by the body's tissues partly by being passed along from systems richer in oxygen to systems poorer in oxygen.*

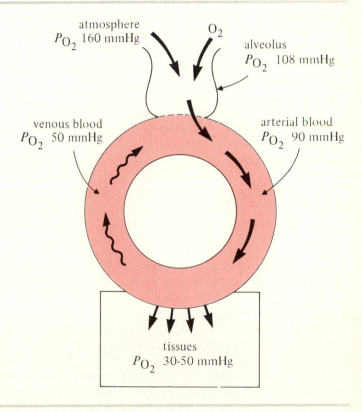

* The partial pressure of a gas over a solution of the gas (such as blood) at atmospheric pressure is more correctly termed *gas tension*. For our purposes, we will continue to use the term partial pressure.

Figure 6.5. *Carbon dioxide is passed from the tissues and returned to the atmosphere partly by its being passed along from CO_2-rich systems to systems less CO_2 rich.*

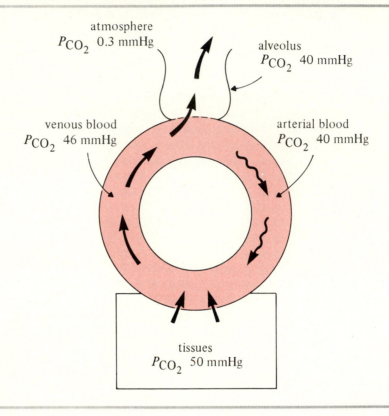

atmosphere
P_{CO_2} 0.3 mmHg

alveolus
P_{CO_2} 40 mmHg

venous blood
P_{CO_2} 46 mmHg

arterial blood
P_{CO_2} 40 mmHg

tissues
P_{CO_2} 50 mmHg

oxyhemoglobin) is now oxygen-rich arterial blood with a P_{O_2} of about 90 mmHg.

The arterial blood is pumped to the tissues of the body (P_{O_2} 30–50 mmHg), where it loses its O_2 to the tissues and becomes oxygen-poor venous blood. The venous blood then returns to the lungs to be oxygenated again. Figure 6.4 shows how the varying partial pressures of O_2 assist the oxygen-transport system.

When the cells of the body oxidize nutrients such as glucose, carbon dioxide is produced, causing the cells to become richer in carbon dioxide than the arterial blood is. Thus, CO_2 is lost to the blood as it changes from arterial blood to venous blood. The blood returns to the alveoli. These sacs collect the CO_2 and return it to the atmosphere, which has an extremely low partial pressure of CO_2. Figure 6.5 is a diagram of the path CO_2 takes from the tissues to the atmosphere.

6.5 SOLUBILITY OF GASES

The quantity of gas that dissolves in a liquid depends partly on the temperature of the liquid. As the temperature of a liquid increases, the solubility of a gas in the liquid decreases. For example, hot water can hold less oxygen than cold water can. "Thermal pollution" of waterways is the introduction of hot water (water from industrial processes or used to remove excess heat in nuclear generating plants, for example) into normally cooler natural water. Thermal pollution causes a decrease of the oxygen level in water and may cause the death of fish and other aquatic organisms that require oxygen.

The quantity of gas that dissolves in a liquid is also dependent on the pressure of the gas. Under greater pressure, more gas dissolves; under lower pressure, less gas dissolves. (See Figure 6.6.) A more formal statement of this relationship is the following:

Henry's Law. At a constant temperature, the solubility of a gas in a liquid is directly proportional to the pressure of the gas above the liquid, or

$$\text{solubility of a gas} = CP$$

where C is a constant and P is the pressure.

When a scuba diver ascends too rapidly from deep water, the diver may develop a painful and sometimes fatal malady known as the "bends." Why? When a scuba diver descends in the water, the pressure of the water on the diver's body increases the pressure of the gases in the lungs. This increased gas pressure, in turn, increases the solubility of air in the blood. When the diver ascends, the pressure of the water decreases, the pressure of the air in the lungs

Figure 6.6. *The solubility of a gas is directly proportional to the pressure of the gas above the liquid.*

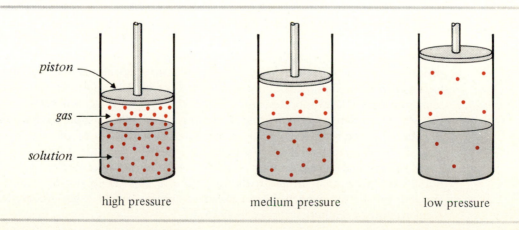

high pressure   medium pressure   low pressure

decreases, and the solubility of air, primarily nitrogen, in the blood decreases. If the ascent is slow, an equilibrium can be maintained, and the excess air can be discharged back into the lungs and exhaled.

On the other hand, if the ascent is rapid, not all the excess air can be discharged quickly enough by way of the lungs, and air bubbles appear in the capillaries. These bubbles cause the symptoms of the bends. Treatment of the bends involves putting the diver under pressure in a decompression chamber so that the pressure on the body can be brought slowly back to atmospheric pressure.

The fizzing of beer or a carbonated drink is another example of a gas behaving in accordance with Henry's law. These beverages contain carbon dioxide maintained in solution by the pressure of the gas above the liquid. When a bottle is opened, the pressure is decreased and some of the dissolved carbon dioxide bubbles out of solution.

Problem 6.5.   A *hyperbaric chamber* is a pressure chamber used to increase the solubility of oxygen in the blood and body tissues. This type of chamber is used in treating victims of carbon monoxide poisoning and in certain types of cancer radiation treatment. Calculate the partial pressure of oxygen in arterial blood (normally about 90 mmHg) when a patient in the chamber is subjected to 2.0 atm of air pressure.

6.6   LIQUID-GAS INTERCONVERSIONS

Evaporation and Condensation

The difference between a gas and a liquid is a difference in the physical state of a substance. A liquid such as water can be converted to a gas such as water vapor. How does this process occur?

Recall from Section 2.4 that the heat of vaporization (ΔH_v) is the number of calories required to convert 1.0 g of liquid at its boiling point to a gas. Even when a liquid is not at its boiling point, some molecules at the surface of a liquid have enough energy to escape from the liquid and become gaseous. This process is called **evaporation**. The rate of evaporation depends on the attractive forces within the liquid and on the temperature of the liquid. At the same temperature, water evaporates more slowly than rubbing alcohol because of the greater attractive forces among water molecules. When a liquid is heated, the energy of the molecules increases, more molecules gain enough energy to escape, and the rate of evaporation increases.

Condensation is the coming together of gaseous molecules to form a liquid. The water that forms on the outside of a glass of ice water is water vapor from the air condensing on the cold surface. When a gas undergoes condensation to a liquid, it loses energy (ΔH_v).

Boiling Point and Pressure

If a container with some liquid in it is closed, some of the liquid evaporates. When a large number of molecules have become gaseous, some of them will strike the surface of the liquid and become liquid again. After a period of time, in a closed container, the system reaches an equilibrium in which the rates of evaporation and condensation are equal.

Because the molecules in the gaseous state collide with the surface of the liquid, they exert a pressure on the surface. This pressure is called the **vapor pressure**. In a closed container at equilibrium, the *equilibrium vapor pressure* is the pressure exerted by escaping molecules, which equals the pressure exerted by returning molecules. A liquid that evaporates readily has a higher vapor pressure than one that evaporates less readily. Thus, rubbing alcohol has a higher vapor pressure than does water. The vapor pressure of a liquid is increased by heat; therefore, warming a liquid causes it to evaporate more rapidly.

As water is heated, it reaches a point where it boils and becomes water vapor. While evaporation is a surface phenomenon, boiling is an internal one. The molecules within the liquid gain enough kinetic energy to become gaseous.

Figure 6.7. *The boiling point is reached when the pressure of the molecules escaping from the liquid is equal to the atmospheric pressure.*

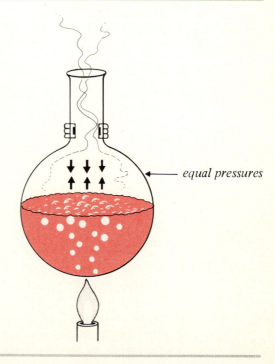

equal pressures

The **boiling point** of a liquid is *the temperature at which the vapor pressure of that liquid equals the atmospheric pressure.* At the boiling point, the molecules leaving the liquid exert as much pressure on the atmosphere as the atmosphere exerts on the liquid. (See Figure 6.7.) If the atmospheric pressure is reduced, the boiling point is reached at a lower temperature. Conversely, if the atmospheric pressure is increased, the temperature of the boiling point increases. Pressure canners are used to can nonacidic foods because the higher boiling point of water under the higher pressure is sufficient to kill *botulinus* (from the Latin *botulus*, "sausage") bacteria, bacteria that multiply in the absence of air and produce the deadly botulism toxin.

SUMMARY

The earth's atmosphere is composed principally of *nitrogen*, N_2 (78%), and *oxygen*, O_2 (21%). Oxygen is used for respiration, combustion, and aerobic decay, all of which produce CO_2. Plants use CO_2 and return O_2 to the air. Nitrogen, which is an essential element in proteins, must be *fixed* (converted from N_2 to a usable nitrogen compound) before it is available to plants. Some gases used in medicine are *nitrous oxide* (N_2O), other anesthetic gases, and oxygen.

The *kinetic molecular theory of gases* states that an *ideal gas* is composed of atoms or molecules that are far apart, move rapidly in straight lines until they collide with other particles or their container walls, and neither attract nor repel one another. The average kinetic energy of molecules of all gases is the same at the same temperature.

Common units of pressure are the *atmosphere* (1.0 atm = average atmospheric pressure at sea level at 0°C) and *mmHg* or *cmHg* (1.0 atm = 76.0 cmHg = 760 mmHg). The following laws relate pressure, volume, temperature, and solubility of a given quantity of ideal gas. (In each case, C is a constant representative of the particular system.)

Boyle's Law: $PV = C$ at constant T

Charles' Law: $V = \dfrac{C}{T}$ at constant P where T is in K (°C + 273)

Dalton's Law of Partial Pressures: $P_{\text{total}} = P_a + P_b + P_c + \cdots$

Henry's Law: solubility of a gas $= C \times P$

The process of a liquid becoming a gas is called *evaporation*; the reverse process (gas to liquid) is called *condensation*. The boiling point of a liquid is reached when its *vapor pressure* equals the atmospheric pressure.

KEY TERMS

*oxygen cycle
*nitrogen cycle
*nitrogen fixation
  kinetic molecular theory of gases
  atmosphere

*Boyle's law
*Charles' law
*Kelvin scale
  partial pressure

Henry's law
evaporation
condensation
vapor pressure

STUDY PROBLEMS

*6.6. Name two natural ways in which nitrogen is made available to plants.

*6.7. (a) List three ways in which ozone is formed.

(b) List two ways in which the ozone layer protects the earth's inhabitants.

*6.8. Why must oxygen be administered to a patient receiving an inhalation anesthetic?

*6.9. Explain how a liquid such as diethyl ether (boiling point 35°C) can be used as an *inhalation* anesthetic.

6.10. Which of the following phrases describe a gas?

(a) may be compressed to a smaller volume

(b) is a type of matter

(c) can hold its shape

(d) is very dense

(e) has atoms or molecules that move in random directions

(f) can undergo a change in temperature

(g) consists of particles that increase in kinetic energy when heated

6.11. When two gases are pumped into a container one after the other, the mixture of gases quickly becomes homogeneous. Explain why.

6.12. Carbon dioxide is heavier than air and can accumulate in mine shafts and old wells. Explain why carbon dioxide does not form a layer on the earth underneath the rest of the atmosphere.

6.13. Convert the following pressures to atm or mmHg, whichever is not given:

(a) 0.5 atm   (b) 1.3 atm

(c) 748 mmHg   (d) 90 mmHg

*6.14. Carbon dioxide (25.0 L at 1.00 atm) has been compressed into a 5.00-L cylinder. At the same temperature, what is the pressure of the carbon dioxide in the cylinder?

*6.15. A balloon contains 3.00 L of helium at 30.0°C. What would be the volume of the helium if it were chilled to 15.0°C without altering its pressure?

*6.16. If 5.0 L of a gas is allowed to expand to 10 L at constant pressure, what will the temperature be if the initial temperature was 23°C? Explain your answer.

6.17. If 1.0 L of O_2 and 1.0 L of N_2, both at 760 mmHg pressure, are compressed together into a 1.0-L tank at constant temperature, what will be the pressure of *each* gas in the tank?

6.18. Explain how nitrogen bubbles form in the capillaries when a person develops the bends.

6.19. Explain why a bottle of carbonated beverage bubbles and fizzes when the cap is removed.

6.20. Some water is placed in a flask, and the flask is corked.

(a) Some of the water molecules become gaseous in a process called _____ .

(b) Some of the gaseous water molecules go back to liquid in a process called _____ .

(c) When equal numbers of molecules are shifting back and forth, the situation is called

_____ .

6.21. (a) Define vapor pressure.

(b) Define boiling point.

(c) Explain why a boiling liquid bubbles.

6.22. At 35°C at sea level, the vapor pressure of water is 42 mmHg and that of diethyl ether is 760 mmHg. Which liquid would evaporate faster? Explain.

6.23. When home-canning tomatoes in a boiling-water bath, cooks are advised to increase the heating time at higher altitudes. Why?

7   Solutions and Their Properties

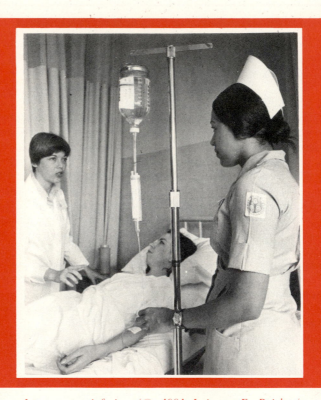

Intravenous infusion. (© 1981 Laimute E. Driskus/ Taurus Photos.) Fluids can be given intravenously for the replenishment of water, electrolytes, blood glucose, or blood itself. For any type of intravenous infusion, even for injections, the fluid used must be isotonic with blood—that is, the fluid must contain the proper concentration of dissolved substances so that blood cells neither lose water and undergo crenation *(shriveling) nor absorb water and undergo* hemolysis *(swelling and rupturing).*

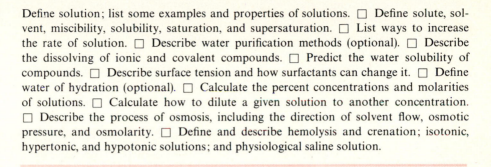

Objectives   Define solution; list some examples and properties of solutions. ☐ Define solute, solvent, miscibility, solubility, saturation, and supersaturation. ☐ List ways to increase the rate of solution. ☐ Describe water purification methods (optional). ☐ Describe the dissolving of ionic and covalent compounds. ☐ Predict the water solubility of compounds. ☐ Describe surface tension and how surfactants can change it. ☐ Define water of hydration (optional). ☐ Calculate the percent concentrations and molarities of solutions. ☐ Calculate how to dilute a given solution to another concentration. ☐ Describe the process of osmosis, including the direction of solvent flow, osmotic pressure, and osmolarity. ☐ Define and describe hemolysis and crenation; isotonic, hypertonic, and hypotonic solutions; and physiological saline solution.

In this chapter, we will survey some types of solutions, then discuss *aqueous solutions* (solutions in water). We will discuss concentrations of solutions and how to dilute a concentrated solution to a weaker solution of a specific concentration. Finally, we will look at the process of osmosis and some of the ways it affects living organisms.

7.1. SOLUTIONS

A **solution** is a *homogeneous, or uniform, mixture of two or more kinds of atoms, molecules, or ions.* (See Figure 7.1.) When a substance goes into solution, or dissolves, the individual molecules or ions separate from one another. These particles are too small to reflect light. Thus, with rare exceptions, solutions are

Figure 7.1.   *A solution is a homogeneous mixture of two or more substances.*

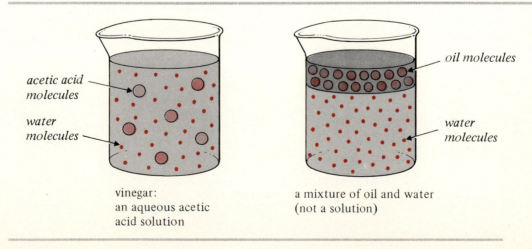

acetic acid
molecules

water
molecules

oil molecules

water
molecules

vinegar:
an aqueous acetic
acid solution

a mixture of oil and water
(not a solution)

transparent. We can see through a solution of salt in water as well as we can see through water. (Although transparent, a solution may be colored, as is red table wine.) If we put some flour in water and mix the two thoroughly, the mixture is cloudy; it is not a solution. The mixture is a colloid, a subject we will discuss in Chapter 8.

Gravity cannot pull dissolved material from a solution. For example, dissolved sugar will not settle out of your morning coffee. However, a solution can be separated into its components by physical means, such as by allowing water to evaporate from a salt solution.

When a particular substance can be dissolved in a liquid, we say that the substance is *soluble* in that liquid. We call the liquid the *solvent* and the material being dissolved the *solute*. When you add sugar to coffee, sugar is the solute and water in the coffee is the solvent.

Substances exhibit varying solubilities in solvents. When a large amount of a substance can dissolve in a given amount of solvent, we say that the substance is *very soluble* in that solvent. Sodium chloride is very soluble in water. When only a small quantity of a substance can dissolve in a given quantity of a solvent, we say that the substance is *slightly soluble* in that solvent. When a substance will not dissolve to any appreciable extent in a solvent, such as glass in water, we say that the substance is *insoluble* in that solvent.

Types of Solutions

We usually think of a solution as a substance dissolved in water, but this is only one type of solution. Other liquid solvents are frequently encountered in chemistry and medicine. For example, ethanol (CH_3CH_2OH) solutions are not uncommon in medical practice. In medicine, ethanol solutions are sometimes called **tinctures**; tincture of iodine is a solution of iodine (I_2) in ethanol. Mercury (Hg) is the only metal that is liquid at room temperature. Solutions of other metals in mercury are called **amalgams**. Silver amalgam (silver metal dissolved in mercury) is used to make dental fillings.

A solution need not be liquid. The atmosphere is a gaseous solution, a homogeneous mixture of a number of different gases. In Table 7.1, we have

Table 7.1.   *Some Common Types of Solutions*

Type	Example
liquid in liquid	vinegar (acetic acid in water)
solid in liquid	brine and saline solution (sodium chloride in water); syrup (sugar in water)
gas in liquid	soda water (carbon dioxide in water)
gas in gas	air (oxygen and other gases in nitrogen)

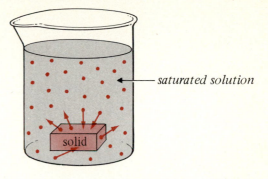

Figure 7.2. *A saturated solution. The number of ions or molecules leaving the solution is the same as the number dissolving.*

saturated solution

solid

listed several types of solutions. In this chapter we will discuss primarily liquid solutions.

Liquid-in-Liquid Solutions.   The definitions of solute and solvent become hazy when we speak of dissolving a liquid in another liquid. For example, ethanol is *infinitely soluble* in water. That is, any amount of ethanol will dissolve in any amount of water, and any amount of water will dissolve in any amount of ethanol. Two liquids that are infinitely soluble in each other are said to be *miscible*. Because diethyl ether ($CH_3CH_2OCH_2CH_3$) is only slightly soluble in water, we say that diethyl ether and water are *partially miscible*. If more ether than can dissolve is added to a quantity of water, the excess ether forms a layer on top of the water. Gasoline and motor oil are insoluble in water. We say that these substances are *immiscible* with water.

Solid-in-Liquid Solutions.   There is a limit to the amount of solid solute that will dissolve in a given amount of liquid solvent. When a solvent has dissolved as much of a solute as is possible, we say that the solution is a **saturated solution**.

By definition, a solute in a saturated solution is in equilibrium with its solid. In a saturated salt solution, salt is dissolving in the water as Na^+ and Cl^-, but an equal amount of salt is coming out of the solution and forming solid material. (See Figure 7.2.) The *number* of sodium ions and chloride ions in the solution remains constant.

The amount of salt that will dissolve in a given amount of water depends on the temperature of the water. Therefore, when we speak of the amount of solute needed to saturate a given amount of solvent, we must specify the temperature. We find that 36.0 g of sodium chloride will dissolve in 100 g of water at 20°C and that 39.8 g will dissolve at 100°C. Although a gas is less soluble at higher temperatures, it is generally true that a greater amount of a solid or liquid solute will dissolve in a solvent as the temperature is increased. The maximum amount of a substance that will dissolve in a given amount of

Table 7.2.   *Some Typical Solubilities in Water*

Solute	Solubility in 100 g of water	
	at 20°C	at 100°C
sodium chloride	36.0 g	39.8 g
sodium hydroxide	109.0	347.0
copper(II) sulfate	20.7	75.4
ethanol (a liquid)	∞^a	∞^a

[a] Infinitely soluble, or miscible.

solvent at a particular temperature is called the **solubility** of that substance. Some typical solubilities of compounds in water are listed in Table 7.2.

Supersaturation. When a hot saturated solution of a substance is cooled, some of the solute may come out of solution. Sometimes, however, the solute does not come out of the solution as we would expect. We call such a solution a **supersaturated solution**. A solution becomes supersaturated because the excess molecules or ions have no crystal lattice present that they can join. For molecules or ions to start a crystal lattice requires energy. To start crystallization from a supersaturated solution, one can add a crystal of the solute, called a *seed crystal*. The seed crystal provides a crystal lattice that can easily be joined by excess solute molecules in solution.

Rate of Solution. Dissolving a solid in a liquid requires a certain amount of time. How fast a solid dissolves depends on a number of factors:

1. *Temperature.* Heating increases the rate of solution of a solid in a liquid.
2. *Surface area.* A powder dissolves faster than large lumps of the same material.
3. *Stirring.* Stirring exposes the solid to fresh solvent and increases the rate of solution.
4. *Solubility in that solvent.* A very soluble substance dissolves faster than a slightly soluble substance.

7.2   WATER

Water is one of the most important compounds on the earth. Life presumably first developed in the oceans, and water is still necessary to sustain life on land because living organisms are composed partly of aqueous solutions. The average person drinks 1–2 L of water per day and ingests almost another liter in the form of water contained in food. An equal amount of water is lost in urine and perspiration. Although we can live a month or more without food, we can live no more than a few days without water.

The water balance in the body is controlled largely by the kidneys, which regulate the quantity of urine. (See Chapter 28.) Kidney malfunction and some other medical problems can lead to retention of excess water in the body's tissues. This condition, which is characterized by puffiness, especially in the extremities, is called *edema*. The opposite condition, *dehydration*, results from too little water in the tissues. Dehydration can result from any number of illnesses or simply from drinking far too little fluid.

■ Purification of Water (optional)

In many medical applications or in other scientific work, water must be pure— that is, free of minerals, microorganisms, and other undesirable substances. Natural water is impure. Rain water contains dissolved gases, dust, and pollutants from the air. Ground water, such as well water or river water, contains dissolved minerals and organic matter.

Water may be purified by *distillation*, or boiling and recondensation, which removes dissolved minerals as well as other undesirable substances. Dissolved minerals can also be removed from water by *softening* or *deionization* (see Section 9.5); however, these two processes may not remove harmful bacteria. Drinking water is usually purified by filtration, perhaps through a bed of sand, to remove undissolved solid material, followed by treatment with chlorine (Cl_2) or ozone (O_3) to kill harmful microorganisms.   ■

Properties of Water

Recall from Section 4.10 that water is a polar molecule and that the oxygen atom carries two pairs of unshared valence electrons. These structural features lead to the formation of *hydrogen bonds* among water molecules.

a hydrogen bond

Hydrogen bonding in water leads to a number of unusual properties, such as a high heat of vaporization (Section 2.4). Let us explore some of the other properties of water in light of its polarity and hydrogen bonding.

Water as a Solvent.   Polar molecules of water attract not only other water molecules, but also other polar molecules or ions. When a crystal of sodium chloride is placed in water, the negative end of a water molecule and a positive sodium ion attract each other. The partially positive hydrogen atom in water and the negative chloride ion also attract each other. Ions are pulled away from the crystal by the attractions of the water, and the compound dissolves.

In solution, both the positive cations and the negative anions become surrounded by a group of water molecules. We say the ions are *hydrated*, or have
undergone *hydration*.

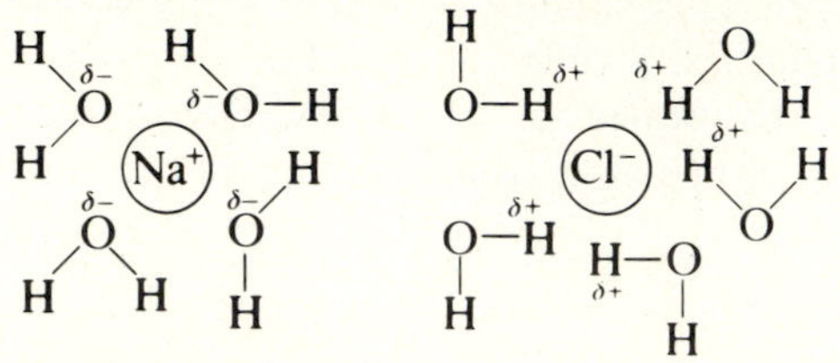

*In the process of hydration, water surrounds
both the cation and the anion, allowing them
to be free from each other in solution.*

Although most covalent compounds do not form ions when they dissolve,
but simply separate into individual molecules, water can dissolve many covalent compounds that contain polar groups.

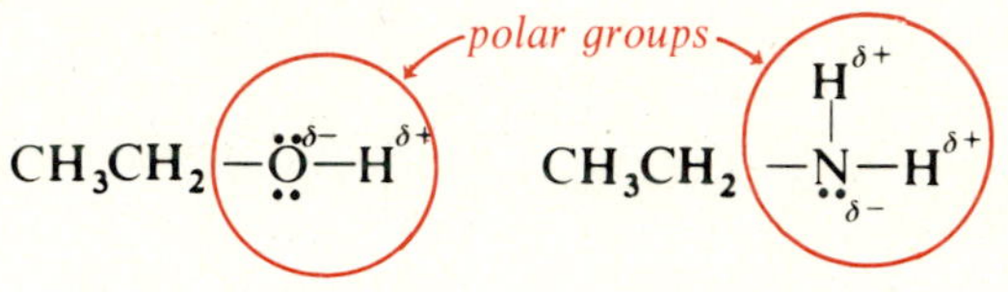

Most polar covalent compounds in living systems contain oxygen or nitrogen.
Compounds containing O or N can form hydrogen bonds with water molecules.
Hydrogen bonds are stronger than most other attractions between covalent
molecules; these hydrogen bonds explain the solubility of many polar compounds in water.

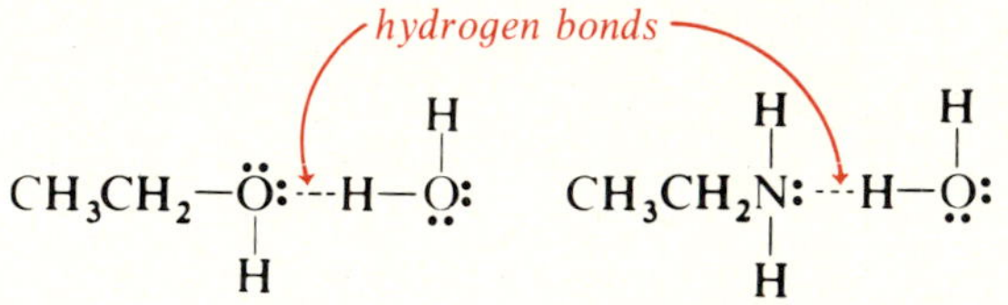

Covalent compounds with nonpolar bonds, such as propane gas
($CH_3CH_2CH_3$), do not dissolve in water because there are few attractions
between their molecules and water. If a compound contains both polar and
nonpolar groups, the size of the nonpolar group determines the solubility.

3 carbons:	*4 carbons:*	*5 carbons:*
$CH_3CH_2CH_2OH$	$CH_3CH_2CH_2CH_2OH$	$CH_3CH_2CH_2CH_2CH_2OH$
water soluble	*slightly soluble*	*water insoluble*

Problem 7.1. Which of the following compounds would you expect to be
soluble in water? Explain your answer.

(a) methanol, CH_3OH

(b) butane, $CH_3CH_2CH_2CH_3$

(c) ethylene glycol, $HOCH_2CH_2OH$

(d) acetic acid, $CH_3\overset{\displaystyle O}{\overset{\displaystyle \|}{C}}OH$

Problem 7.2.   We have shown an ethanol molecule forming a hydrogen bond with a water molecule, but ethanol can form another type of hydrogen bond with water. Write formulas for these two molecules showing the other type.

Surface Tension.   Surface tension is a liquid's resistance to surface expansion and is the phenomenon that causes a drop of liquid to hold its shape rather than flatten out on a surface. The surface tension of water is unusually high compared to that of other liquids because of the exceptionally strong attractions between H_2O molecules. The H_2O molecules on the surface of the liquid are pulled toward the bulk of the liquid and become packed more closely together than the molecules in the bulk of the liquid, which are surrounded on *all* sides by other attractive molecules. (See Figure 7.3.)

Capillary Action.   A glass tube with a small inside diameter is called a *capillary tube*. When the open tip of a capillary tube is placed in water, the water rises in the tube. Similarly, when a small amount of blood is needed for laboratory testing, the blood may be collected in a capillary tube. The rising of the liquid, called *capillary action*, is the result of both surface tension and attractions between the glass and liquid molecules.

Surfactants.   A **surfactant** (from "surface active agent") is a substance that lowers the surface tension of water. Water droplets containing a surfactant will flatten and spread out on a surface rather than holding their shape. Soaps and detergents are familiar examples of surfactants. A molecule of soap or detergent has an ionic end that is attracted to water molecules and a large nonpolar end that is insoluble in water. Therefore, molecules of surfactant concentrate at

Figure 7.3.   *An illustration of the origin of surface tension.*

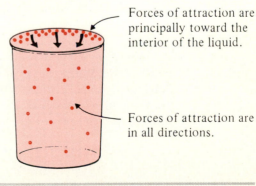

Forces of attraction are principally toward the interior of the liquid.

Forces of attraction are in all directions.

the surface of a water droplet with their ionic "heads" in among the water molecules and their nonpolar "tails" protruding from the surface of the drop. In this way, the surfactant interferes with the water molecules' attractions for one another and thus lowers the surface tension.

nonpolar tail   *ionic head*

$$CH_3CH_2CH_2CH_2CH_2CH_2CH_2CH_2CH_2CH_2CH_2 \overset{\overset{O}{\|}}{C}O^- Na^+$$

an example of a soap molecule

The alveoli (air sacs in the lungs) are coated with a surfactant and a very thin film of water. This surfactant keeps the surface tension of the water low so that the alveoli do not collapse. Premature infants, whose lungs are not always fully developed, may have an insufficient quantity of this lung surfactant, a condition that leads to *respiratory distress syndrome* (RDS), also called *hyaline membrane disease.*

■ **Water of Hydration (optional).**   When water is allowed to evaporate from a sodium chloride solution, the pure, crystalline sodium chloride can be recovered. Some salts, however, crystallize from water with water molecules incorporated into their crystalline structure in a definite proportion by weight. The water that is part of the crystalline structure is called **water of hydration**, and the salt is called a **hydrate**. *Epsom salt* ($MgSO_4 \cdot 7\ H_2O$) is an example. The dot in the formula is used to show that the water molecules and $MgSO_4$ are closely associated with each other but retain their identities. If Epsom salt is heated to 200°C, the water is driven off, leaving anhydrous $MgSO_4$ behind. *Anhydrous* means "without water."

$$MgSO_4 \cdot 7\ H_2O \xrightarrow{200°C} \quad MgSO_4 \quad + \quad 7\ H_2O$$

| magnesium sulfate | anhydrous | steam |
| heptahydrate | magnesium sulfate | |

Epsom salt
a laxative

Plaster of Paris, a hydrate of calcium sulfate ($CaSO_4$), is another important hydrate. When plaster of Paris is mixed with water, it rapidly turns into a different hydrate, *gypsum*, which is much harder than plaster of Paris. This property allows plaster of Paris to be molded into plaster casts for immobilizing an injured arm or leg.

$$2\ CaSO_4 \cdot H_2O \quad + \quad 2\ H_2O \longrightarrow 2\ [CaSO_4 \cdot 2\ H_2O]$$

plaster of Paris   gypsum
soft   *hard*

7.3 CONCENTRATIONS OF SOLUTIONS

We may say that a solution is *dilute* (of low concentration) or that it is *concentrated* (of high concentration). For example, we can describe vinegar as a dilute solution of acetic acid and shampoo as a concentrated solution of soap. However, we also need *quantitative* methods of expressing concentration. A number of methods of expressing concentration are in common use.

Percents

Percent (weight/volume). A widely used method of expressing concentrations is a combination of weight and volume. In work with a solid solute and a liquid solvent, it is convenient to weigh the solute in grams and to measure the volume of the solution in milliliters. In this case, we speak of concentration in terms of *percent* (*weight/volume*), or % (w/v). (Note that the volume in this formula is the *total volume* of the final solution and not the amount of solvent added.) In clinical reports, the term *grams/deciliter* (g/dL), where 1.0 dL = 100 mL, is sometimes encountered. This term is synonymous with % (w/v).

$$\% \text{ (weight/volume)} = \frac{\text{g of solute}}{100 \text{ mL of solution}} = \frac{\text{g}}{\text{dL}}$$

Example   If 2.0 g of NaCl is dissolved, then diluted with water to 200 mL, what is the percent (weight/volume) of the NaCl in the solution?

Solution
$$\text{concentration} = \frac{2.0 \text{ g}}{200 \text{ mL}} = \frac{0.010 \text{ g}}{\text{mL}}$$

$$\% \text{ (w/v)} = \frac{\text{g}}{100 \text{ mL}}$$

$$= \frac{0.010 \text{ g}}{\text{mL}} \times 100$$

$$= \frac{1.0 \text{ g}}{\text{mL}}$$

$$= 1.0\%$$

Example   How many grams of NaCl are necessary to prepare 500 mL of a 10% (w/v) solution?

Solution
$$\% \text{ (w/v)} = \frac{\text{g of solute}}{100 \text{ mL of solution}}$$

$$10\% \text{ (w/v)} = \frac{10 \text{ g NaCl}}{100 \text{ mL}} = \frac{0.10 \text{ g NaCl}}{\text{mL}}$$

For 500 mL of solution,

$$\frac{0.10 \text{ g NaCl}}{\text{mL}} \times 500 \text{ mL} = 50 \text{ g NaCl}$$

Problem 7.3.   What is the % (w/v) of 1.0 L of solution that contains 5.0 g of NaCl?

Problem 7.4.   How many grams of glucose would be required in the preparation of 3.0 L of a 5% (w/v) solution?

Percent (weight/weight).   Expressing a concentration in *percent by weight*, *percent* (*weight/weight*), or % (w/w) is similar to expressing a concentration in % (w/v) except that grams of solution are used instead of milliliters.

$$\% \text{ by weight} = \% \text{ (w/w)} = \frac{\text{g of solute}}{100 \text{ g of solution}}$$

Percent (volume/volume).   One convenient way to mix a liquid-liquid solution is to measure the volume of the solute, then dilute it with solvent to the desired volume. The concentration of the solute then can be expressed as a *percent by volume*, or % (v/v).

$$\% \text{ by volume} = \% \text{ (v/v)} = \frac{\text{mL of solute}}{100 \text{ mL of solution}}$$

Problem 7.5.   How many grams of sodium chloride are needed to prepare 60 g of 5.0% (w/w) solution?

Problem 7.6.   Find the percent by volume of ethanol in a solution prepared by diluting 30.0 mL of ethanol to 250 mL.

Concentration Expressions for Very Dilute Solutions

When we deal with very dilute solutions, as is often the case in biological sciences and medicine, we need other expressions for concentration. After all, it is not very convenient to say that drinking water should contain no more than 0.000195 percent by weight of sodium fluoride (the additive used to prevent tooth decay).

   Milligram percent (mg%) is the number of milligrams present in 100 mL of solution. The term *milligrams/deciliter* (mg/dL) is synonymous with mg%.

$$\text{mg\%} = \frac{\text{mg of solute}}{100 \text{ mL of solution}} = \frac{\text{mg}}{\text{dL}}$$

Parts per million (ppm) is frequently encountered in pollution-control work and refers to the number of parts of solute per million parts of solution. There are two ways of calculating and interpreting the ppm expression, depending on whether the solvent is a liquid or a gas. In liquids, ppm means the number of milligrams of solute per liter of solution. For example, 1.0 L of solution that contains 5 ppm of cadmium chloride ($CdCl_2$) contains 5 mg of $CdCl_2$.

$$\text{ppm (liquids)} = \frac{\text{mg of solute}}{\text{L of solution}}$$

In air-pollution control, the term ppm is based on volume measurements rather than weight measurements, and may be expressed as microliters (μL) of solute (where 1.0 μL = 0.001 mL, or 1.0×10^{-6} L) per liter of air. For example, 5 ppm of sulfur dioxide (SO_2) in the air means 5 μL of SO_2 per liter of air.

$$\text{ppm (gases)} = \frac{\mu\text{L of solute}}{\text{L of air}}$$

Table 7.3 summarizes the units of concentration we have presented here.

Table 7.3.   *Some Units Used for Concentrations of Solutions*

Unit	Definition	Equivalent
% (w/v)	$\dfrac{\text{g of solute}}{100 \text{ mL of solution}}$	g/dL
% (w/w)	$\dfrac{\text{g of solute}}{100 \text{ g of solution}}$	—
% (v/v)	$\dfrac{\text{mL of solute}}{100 \text{ mL of solution}}$	—
mg%	$\dfrac{\text{mg of solute}}{100 \text{ mL of solution}}$	mg/dL
ppm	$\dfrac{\text{parts of solute}}{10^6 \text{ parts of solution}}$	$\dfrac{\text{mg or } \mu\text{L}}{1.0 \text{ L of solution}}$

Problem 7.7.   Carry out the following calculations:
(a) What is the concentration in mg% of a solution that contains 5.0 mg of NaCl in 500 mL of water?
(b) How many ppm of NaCl are present in the solution in (a)?

Molarity

Chemical reactions occur in definite proportions. When HCl undergoes reaction with NaOH, one mole of HCl (36.5 g) undergoes reaction with one mole of NaOH (40.0 g) to yield one mole of NaCl (58.5 g) and one mole of water (18.0 g).

$$HCl + NaOH \longrightarrow NaCl + H_2O$$

If less than a mole or more than a mole of one reactant is used, the amounts of the other reactants and of the expected products must be adjusted accordingly to keep the molar ratio in the proper proportions. Otherwise, any excess reactant will not undergo reaction and will just be left over when the reaction is completed.

Chemical reactions are often carried out in solutions. It would be inconvenient to treat pure sodium hydroxide, which we buy as pellets, with pure hydrogen chloride, which is a gas, to effect the preceding reaction. It is far easier to mix an aqueous solution of sodium hydroxide with an aqueous solution of hydrochloric acid.

When we carry out a chemical reaction using solutions, it is convenient to express the concentration of the solutions in terms of the number of moles of solute present in a given volume of solution. The expression we use is **molarity**, which is defined as *the number of moles of solute per liter of solution* and can be calculated by the following equation:

$$\text{molarity } (M) = \frac{\text{number of moles of solute}}{\text{number of liters of solution}}$$

A solution that contains 1.0 mole of NaOH in 1.0 L of solution would therefore be a 1.0 molar (1.0 M) solution of NaOH. A solution that contains 2.0 moles of NaOH in 1.0 L of solution would be a 2.0 molar (2.0 M) solution of NaOH.

Example   What is the molarity of a solution prepared by dissolving 4.0 g NaOH in water and diluting to 1.0 L?

Solution

formula weight of NaOH = 23.0 + 16.0 + 1.0 = 40.0

$$\text{number of moles NaOH} = \frac{\text{weight of NaOH}}{\text{weight of 1.0 mole of NaOH}}$$

$$= \frac{4.0 \text{ g}}{40.0 \text{ g}}$$

$$= 0.10$$

$$M = \frac{\text{number of moles of solute}}{\text{number of liters of solution}}$$

$$= \frac{0.10 \text{ mole}}{1.0 \text{ L}}$$

$$= 0.10 \text{ mole/L}$$

Example   What is the molarity of a solution that contains 2.0 moles in 2.0 L?

Solution

$$M = \frac{\text{number of moles of solute}}{\text{number of liters of solution}}$$

$$= \frac{2.0 \text{ moles}}{2.0 \text{ L}}$$

$$= 1.0 \text{ mole/L}$$

Rearranging the equation that defines molarity allows us to calculate the *number of moles of solute* in a specific volume of solution of known molarity.

number of moles of solute = M × number of liters of solution

or

number of moles = MV

Example   How many moles of NaCl are present in 150 mL of a 1.5 M solution?

Solution   number of moles = MV

where

$$V = \frac{150 \text{ mL}}{1000 \text{ mL/L}} = 0.150 \text{ L} \quad \text{and} \quad M = 1.5 \text{ moles/L}.$$

number of moles = 1.5 moles/L × 0.150 L

$$= 0.225 \text{ mole}$$

Problem 7.8. (a) How many moles of NaCl are required to prepare 100 mL of a 1.6 M solution? (b) How many *grams* of NaCl are required? (Remember to change milliliters to liters.)

Problem 7.9. How many moles of HCl are present in 50 mL of 3.0 M HCl solution?

7.4   DILUTION PROBLEMS

A common laboratory task is converting a concentrated solution of a substance to a more dilute solution. For example, let us assume that we need 100 mL of 1.00 M NaCl solution but can find only a 1.50 M solution. How much of this 1.50 M solution should we use to mix the 1.00 M solution? This type of problem, called a *dilution problem*, can be solved with the following equation:

$$M_{old} V_{old} = M_{new} V_{new}$$

or

$$M_1 V_1 = M_2 V_2$$

where M = molarity and V = liters of solution. This equation is simply an expression that tells us that the *total number of moles in the initial solution* ($M_1 V_1$) equals *the total number of moles in the final solution* ($M_2 V_2$). (Because only water is added to the first solution, this expression must be true.)

The value of this equation is that *any concentration units* (for example, percent by weight instead of M) can be used as long as the same units are used on both sides of the equation. Also, *any volume units* (milliliters, liters, or even quarts) can be used as long as the same units are used on both sides of the equation. Therefore, a more general equation can be written.

$$C_1 V_1 = C_2 V_2$$

where C_1 and C_2 are concentrations and V_1 and V_2 are volumes.

Example   What volume of 1.50 M NaCl should we use to prepare 100 mL of 1.00 M NaCl?

Solution   Define terms:

$$C_1 = 1.50 \ M \qquad C_2 = 1.00 \ M$$

$$V_1 = ? \qquad V_2 = 100 \ mL$$

Write the equation:

$$C_1 V_1 = C_2 V_2$$

Substitute:

$$(1.50 \ M)(V_1) = (1.00 \ M)(100 \ mL)$$

Rearrange and solve:

$$V_1 = \frac{(1.00 \ M)(100 \ mL)}{1.50 \ M}$$

$$= \frac{100 \ mL}{1.50} = 66.7 \ mL$$

In words, the solution to this equation tells us that we should take 66.7 mL of 1.50 *M* NaCl solution, then *dilute this solution to 100 mL* to obtain the desired 100 mL of 1.00 *M* NaCl.

Problem 7.10.   (a) How would you prepare 200 mL of a 1.0% (w/v) solution of aluminum chloride (which, in dilute solution, is used as an antiperspirant) from a 6.0% (w/v) solution? (b) How would you prepare 2.5 L of 2% (w/v) glucose solution from 5% (w/v) aqueous glucose?

7.5   COLLIGATIVE PROPERTIES OF SOLUTIONS

The presence of a solute in a liquid solvent changes the physical properties of the solvent from those of pure solvent. The properties that we will discuss are referred to as the **colligative properties** of solutions: properties that depend on the *number of dissolved particles* (ions or molecules) in a given mass of solvent, regardless of the identity of the solute (or solutes).

One mole of any substance contains 6×10^{23} molecules or formula units. For example, one mole of glucose contains 6×10^{23} molecules of glucose, and one mole of sodium chloride contains 6×10^{23} units of $Na^+ Cl^-$. However, the number of particles resulting from the dissolving of ionic compounds differs from the number resulting from the dissolving of covalent compounds because ionic compounds dissociate into ions (different particles) when they dissolve.

$$1.0 \text{ mole glucose} = 6 \times 10^{23} \text{ particles (molecules)}$$

$$1.0 \text{ mole } Na^+ Cl^- = 6 \times 10^{23} \; Na^+ Cl^-$$

$$= 6 \times 10^{23} \; Na^+ + 6 \times 10^{23} \; Cl^-$$

$$= 12 \times 10^{23} \text{ particles (ions)}$$

Because the colligative properties of solutions depend on the number of particles in solution, ionic compounds have a greater effect on these properties than do covalent compounds if they are present in the same molar concentration. For example, one mole of NaCl, with two ions per formula unit, has twice the effect on the colligative properties of a solution as does one mole of a covalent compound.

One of the colligative properties of a solution is its **freezing point depression**. A solution freezes at a temperature *lower* than the freezing point of the pure solvent. The freezing point of pure water is $0°C$. For each mole of dissolved glucose (or other nonionic substance) in 1000 g of water, the freezing point is depressed $1.86°C$. However, each mole of NaCl in the same amount of water will depress the freezing point by $3.72°C$ (twice as much).

A practical application of freezing point depression is the use of antifreeze in automobile radiators. Antifreezes are nonvolatile (high-boiling) covalent organic compounds that are added to the radiator to prevent the water from freezing. Ethylene glycol ($HOCH_2CH_2OH$) is commonly used as an antifreeze. (Ethylene glycol tastes sweet but is toxic; therefore, care should be taken when it is used around children or pets.)

Another colligative property is **boiling point elevation**. The addition of a solid solute to a solvent *raises* the boiling point of the solution over that of pure solvent. One mole of the covalent solute table sugar (sucrose) in 1000 g of water raises the boiling point by 0.52°C. Boiling jelly or jam, which may be 50% sugar, is much hotter than boiling water.

7.6   OSMOSIS

One very important colligative property of solutions is called *osmosis*. Before defining osmosis, we first must consider diffusion and semipermeable membranes. Liquids are capable of *diffusion*, or mixing, because their molecules are in constant motion. If you place a drop of red food coloring in a glass of water and set the glass aside without stirring, the liquid will soon become a homogeneous pink; this is a simple example of diffusion.

When solutions of different concentrations are mixed, the natural movement of solute and solvent molecules by **diffusion** results in an equal concentration throughout the sample.

A **semipermeable membrane** is a thin sheet of material that contains microscopic holes, or pores, that are small enough to block the passage of large molecules but large enough that smaller molecules are able to pass through. There are several types of semipermeable membranes. *Dialyzing membranes* (Section 8.4) are semipermeable membranes of such porosity that water, small molecules, and ions can pass through. However, larger ions and molecules cannot pass. *Cell membranes* (Section 22.2) are dialyzing membranes through which water, ions, and small molecules (such as nutrients and waste molecules) are *selectively* passed. *Osmotic membranes* are membranes with pores so small that only water can pass through. Even ions (which are relatively large because of hydration) cannot pass through an osmotic membrane, or do so at a relatively slow rate.

When a concentrated solution of sodium chloride or other salt is placed on one side of an osmotic membrane and a dilute solution is placed on the other side, water molecules pass back and forth freely through the membrane while the ions remain behind. Figure 7.4 illustrates this phenomenon.

Although the solvent molecules can flow back and forth through the membrane, a greater number will flow in the direction of the more concentrated solution in an attempt to make the solutions on both sides of the

Figure 7.4.   *An osmotic membrane contains pores of such a size that water molecules are permitted to pass through, but not larger molecules or solvated ions.*

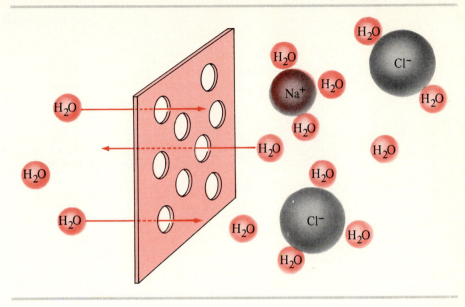

membrane of equal concentration. The net result is a flow of solvent from the dilute side to the concentrated side of the membrane. This is the process called *osmosis.*

Osmosis is the net flow of solvent molecules through an osmotic membrane from a dilute solution to a more concentrated solution, so as to make the concentrations more nearly equal.

Figure 7.5 depicts the process of osmosis. As you can see from the figure, as water moves from the dilute solution to the concentrated solution, the volume of the more concentrated solution must increase. The level of solution in the tube rises, and the level of the dilute solution drops. Eventually, a point is reached where the pressure forcing water into the tube (to equalize the concentrations) is counteracted by the pressure from the force of gravity, which prevents the liquid in the tube from rising higher. At this point, the system is at equilibrium, and water molecules pass back and forth through the membrane at equal rates.

Osmotic Pressure and Osmolarity

Osmosis of a system such as the one shown in Figure 7.5 can be prevented by applying pressure to the liquid in the tube. The amount of pressure that must

Figure 7.5.   *In the process of osmosis, solvent flows through an osmotic membrane from the dilute solution toward the more concentrated solution. The flow continues until the pressure of the increased volume of the concentrated solution is equal to the pressure of the solvent flow.*

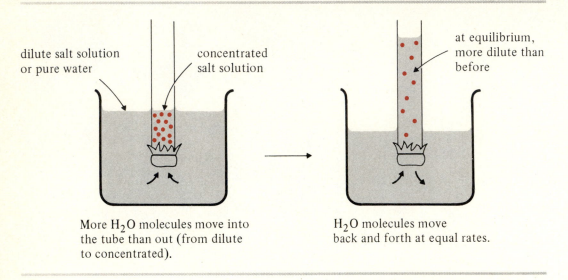

dilute salt solution
or pure water

concentrated
salt solution

at equilibrium,
more dilute than
before

More H_2O molecules move into
the tube than out (from dilute
to concentrated).

H_2O molecules move
back and forth at equal rates.

be applied to prevent the osmotic flow of pure water into a particular solution is called the **osmotic pressure** of that solution. A solution with high osmotic pressure will take up more water than a solution with lower osmotic pressure; thus, more pressure must be applied to prevent osmosis. Figure 7.6 depicts the concept of measuring osmotic pressure.

Figure 7.6.   *Osmotic pressure is the amount of pressure required to prevent the solution in the tube from rising.*

P (osmotic pressure)

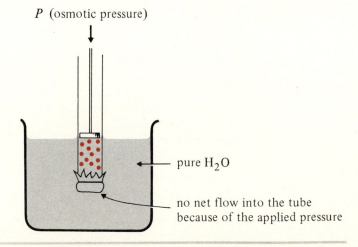

pure H_2O

no net flow into the tube
because of the applied pressure

The osmotic pressure of a solution is a colligative property and is proportional to the concentration of the solute and the number of particles formed when the solute dissolves, regardless of what the actual solute is. The term that takes both these variables into account is called **osmolarity**.

$$\text{osmolarity} = M \times \text{number of particles from one molecule}$$
$$\text{(or formula unit) of solute}$$

$$\text{osmotic pressure} = \text{osmolarity} \times \text{constant}$$

Example   (a) What is the osmolarity of a 2 *M* solution of glucose? (b) What is the osmolarity of a 2 *M* solution of $Na^+ Cl^-$?

Solution   (a) osmolarity = 2 *M* × 1 particle = 2
(b) osmolarity = 2 *M* × 2 particles = 4

Example   Let us set up two identical apparatuses to measure osmotic pressure. In the tube of one apparatus, we place a 1.0 *M* solution of NaCl. In the tube of the other apparatus, we place a 1.0 *M* solution of glucose. Which solution would exhibit the greater osmotic pressure and by how much?

Solution   The osmotic pressure of the 1.0 *M* NaCl solution is *twice* the osmotic pressure of the 1.0 *M* glucose solution, because it has twice as many particles in any given volume. Phrased in another way, the osmolarity of a NaCl solution is twice that of a glucose solution of the same molarity; therefore, the osmotic pressure is twice that of the glucose solution.

Problem 7.11.   If body fluids have an osmolarity of 0.300, what is the approximate total molarity of compounds in these fluids (assuming that most of the dissolved compounds are ionic and yield two particles when dissolved)?

Problem 7.12.   The *milliosmolarity* of a solution is defined as its osmolarity × 1000. What is the approximate milliosmolarity of body fluids? (See Problem 7.11.)

Problem 7.13.   Compare the osmotic pressures of the following solutions:
(a) a 0.50 *M* NaCl solution and a 1.0 *M* glucose solution
(b) a 0.20 *M* $CaCl_2$ solution and a 0.20 *M* NaCl solution

Hemolysis and Crenation

An erythrocyte (red blood cell) is surrounded by a semipermeable membrane that allows the passage of water, as well as of certain ions and small molecules. Both erythrocytes and blood contain fluids that are equal in osmolarity to

Figure 7.7.   *The effects of bathing a red blood cell in hypotonic, hypertonic, and isotonic solutions.*

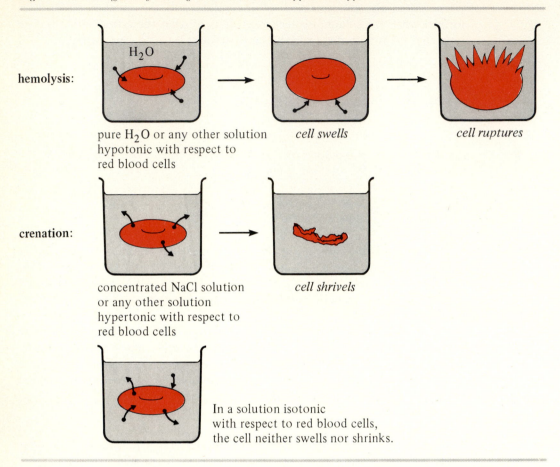

hemolysis:

pure H$_2$O or any other solution
hypotonic with respect to
red blood cells

cell swells

cell ruptures

crenation:

concentrated NaCl solution
or any other solution
hypertonic with respect to
red blood cells

cell shrivels

In a solution isotonic
with respect to red blood cells,
the cell neither swells nor shrinks.

0.9% (w/v) aqueous NaCl. Osmosis does not occur between blood and a red blood cell because water molecules pass in and out of the cell at equal rates.

If we place red blood cells in pure water, however, more water enters the cell by osmosis than leaves (dilute solution to concentrated). The cell swells up and eventually bursts. The rupturing of red blood cells by osmosis is called **hemolysis**.

By contrast, if red blood cells are placed in a water solution containing more than 0.9% NaCl, water *leaves* the cell. The cell becomes shriveled. This process is called **crenation**. (A simplified diagram of hemolysis and crenation is shown in Figure 7.7.)

Physiological Saline Solution.   What if red blood cells are placed in a solution containing 0.9% (w/v) NaCl? No osmosis is observed. A 0.9% solution of

NaCl is called *physiological saline solution.* Physiological saline solution (instead of water) is used for intravenous preparations to prevent dilution of the blood and subsequent hemolysis.

Two solutions that have the same osmotic pressure (the same osmolarities) are said to be *isotonic* (from *iso-,* meaning "equal") with each other. Blood and physiological saline solution are isotonic solutions. By contrast, a *hypertonic solution* (*hyper-,* meaning "over" or "more than normal") is one in which the osmolarity is *greater* than that of a second solution. A concentrated NaCl solution is hypertonic with respect to blood. A *hypotonic solution* (*hypo-,* "under") is one in which the osmolarity is *lower* than that of a second solution. Pure water is hypotonic compared to blood. The effects of bathing red blood cells in these three types of solution are illustrated in Figure 7.7.

isotonic solutions: equal osmolarities; no osmosis occurs
hypertonic solution: higher osmolarity; this solution gains water by osmosis
hypotonic solution: lower osmolarity; this solution loses water by osmosis

Epsom salt ($MgSO_4 \cdot 7H_2O$) and some other cathartics act as hypertonic solutions with respect to fluids in the intestinal wall and cause water to be drawn by osmosis into the intestine. Consequently, the contents of the intestine become more liquid and are more readily evacuated.

Problem 7.14.   As you are aware, humans must drink fresh water, not sea water. Using the proper terminology, explain why this is so.

SUMMARY

A *solution* is a microscopically homogeneous mixture of a *solute* in a *solvent.* A *tincture* is a solution in ethanol. An *amalgam* is a solution in mercury. *Miscible liquids* are liquids infinitely soluble in each other. A *saturated solution* is a solution in equilibrium with pure solute; at a given temperature, the solvent can dissolve no more solute.

Water is a good solvent for ionic compounds and polar covalent compounds because water molecules are polar. A *hydrogen bond* is a strong attraction between an H bonded to an N or O and the N or O of another molecule. For example:

$$\}{-}OH{-}{-}{-}{:}\ddot{O}{-}\} \qquad or \qquad \}{-}\ddot{N}H{-}{-}{-}{:}\ddot{O}{-}\}$$

The *surface tension* of water is unusually high because of these attractions. A *surfactant* is an agent that lowers the surface tension.

A *hydrate* is a solid salt that contains *water of hydration* in its crystals in a definite proportion by weight.

Some important units of concentration are summarized in Table 7.3. In addition to those units, concentration may be expressed in *molarity*, M:

$$M = \frac{\text{number of moles of solute}}{\text{number of liters of solution}}$$

The number of moles present in a solution equals *molarity* × *volume* in liters, or MV. In dilution problems, the following equations are used.

$$M_1 V_1 = M_2 V_2 \qquad \text{or} \qquad C_1 V_1 = C_2 V_2$$

Typical *colligative properties* of solutions are freezing point depression, boiling point elevation, and *osmosis*, the passage of solvent through an osmotic semipermeable membrane from a more dilute solution to a more concentrated solution. *Osmotic pressure*, the pressure required to prevent osmosis, is proportional to *osmolarity* (M × number of particles per molecule or formula unit).

Hemolysis is the swelling and bursting of blood cells in a *hypotonic* (lower osmolarity) solution. *Crenation* is the shriveling of blood cells in a *hypertonic* (higher osmolarity) solution. *Physiological saline solution* is *isotonic* with blood (same osmolarity) and causes neither hemolysis nor crenation.

KEY TERMS

solution	surfactant	osmosis
solute	*hydrate	osmotic pressure
solvent	*water of hydration	osmolarity
solubility	percent concentrations	hemolysis
saturated solution	parts per million	crenation
hydration	molarity	physiological saline solution
hydrogen bond	colligative property	isotonic solution
surface tension	diffusion	hypertonic solution
capillary action	semipermeable membrane	hypotonic solution

STUDY PROBLEMS

7.15. A chemist heated the following mixture in a beaker: 2.0 g sugar, 100 mL H_2O, and 10.0 g ground coffee. Tell which is the solvent and which is the solute (or solutes). Tell which of the ingredients would be considered soluble and which only partially soluble.

7.16. Designate each of the following sets of liquids as miscible or immiscible:
(a) salad oil and vinegar   (b) wine and water

7.17. The following mixtures have been heated, then cooled to 20°C. Tell whether each is a saturated or unsaturated solution (see Table 7.2).
(a) 38.0 g NaCl in 100 g H_2O
(b) 60.0 g $CuSO_4$ in 200 g H_2O

7.18. Suggest a reason why droplets of water remain distinct on a waxed or greasy glass or metal surface (such as an automobile body) but spread out when the surface is clean.

7.19. Which of the following compounds could act as a surfactant? Explain your answers.

(a) the first two phosphorus compounds shown in Section 19.3

(b) decane (Table 14.1)

(c) $CH_3CH_2CH_2CH_2CH_2CH_2CH_2CH_2OSO_3^-$
Na^+

7.20. Tell how you would prepare the following solutions from pure compounds:

(a) 250 mL of 5.00% (w/v) aqueous sodium bicarbonate

(b) 3 L of 0.9% (w/v) NaCl (physiological saline solution)

(c) 2 L of 70% (v/v) aqueous isopropyl alcohol (rubbing alcohol)

(d) 10 g of a 5% (w/w) solution of NaOH in water

7.21. Find the percent by volume of ethanol in an aqueous solution prepared by diluting 15 mL of pure ethanol to 70 mL.

7.22. Calculate the number of moles of solute in each of the following solutions:

(a) 100 mL of a 0.5 M solution of $CdCl_2$

(b) 50 mL of a 1.5 M solution of $Na_2Cr_2O_7$

7.23. How many grams of KOH would be needed to prepare 100 mL of a 2.00 M solution?

7.24. What is the molarity of each of the following solutions?

(a) 20.0 g of $FeCl_3$ in 100 mL of H_2O solution

(b) 0.050 g of KCl in 500 mL of H_2O solution

7.25. How many milliliters of each of the following solutions would be required to give 1.5 moles?

(a) 12 M H_2SO_4 (b) 0.05 M NH_4Cl

7.26. We want to prepare a 1.0% (w/v) solution of NaCl from 25 mL of a 2.5% (w/v) solu-

tion. How much water should we add to the first solution (assuming the volumes are additive)?

7.27. A chemist diluted 100 mL of 3 M H_2SO_4 to 600 mL with water. What is the molarity of the new solution?

7.28. For each of the following solutions, how many mL of H_2O should be added to yield a solution that has a concentration of 1.5 M (assuming that the volumes are additive)?

(a) 25 mL of 6 M HCl

(b) 75 mL of 2 M NaOH

7.29. Is it possible to dilute 10 mL of a 0.50 M NaCl solution to obtain a 1.0 M solution? (Solve the problem by the dilution equation and interpret the result.)

7.30. Give the number of particles of solute that would be formed if each of the following samples were dissolved in water:

(a) 0.5 mole glucose   (b) 0.5 mole $MgBr_2$

7.31. Calculate the osmolarity of each of the following solutions:

(a) 0.5 M NaCl   (b) 0.5 M $MgBr_2$

7.32. Calculate the freezing point of (a) 1.5 M aqueous glucose and (b) 1.5 M NaCl.

7.33. If pure water is placed inside a bag made of an osmotic membrane and the bag is then suspended in an ionic solution,

(a) in which direction will the water flow?

(b) will the bag expand or contract?

(c) will the ionic solution increase or decrease in ionic strength?

7.34. What is the ratio of the osmotic pressures of (a) 0.25 M glucose and (b) 0.25 M $MgCl_2$?

8 Colloids

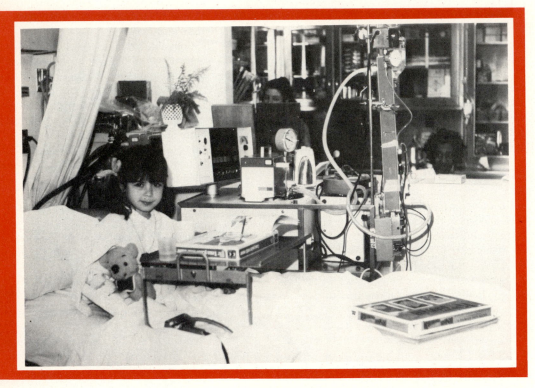

A child undergoing hemodialysis with an artificial kidney machine. (Reprinted with permission of the National Kidney Foundation, Inc., New York, N.Y., and Walter Reed Army Medical Center, Washington, D.C.) Not too many decades ago, kidney failure meant death due to uremic poisoning, the build-up of toxic waste products in the blood and tissues. Today, artificial kidney machines allow the removal of these waste products from the blood by hemodialysis. Early kidney machines were large and expensive. In recent years, small portable units have become available.

Objectives   Define colloid and list some types of colloids. ☐ List the similarities and differences among solutions, colloids, and suspensions. ☐ Describe the Tyndall effect and Brownian movement. ☐ Describe how ions can stabilize a colloid. ☐ Relate the structure of an emulsifying agent to emulsion stabilization. ☐ Define and describe dialysis and hemodialysis, including what types of substances can pass through a dialyzing membrane.

The terms **colloid** and **colloidal dispersion** are synonymous and refer to a dispersion of small particles of one substance (the *dispersed phase*) in another substance (the *dispersion medium*). If we mix a small amount of flour in water and shake up the mixture, the flour does not dissolve, nor does all the flour settle out after the mixture is allowed to stand undisturbed. The water will have a cloudy appearance due to dispersed flour particles. In this example, flour is the dispersed phase and water is the dispersion medium. The homogeneous cloudy liquid is the colloidal dispersion. The differences between the properties of a colloidal dispersion and those of a solution arise from the fact that colloidal particles are larger than solute particles.

The word *colloid* comes from the Greek words *kolla*, meaning "glue," and *eidos*, meaning "looks like." Animal glue itself is a colloid formed from solid protein material dispersed in a solvent. Many other colloids resemble glue, but some do not. For example, smoke and marshmallows are colloids whose physical resemblance to glue is remote.

Some types of colloids and their properties are discussed in this chapter. We will also introduce *dialysis*, a process similar to osmosis, and *hemodialysis*, dialysis of the blood in a kidney machine.

8.1   TYPES OF COLLOIDS

A mixture of flour in water is an example of a colloidal dispersion of a *solid in a liquid*; this type of colloid is called a **sol**. Smoke is a dispersion of a *solid in a gas*. Some common types of colloids are listed in Table 8.1.

Gelatin desserts belong to a special class of jellylike solid-in-liquid colloids called **gels**. Gelatin is protein material obtained by boiling animal bones and skins in water. The protein material that forms connective tissue (collagen) occurs as twisted, ropelike strands containing three individual protein molecules. Boiling causes the strands to unravel; gelatin is composed of these individual unraveled protein molecules. Dried and powdered gelatin is soluble in hot water but not soluble in cold water. When a hot solution of gelatin is chilled, gelatin molecules do not form crystals because of their strong attractions for water molecules. Instead, a three-dimensional network of gelatin and water molecules is formed, leading to the familiar jellylike dessert.

Table 8.1. *Types of Colloids*

Type	Name	Example
gas in liquid	foam	whipped cream, shaving cream
gas in solid		marshmallows, foam rubber
liquid in gas	aerosol	fog, aerosol sprays (insecticide, deodorant, and hair sprays)
liquid in liquid	emulsion	milk, mayonnaise
liquid in solid		cheese
solid in gas		smoke
solid in liquid	sol; gel	paint, flour in water, ink, milk of magnesia, hot chocolate; gelatin
solid in solid		colored glass

An **emulsion** is a colloidal dispersion of one liquid in another liquid. An emulsion is a fairly homogeneous mixture, and the two liquids do not separate. Although mayonnaise is principally an emulsion of vinegar in vegetable oil, it does not resemble vinegar-and-oil salad dressing because it also contains an emulsifying agent, which we will discuss later in this chapter.

8.2   PROPERTIES OF COLLOIDS

Particle Size

The particle size of the dispersed phase of a colloidal dispersion is larger than that of a solute in a solution. Solute particles have diameters under 1 nanometer, or nm (where 1 nm = 10^{-9} m). Colloidal particle sizes fall in the 1–100 nm range. Like solute particles, colloidal particles can pass through filter paper and do not settle out of the dispersion.

If the particles of a dispersed material are greater than 100 nanometers in diameter, the particles may be suspended in water temporarily to yield a **suspension**. The particles of a suspension cannot pass through filter paper and will eventually settle, Many oral medications are suspensions and therefore should be shaken before use.

Table 8.2 summarizes the properties of colloids and compares them to the properties of solutions and suspensions.

Tyndall Effect

We cannot actually see a beam of light unless it is directed into our eyes. When we see a shaft of sunlight coming through the window of a dimly lit room, what we really observe is the reflection and scattering of the light by minute dust particles floating in the air—dust particles that would also be invisible if it were not for the beam of light.

Table 8.2. *Comparison of the Properties of Solutions, Colloids, and Suspensions*

Property	Solution	Colloid	Suspension
particle size	<1 nm	1–100 nm	>100 nm
appearance	clear, homogeneous under a microscope	usually cloudy, homogeneous to the eye	cloudy, not homogeneous
separation	does not separate	does not separate	separates, or settles
filterability	passes through filter paper	passes through filter paper	does not pass through filter paper
	can pass through a semipermeable membrane of suitable porosity	does not pass through a semipermeable membrane	does not pass through a semipermeable membrane
Tyndall effect	no Tyndall effect	Tyndall effect (if not opaque)	depends on particle size
Brownian movement	particles invisible under microscope	exhibits Brownian movement	exhibits little or no Brownian movement
adsorptive properties	not adsorptive	can adsorb ions	usually not adsorptive

If we shine a beam of light through a true solution in a jar, we cannot see the track of the light. (See Figure 8.1.) However, if we shine a beam of light through a colloidal dispersion, even one that does not appear cloudy to us, we can see the path of the light just as we can see a sunbeam in a dusty room, because the light is scattered by the particles. The ability of a colloid to scatter

Figure 8.1. *The Tyndall effect. A beam of light passes through a solution without being reflected. A beam of light passing through a colloidal dispersion is reflected to our eyes.*

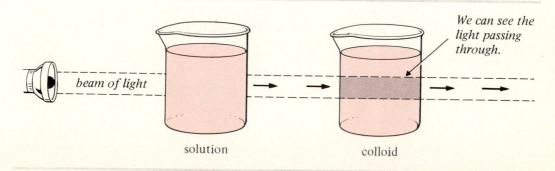

We can see the light passing through.

beam of light

solution   colloid

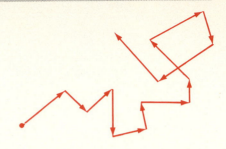

Figure 8.2. *Brownian movement is the erratic motion of a colloidal particle.*

light rays so that we can see a beam of light passing through it is known as the **Tyndall effect**.

Brownian Movement

If they are large enough, the particles of colloidal dispersions can be seen under a microscope. The principal property observed is that colloidal particles move continuously in an erratic rather than a smooth fashion. This erratic motion of colloidal particles is called **Brownian movement**, after its discoverer, Robert Brown (1773–1858). Brown suggested that the colloidal particles might be alive, but we now know that their dancing motion is caused by the small colloidal particles responding to collisions with the even smaller molecules of the medium. (See Figure 8.2.) The fact that colloidal particles are small enough to be buffeted about by molecular collisions is one reason why the particles do not settle out of the dispersion medium. The collisions exert a greater force on the particles than does gravity.

Adsorption of Ions

Colloidal particles can adsorb (attract to their surface) ions present in the dispersion medium. Depending on the type of colloid, the particles attract either positive ions or negative ions, but not both. Thus, each colloidal particle is blanketed in ions and repels other similar colloidal particles. This is a principal reason that colloidal particles do not aggregate and precipitate.

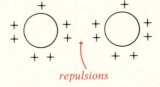

repulsions

Problem 8.1.   An *electrostatic precipitator* is a device with charged plates that removes colloidal particles from air. In industry, large electrostatic precipitators are used to remove colloidal particles from smoke before it passes out of a stack. Describe *how* and *why* these devices are able to remove colloidal particles from the smoke.

An **emulsifying agent** is a substance that stabilizes an emulsion so that it does not separate into its components. The emulsifying agent in mayonnaise is egg yolk, which is rich in *lecithins* (see Section 19.3). Soaps (see Section 17.6) and detergents are also emulsifying agents. A molecule of each of these types of compounds contains a large nonpolar end plus an ionic end. The structures of emulsifying agents are similar to those of surfactants (Section 7.2); many surfactants are emulsifying agents and vice versa.

Figure 8.3.   *Emulsification of an oil droplet by soap molecules.*

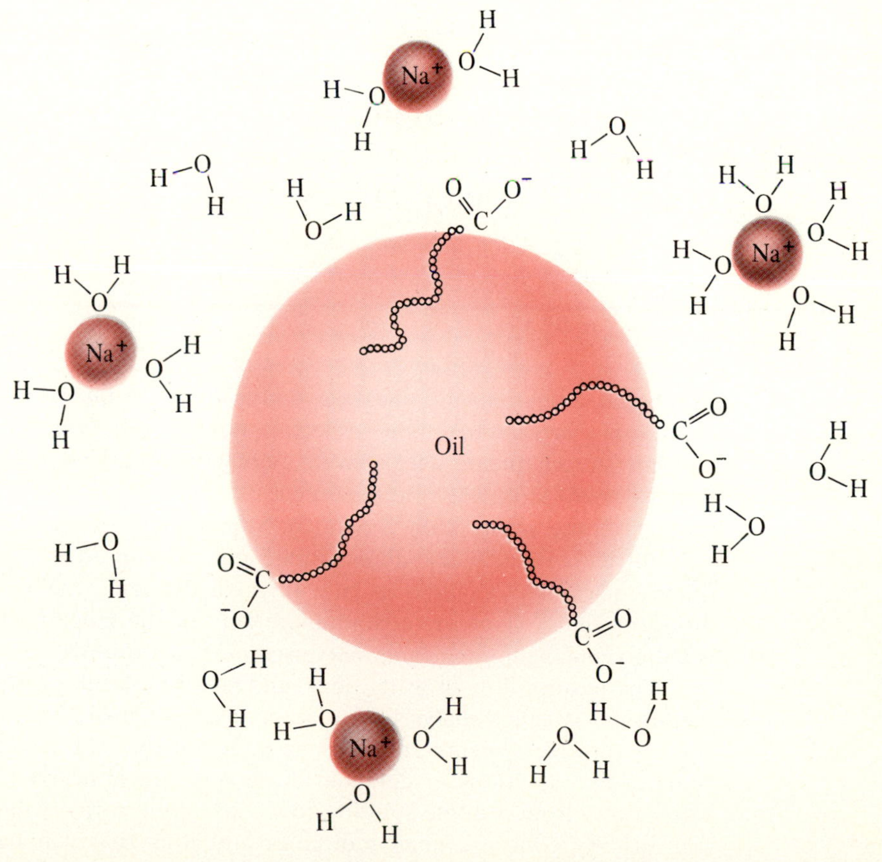

Let us consider a soap as an example of an emulsifying agent. Most large organic molecules are insoluble in water. However, because the ionic portion of a soap molecule is attracted to water, a soap readily forms a colloidal dispersion in water. Oils and greases (large organic molecules) are not soluble in water, but they are attracted to the organic portion of a soap molecule. The result of these attractions is that the nonpolar portions of soap molecules hold onto small droplets of oil and grease, which thus become emulsified in water. The soap-oil droplets do not coalesce because each is "coated" with negative ionic charges from the soap; thus, the droplets repel one another. Once the oil droplets are dispersed in the water by the soap, they can be washed away. (See Figure 8.3.) In mayonnaise, the lecithins from the egg yolk behave similarly. The nonpolar tails of the molecules dissolve in oil droplets, while the ionic heads protrude and are attracted to the water molecules in the vinegar.

Problem 8.2.   Which of the following compounds could act as an emulsifying agent for oils in water? Explain your answers.

(a) $CH_3CH_2CH_2CH_2CH_2CH_3$

(b) $CH_3CH_2CH_2CH_2CH_2CH_2OSO_3^- \ Na^+$

(c) $CH_3CH_2CH_2CH_2CH_2CH_2\overset{\overset{\displaystyle CH_3}{\displaystyle |}}{\underset{\underset{\displaystyle CH_3}{\displaystyle |}}{N^+}}-CH_3 \ Cl^-$

(d) $CH_3CH_2CH_2CH_2CH_2CH_2OPO_3^- \ Na^+$

8.4   DIALYSIS

Plant and animal membranes are dialyzing membranes similar to osmotic membranes in that they allow selective flow of substances from one side to another. A cell membrane is more permeable than an osmotic membrane because small molecules and ions must be able to pass through. For example, the membrane surrounding a plant or animal cell (Section 22.2) allows water and nutrients to enter and allows water and waste products to escape. Colloidal particles are too large to pass through dialyzing membranes. Therefore blood proteins, which are colloidal, are kept in the blood stream. Cellophane is a dialyzing membrane commonly used in the laboratory.

The natural flow of water, ions, and small molecules through a dialyzing membrane is called **dialysis**. In dialysis, as in osmosis, the net flow of substances through a membrane is in the direction that will make the concentrations of two solutions or colloidal dispersions more nearly equal. Therefore, water flows from a dilute solution to a more concentrated solution, just as in osmosis. By contrast, small molecules and ions that can pass through the membrane move from the more concentrated solution to the more dilute

Figure 8.4. *Dialysis. Small ions and molecules can pass through the dialyzing membrane; the larger particles, such as colloids, cannot.*

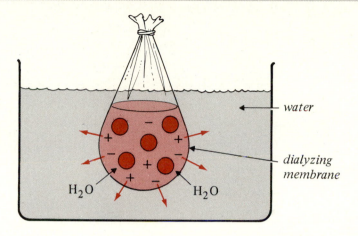

solution—again in the direction that will tend to equalize concentrations. Figure 8.4 shows the net movements of these substances. (You will learn in Chapter 22 that living cell membranes, unlike laboratory dialyzing membranes, can *force* ions and small molecules through in the direction that opposes the natural flow of dialysis.)

Hemodialysis

The kidneys remove waste products (which are small molecules) and excess ions from the blood so that they can be eliminated in the urine. If the kidneys are not functioning normally, waste products and toxins accumulate in the blood stream, causing a condition called *uremic poisoning.* When enough of these substances accumulate, the person so afflicted may die. However, the toxins and waste products can be removed from the blood artificially by a kidney machine in a process called **hemodialysis**, which means "dialysis of the blood."

A kidney machine (Figure 8.5) contains a long cellophane tube (coiled so as to be compact) through which a patient's blood can be circulated. The coiled tube is immersed in a special, temperature-controlled water solution, which must be free of microorganisms and other harmful substances. Also, the solution must be isotonic with respect to normal blood and must contain the proper concentrations of sodium ions, chloride ions, calcium ions, and all other desirable ions and small molecules.

When the patient's blood is pumped through the dialysis tube, blood cells, protein molecules, and other colloidal components remain in the blood

Figure 8.5.   *Diagram of a kidney machine.*

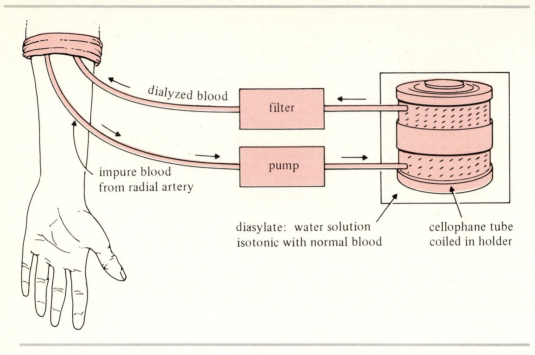

because they cannot pass through the pores of the cellophane membrane. The smaller molecules of waste products can pass through the membrane and thus are removed from the blood. Ions also can pass through the membrane. However, the ions that have been added to the water solution prior to dialysis pass back and forth at equal rates. Thus, there is no net loss of desirable ions. As the concentration of undesirable compounds in the water solution increases, these too can return to the blood. In order to prevent this from occurring, the water solution is replaced with fresh solution periodically.

SUMMARY

Colloids are dispersions of colloidal particles in a dispersion medium. Colloidal particles are intermediate in size between solute molecules or ions in a solution and particles in a suspension. A *gel* is one type of solid-in-liquid colloid. An *emulsion* (liquid-in-liquid colloid) can be stabilized by an *emulsifying agent*, generally an ionic compound with a long, nonpolar chain. Colloids exhibit the *Tyndall effect*. *Brownian movement* and *adsorbed ions* keep colloidal particles from settling.

Dialysis is similar to osmosis, except that a dialyzing membrane is sufficiently porous to allow the passage of ions and small molecules, as well as water, but not colloidal particles. *Hemodialysis* is the dialysis of blood in a kidney machine.

KEY TERMS

colloid   Tyndall effect   dialysis
emulsion   Brownian movement   hemodialysis
suspension   emulsifying agent

STUDY PROBLEMS

8.3. Identify the dispersed phase (solid, liquid, or gas) and the dispersion medium for each of the following colloids or suspensions:
(a) gravy   (b) aerosol dessert topping
(c) vinegar-and-oil dressing

8.4. Each of the following properties applies to a solution, a colloid, or a suspension. Identify which type (or types) of mixture is indicated by each property.
(a) particles settle out
(b) homogeneous appearing, but opaque
(c) homogeneous and transparent
(d) completely passes through filter paper
(e) does not completely pass through a dialyzing membrane

8.5. Which of the following mixtures would exhibit the Tyndall effect? Brownian movement?
(a) gelatin dessert   (b) smoke
(c) pure air   (d) diluted ink
(e) salt solution

8.6. Give two reasons for the ability of colloidal particles to stay suspended in the dispersion medium.

8.7. Explain why the addition of a large quantity of an ionic substance may cause a colloidal suspension to precipitate.

8.8. Bile salts are substances produced in the liver and transported to the intestines, where they act as emulsifying agents for ingested fats and oils and expedite the digestion of these compounds.

$$R-\overset{\overset{\displaystyle O}{\|}}{C}NHCH_2\overset{\overset{\displaystyle O}{\|}}{C}O^-\ Na^+$$

a bile salt

where R = a complex organic structure called a *steroid* (see Section 19.5). Explain how bile salts can act as emulsifying agents.

8.9. A colloidal dispersion of flour and water is placed on side 1 of a dialyzing membrane. A concentrated aqueous solution of NaCl is placed on side 2 of the membrane. Describe the movement across the membrane (if any) of each of the following species:
(a) H_2O (b) Na^+ (c) Cl^- (d) flour

8.10. If the membrane in Problem 8.9 were an osmotic membrane instead of a dialyzing membrane, what would be observed?

8.11. If distilled water were used as a hemodialysis solution, what would be the effects on the blood of the patient in terms of (a) waste products, (b) glucose (water soluble), (c) ions, and (d) colloidal proteins?

9 Ionic Compounds in Solution

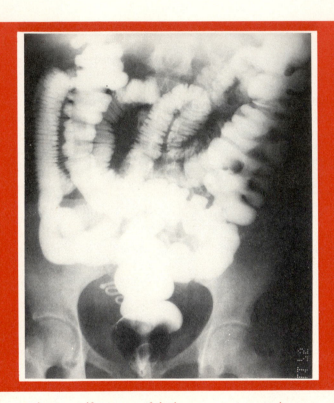

A barium sulfate x ray of the lower gastrointestinal tract. (© 1981 Martin M. Rotker/Taurus Photos.) The body's soft tissues are relatively transparent to x radiation, but the water-insoluble salt barium sulfate, which can be administered orally, renders the gastrointestinal tract opaque to this type of radiation. This radiograph shows the small intestines toward the top; the large intestine and pelvic bones are visible toward the bottom.

Objectives   Define and give examples of electrolytes and nonelectrolytes. □ Define acid and write equations that show how acids yield H_3O^+ or H^+. □ Define base and write equations for acid-base reactions. □ Define and give examples of salts. □ Differentiate strong and weak electrolytes. □ List some electrolytes in body fluids and explain their importance (optional). □ List ions that impart water-solubility to a salt. □ Construct net ionic equations from molecular equations. □ List and give examples of the phenomena that drive ionic reactions to completion. □ Define hard water and explain how it can be softened (optional).

As we mentioned in Chapters 3 and 7, an ionic compound dissociates into individual ions when it is dissolved in water. In this chapter, we will discuss some types of electrolytes (compounds that yield ions in water) and some of their chemical reactions. Because the human body is primarily water, many of these reactions can occur in the body fluids as well as in a test tube. We will also consider the differences between hard and soft water. In Chapter 10, we will expand our discussion of acids and bases.

9.1   ELECTROLYTES

Pure water cannot conduct electricity to any measurable extent. How then can a person be electrocuted in the bathtub? Water containing ions is an excellent conductor of electricity. Bath water contains ionic compounds dissolved from soil, from plumbing pipes, from skin, and from soap; therefore, bath water can conduct electricity.

Because a solution of ions in water can conduct electricity, an ionic compound is said to be an **electrolyte**. A solution of a covalently bonded compound, such as table sugar (sucrose), does not conduct electricity because sucrose does not form ions in water. Sucrose and other nonionic compounds are called **nonelectrolytes**.

some electrolytes:

$Na^+ Cl^-$, $Na^+ OH^-$, $K^+ NO_3^-$

some nonelectrolytes:

$C_{12}H_{22}O_{11}$ (sucrose), CH_3CH_2OH (ethanol)

Ions in Water

Because an ionic compound dissociates into separate ions in solution, its dissolving is called **dissociation**.

dissociation of ionic compounds in water:

$$NaCl_{solid} \xrightarrow{H_2O} Na^+ \quad + \quad Cl^-$$

sodium chloride   sodium ion   chloride ion
                        a cation   *an anion*

$$KOH_{solid} \xrightarrow{H_2O} K^+ \quad + \quad OH^-$$

potassium hydroxide   potassium ion   hydroxide ion
                        a cation   *an anion*

Some highly polar covalent compounds can yield ions, or undergo **ionization**, when they are dissolved in water. For example, hydrogen chloride gas (HCl) is a polar covalent compound that undergoes ionization in water to yield H^+ and Cl^-. A water solution of hydrogen chloride is called *hydrochloric acid.*

ionization of a covalent compound in water:

$$H-\overset{\cdot\cdot}{\underset{\cdot\cdot}{Cl}}{:}_{gas} \xrightarrow{H_2O} H^+ \quad + \quad {:}\overset{\cdot\cdot}{\underset{\cdot\cdot}{Cl}}{:}^-$$

hydrogen   hydrogen   chloride
chloride   ion   ion

hydrochloric acid

When HCl ionizes in water, it does not actually form a hydrogen ion as we have shown. When a hydrogen ion is formed in water, it is attracted to the partially negative oxygen of water. This oxygen has unshared electrons that form a covalent bond with the H^+ to yield H_3O^+, called the **hydronium ion**.

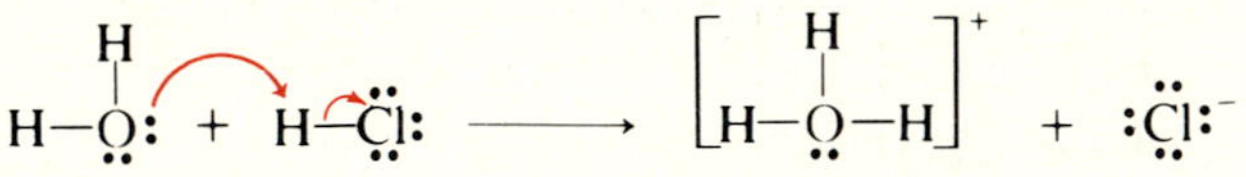

$$H-\overset{H}{\underset{\cdot\cdot}{O}}{:} \; + \; H-\overset{\cdot\cdot}{\underset{\cdot\cdot}{Cl}}{:} \longrightarrow \left[H-\overset{H}{\underset{\cdot\cdot}{O}}-H \right]^+ \; + \; {:}\overset{\cdot\cdot}{\underset{\cdot\cdot}{Cl}}{:}^-$$

hydronium ion

or simply

$$H_2O + HCl \longrightarrow H_3O^+ + Cl^-$$

In water solution, a hydrogen ion always exists as a hydronium ion rather than as H^+. However, for the sake of simplicity, acids are usually represented as ionizing to H^+ and an anion, even though this is not strictly correct.

Example   Write two equations for the ionization of nitric acid (HNO_3) in water. In one equation, show the hydronium ion product; in the other, show H^+ as the product.

Solution   (1) $HNO_3 + H_2O \longrightarrow H_3O^+ + NO_3^-$

(2) $HNO_3 \xrightarrow{H_2O} H^+ + NO_3^-$

Problem 9.1.   Write equations that show what happens when the following compounds are dissolved in water:
(a) calcium sulfate, $CaSO_4$
(b) hydrogen bromide, HBr
(c) lithium hydroxide, LiOH

Types of Electrolytes

Acids.   The three fundamental types of electrolytes are *acids*, *bases*, and *salts*. An **acid** is defined as a compound that can *lose hydrogen ions* (H^+) in chemical reactions.

two typical acids:

$$HCl \xrightarrow{H_2O} H^+ + Cl^-$$

hydrogen chloride

$$H_2SO_4 \xrightarrow{H_2O} H^+ + HSO_4^-$$

sulfuric acid

One of the properties common to acids is that they have a sour, tart taste. The sour taste of lemons and grapefruits is due to the presence of citric acid. Vinegar contains about 5% acetic acid. Gastric juices contain very dilute hydrochloric acid (about 0.1 *M*, or 0.5%).

Bases.   A **base** is a compound that can *accept hydrogen ions* in chemical reactions. Typical bases are compounds that yield hydroxide ions (OH^-) upon dissociation.

two typical bases:

$$NaOH \xrightarrow{H_2O} Na^+ + OH^-$$

sodium hydroxide
(lye)

$$Ca(OH)_2 \xrightarrow{H_2O} Ca^{2+} + 2 OH^-$$

calcium hydroxide
(lime)

A compound containing hydroxide ions can accept hydrogen ions from an acid in a reaction called an **acid-base reaction**, to yield a salt plus water.

an acid-base reaction:

$$Na^+ + OH^- + H^+ + Cl^- \longrightarrow Na^+ + Cl^- + H_2O$$

sodium hydroxide   hydrochloric acid   sodium chloride
a base   *an acid*   *a salt*

or simply

$$NaOH + HCl \longrightarrow NaCl + H_2O$$

We will discuss acids and bases in more detail in Chapter 10.

Salts.   A **salt** is an ionic compound that contains a cation other than H^+ and an anion other than OH^-. Pure salts are generally high-melting solid compounds. For example, sodium chloride melts at $801°C$.

some typical salts:

sodium chloride $Na^+ Cl^-$

potassium nitrate $K^+ NO_3^-$

potassium chloride $K^+ Cl^-$

A salt is the other product that is formed when an acid undergoes reaction with a hydroxide to yield water. If the salt is soluble, it can be recovered from solution by boiling off the water. An insoluble salt can be isolated by filtration.

$$Mg(OH)_2 + H_2SO_4 \longrightarrow MgSO_4 + 2 H_2O$$

magnesium   sulfuric   magnesium
hydroxide   acid   sulfate
                               a salt

Problem 9.2.   Classify each of the following compounds as an acid, a base, or a salt:
(a) LiBr   (b) $Ba(OH)_2$ (c) H_3PO_4

Problem 9.3.   Complete and balance the following equations:
(a) $KOH + HNO_3 \longrightarrow$ (b) $HCl + Ca(OH)_2 \longrightarrow$
(c) $H_2SO_4 + NaOH \longrightarrow$

Strong and Weak Electrolytes

The existence of ions explains why some aqueous solutions can conduct electricity. When an aqueous solution of an electrolyte conducts electricity, the ions in solution migrate, or move, to the electrode of opposite charge and

Figure 9.1. *A solution containing ions can conduct electricity. Part of the process involves ions moving to the electrode of opposite charge.*

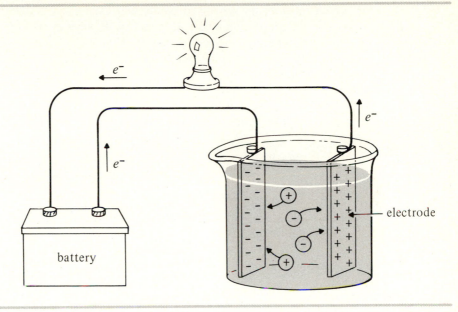

undergo reaction there. The cations (+) migrate toward the negative electrode, and the anions (−) migrate toward the positive electrode (see Figure 9.1). Therefore, the migrating ions actually carry the electrical charges through the solution. A nonionic compound, or *nonelectrolyte*, carries no ionic charge, and thus its molecules do not migrate toward either electrode. This is why electricity is not conducted through a solution of a nonelectrolyte.

Soluble salts and many acids and bases are completely dissociated or ionized when dissolved in water. Their ionic solutions are excellent conductors of electricity, and thus they are called **strong electrolytes**.

some typical strong electrolytes (completely dissociated or ionized in water):

soluble salts	$NaCl$, $CaCl_2$, $NaHCO_3$
many inorganic acids	HCl, HNO_3, H_2SO_4
many inorganic bases	$NaOH$, KOH

Other compounds, called **weak electrolytes**, are intermediate between strong electrolytes and nonelectrolytes in their ability to conduct an electric current when in aqueous solution. Solutions of these compounds can conduct a current, but not to the extent that a strong electrolyte can. (The light bulb in Figure 9.1 would glow only feebly.)

The reason for this intermediate behavior is that a weak electrolyte is only partially ionized when it is dissolved in water. An equilibrium is achieved in which most of the dissolved compound exists as covalent molecules and only a small proportion exists as ions. Acetic acid is an example of a compound that undergoes partial ionization. When dissolved in water, this compound yields some acetate ions and some hydrogen ions in a reversible reaction with water.

$$\underset{\substack{\text{acetic acid} \\ 99.5\%}}{\text{CH}_3\overset{\overset{\text{O}}{\|}}{\text{C}}\text{OH}} \underset{\text{H}_2\text{O}}{\longleftrightarrow} \underset{\text{acetate ion}}{\text{CH}_3\overset{\overset{\text{O}}{\|}}{\text{C}}\text{O}^-} + \underset{\text{hydrogen ion}}{\text{H}^+}$$

Only 0.5% exists as ions.

some typical weak electrolytes (only partially ionized in water):

most soluble organic acids $\text{CH}_3\overset{\overset{\text{O}}{\|}}{\text{C}}\text{OH}, \text{CH}_3\text{CH}_2\overset{\overset{\text{O}}{\|}}{\text{C}}\text{OH}$

carbonic acid $\text{HO}\overset{\overset{\text{O}}{\|}}{\text{C}}\text{OH}$

ammonia NH_3

◼ 9.2   ELECTROLYTES IN THE BODY FLUIDS (optional)

The following lists show the principal ions found in the blood and interstitial fluids (fluids in the spaces, or interstices, *between* cells) and within the cells themselves. Note that sodium chloride ($\text{Na}^+ + \text{Cl}^-$) is the principal extracellular (outside of cells) electrolyte, while potassium hydrogen phosphate ($2\text{ K}^+ + \text{HPO}_4^{2-}$) is the principal intracellular (within cells) electrolyte.

principal ions in the extracellular body fluids (blood and interstitial fluids) Na^+, Cl^-, and small amounts of Ca^{2+}, K^+, Mg^{2+}, HCO_3^-, HPO_4^{2-}, and organic anions

principal ions within the cells K^+, HPO_4^{2-}, and small amounts of Mg^{2+}, Na^+, Ca^{2+}, SO_4^{2-}, HCO_3^-, and Cl^-

The sodium ions in the blood and interstitial fluids help maintain the osmotic pressure of these fluids, help maintain blood pressure, aid in water retention between cells, and provide a cationic balance for the important anions. Chloride ions are present to keep an ionic balance and are necessary in gastric juices, which contain very dilute HCl.

Although other ions in the blood and interstitial fluids are present only in relatively small amounts, these ions are nonetheless important. Calcium ions (Ca^{2+}), for example, are present to the extent of about 0.0025 M but play an important role in the blood-clotting mechanism and in the regulation of membrane permeability, especially of the nerve cell membranes. A severe lack of calcium ions allows the nerves to be overstimulated, resulting in spasms or even convulsions. Magnesium ions (Mg^{2+}), present in even lower concentrations, are required to activate many of the body's enzymes. Bicarbonate ions (HCO_3^-) help regulate the acidity of the blood and are an important part of the O_2-CO_2 transfer system.

Within the cell, potassium ions (K^+) perform functions similar to those performed by Na^+ in the extracellular fluids. Hydrogen phosphate ions (HPO_4^{2-}) help control the acid-base balance in the cells and play a principal role in cell respiration, the oxidation of nutrients by a cell to yield energy.

Other electrolytes besides those found in the principal body fluids are also important. For example, traces of iodide ion (I^-) are necessary in human nutrition for the proper functioning of the thyroid gland. Lack of iodides results in *goiter*, an enlarged thyroid gland. In coastal areas, iodides are found in the soil and in the fruits and vegetables growing there. Salt-water fish also contain iodides. However, over the eons, most of the iodides in the soil of inland areas have been leached out by rain water, and natural food from these inland areas does not contain a sufficient amount of iodides. The introduction and widespread use of *iodized salt*, salt to which a trace of sodium iodide or potassium iodide has been added, has alleviated the problem of iodide deficiency in inland areas. ■

9.3 SOLUBILITIES OF SALTS IN WATER

Salts differ from one another in their solubilities in water. In general, a very soluble salt is one in which the attraction of the ions for water molecules (and water molecules for the ions) is greater than the attraction of the two ions for each other. The ions that almost always form water-soluble salts are listed in Table 9.1. If an ionic compound contains one (not necessarily both) of the ions on this list, it is very probably a soluble salt.

From this table, we would deduce correctly that Na_2CO_3 is water soluble because it contains Na^+, but that $CaCO_3$ is insoluble because neither Ca^{2+} nor CO_3^{2-} is on the list of soluble-salt ions. Calcium carbonate ($CaCO_3$) is indeed insoluble and is found in nature as limestone, marble, and pearls. Calcium carbonate mixed with magnesium carbonate ($MgCO_3$) is also found in chalk, coral, and eggshells.

Problem 9.4. Predict whether or not the following compounds are water soluble:

(a) sodium bicarbonate, $NaHCO_3$ (baking soda)

Table 9.1.   *Ions That Form Water-Soluble Salts*

Soluble if cation is	Soluble if anion is
sodium ion, Na^+	nitrate ion, NO_3^-
potassium ion, K^+	
ammonium ion, NH_4^+	acetate ion, $CH_3\overset{\overset{\textstyle O}{\|}}{C}O^-$
	chloride ion, Cl^-
	(except for $AgCl$, $PbCl_2$, Hg_2Cl_2)
	sulfate ion, SO_4^{2-}
	(except for $BaSO_4$ and $PbSO_4$;
	$CaSO_4$ is only slightly soluble)

(b) magnesium sulfate, $MgSO_4$ (Epsom salt)

(c) calcium chloride, $CaCl_2$ (used to decrease blood-clotting time)

(d) ferrous sulfate, or iron(II) sulfate, $FeSO_4$ (used to treat iron-deficiency anemia)

(e) magnesium carbonate, $MgCO_3$ (used as a cathartic)

Saying that a salt is soluble or insoluble is a qualitative way of expressing solubility. Barium sulfate ($BaSO_4$) is an insoluble salt that is opaque to x rays. It is sometimes administered as an enema prior to x rays of the gastro-intestinal (GI) tract. Because of the opacity of $BaSO_4$, the intestines show up clearly on the x-ray film. Any insoluble salt has a slight solubility. In the case of $BaSO_4$, 0.0002 g dissolves in 100 g of water. Even though barium compounds are poisonous, the very low solubility of barium sulfate means that no appreciable amount is absorbed into the body's tissues.

9.4   HOW IONIC REACTIONS PROCEED

Precipitation and Net Ionic Equations

What happens if sodium chloride (NaCl) and potassium nitrate (KNO_3) are dissolved in water? The NaCl dissociates into Na^+ and Cl^- ions, and the KNO_3 dissociates into K^+ and NO_3^- ions. The result is a mixture of four ions—Na^+, K^+, Cl^-, and NO_3^-—in solution. In this case, we say that *no reaction* takes place.

Now let us mix a solution of NaCl with a solution of silver nitrate ($AgNO_3$). When we mix these two solutions, we observe a heavy white *precipitate*, or solid material, coming out of solution. What has happened? The situation is very similar to the previous example, but with one important difference. When Na^+, K^+, Cl^-, and NO_3^- are in solution, no insoluble salt can be formed. The four possible combinations—NaCl, $NaNO_3$, KCl, and KNO_3—are all soluble. But when Ag^+ ions and Cl^- ions are in solution, silver chloride (AgCl) can be formed. Silver chloride is insoluble and does not redissociate. Instead, the ions combine and come out of solution as a precipitate of AgCl. A precipitate may be indicated by a down-pointing arrow (↓), as in the following equation. The Na^+ and NO_3^- ions remain behind in solution. Note that this reaction is an example of a *double displacement reaction* (Section 5.1): $AB + CD \rightarrow AD + CB$.

$$Ag^+ + NO_3^- + Na^+ + Cl^- \xrightarrow{\ H_2O\ } \underset{\substack{\text{silver chloride} \\ \text{a white} \\ \text{precipitate}}}{AgCl\downarrow} + \underset{\substack{\text{sodium nitrate} \\ \text{in solution}}}{Na^+ + NO_3^-}$$

Because the Na^+ and NO_3^- ions are common to both sides of this equation, we may cancel them.

cancellation:

$$\cancel{Na^+} + Ag^+ + \cancel{NO_3^-} + Cl^- \xrightarrow{\ H_2O\ } AgCl\downarrow + \cancel{Na^+} + \cancel{NO_3^-}$$

net ionic equation:

$$Ag^+ + Cl^- \xrightarrow{\ H_2O\ } AgCl\downarrow$$

The portion of the equation that is left is called a **net ionic equation**. This equation tells us that any solution that contains Ag^+ ions and Cl^- ions will give a precipitate of AgCl. The other ions that might be present are of no importance in this ionic reaction. We could have started with KCl or $CaCl_2$ instead of NaCl; the net ionic equation would have been the same.

In a net ionic equation, we show ions in solution as separate entities, such as $Na^+ + Cl^-$. We indicate solids or compounds that do not ionize (or ionize only slightly) by writing their entire formulas. In the preceding equation, the formula AgCl was used for the precipitate. Similarly, H_2O, CO_2, and CH_3CO_2H would be represented as formulas in net ionic equations for other reactions.

In contrast to a net ionic equation, a **molecular equation** shows the actual compounds in the reaction mixture as formulas, rather than showing just the ions. Throughout this book, we will use both net ionic and molecular equations.

molecular equation:

$$NaCl + AgNO_3 \xrightarrow{\ H_2O\ } AgCl\downarrow + NaNO_3$$

Why Ionic Reactions Occur

There must be a driving force for an ionic reaction to yield a product, or *proceed to completion*. A reaction that proceeds to completion is essentially an irreversible reaction. In our last example, the precipitation of an insoluble salt forced the ionic reaction to completion. Precipitation is just one of three occurrences that can drive an ionic reaction to completion.

1. The **precipitation of an insoluble salt** as a driving force has been exemplified with the precipitation of silver chloride. Any mixture of ions that leads to an insoluble salt will lead to an ionic reaction. For example, if we mix solutions of sodium sulfide, Na_2S, and lead(II) nitrate, $Pb(NO_3)_2$, the insoluble salt lead(II) sulfide, PbS, will be formed.

$$2\ Na^+ + S^{2-} + Pb^{2+} + 2\ NO_3^- \xrightarrow{\ H_2O\ } PbS\downarrow + 2\ Na^+ + 2\ NO_3^-$$

net ionic equation:

$$Pb^{2+} + S^{2-} \xrightarrow{\ H_2O\ } PbS\downarrow$$

2. The **evolution of a gas** also will drive an ionic reaction to completion. In an equation, we represent a gas coming out of a solution by an up-pointing arrow ($\uparrow$). We have already mentioned that carbonic acid (H_2CO_3) yields H_2O and the gas CO_2. Therefore, if a solution of sulfuric acid (H_2SO_4) is mixed with a solution of sodium carbonate (Na_2CO_3), the product will be H_2CO_3, which, in turn, gives carbon dioxide, a gas. If the reaction is carried out in an open container, the gas will bubble off and thus be removed from the reaction site.

$$2\ H^+ + SO_4^{2-} + 2\ Na^+ + CO_3^{2-} \xrightarrow{\ H_2O\ } SO_4^{2-} + 2\ Na^+ + H_2O + CO_2\uparrow$$

net ionic equation:

$$2\ H^+ + CO_3^{2-} \xrightarrow{\ H_2O\ } H_2O + CO_2\uparrow$$

3. The **formation of a nonelectrolyte (un-ionized) or weak electrolyte (partially ionized)** will also drive a reaction to completion. For example, we have seen that a hydrogen ion will react with a hydroxide ion to yield water, which is essentially an un-ionized compound. If we mix a solution of potassium hydroxide (KOH) with nitric acid (HNO_3), water is formed and the reaction proceeds to completion.

$$H^+ + NO_3^- + K^+ + OH^- \xrightarrow{\ H_2O\ } H_2O + NO_3^- + K^+$$

net ionic equation:

$$H^+ + OH^- \xrightarrow{\ H_2O\ } H_2O$$

Example   Write the net ionic equation for the reaction that occurs in the stomach (which contains dilute HCl) when a person takes the antacid sodium bicarbonate ($Na^+\ {}^-HCO_3$).

Solution   (1) Write the formulas for the reactant ions.

$$H^+ + Cl^- + Na^+ + HCO_3{}^- \longrightarrow$$

Mentally, figure the possible combinations of ions and decide if an insoluble salt, a gas, or an un-ionized compound can be formed from these ions. In this case, H_2CO_3 can be formed. We have already seen that H_2CO_3 yields water (H_2O) and carbon dioxide gas (CO_2).

(2) Write the formulas for the products.

$$\longrightarrow H_2O + CO_2\uparrow + Cl^- + Na^+$$

(3) Cancel the ions that are common to both sides of the equation.

net ionic equation:

$$H^+ + HCO_3{}^- \longrightarrow H_2O + CO_2\uparrow$$

Problem 9.5.   Write net ionic equations for the reactions that would occur if an aqueous solution of each of the following compounds were mixed with hydrochloric acid (aqueous HCl):

(a) sodium acetate ($CH_3\overset{\displaystyle O}{\overset{\displaystyle \|}{C}}O^-\ Na^+$)

(b) silver nitrate ($AgNO_3$)

(c) sodium carbonate (Na_2CO_3)

■ 9.5   SALTS IN HARD AND SOFT WATER (optional)

Ground water and well water always contain some dissolved minerals. If water contains metal ions other than sodium ions (Na^+) or potassium ions (K^+), it is said to be **hard water.** Water containing no metal ions, such as distilled water or rain water, or water containing only Na^+ and K^+ as metal ions is called **soft water.** The principal metal ions found in hard water are calcium ions (Ca^{2+}), magnesium ions (Mg^{2+}), and iron(III) ions (Fe^{3+}). These ions are paired with anions such as sulfate ions ($SO_4{}^{2-}$) or bicarbonate ions ($HCO_3{}^-$).

One of the problems with hard water is that metal ions (except for Na^+ and K^+) form precipitates with the anions of soap; we see this precipitate as a scum in the bath water or as a ring around the tub. Detergents contain anions that do not form precipitates with calcium or magnesium ions.

$$Ca^{2+} + 2\ CH_3(CH_2)_{16}\overset{\displaystyle O}{\overset{\displaystyle \|}{C}}O^- \xrightarrow{\ H_2O\ } [CH_3(CH_2)_{16}\overset{\displaystyle O}{\overset{\displaystyle \|}{C}}O]_2Ca\downarrow$$

from a soap

$$Ca^{2+} + CH_3(CH_2)_{16}CH_2OSO_3{}^- \xrightarrow{\ H_2O\ } \text{no reaction}$$

from a detergent

Problem 9.6.   Write an equation to show the precipitation of Mg^{2+} with the soap anion shown in the preceding equation.

If hard water contains principally *metal bicarbonates*, such as $Ca(HCO_3)_2$, the metal ions can be precipitated simply by heating the water. This type of hard water is called **temporary hard water.** By contrast, dissolved metal salts other than bicarbonates, such as $CaSO_4$, do not precipitate from water upon heating. This water is said to be *permanently hard*. Metal ions can be precipitated from temporary hard water by heating because bicarbonate ions yield carbon dioxide gas (CO_2), carbonate ions ($CO_3{}^{2-}$), and water when heated. Because most metal carbonates are insoluble, they precipitate.

$$\underbrace{Ca^{2+} + 2\,HCO_3^-}_{\textit{in temporary hard water}} \xrightarrow{\text{heat}} CaCO_3\downarrow + CO_2\uparrow + H_2O$$

The precipitation of metal carbonates from temporary hard water can occur in hot-water heaters, boilers, hot-water pipes, and teakettles. The resultant scaly deposit of metal carbonates cuts heating efficiency and can actually clog pipes.

Softening Hard Water

A process called *softening the water* is used to remove the hard-water metal ions from water. Softening hard water prevents metal salts from precipitating, either in pipes or by combination with soap.

Precipitating the Ions.   One way to prevent the precipitation of soap by metal ions is to add a compound that yields a precipitate with the metal ions before they can react with soap. Aqueous solutions of ammonia (NH_3), which contain hydroxide ions, washing soda (Na_2CO_3), and borax ($Na_2B_4O_7$) all contain anions that will precipitate the hard-water ions.

Problem 9.7.   Write a net ionic equation to show the reaction that occurs when
(a) Na_2CO_3 is added to water containing dissolved $CaSO_4$
(b) $Na_2B_4O_7$ is added to water containing dissolved $MgCl_2$

Ion-Exchange Water Softeners.   A popular way of softening water is with an *ion-exchange device*, in which the hard-water ions are exchanged for sodium ions. The most common method of accomplishing this is by treating the water in a water softener containing sodium aluminum silicate (zeolite), an insoluble compound that traps calcium ions and releases sodium ions.

Figure 9.2.   *Diagram of a water softener.*

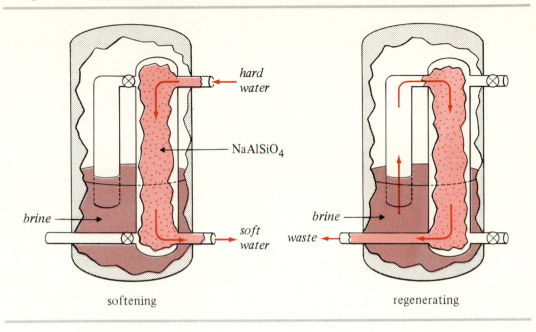

softening   regenerating

softening:

$$2 \, NaAlSiO_4 + Ca^{2+} \longrightarrow Ca(AlSiO_4)_2\downarrow + 2 \, Na^+$$

removed from   *released into*
the water   *the water*

When the silicate salt has been used up, it is regenerated by being flooded with brine (sodium chloride solution). The calcium ions are then flushed away as soluble calcium chloride. (See Figure 9.2.)

regenerating:

$$Ca(AlSiO_4)_2 + 2 \, NaCl \longrightarrow 2 \, NaAlSiO_4 + CaCl_2$$

flushed away

Water softened in this type of ion-exchange device contains a fairly high concentration of sodium ions. Because an excess of sodium ions can lead to water retention in the body's tissues and adds to the work of the kidneys, persons with hypertension (high blood pressure) or kidney problems are often advised to avoid drinking artificially softened water.

Deionized water is softened water in which the dissolved minerals have actually been removed, not simply exchanged for sodium compounds. The water is first passed through an acidic resin that replaces metal cations with

H^+. Then the water is passed through a basic resin that removes the acid. The salts are effectively replaced by H^+ and OH^-, or H_2O.   ■

SUMMARY

An aqueous solution containing ions conducts electricity; therefore, ionic compounds are called *electrolytes*. Ionic solutions can arise from the *dissociation* of soluble ionic compounds or from the *ionization* of certain polar covalent molecules in water.

An *acid* is a compound that can lose H^+. In water solution, the H^+ is bonded to a water molecule as a *hydronium ion* (H_3O^+). A *base* is a compound that can accept H^+. Hydroxides are typical bases ($H^+ + {}^-OH \rightarrow H_2O$). A *salt* is an ionic compound containing a cation other than H^+ and an anion other than OH^-. A salt can be obtained from the reaction of an acid and a base ($NaOH + HCl \rightarrow NaCl + H_2O$).

Salts, some acids, and some bases are *strong electrolytes* (completely dissociated or ionized). Organic acids, carbonic acid, and ammonia are *weak electrolytes* (only partially ionized). Some electrolytes are important in body chemistry—notably NaCl, the principal electrolyte of the extracellular body fluids, and K_2HPO_4, the principal electrolyte of the intracellular fluid.

Salts containing

$$Na^+, K^+, NH_4^+, NO_3^-, CH_3\overset{\overset{\displaystyle O}{\|}}{C}O^-, Cl^-, \text{ or } SO_4^{2-}$$

are generally soluble. (Exceptions are noted in Table 9.1.)

Ionic reactions can be represented by *net ionic equations* or by *molecular equations*. Ionic reactions are caused by the precipitation of an insoluble salt, the evolution of a gas, or the formation of an un-ionized or partially ionized product.

precipitation:

$$Ag^+ + Cl^- \longrightarrow AgCl\downarrow$$

gas evolution:

$$2\,H^+ + CO_3^{2-} \longrightarrow H_2CO_3 \longrightarrow H_2O + CO_2\uparrow$$

formation of weak electrolytes:

$$CH_3\overset{\overset{\displaystyle O}{\|}}{C}O^- + H^+ \longrightarrow CH_3\overset{\overset{\displaystyle O}{\|}}{C}OH$$

$$H^+ + OH^- \longrightarrow H_2O$$

Hard water contains metal ions other than Na^+ or K^+. These metal ions precipitate soap and may form scaly carbonate deposits. Water can be softened in three ways: by removing these metal ions as precipitates, with sodium zeolite, or with deionizing resins. *Temporary hardness* (caused by the presence of bicarbonates) can be removed by heat.

KEY TERMS

electrolyte	base	net ionic equation
dissociation	salt	molecular equation
ionization	acid-base reaction	*hard water
hydronium ion	weak electrolyte	*soft water
acid	precipitation	

STUDY PROBLEMS

9.8. Which of the following compounds would you expect to be electrolytes?
(a) $MgSO_4$ (b) H_2O
(c) CH_4 (d) H_2SO_4

9.9. Classify each of the following electrolytes as an acid, a base, or a salt. Then write balanced equations to show what happens when each dissolves in water.
(a) $Mg(NO_3)_2$ (b) HI   (c) NaOH

9.10. Complete and balance the following equations for acid-base reactions:

(a) $KOH + HCl \xrightarrow{H_2O}$

(b) $Ca(OH)_2 + HNO_3 \xrightarrow{H_2O}$

9.11. Write an equation for the reaction of baking soda ($NaHCO_3$) with stomach acid (dilute HCl).

9.12. Classify each of the following compounds as a strong or weak electrolyte:

(a) $CH_3\overset{O}{\overset{\|}{C}}-\overset{O}{\overset{\|}{C}}OH$ (b) H_2CO_3
(c) H_2SO_4 (d) KCl

9.13. (a) Write an equation that shows the ionization of carbonic acid (using the structural formula shown at the end of Section 9.1) in water.
(b) List *all* the molecules and ions that would be found in an aqueous solution of CO_2.

9.14. Using Table 9.1, predict the solubilities of the following compounds in water:
(a) NH_4Cl (b) $AgNO_3$

(c) $FeCO_3$ (d) Na_2SO_4
(e) $CoCl_2$ (f) $CaCl_2$

9.15. Write (and balance) the following equations as net ionic equations:

(a) $MgCl_2 + NaOH \xrightarrow{H_2O} Mg(OH)_2 + NaCl$

(b) $Na_2S + CoCl_2 \xrightarrow{H_2O} CoS + NaCl$

(c) $NaHCO_3 + CH_3\overset{O}{\overset{\|}{C}}OH \xrightarrow{H_2O}$

$CH_3\overset{O}{\overset{\|}{C}}O^- Na^+ + H_2O + CO_2$

9.16. Write the net ionic equation and the molecular equation for the reaction that would occur if solutions of sodium sulfide (Na_2S) and copper(II) chloride ($CuCl_2$) were mixed.

9.17. Complete and balance the following molecular equations:

(a) $HNO_3 + CH_3\overset{O}{\overset{\|}{C}}O^- Na^+ \xrightarrow{H_2O}$

(b) $Ba(NO_3)_2 + Na_2SO_4 \xrightarrow{H_2O}$

(c) $Pb(NO_3)_2 + HCl \xrightarrow{H_2O}$

*9.18. Write an equation that shows the formation of boiler scale from water containing $Mg(HCO_3)_2$.

*9.19. (a) Explain why a person with high blood pressure should not drink water softened by the sodium zeolite process.
(b) Would drinking hard water containing $CaSO_4$ cause any problems for this person?

10 Acids and Bases

Granite blocks eaten away by acid rain. (Mark Antman/The Image Works.) One of the most damaging air pollutants is the gas sulfur dioxide (SO_2), which is formed when sulfur is burned. The burning of high-sulfur coal is a primary source of atmospheric sulfur dioxide. In the atmosphere, sulfur dioxide is converted to sulfuric acid, which is washed to the earth by rain. The acid rain can kill plants and fish, corrode metals, and eat away building materials. This photograph shows the effects of acid rain on granite, a type of rock generally thought of as indestructible.

Objectives   Define and give examples of strong and weak acids. ☐ Write equations to show the formation of acids and the reactions of acids with certain metals and bases. ☐ Define and give examples of strong and weak bases. ☐ Write an equation for the ionization of water; relate the concentrations of H^+ and OH^- using K_w. ☐ Determine pH values; relate acidity, neutrality, and alkalinity to the pH scale. ☐ Calculate equivalent weights. ☐ Define normality; calculate a normality from titration results (optional). ☐ Give examples of equations depicting the ionization of a diprotic acid. ☐ With equations, describe amphoterism, buffer solutions, and the role of buffers in blood. ☐ With equations, describe acidosis and alkalosis (optional).

Acids and bases are classes of compounds that are encountered in every walk of life. Vinegar is a dilute aqueous solution of acetic acid. Aspirin is acetylsalicylic acid. Hydrochloric acid (called *muriatic acid* in industry) is used to clean rust and paint from metals and concrete. In very dilute solution, hydrochloric acid may be given to patients whose own gastric juices are not sufficiently acidic for efficient digestion. Nitric acid can be used to test the urine for albumins because it causes these colloidal proteins to coagulate; industrially, nitric acid is used to make fertilizers and explosives. Bases, also called *alkalis* or *alkaline substances,* range from the extremely caustic sodium hydroxide (lye), which is used to dissolve grease and hair in clogged drains, to the mild stomach antacids.

In this chapter, we will redefine acids and bases and will consider some reactions of these compounds. We will also consider the relative strengths of acids and bases and will introduce equivalent weights of substances, normality as a unit of concentration, and the pH scale as a measure of acidity and basicity. Finally, we will discuss how the pH of blood is regulated.

10.1   ACIDS

Types of Acids

In Chapter 9, we defined an acid as a compound that can donate H^+, called a *hydrogen ion* or a *proton.* In aqueous solution, H^+ becomes bonded to water to yield a *hydronium ion,* H_3O^+. Some acids, like salts, are almost completely ionized in water. We call these acids **strong acids**. Hydrochloric acid (HCl), sulfuric acid (H_2SO_4), and nitric acid (HNO_3) are all strong acids. (Sulfuric acid, which can give up two hydrogen ions, will be discussed later in this chapter.) Since they do not contain carbon, these acids are referred to as *mineral acids.*

strong acids are highly ionized in H_2O:

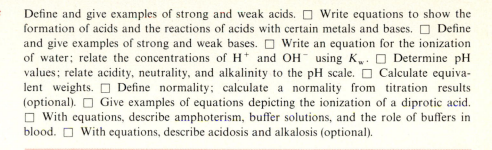

or simply

$$HCl \xrightarrow{H_2O} H^+ + Cl^-$$

$$H_2\ddot{O}: + H\overset{O}{\underset{O}{\overset{\|}{\ddot{O}}}}SOH \longrightarrow H_3\ddot{O}: + \overset{O}{\underset{O}{\overset{\|}{^-:\ddot{O}}}}SOH$$

or simply

$$H_2SO_4 \xrightarrow{H_2O} H^+ + HSO_4$$

Except in very dilute solutions, strong acids are corrosive compounds that can react with skin, clothing, metals, and other substances. We dare not taste a 1.0 M solution of a strong acid such as HCl. The concentration of hydrogen ions would be high enough to destroy the tissues of the mouth. Yet vinegar and lemon juice contain acetic acid and citric acid, respectively, in almost the same concentration. Why can a person eat pickles or drink lemonade without suffering harm? The reason is that, unlike HCl, acetic acid and citric acid do not ionize to a large extent in solution. Acids that undergo only partial ionization, such as acetic acid and citric acid, are called **weak acids**.

weak acids are partially ionized in H_2O:

$$\underset{\text{acetic acid}}{CH_3\overset{O}{\overset{\|}{C}}\ddot{O}-H} + :\ddot{O}H \rightleftharpoons CH_3\overset{O}{\overset{\|}{C}}\ddot{O}:^- + H_3\overset{+}{O}:$$

or simply

$$CH_3\overset{O}{\overset{\|}{C}}OH \rightleftharpoons CH_3\overset{O}{\overset{\|}{C}}O^- + H^+$$

Keep in mind that the terms *strong acid* and *weak acid* refer to the extent of ionization of an acid and not to its concentration. We may talk about a very dilute solution of a strong acid. Our gastric juices are about 0.5% hydrochloric acid. We may also speak of a highly concentrated solution of a weak acid. See Figure 10.1 for some typical strong and weak acids.

Formation of Acids; Acid Rain

Oxides of metals undergo reaction with water to yield hydroxides. (Many metal oxides are water insoluble and do not undergo this reaction.) By contrast, oxides of *nonmetals* undergo reaction with water to yield *acidic solutions*.

$$\underset{\substack{\text{calcium oxide} \\ \textit{a metal oxide}}}{CaO} + H_2O \longrightarrow \underset{\text{calcium hydroxide}}{Ca(OH)_2}$$

— *a base*

Figure 10.1. *The relative strengths of some common acids in aqueous solution.*

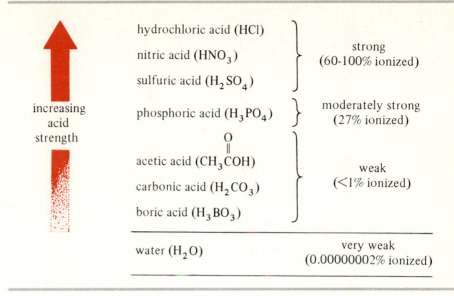

increasing acid strength	hydrochloric acid (HCl) nitric acid (HNO$_3$) sulfuric acid (H$_2$SO$_4$)	strong (60-100% ionized)
	phosphoric acid (H$_3$PO$_4$)	moderately strong (27% ionized)
	acetic acid (CH$_3$COH) carbonic acid (H$_2$CO$_3$) boric acid (H$_3$BO$_3$)	weak (<1% ionized)
	water (H$_2$O)	very weak (0.00000002% ionized)

$$H_2O \quad + \quad SO_3 \quad \longrightarrow \quad H_2SO_4$$

sulfur trioxide   sulfuric acid
a nonmetal oxide

an acid

Most coal contains sulfur. When coal is burned, the sulfur is converted to sulfur dioxide (SO$_2$) which, in the presence of air and sunlight, is further oxidized to SO$_3$.

$$2\ SO_2 + O_2 \xrightarrow{\ sunlight\ } 2\ SO_3$$

Rain washes the SO$_3$ out of the air as **acid rain**, which can corrode automobiles and buildings and can cause streams and lakes to become too acidic for fish. Rainfall as acidic as lemon juice and vinegar has been reported.

10.2   REACTIONS OF ACIDS

Reactions of Acids with Metals

Many metals undergo displacement reactions with acids to yield metal salts and hydrogen gas (H$_2$). This is the reason that acid rain promotes corrosion of automobile parts. Two types of equations for the reaction of zinc metal (Zn) with hydrochloric acid follow.

Figure 10.2.   *Reactivity of metals toward acids.*

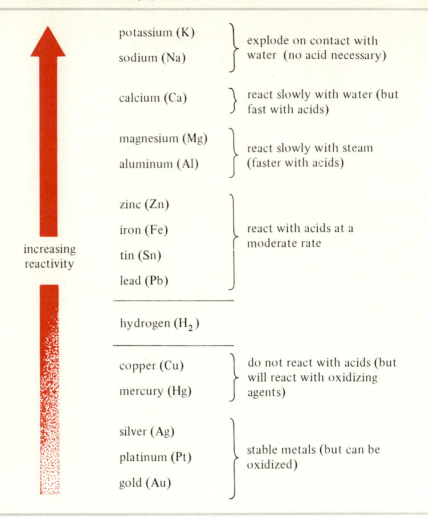

increasing
reactivity

potassium (K)	explode on contact with
sodium (Na)	water (no acid necessary)
calcium (Ca)	react slowly with water (but fast with acids)
magnesium (Mg)	react slowly with steam
aluminum (Al)	(faster with acids)
zinc (Zn)	
iron (Fe)	react with acids at a
tin (Sn)	moderate rate
lead (Pb)	
hydrogen (H$_2$)	
copper (Cu)	do not react with acids (but
mercury (Hg)	will react with oxidizing agents)
silver (Ag)	
platinum (Pt)	stable metals (but can be oxidized)
gold (Au)	

molecular equation:

$$\text{Zn} \; + \; 2\,\text{HCl} \; \xrightarrow{\text{H}_2\text{O}} \; \text{ZnCl}_2 \; + \; \text{H}_2\uparrow$$

zinc metal   hydrochloric acid   zinc chloride   hydrogen gas

net ionic equation:

$$\text{Zn} + 2\,\text{H}^+ \xrightarrow{\text{H}_2\text{O}} \text{Zn}^{2+} + \text{H}_2\uparrow$$

The list in Figure 10.2, called the **activity series of the metals**, indicates the relative reactivities of some metals toward acids. The metals above H$_2$ in the series all react with acids to yield salts and H_2. The most reactive metals are at the top of the list. Sodium and potassium metals require special handling techniques because these metals explode on contact with water. They must be stored in containers of mineral oil or other inert liquid to exclude air and moisture. Calcium metal, too, is fairly reactive, although not as reactive as sodium and potassium. The other metals listed are relatively stable and are used commercially ("mag" hub caps, aluminum pans, etc.), either as pure metals or as alloys (solid solutions) of mixtures of metals.

The metals on the list from magnesium through lead will all react with acids. For this reason, acidic substances should be stored in glass or plastic bottles, but never in metal containers. Some metals, such as aluminum, form oxide coatings that protect the metal, to an extent, from further corrosion. Although one would not make pickles in an aluminum pan, cooking mildly acidic foods such as tomatoes in aluminum pans is not considered harmful, according to the American Medical Association.

Note from Figure 10.2 that the most reactive metals are those in Group IA of the periodic table, followed by the metals in Group IIA, and finally by the metals in Group IIIA and the transition metals. One reason for this order of reactivity is that electronegativity (attraction for outer electrons) increases as we proceed from left to right in the periodic table. Each of the Group IA metals has little attraction for its outer electron and loses it readily.

The more reactive metals yield extremely stable salts. In nature, these metals are always found in ionic compounds ($NaCl$, $CaCO_3$, etc.), never in their elemental states. In fact, the only metals found in nature in their free elemental state are the extremely unreactive ones such as silver and gold.

Problem 10.1.   Write equations for the reactions of (a) calcium metal with HNO_3 and (b) magnesium metal with HCl.

Reactions of Acids with Hydroxides, Oxides, and Carbonates

Acids can undergo reaction with a variety of substances besides metals. We have already mentioned in Chapter 9 that acids undergo irreversible reactions with hydroxides to yield salts and water. Both strong and weak acids undergo this type of reaction. These reactions can occur with pure substances or in aqueous solution. Examples of the reactions of acids with various compounds follow.

reaction with hydroxides:

$$HCl + NaOH \longrightarrow NaCl + H_2O$$

$$\underset{\text{CH}_3\text{COH}}{\overset{\overset{\text{O}}{\|}}{}} + NaOH \longrightarrow \underset{\text{CH}_3\text{CO}^-\text{Na}^+}{\overset{\overset{\text{O}}{\|}}{}} + H_2O$$

reaction with metal oxides:

$$6\,HCl + Fe_2O_3 \longrightarrow 2\,FeCl_3 + 3\,H_2O$$

reaction with carbonates:

$$2 \text{ HCl} + \text{Na}_2\text{CO}_3 \longrightarrow 2 \text{ NaCl} + \text{H}_2\text{O} + \text{CO}_2\uparrow$$

reaction with bicarbonates:

$$\text{HCl} + \text{NaHCO}_3 \longrightarrow \text{NaCl} + \text{H}_2\text{O} + \text{CO}_2\uparrow$$

Problem 10.2. Write a molecular equation to show how the H_2SO_4 in acid rain corrodes statuary and buildings made of limestone ($CaCO_3$).

Problem 10.3. When vinegar is mixed with sodium bicarbonate (baking soda), the mixture effervesces, or fizzes. Write an equation that explains this behavior.

Because they react with acids to destroy, or neutralize, acids, many representatives of the preceding classes of compounds are used as stomach antacids. If you read the labels of commercial antacid preparations, you will probably find calcium carbonate ($CaCO_3$), magnesium carbonate ($MgCO_3$), magnesium hydroxide [$Mg(OH)_2$], aluminum hydroxide [$Al(OH)_3$], or some combination of these. The reaction of magnesium hydroxide (in milk of magnesia) with stomach acid is typical:

in gastric juices

$$\text{Mg(OH)}_2 \qquad + 2 \text{ HCl} \longrightarrow \text{MgCl}_2 + 2 \text{ H}_2\text{O}$$

magnesium hydroxide
as water suspension

Sodium bicarbonate finds use as baking soda because the CO_2 liberated when this compound comes into contact with acid (from sour milk or lemon juice, for example) causes baked goods to be light and fluffy instead of hard and flat. Yeast serves the same purpose because yeast organisms give off CO_2 when their cells metabolize, or break down, carbohydrates. At one time, sodium bicarbonate was very widely used as a stomach antacid. It has been largely replaced as an antacid because of the problems that excessive sodium ion intake can cause and because it sometimes causes a rebound effect, which results in greater stomach acidity than before.

10.3   BASES

Svante Arrhenius, a Swedish chemist who proposed the ionic theory in 1887, defined a base as a substance that produces hydroxide ions upon dissociation. We mentioned in the previous section that hydroxides undergo reaction with acids to yield salts and water. But what of the other substances (metal oxides, carbonates, and bicarbonates) that also react with acids to yield salts and water? In 1923, the Danish chemist Johannes Brønsted proposed that the definition of base be broadened to include *any proton acceptor.* Thus, metal oxides, carbonates, and bicarbonates, as well as hydroxides, are considered to be bases by the Brønsted definition.

A **Brønsted base** is any ion or compound that accepts H^+ from an acid.

This is the definition of a base that we will use in this text. Sometimes, we will refer to an *ion* (such as the hydroxide ion) as a base. Because the OH^- ion must always be paired with a cation, we will also sometimes refer to a *compound* (such as NaOH or KOH) as a base.

Note that a base must contain ions or molecules with at least one pair of unshared valence electrons, which are used to form a covalent bond with H^+.

$$H^+ \; + \; {}^-\!:\!\ddot{O}\!-\!H \longrightarrow \overset{H}{\underset{}{:\!\ddot{O}\!-\!H}}$$

Bases, like acids, can be categorized as *strong bases* or *weak bases.* (See Figure 10.3.) Sodium hydroxide and potassium hydroxide are examples of strong bases, completely dissociated in water solution.

$$KOH \xrightarrow{\;H_2O\;} K^+ + OH^-$$

potassium hydroxide
a strong base

Figure 10.3. *The relative base strengths of some common bases.*

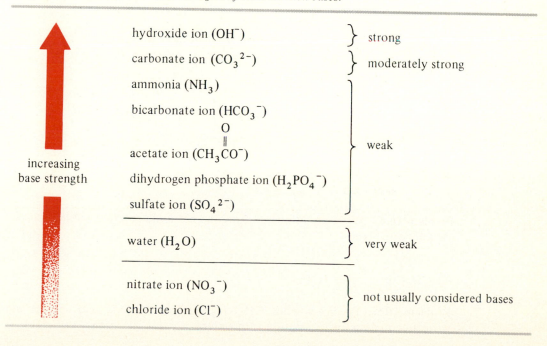

hydroxide ion (OH^-)	strong
carbonate ion ($CO_3{}^{2-}$)	moderately strong
ammonia (NH_3)	
bicarbonate ion ($HCO_3{}^-$)	
acetate ion ($CH_3\overset{O}{\overset{\|}{C}}O^-$)	weak
dihydrogen phosphate ion ($H_2PO_4{}^-$)	
sulfate ion ($SO_4{}^{2-}$)	
water (H_2O)	very weak
nitrate ion ($NO_3{}^-$)	not usually considered bases
chloride ion (Cl^-)	

increasing base strength

Calcium hydroxide and magnesium hydroxide are not very soluble in water, but the portion that does dissolve is completely dissociated. These compounds are also considered strong bases. Because they are less soluble, these compounds are not as caustic as NaOH or KOH.

$$Ca(OH)_2 \xrightarrow{\;H_2O\;} Ca^{2+} + 2\,OH^-$$

calcium hydroxide
(lime)

Compounds that contain covalently bonded —OH groups, such as ethanol (CH_3CH_2OH), do not undergo ionization and thus are not considered bases in the ordinary sense.

Ammonia gas (NH_3) dissolves in water to form a basic, or alkaline, solution. This reaction is somewhat different from simple dissociation, in that ammonia actually undergoes a reaction with water to form an ammonium ion and a hydroxide ion. Because only a small percentage (about 0.5%) of the ammonia molecules in solution undergo this reaction, an aqueous solution of ammonia is considered a *weak base*. Aqueous ammonia is sometimes called ammonium hydroxide, a misnomer because this name implies complete ionization.

$$:NH_3 \;+\; :\overset{..}{O}H \;\rightleftharpoons\; NH_4^+ \;+\; {}^-:\overset{..}{O}H$$

ammonia gas   ammonium   hydroxide
a weak base   ion   ion

Some salts also undergo reversible reactions with water to form alkaline solutions. For example, soluble carbonates or bicarbonates accept H^+ from water to yield OH^- ions.

alkaline solution

$$CO_3{}^{2-} \;+\; H_2O \;\rightleftharpoons\; HCO_3{}^- \;+\; OH^-$$

carbonate ion

$$HCO_3{}^- \;+\; H_2O \;\rightleftharpoons\; H_2CO_3 \;+\; OH^-$$

bicarbonate ion

Problem 10.4.   Write equations to show the net ionic reactions of the following ions with water:

(a) $CH_3\overset{\displaystyle O}{\overset{\|}{C}}O^-$ (b) $H_2PO_4{}^-$

Problem 10.5.   Predict whether CH_3NH_2 would be a weak acid or base and write an equation for its reaction with water.

Problem 10.6.   Identify the acid and the base in each of the following acid-base reactions:

(a) $Na_2O + H_2O \longrightarrow 2\ NaOH$

(b) $Na_2CO_3 + H_2CO_3 \longrightarrow 2\ NaHCO_3$

(c)
$$CH_3\overset{\overset{\displaystyle O}{\displaystyle \|}}{C}OH + NH_3 \longrightarrow NH_4{}^+ + CH_3\overset{\overset{\displaystyle O}{\displaystyle \|}}{C}O{}^-$$

10.4.   REACTIONS OF BASES

As we have mentioned, strong bases such as sodium hydroxide are extremely caustic. Either in their pure form or in solution, they are capable of destroying protein molecules and thus can cause extremely severe burns. Because bases are able to react with a variety of compounds including grease, laundry products are often very alkaline. Liquid chlorine bleaches are solutions of the toxic gas chlorine (Cl_2) in aqueous sodium hydroxide; this type of bleach is discussed in Section 11.5. Some laundry soaps contain sodium carbonate $(Na_2CO_3$, washing soda). Clothes that have been washed with such products and not thoroughly rinsed can cause skin irritation. Strong bases can oxidize some metals to their salts. Therefore bases, like acids, should be stored in glass or plastic containers.

By their very definition, bases (proton acceptors) undergo reaction with acids (proton donors). In these acid-base reactions, the driving force is always *the conversion of a stronger acid to a weaker acid and a stronger base to a weaker base*:

$$HCl \quad + \quad H_2O \quad \longrightarrow H_3O^+ + Cl^-$$

stronger acid   *stronger base*
than H_3O^+   *than Cl^-*

$$HCl \quad + \quad NaOH \quad \longrightarrow H_2O + NaCl$$

stronger acid   *stronger base*
than H_2O   *than NaCl*

Example   Aqueous ammonia (NH_3) and hydrochloric acid (aqueous HCl) emit fumes of gaseous NH_3 and HCl, respectively. When their containers are opened near each other, a white cloud of finely divided solid is observed over the containers. What is this cloud? Write the equation for the reaction.

Solution   HCl is an acid, while NH_3 is a base. These two gases undergo an acid-base reaction to yield the solid salt ammonium chloride, NH_4Cl.

$$H-\overset{..}{\underset{..}{Cl}}: \;+\; :NH_3 \;\longrightarrow\; NH_4^+\; :\overset{..}{\underset{..}{Cl}}:^-$$

stronger acid   *stronger base*   ammonium chloride
than NH_4^+ *than* Cl^- *a salt*

Problem 10.7. Complete the following equations for acid-base reactions. Indicate the relative acid strengths and base strengths, as we have done in the preceding equations.
(a) $Mg(OH)_2 + HCl \longrightarrow$
(b) $NH_3 + HNO_3 \longrightarrow$

10.5. IONIZATION OF WATER AND pH

If two water molecules collide with enough energy, a water molecule can be broken into H^+ and OH^-. In other words, water itself can ionize.

$$H-\overset{..}{\underset{..}{O}}-H \;+\; H-\overset{..}{\underset{..}{O}}-H \;\rightleftharpoons\; H_3\overset{+}{O} \;+\; ^-\overset{..}{\underset{..}{O}}H$$

or simply

$$H_2O \;\rightleftharpoons\; H^+ + OH^-$$

All water, no matter how pure, contains a few hydrogen ions (as hydronium ions) and a few hydroxide ions. The concentration of the ions in pure water is far too low for pure water to conduct electricity to any extent. It has been determined that at 25°C the hydrogen ion and the hydroxide ion are each present in pure water to the extent of only 1×10^{-7} M, or 0.0000001 M. Stated in another way, only about $2 \times 10^{-8}\%$ (0.00000002%) of the water molecules are ionized.

Ion Product Constant

In pure water or in any dilute aqueous solution, the multiplication product of the concentration of H^+ and the concentration of OH^- is a constant. This multiplication product is called the **ion product constant for water** and is represented as K_w (K for *equilibrium constant*, subscript w for *water*).

for pure water at 25°C:

$$K_w = [H^+][OH^-]$$

where

$$K_w = \text{ion product constant for } H_2O$$

$$[H^+] = \text{molarity of } H^+ = 1 \times 10^{-7}\ M$$

$$[OH^-] = \text{molarity of } OH^- = 1 \times 10^{-7}\ M$$

Substituting,

$$K_w = (1 \times 10^{-7})(1 \times 10^{-7})$$

$$= 1 \times 10^{-14}$$

(Note that when terms containing exponents are multiplied, the exponents are added. See the appendix for further discussion.)

For any dilute aqueous solution at 25°C, regardless of the solute, $K_w = 1 \times 10^{-14}$. Thus, for any solution that contains equal amounts of H^+ and OH^-, $[H^+]$ and $[OH^-]$ are both equal to $1 \times 10^{-7}\ M$. A solution with equal amounts of H^+ and OH^- is said to be a **neutral solution**—neither acidic nor alkaline.

A solution that contains more hydrogen ions than hydroxide ions is said to be an **acidic solution**. A solution with more hydroxide ions than hydrogen ions is said to be a **basic solution**, or an **alkaline solution**. The multiplication product of the concentrations of H^+ ions and OH^- ions in an acidic solution still must equal 1×10^{-14}. Consequently, even an acidic solution contains some OH^- ions; the equilibrium is just shifted (Le Châtelier's principle, Section 5.6). For example, if the concentration of H^+ is $1 \times 10^{-2}\ M$, then the concentration of OH^- ions is $1 \times 10^{-12}\ M$.

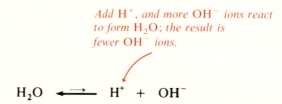

Add H^+, and more OH^- ions react to form H_2O; the result is fewer OH^- ions.

$$H_2O \;\longleftrightarrow\; H^+ \;+\; OH^-$$

The pH Scale

The concentration of H^+ in most biological substances is quite low. Water contains $1.0 \times 10^{-7}\ M\ H^+$, while blood contains $4.0 \times 10^{-8}\ M\ H^+$. It is inconvenient to work with exponential numbers like these; yet we dare not express these concentrations as decimals (for example, 0.00000010 M), because it is too easy to miscount the number of zeros. For this reason, the **pH scale** was developed.

The pH of a solution is defined as the negative log of the hydrogen ion concentration: $pH = -\log[H^+]$. A logarithm is an exponent to the base 10;

Table 10.1. *Hydrogen Ion Concentration and pH Values*

	$[H^+]$ expressed as a decimal	$[H^+]$ expressed with an exponent	pH Value
↑	1.0 M	$1 \times 10^0\ M$	0
	0.1 M	$1 \times 10^{-1}\ M$	1
increasing	0.001 M	$1 \times 10^{-3}\ M$	3
acidity	0.00001 M	$1 \times 10^{-5}\ M$	5
	0.0000001 M	$1 \times 10^{-7}\ M$	7
	0.00000001 M	$1 \times 10^{-8}\ M$	8

therefore, it is easier to think of pH as the *negative exponent of the hydrogen ion concentration.*

$$[H^+] = 1 \times 10^{-6}\ M \qquad \overset{pH = 6}{\nearrow} \qquad\qquad [H^+] = 1 \times 10^{-3}\ M \qquad \overset{pH = 3}{\nearrow}$$

Table 10.1 shows some examples of hydrogen ion concentrations and the corresponding pH values. Note that the higher the concentration of hydrogen ion, the smaller is the pH value. Also note that a difference of 1.0 in pH means a *tenfold* difference in the concentration of H^+.

difference of 1 $\begin{cases} \text{pH} = 0, \text{ then } [H^+] = 1.0\ M \\ \text{pH} = 1, \text{ then } [H^+] = 0.1\ M \end{cases}$ *tenfold difference*

Example   What is the pH of a solution whose hydrogen ion concentration is 0.01 M?

Solution

$$[H^+] = 0.01\ M \qquad \overset{pH = 2}{\nearrow}$$
$$= 1 \times 10^{-2}\ M$$

Problem 10.8. The pH of a solution is 4.0. Express the hydrogen ion concentration of this solution as an exponential function and as a decimal.

In terms of acidity and basicity:

1. A neutral solution, neither acidic nor basic, has pH 7.
2. Acidic solutions have pH values *less than 7.*
3. Basic, or alkaline, solutions have pH values *greater than 7.*

Figure 10.4 shows the pH scale along with the pH values of a few biological substances. Note that some of the pH values are expressed as decimals.

Figure 10.4.   *The pH scale, showing the relative pH values of some commom substances.*

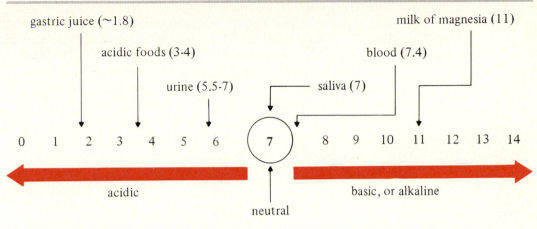

For example, blood has a pH around 7.4. The decimal indicates that the pH is between two whole numbers; in the case of blood, the pH is between 7 and 8—very slightly alkaline.

10.6   NEUTRALIZATION REACTIONS AND EQUIVALENTS

The reaction of equivalent amounts of an acid and a base results in the destruction of both the acid and the base and yields a *neutral* product, neither acidic nor basic. The reaction of equivalent amounts of an acid and a base is called a **neutralization reaction**.

a neutralization reaction:

$$HCl \quad + \quad NaOH \quad \longrightarrow \quad NaCl \quad + \quad H_2O$$

1.0 mole   1.0 mole   *a neutral solution*
(36.5 g)   (40.0 g)

A **diprotic acid**, one that can donate *two* protons to a base, can also be used in a neutralization reaction. Note that although 1.0 mole of a monoprotic acid (one H^+) is needed to neutralize 1.0 mole of NaOH, only 0.5 mole of a diprotic acid, such as sulfuric acid, is needed.

a diprotic acid

$$H_2SO_4 \quad + \quad 2\,NaOH \quad \longrightarrow \quad Na_2SO_4 \quad + \quad 2\,H_2O$$

0.5 mole   1.0 mole   *a neutral solution*
(49.0 g)   (40.0 g)

Equivalent Weight

Acid-base reactions do not always occur in a 1:1 molar ratio. We have just shown that 0.5 mole of sulfuric acid (H_2SO_4) can supply 1.0 mole of H^+ and thus neutralize 1.0 mole of NaOH. Thus, we can say that half a mole of H_2SO_4 is *equivalent* to one mole of HCl or to one mole of NaOH. Because of this equivalency, the terms *equivalent* and *equivalent weight* were coined.

In a reaction involving H^+ and OH^-, one **equivalent** of a compound is the number of moles of that compound required to supply one mole of H^+ or OH^-. The **equivalent weight** of a compound is the weight in grams of one equivalent, or

$$\text{equivalent wt} = \frac{\text{formula wt in g}}{\text{number of } H^+ \text{ or } OH^- \text{ per formula unit}}$$

Example   What are the equivalent weights of (a) HCl and (b) H_2SO_4?

Solution   (a) formula wt of HCl = 36.5 amu

$$\text{equivalent wt} = \frac{36.5 \text{ g}}{1} = 36.5 \text{ g}$$

(b) formula wt of H_2SO_4 = 98.0 amu

$$\text{equivalent wt} = \frac{98.0 \text{ g}}{2} = 49.0 \text{ g}$$

Problem 10.9.   The formula for citric acid is

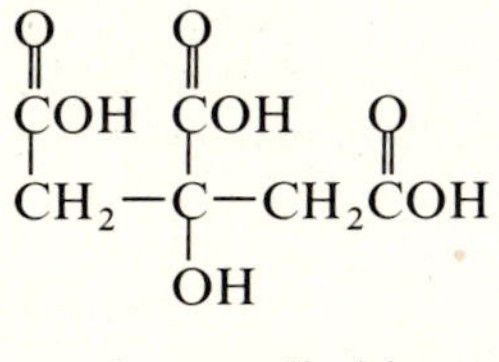

$$\begin{array}{c} \overset{O}{\underset{\|}{}} \quad \overset{O}{\underset{\|}{}} \\ \text{COH} \quad \text{COH} \quad O \\ | \qquad | \qquad \| \\ \text{CH}_2-\text{C}-\text{CH}_2\text{COH} \\ | \\ \text{OH} \end{array}$$

(a) How many moles of H^+ can be supplied by one mole of citric acid? (*Hint:*

In an organic compound of this type, only a $-\overset{O}{\overset{\|}{C}}\text{OH}$ group can lose H^+.)
(b) What is the equivalent weight of citric acid?

Equivalent Weight in Medicine

In medical practice, the terms *equivalent weight* and *equivalent* are often used to denote *the combining power of ions*, rather than just acids and bases. The equivalent weight of an ion is its formula weight (expressed in g) divided by the ionic charge.

$$\text{equivalent wt of an ion} = \frac{\text{formula wt in g}}{\text{ionic charge}}$$

Therefore, the equivalent weight of Na^+ is 23.0 g/1, or 23.0 g. The equivalent weight of Ca^{2+} is 40.1 g/2, or 20.0 g.

Because the body fluids are very dilute solutions, the term *milliequivalent* (mEq) is more frequently encountered in medicine than the term equivalent (equiv).

1 milliequivalent (mEq) = 0.001 equivalent (equiv)

Thus, 1 mEq of Na^+ is 0.001×23.0 g, or 0.0230 g.

Electrolyte analyses of the body fluids or of intravenous solutions are often reported in mEq/L. In Table 10.2 the concentrations of some ions found in the extracellular fluids are expressed in this way.

Table 10.2.   *Electrolytes in Extracellular Body Fluids*

Ion	Normal concentration in mEq/L
cations:	
sodium (Na^+)	142 ⎱
potassium (K^+)	5 ⎰ total of
calcium (Ca^{2+})	5 ⎱ 155 mEq/L
magnesium (Mg^{2+})	3 ⎰
anions:	
chloride (Cl^-)	104 ⎱
bicarbonate (HCO_3^-)	27 ⎰ total of
hydrogen phosphate (HPO_4^{2-})	2 ⎱ 155 mEq/L
sulfate (SO_4^{2-})	1 ⎰
organic anions	21 ⎰

Example   If we wanted to mix 1.0 L of a solution that contains 142 mEq of Na^+, how many grams of Na^+ would we need?

Solution   (1) wt of 1 equiv of Na^+ = 23.0 g

(2) wt of 1 mEq $= \dfrac{23.0 \text{ g}}{\text{equiv}} \times \dfrac{0.001 \text{ equiv}}{\text{mEq}}$

$= 0.0230 \dfrac{\text{g}}{\text{mEq}}$

(3) wt of 142 mEq = 0.0230 g/mEq × 142 mEq

$$= 3.27 \text{ g}$$

Problem 10.10.   How many grams of NaCl would be needed to supply 3.27 g of Na^+?

Problem 10.11.   (a) Using Table 10.2, determine the weight of calcium ions in 1.0 L of extracellular body fluids. (b) What weight of sulfate ion is found in 1.0 L of these fluids?

■ **10.7   NORMALITY AND TITRATION (optional)**

Now that we have defined equivalent weight, we can discuss a commonly used term for expressing concentrations of acids and bases—normality. **Normality** is defined as *the number of equivalents of an acid or base in 1.0 L of solution* and is calculated by the following equation:

$$\text{normality } (N) = \frac{\text{equivalents of solute}}{\text{number of liters of solution}}$$

For compounds that yield only one H^+ or OH^- (such as HCl, NaOH, or KOH), normality and molarity are identical. Only when one mole of an acid or a base can yield two or more moles of H^+ or OH^- is normality different from molarity. The most commonly encountered cases are H_2SO_4 and H_3PO_4. In these cases, the *normality of the solution is always larger than its molarity.* For example, a 0.5 *M* solution of H_2SO_4 (49 g in 1.0 L) is 1.0 *N* (1.0 equiv in 1.0 L).

Example   What is the normality of 1.0 L of an aqueous solution of 98.0 g of H_2SO_4?

Solution   (1) equiv wt of H_2SO_4 = 49.0 g

(2) number of equiv of $H_2SO_4 = \dfrac{98.0 \text{ g}}{49.0 \text{ g}} = 2.0$

(3) $N = \dfrac{2.0 \text{ equiv}}{1.0 \text{ L}}$

$$= 2.0$$

Problem 10.12.   What are the normalities of (a) 0.213 *M* HCl and (b) 0.010 *M* $Ca(OH)_2$?

Figure 10.5.   *A titration apparatus.*

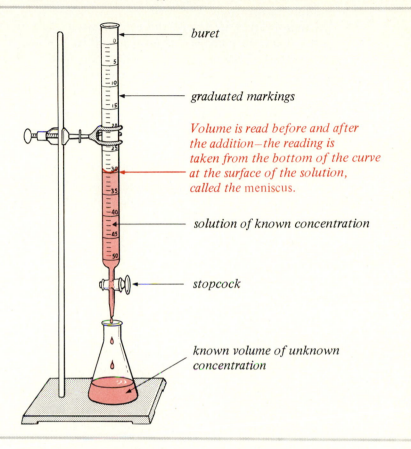

buret

graduated markings

*Volume is read before and after
the addition—the reading is
taken from the bottom of the curve
at the surface of the solution,
called the* meniscus.

solution of known concentration

stopcock

known volume of unknown
concentration

Problem 10.13.   What are the molarities of (a) 6.0 N H_2SO_4 and (b) 0.5 N H_3PO_4?

A common laboratory problem is determining the exact concentration of an acid or a base in a given sample. The procedure that is used to solve this problem is called titration. **Titration** involves measuring the volume of a sample solution (acid or base), then slowly adding another solution (base or acid) of known concentration (a "standardized" solution). The standardized solution is added from a specially calibrated tube called a *buret* until the mixture is neutral. (See Figure 10.5.)

An acid-base reaction gives no outward sign that the reaction is complete, so we need something to tell us to stop when the neutral point is reached. For this purpose, an indicator may be added to the mixture. An *indicator* is an organic compound that is one color in acidic solution and another color in alkaline solution. At the point of neutrality, or *end point*, the indicator changes color. For exact work, a *pH meter*, an instrument that measures the pH of a solution electrically, may be used.

In a titration, we carry out a neutralization reaction just to the end point. At this point, the number of equivalents of the acid (or base) added just equals the number of equivalents of the base (or acid) in the unknown solution. Therefore, we can set up a convenient equation with which to make calculations. (Compare this equation to that used in dilution problems, Section 7.4.)

number of equiv of acid = number of equiv of base

or

$$N_{acid} V_{acid} = N_{base} V_{base}$$

In this equation, we can use milliliters or liters, as long as we are consistent. The use of this equation is shown in the following example.

Example   A solution of 15.00 mL of a base requires 30.00 mL of 0.2000 N HCl to be neutralized. What is the normality of the basic solution?

Solution

$$N_{acid} V_{acid} = N_{base} V_{base}$$

$$(0.2000\ N)(30.00\ mL) = (N_{base})(15.00\ mL)$$

$$N_{base} = \frac{(0.2000\ N)(30.00\ mL)}{(15.00\ mL)}$$

$$= 0.4000$$

Problem 10.14.   A solution of 10.00 mL of an acid requires 23.00 mL of 0.1000 N NaOH solution to reach neutrality.
(a) What is the concentration in N of the acidic solution?
(b) How many equivalents of acid are present?   ■

10.8   DIPROTIC ACIDS AND THEIR ROLE IN BUFFER SYSTEMS

Diprotic Acids

When a diprotic acid, such as H_2SO_4 or H_2CO_3, ionizes in water, it does so in steps. First one hydrogen ion ionizes; then the second one ionizes. The second ionization always occurs to a lesser extent than the first.

Let us use sulfuric acid as an example. The first hydrogen ion separates completely from the sulfuric acid molecule in water solution. Sulfuric acid is a strong acid. The bisulfate, or hydrogen sulfate, ion (HSO_4^-) that is formed can lose yet another hydrogen ion, but this second ionization is incomplete. The bisulfate ion is a weak acid.

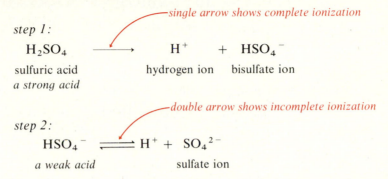

step 1:

$$H_2SO_4 \longrightarrow H^+ + HSO_4^-$$

sulfuric acid   hydrogen ion bisulfate ion
a strong acid

double arrow shows incomplete ionization

step 2:

$$HSO_4^- \rightleftharpoons H^+ + SO_4^{2-}$$

a weak acid   sulfate ion

Carbonic acid is a weak acid that ionizes only partially to hydrogen ions and bicarbonate ions (HCO_3^-). The bicarbonate ion is a weaker acid yet and ionizes scarcely at all. In fact, the bicarbonate ion generally acts like a base, not an acid.

step 1:

$$H_2CO_3 \rightleftharpoons H^+ + HCO_3^-$$

carbonic acid   bicarbonate ion
a weak acid

step 2:

$$HCO_3^- \longleftarrow H^+ + CO_3^{2-}$$

       carbonate ion

Even though a diprotic acid does not lose both protons to water readily, the addition of a base stronger than the final product (sulfate ion or carbonate ion, in our examples) will remove the second proton.

single arrow shows complete reaction

$$H_2SO_4 + 2\,OH^- \longrightarrow SO_4^{2-} + 2\,H_2O$$

Problem 10.15.   Calcium carbonate ($CaCO_3$) is insoluble in water. Show by an equation how you could prepare this compound from aqueous sodium bicarbonate and another reagent.

Amphoteric Substances

Certain compounds or ions can act as either acids or bases. These are called **amphoteric compounds** or **amphoteric ions**. Sodium bicarbonate ($NaHCO_3$) is an amphoteric compound. Sodium bicarbonate can act as a base (a proton acceptor) in its reaction with HCl; it can also act as an acid (a proton donor) in its reaction with NaOH. The anions of other diprotic or triprotic acids show similar behavior.

sodium bicarbonate as a base:

molecular equation $HCl + NaHCO_3 \longrightarrow NaCl + H_2CO_3$

net ionic equation $H^+ + HCO_3^- \longrightarrow H_2CO_3$ (which, in turn,
goes to $H_2O + CO_2$)

sodium bicarbonate as an acid:

molecular equation $NaOH + NaHCO_3 \longrightarrow Na_2CO_3 + H_2O$

net ionic equation $OH^- + HCO_3^- \longrightarrow CO_3^{2-} + H_2O$

Buffer Solutions

If an acid is added to pure water, the pH of the water solution is lowered. Conversely, if a base is dissolved in water, the pH rises. A **buffer solution** is a solution that resists a change in pH when either an acid or a base is added. The solutes that help a solution resist a change in pH are called **buffers**.

A buffer solution is simply a solution of substances that can react with either H^+ or OH^- ions to remove them from solution. A solution of HCO_3^- in equilibrium with H_2CO_3 is a buffer solution:

removed from solution

add H^+:

$$HCO_3^- + \boxed{H^+} \longrightarrow H_2CO_3$$

removed from solution

add OH^-:

$$H_2CO_3 + \boxed{OH^-} \longrightarrow HCO_3^- + H_2O$$

Similarly, a mixture of HPO_4^{2-} and $H_2PO_4^-$ can act as a buffer:

add H^+:

$$HPO_4^{2-} + H^+ \longrightarrow H_2PO_4^-$$

add OH^-:

$$H_2PO_4^- + OH^- \longrightarrow HPO_4^{2-} + H_2O$$

Any buffer system can be overwhelmed by the addition of large amounts of acid or base because the acid-base equilibrium is completely shifted to one side. For example, if large amounts of acid were added to a bicarbonate–carbonic acid buffer, the solution would contain no bicarbonate ions and could dispose of no more acid.

removed from solution

$$\text{excess } H^+ \; + \; \boxed{HCO_3^-} \; \longrightarrow \; H_2CO_3$$

Buffers in Blood

The pH of the blood is critical to human life. Hemoglobin is the substance in red blood cells that carries O_2 from the lungs to the cells. The following equation represents hemoglobin without O_2 (in venous blood) as HHb. Oxygenated hemoglobin, or oxyhemoglobin (in arterial blood), is represented as HbO_2^-. The equation shows that H^+ is lost from HHb as it is converted to HbO_2^-.

In normal blood, slight alkalinity removes H^+ and helps drive reaction to the right.

$$\text{HHb} \;\; + \;\; O_2 \;\; \underset{\text{in lungs}}{\xrightleftharpoons{\quad}} \;\; HbO_2^- \;\; + \;\; \boxed{H^+}$$

hemoglobin   oxyhemoglobin
in veins   *in arteries*

From the equation, you can see that oxygen pickup would not be favored by too-acidic blood because an excess of H^+ would drive the reaction to the left. (For more details about gas exchange in the blood, see Sections 6.4 and 28.4.)

Most foodstuffs are oxidized by the cells to acidic products, yet the blood of a normal person is maintained at a very narrow pH range (7.35–7.45). This is accomplished primarily by the bicarbonate–carbonic acid buffer system. Normal blood contains about 24–28 mEq/L of HCO_3^- and 1.2 mEq/L of H_2CO_3. Excess acid is neutralized by HCO_3^- in the blood. The product H_2CO_3 is in equilibrium with H_2O and CO_2. Excess CO_2 can be eliminated by the lungs.

destruction of excess acid by the blood and respiratory system:

$$\text{excess } H^+ + HCO_3^- \;\; \underset{\text{blood}}{\xrightleftharpoons{\quad}} \;\; H_2CO_3 \;\; \underset{\text{lungs}}{\xrightleftharpoons{\quad}} \;\; H_2O + CO_2\uparrow$$

exhaled

The kidneys also help regulate the blood pH by removing excess hydrogen ions from H_2CO_3 in the blood. Thus, the kidneys help maintain the HCO_3^-–H_2CO_3 balance in the blood.

$$H_2CO_3 \;\; \xrightarrow{\text{kidneys}} \;\; H^+ \;\; + \;\; HCO_3^-$$

to urine   *to blood*

The blood contains other buffers, but these are not as important as the bicarbonate–carbonic acid buffer. One such buffer is the phosphate buffer shown previously in this section.

■ **10.9   ACIDOSIS AND ALKALOSIS (optional)**

The normal pH of blood is 7.35–7.45; blood pH lower than 7.35 results in a condition called **acidosis**. Acidosis can occur in patients with untreated diabetes mellitus. These individuals lack insulin and thus cannot utilize carbohydrates. Acidosis also can result from starvation, fasting, or a low-carbohydrate diet. In these cases, either improper metabolism of carbohydrates or lack of carbohydrates in the diet means that the body must metabolize fats for energy. When broken down by the body, fats produce larger quantities of acidic end-products than do carbohydrates; these acids are removed from the cells by the blood, which then becomes overly acidic.

This type of acidosis, called metabolic acidosis (because it is a result of cellular reactions, or metabolism), is characterized by low HCO_3^- levels in the blood. A patient suffering metabolic acidosis may *hyperventilate* (overbreathe) to remove excess CO_2 from the blood and to obtain more O_2.

depleted by exhalation

$$H^+ \; + \; HCO_3^- \; \rightleftharpoons \; H_2CO_3 \; \rightleftharpoons \; H_2O \; + \; \boxed{CO_2}$$

*in overly acidic   depleted by
blood   reaction with* H^+

Another type of acidosis, called respiratory acidosis, results from *hypoventilation* (underbreathing), which can arise from pulmonary (lung) disease. In this case, CO_2 buildup in the blood results in excessive amounts of H_2CO_3, H^+, and HCO_3^-.

$$H_2O \; + \; CO_2 \; \rightleftharpoons \; H_2CO_3 \; \rightleftharpoons \; H^+ \; + \; HCO_3^-$$

not exhaled   *concentrations in blood
increase*

If the pH of blood is greater than 7.45, the person is said to suffer from **alkalosis**. Alkalosis is not as common as acidosis. One way alkalosis occurs is if a person has suffered from prolonged vomiting and thus has lost an unusual amount of acid from the stomach. Excessive use of antacid preparations can also lead to alkalosis. Hyperventilation is another cause of alkalosis (called, in this case, respiratory alkalosis):

depleted by exhalation

$$HCO_3^- \; + \; H^+ \; \rightleftharpoons \; H_2CO_3 \; \rightleftharpoons \; H_2O \; + \; \boxed{CO_2}$$

*depleted
by loss of* CO_2

Hyperventilation itself can be caused by disorders of the central nervous system, neurotic disorders, or high altitudes. ■

SUMMARY

An *acid* is a proton donor. *Strong acids* are almost completely ionized in water, while *weak acids* are only partially ionized. Figure 10.1 lists some strong and weak acids. Acids can be formed by the reaction of water with nonmetal oxides, such as SO_3. Burning sulfur-containing fuels can lead to SO_3 and *acid rain*.

Acids react with some metals to yield salts and hydrogen gas:

$$M + 2\,H^+ \rightarrow M^{2+} + H_2\uparrow$$

A list of metals in terms of their reactivity toward acids is called the *activity series of the metals* (see Figure 10.2). Acids also react with *bases* (proton acceptors), such as hydroxides, metal oxides, carbonates, bicarbonates, and ammonia.

Strong bases are completely ionized in water, while *weak bases* are only partially ionized. Figure 10.3 lists some strong and weak bases. Acid-base reactions proceed from stronger acids and bases to weaker acids and bases.

Pure water contains $1 \times 10^{-7}\,M\,H^+$ and an equal concentration of OH^-. The *ion product constant of water* is represented as K_w. $K_w = [H^+]\,[OH^-] = 1 \times 10^{-14}$ at $25°C$, regardless of the concentration of H^+. The *pH* is a measure of acidity or alkalinity. At neutrality, $[H^+] = 1 \times 10^{-7}\,M$, and pH = 7. Acidic solutions have pH values *smaller than 7*, while alkaline solutions have pH values *larger than 7*.

A *neutralization reaction* is one in which equivalent amounts of acid and base react to yield a neutral solution. One *equivalent* of a substance is the number of moles required to provide 1.0 mole of H^+ or OH^-. The *equivalent weight* is the weight in g that will provide 1.0 mole of H^+ or OH^-. A *milliequivalent* is 0.001 equivalent. The *normality* (N) of a solution is the number of equivalents of solute per liter of solution. *Titration* involves measuring acidity or basicity by adding a solution of known concentration until neutrality is reached. The unknown concentration can be calculated by the equation $N_a V_a = N_b V_b$.

The second ionization of a diprotic acid in water occurs to a lesser extent than the first, although both protons can be removed by a sufficiently strong base.

$$H_2A \underset{\xrightarrow{\hspace{1cm}}}{\overset{H_2O}{\longleftarrow}} H_3O^+ + HA^- \underset{\overset{H_2O}{\xrightarrow{\hspace{1cm}}}}{\longleftarrow} H_3O^+ + A^-$$

stronger acid   *weaker acid*

but

$$H_2A + 2\,OH^- \longrightarrow A^{2-} + 2\,H_2O$$

An *amphoteric substance* (such as HCO_3^-) can react with either an acid or a base. A *buffer solution* is a solution that resists changes in pH. The HCO_3^--H_2CO_3 buffer is the principal buffer in blood.

$$HCO_3^- \underset{OH^-}{\overset{H^+}{\rightleftharpoons}} H_2CO_3 \rightleftharpoons H_2O + CO_2\uparrow$$

Acidosis (overly acidic blood) can result from a deficiency in carbohydrate metabolism or from lung disease. *Alkalosis* (overly alkaline blood) can result from direct loss of acid from the stomach or from hyperventilation.

KEY TERMS

strong acid	weak base	*normality
weak acid	pH scale	*titration
activity series	neutralization	amphoteric substance
Brønsted base	diprotic acid	buffer solution
alkali	equivalent	acidosis
strong base	equivalent weight	alkalosis

STUDY PROBLEMS

10.16.   State whether each of the following acids is strong or weak:

(a) H_2SO_4 (b) $CH_3CH\overset{OH}{\underset{|}{}}\overset{O}{\underset{\|}{}}COH$ (lactic acid)

(c) HSO_4^- (d) $H_2PO_4^-$

10.17.   Write equations for (a) the ionization of lactic acid and (b) the reaction of SO_3 with water.

10.18.   Referring to Figure 10.2, complete and balance the following equations. If no reaction occurs, write "no reaction." (Refer to the periodic table inside the cover of this book to determine the oxidation numbers of the product metal ions.)

(a) $Na + H_2O \longrightarrow$
(b) $Al + HCl \longrightarrow$
(c) $Ag + HCl \longrightarrow$

10.19.   Vinegar is often used to remove hard-water deposits (such as $CaCO_3$) from the inside of automatic drip coffee makers. Write the equation that shows how $CaCO_3$ can be dissolved by vinegar.

10.20.   Toilet-bowl cleaners generally contain either sodium hydrogen sulfate ($NaHSO_4$) or dilute hydrochloric acid (HCl), which removes rust stains and hard-water deposits. Complete and balance the following equations that exemplify these reactions:

(a) $\underset{\text{rust}}{Fe_2O_3} + NaHSO_4 \longrightarrow$

(b) $CaCO_3 + HCl \longrightarrow$

10.21.   Which of the following compounds can act as a base? Would each be a strong or weak base?

(a) $Mg(OH)_2$ (b) $CH_3\overset{..}{N}CH_3$
                                  $\underset{H}{|}$

(c) $Ca(HCO_3)_2$ (d) NH_4Cl

(e) Na_2CO_3

10.22. Complete and balance the following equations. (If no reaction occurs, write "no reaction.") Also, indicate which of the reactants and products are stronger and weaker acids and bases.

(a) $CH_3\overset{\overset{\displaystyle O}{\|}}{C}O^- + HCl \longrightarrow$

(b) $NH_4Cl + NaOH \longrightarrow$

(c) $CH_3\overset{\overset{\displaystyle O}{\|}}{C}OH + NH_3 \longrightarrow$

10.23. Using K_w, determine the concentration of H^+ in each of the following solutions:
(a) $0.001\ M$ NaOH
(b) $1 \times 10^{-9}\ M$ KOH

10.24. Calculate the pH of each of the following solutions:
(a) $1.0\ M$ HCl   (b) $0.1\ M$ HCl
(c) $1 \times 10^{-9}\ M$ KOH

10.25. The pH of urine is usually about 6.0.
(a) Is urine acidic or alkaline?
(b) What is the hydrogen-ion concentration of urine?
(c) What is the hydroxide-ion concentration?

10.26. Which of the following equations represent neutralization reactions?

(a) $2\ HNO_3 + K_2CO_3 \longrightarrow$
$$CO_2\uparrow + H_2O + 2\ KNO_3$$
(b) $AgNO_3 + NaCl \longrightarrow AgCl\downarrow + NaNO_3$

10.27. Give the number of equivalents present in each of the following samples:
(a) 8.0 moles H_2SO_4
(b) 2.0 moles HCl
(c) 0.4 mole $Ca(OH)_2$

10.28. Calculate the equivalent weight of each of the following compounds:
(a) NaOH   (b) $Ba(OH)_2$

10.29. (a) Determine the number of grams of $CaCl_2$ needed to prepare 1.0 L of a solution containing 5 mEq of Ca^{2+}.
(b) How many mEq of Cl^- would be present in this solution?

*****10.30.** (a) What is the normality of $2\ M$ H_2SO_4 solution?
(b) Calculate the normality of 250 mL of a solution that was prepared from 49 g pure H_2SO_4 and water.

*****10.31.** (a) If a solution of 25.00 mL of an acid requires 15.00 mL of $0.500\ N$ sodium hydroxide solution for neutralization, what was the normality of the original acid solution?
(b) How many equivalents of acid were present?

10.32. Write equations for the stepwise ionization of phosphoric acid (H_3PO_4) in water. Show by arrow directions and lengths the relative extent of each step.

10.33. Write equations for the reaction of each of the following ions first with H^+ and then with OH^-:
(a) HCO_3^- (b) $H_2PO_4^-$

10.34. (a) Explain with an equation why the alkalinity of blood favors O_2-uptake by hemoglobin.
(b) Because cells give off CO_2, the fluids surrounding cells are more acidic than is arterial blood. Write equations to show why the fluids are more acidic and how the increased acidity favors O_2-release to the cells.

*****10.35.** Suggest a reason for the fact that re-breathing (breathing into a bag) can be used to treat respiratory alkalosis caused by neurotic hyperventilation.

11 Oxidation and Reduction Reactions

The burning of the German dirigible Hindenburg. *(Courtesy of the National Air and Space Museum, Smithsonian Institution.) In the early twentieth century, lighter-than-air craft were filled with the lightest gas—hydrogen. When ignited, hydrogen can undergo a rapid, explosive oxidation-reduction reaction with oxygen. In 1937, an explosion and fire destroyed the German dirigible* Hindenburg *in Lakehurst, New Jersey, and claimed 35 lives. This tragedy ended the use of hydrogen in balloons, blimps, and other such crafts. Today, the more expensive, but nonflammable, gas helium is used.*

Objectives   Determine oxidation numbers of simple ions and atoms in small molecules. ☐ Define oxidation and reduction in terms of electron transfer and addition or subtraction of H or O atoms. ☐ Define and identify oxidizing and reducing agents. ☐ Define combustion and incomplete combustion. ☐ List some common oxidizing agents and their uses (optional).

The burning of gasoline in an automobile engine, the digestion of food, the bleaching of clothes, and the corrosion of metals—all of these reactions and many others involve oxidation and reduction reactions.

$$2\ C_8H_{18} + 25\ O_2 \xrightarrow{\text{spark}} 16\ CO_2\uparrow + 18\ H_2O\uparrow + \text{energy}$$

octane
in gasoline

$$C_6H_{12}O_6 + 6\ O_2 \xrightarrow{\substack{\text{many} \\ \text{steps}}} 6\ CO_2\uparrow + 6\ H_2O + \text{energy}$$

glucose
from carbohydrates
in the diet

$$2\ Fe + 3\ O_2 \longrightarrow 2\ Fe_2O_3$$

iron metal   iron(III) oxide
                                    (rust)

In this chapter, we will redefine oxidation and reduction, and we will show how to determine whether a covalently bonded compound is oxidized or reduced in a reaction. We will also consider combustion and some commonly encountered antiseptics and bleaches.

11.1   CHANGES IN OXIDATION NUMBER

Before we consider oxidation and reduction reactions, we must discuss a few of the conventions that have been devised to handle these reactions. Recall that the *oxidation number* is the electrical charge of an ion or a covalently bonded atom, a charge that results from an unequal number of protons and electrons. An atom of a substance in the elemental state has no electrical charge, and its oxidation number is 0. For example, a neutral atom of calcium metal contains 20 protons and 20 electrons; the electrical charge and the oxidation number are both 0. A calcium ion, however, contains 20 protons and only 18 electrons; its oxidation number is $+2$.

oxidation number of 0

$$Ca^0$$

20 protons
20 electrons

oxidation number of +2

$$Ca^{2+}$$

20 protons
18 electrons

oxidation number of 0

$$Cl_2{}^0$$

a total of 34 protons and 34 electrons
for the two chlorine atoms

oxidation number of −1

$$Cl^-$$

17 protons
18 electrons

Oxidation numbers can be changed in chemical reactions. An *increase in the oxidation number* of an atom is defined as **oxidation**. Oxidation of an atom occurs through the loss of electrons. Thus, when Ca^0 loses two electrons to become Ca^{2+}, its oxidation number increases from 0 to $+2$, and an oxidation reaction takes place.

an oxidation reaction (a loss of electrons):

an increase from 0 to +2 in oxidation number

$$Ca:^0 \longrightarrow Ca^{2+} + 2\ e^-$$

Reduction is defined as a *decrease in the oxidation number* of an atom. In other words, for an atom to be reduced, it must *gain* electrons.

a reduction reaction (a gain of electrons):

a decrease in oxidation number from 0 to −1

$$:\ddot{C}l:\ddot{C}l: \ + \ 2\ e^- \longrightarrow 2\ :\ddot{C}l:^-$$

An **oxidation-reduction reaction**, sometimes called a **redox reaction**, is one in which there is *electron transfer*. Any time we observe an oxidation, there must be a corresponding reduction of something else. If one substance gains electrons, another substance must have lost the same number of electrons.

$$Ca + Cl_2 \longrightarrow CaCl_2$$

Ca is oxidized to Ca^{2+} (two electrons lost).
Cl_2 is reduced to two Cl^- (two electrons gained).

In the reaction of calcium metal with chlorine gas, one calcium atom loses two electrons and is oxidized. Each chlorine atom gains one electron and is reduced. The number of electrons lost by one calcium atom is the same as the total number of electrons gained by the two chlorine atoms.

Recall from Section 10.2 that acids undergo reaction with active metals, such as magnesium, to yield salts and hydrogen gas. These reactions are typical oxidation-reduction reactions.

$$\text{Mg:}^0 + 2 \text{ H}^+ \longrightarrow \text{Mg}^{2+} + \text{H:H}$$

Mg *is oxidized to* Mg^{2+} *(two electrons lost).*
Two H^+ *are reduced to* H_2 *(two electrons gained).*

Problem 11.1. Refer to Sections 4.3 and 4.13 for the rules for determining oxidation numbers, then show by oxidation numbers which atoms are oxidized and which are reduced in the following equation:

$$3 \text{ Cu} + 8 \text{ HNO}_3 \longrightarrow 3 \text{ Cu(NO}_3)_2 + 2 \text{ NO} + 4 \text{ H}_2\text{O}$$

Problem 11.2. The salt ammonium nitrate (NH_4NO_3) is useful as a fertilizer because it is rich in nitrogen. It is also an explosive that is used in dynamite. In the following equation for the explosion of ammonium nitrate, determine which atoms are oxidized and which are reduced.

$$2 \text{ NH}_4\text{NO}_3 \xrightarrow{\text{pressure}} 2 \text{ N}_2\uparrow + \text{O}_2\uparrow + 4 \text{ H}_2\text{O}\uparrow$$

11.2 OXIDIZING AGENTS AND REDUCING AGENTS

Recall that an **oxidizing agent** is a substance that causes the oxidation of something else. While it is causing an oxidation, the oxidizing agent is reduced. In the following equation, bromine (Br_2) causes the oxidation of sodium (Na). Therefore, bromine acts as an oxidizing agent and, in the process, is reduced.

A **reducing agent** is a substance that causes the reduction of something else. In the process, the reducing agent is oxidized. The definitions and conventions used in oxidation and reduction are summarized in Table 11.1.

Table 11.1. *Conventions Used for Oxidation and Reduction Reactions*

$$2 \text{ Na}^0 + \text{Br}_2{}^0 \longrightarrow 2 \text{ Na}^+ \text{Br}^-$$

$2 \text{ Na}^0 \longrightarrow 2 \text{ Na}^+ + 2 \text{ } e^-$	$\text{Br}_2{}^0 + 2 \text{ } e^- \longrightarrow 2 \text{ Br}^-$
Oxidation reaction	*Reduction reaction*
Electrons are lost.	Electrons are gained.
Oxidation number increases.	Oxidation number decreases.
Na is oxidized.	Br_2 is reduced.
Na acts as a reducing agent.	Br_2 acts as an oxidizing agent.

Problem 11.3.   Label the oxidizing agent and the reducing agent in the equation in Problem 11.1.

11.3   OXIDATION AND REDUCTION OF COVALENT COMPOUNDS

In the later chapters of this book, we will discuss how nutrients, which are covalently bonded organic compounds, are oxidized to provide energy for organisms. Generally, we say that a covalently bonded compound is oxidized if it *gains oxygen* or *loses hydrogen* (or both). Conversely, reduction is the loss of oxygen or the gain of hydrogen.

oxidation of organic compounds:

gain of two O,
loss of four H

$$CH_4 \ + 2\,O_2 \xrightarrow{\text{spark}} CO_2 \ + 2\,H_2O$$

methane   carbon
                 dioxide

gain of O,
loss of two H

$$\underset{\text{ethanol}}{CH_3CH_2OH} + O_2 \xrightarrow{\text{enzymes}} \underset{\substack{\text{acetic acid}\\ \text{(vinegar)}}}{CH_3\overset{\displaystyle O}{\overset{\|}{C}}OH} + H_2O$$

reduction of organic compounds:

gain of two H

$$\underset{\text{ethylene}}{CH_2{=}CH_2} + H_2 \xrightarrow{\text{catalyst}} \underset{\text{ethane}}{CH_3CH_3}$$

gain of two H

$$\underset{\text{acetone}}{CH_3\overset{\displaystyle O}{\overset{\|}{C}}CH_3} + H_2 \xrightarrow[\text{heat, pressure}]{\text{catalyst}} \underset{\substack{\text{isopropyl alcohol}\\ \text{(rubbing alcohol)}}}{CH_3\overset{\displaystyle OH}{\overset{|}{C}}HCH_3}$$

Example   Label each of the following conversions as oxidation or reduction:

(a) $CH_3CH_2OH \longrightarrow CH_3\overset{\displaystyle O}{\overset{\displaystyle \|}{C}}H$

(b) $CH \equiv CH \longrightarrow CH_2 = CH_2$

(c) $HO\overset{\displaystyle O}{\overset{\displaystyle \|}{C}}CH = CH\overset{\displaystyle O}{\overset{\displaystyle \|}{C}}OH \longrightarrow HO\overset{\displaystyle O}{\overset{\displaystyle \|}{C}}CH_2CH_2\overset{\displaystyle O}{\overset{\displaystyle \|}{C}}OH$

Solution   (a) is an oxidation (loss of hydrogen); (b) and (c) are reductions (gain of hydrogen).

Problem 11.4.   When methanol (CH_3OH, wood alcohol) is ingested, the body converts it to the toxic compound formaldehyde ($H_2C = O$). The result may be blindness or death. (a) Write a partial equation for this reaction. (b) State whether the methanol has been oxidized or reduced.

11.4   COMBUSTION

Oxygen gas is a potent oxidizing agent. The burning of a substance in air, called **combustion**, is the rapid oxidation of that substance by oxygen. During combustion, energy is given off in the forms of heat and light, which we often harness for power and heat.

$$CH_4 \quad + \quad 2\,O_2 \longrightarrow CO_2 + 2\,H_2O + energy$$
methane
in natural gas

$$2\,Mg \quad + \quad O_2 \longrightarrow 2\,MgO + energy$$
magnesium
used in flares

A substance that can undergo combustion is said to be *combustible*, *flammable*, or *inflammable*. (Although the prefix *in-* usually means *not*, the word *inflammable* means the same as the word *flammable*—the substance can burn.) The term *nonflammable* is used to describe a substance that does not burn. Note that oxygen itself is not combustible; oxygen causes the combustion of other substances.

As we mentioned in Section 6.2, compounds such as those in wood that can undergo combustion readily in air, which is only 21% O_2, burn extremely rapidly in pure oxygen or oxygen-enriched air. Therefore, great care must be taken to avoid sparks when oxygen gas is being used to aid a patient's respiration.

Incomplete Combustion

Incomplete combustion is the incomplete reaction of a substance (generally a carbon compound) with oxygen. The complete combustion of carbon (from coal or charcoal, for example) or a carbon compound gives carbon dioxide as a product. Carbon cannot be oxidized any further than to carbon dioxide.

complete combustion:

$$CH_4 + 2 O_2 \longrightarrow CO_2 + 2 H_2O$$
$$C + O_2 \longrightarrow CO_2$$

incomplete combustion leading to carbon monoxide:

$$2 CH_4 + 3 O_2 \longrightarrow 2 CO + 4 H_2O$$
$$2 C + O_2 \longrightarrow 2 CO$$

Incomplete combustion produces carbon products with carbon in an oxidation state lower than $+4$. For example, the poisonous gas carbon monoxide can be formed by incomplete combustion. Virtually any device that uses carbon fuels, such as an automobile engine, a gas furnace, or even a cigarette, produces some carbon monoxide. Another product of incomplete combustion may be carbon itself. Black, sooty smoke contains particles of carbon. If the oxygen supply is inadequate, incomplete combustion is an especially prevalent reaction.

■ 11.5   SOME COMMON OXIDIZING AGENTS (optional)

In medicine, in the home, and in industry, oxidizing agents are commonly used as bleaches, as antiseptics for wounds, and as disinfectants for clothes, bedding, and hard surfaces. Let us consider a few examples.

Hydrogen peroxide (H_2O_2, or H—O—O—H) is an unusual compound because the oxygen has an oxidation number of -1 instead of its usual -2. This compound is readily reduced to H_2O or oxidized to O_2. A 3% solution of hydrogen peroxide in water is used as a mild antiseptic for minor wounds or as a mouthwash. A 10% solution is used to bleach hair. In more concentrated solutions, hydrogen peroxide is likely to explode in contact with material that can be oxidized.

Chlorine (Cl_2) is an oxidizing agent that is used by many cities to kill bacteria in water supplies. Chlorine laundry bleaches are made by dissolving chlorine gas in dilute, aqueous sodium hydroxide solution. Chlorine undergoes reaction with the sodium hydroxide to form *sodium hypochlorite*, NaOCl, in the following reaction:

$$Cl_2 + 2\ NaOH \rightleftharpoons \quad NaOCl \quad + NaCl + H_2O$$

sodium
hypochlorite
an oxidizing agent

The Lewis formula for NaOCl *is*

$$Na^+ \quad {}^- :\!\overset{..}{\underset{..}{O}}\!:\!\overset{..}{\underset{..}{Cl}}\!:$$

Dakin's solution, a more dilute solution of sodium hypochlorite that is less alkaline than laundry bleach, is used to treat wounds.

In neutral or acidic solution, the equilibrium between the hypochlorite ions and chlorine gas is shifted to the left, and toxic chlorine gas is given off. For this reason, chlorine bleaches should never be used in conjunction with acidic cleaners, such as toilet-bowl cleaners, which usually contain hydrochloric acid or the acidic salt sodium bisulfate ($NaHSO_4$).

Problem 11.5. (a) If oxygen is considered to have an oxidation number of -2, what is the oxidation number of chlorine in NaOCl?
(b) Complete the following net ionic equation for the acidic decomposition of the hypochlorite ion to yield chlorine gas.

$$OCl^- + Cl^- + 2\ H^+ \longrightarrow$$

Iodine (I_2) is a purple, crystalline solid that readily sublimes (is converted from the solid state directly to the gaseous state). In solution, it is used as an antiseptic. The old-fashioned tincture of iodine (an ethanol solution of I_2) is no longer used routinely for cuts and scrapes in the home because iodine is capable of damaging tissues as well as killing bacteria. Also, as the ethanol evaporates over a period of time, the increased concentration of I_2 can lead to an even greater amount of tissue damage.

Table 11.2.   *Some Common Oxidizing Agents and Their Uses*

Name	Formula	Use
oxygen gas	O_2	coupled with ultraviolet radiation, acts as a germicide and bleach
3% hydrogen peroxide	H_2O_2	used as a mild antiseptic
ozone gas	O_3	used to kill microorganisms in drinking water supplies
chlorine gas	Cl_2	used to kill microorganisms in drinking water supplies and swimming pools
sodium hypochlorite	NaOCl	used as a laundry bleach or an antiseptic
iodine	I_2	in solution, used as an antiseptic

Atmospheric oxygen gas (O_2), in conjunction with ultraviolet light from the sun or from uv lamps, is capable of killing bacteria. This is why uv lamps are often found in examining rooms and other places where air free of disease-causing organisms is desired.

Ozone gas (O_3) is a toxic gas that forms a protective layer in the upper atmosphere and is also found in smog. (See Section 6.1.) Ozone is used commercially as a bleach for oils, waxes, fabric, and starch. Ozone, like chlorine, is used by some cities to kill bacteria in drinking water. Ozone has the advantage over chlorine of not forming toxic chlorinated organic compounds when it reacts with organic material in the water.   ■

SUMMARY

Oxidation can be defined as:

1. a loss of electrons and an increase in oxidation number
2. a gain of oxygen atoms
3. a loss of hydrogen atoms

Reduction can be defined as:

1. a gain of electrons and a decrease in oxidation number
2. a loss of oxygen atoms
3. a gain of hydrogen atoms

An *oxidizing agent* causes the oxidation of another substance, while a *reducing agent* causes the reduction of another substance. Some oxidizing agents are used as bleaches and germicides, as shown in Table 11.2.

KEY TERMS

oxidation number   oxidation-reduction reaction   combustion
oxidation   oxidizing agent
reduction   reducing agent

STUDY PROBLEMS

11.6. What is the oxidation number of each atom in the following ions or molecules? (Assume that O is -2 and H is $+1$ in each case.)

(a) Fe^{3+} (b) I_2 (c) H_2S
(d) SO_2 (e) $H_2PO_4^-$

11.7. Tell which of the following are oxidations and which are reductions, and add the proper number of electrons to the proper side of each equation.

(a) $Fe^{3+} \longrightarrow Fe^{2+}$
(b) $O_2 \longrightarrow 2\,O^{2-}$
(c) $Na \longrightarrow Na^+$
(d) $Cu^+ \longrightarrow Cu^{2+}$

11.8. In the following equations, which atoms are oxidized and which are reduced?

(a) $Cl_2 + 2\ Br^- \longrightarrow Br_2 + 2\ Cl^-$

(b) $2\ Na + 2\ H_2O \longrightarrow 2\ NaOH + H_2$

(c) $CH_4 + 2\ O_2 \longrightarrow CO_2 + 2\ H_2O$

11.9. State which is the reducing agent and which is the oxidizing agent in the equations shown in problem 11.8.

11.10. Research has been aimed at development of fuel cells to provide the energy for the operation of automobiles. One such fuel cell would use the reaction between lithium metal and chlorine. Write the equation for this reaction.

11.11. The following reactions all occur in animal cells. Tell whether each is an oxidation or reduction of the organic reactant, or neither.

(a)
$$\underset{\displaystyle}{HOCH_2\overset{\textstyle O}{\overset{\|}{C}}CH_2OH}$$

$$\longrightarrow HOCH_2\overset{OH\ \ \ O}{\underset{}{\overset{|\ \ \ \ \|}{CH-CH}}}$$

(b)
$$HOCH_2\overset{OH\ \ \ O}{\underset{}{\overset{|\ \ \ \ \|}{CH-CH}}}$$

$$\longrightarrow HOCH_2\overset{OH\ \ \ O}{\underset{}{\overset{|\ \ \ \ \|}{CH-COH}}}$$

(c)
$$\underset{\text{pyruvic acid}}{CH_3\overset{O\ \ \ O}{\overset{\|\ \ \ \|}{C-COH}}} \longrightarrow \underset{\text{lactic acid}}{CH_3\overset{OH\ \ \ O}{\overset{|\ \ \ \ \|}{CH-COH}}}$$

(d)
$$\underset{\text{carbonic acid}}{HO\overset{O}{\overset{\|}{C}}OH} \longrightarrow CO_2 + H_2O$$

11.12. Two important compounds that work in conjunction with enzymes in biological oxidations and reductions are *nicotinamide adenine dinucleotide* (NAD$^+$) and *flavin adenine dinucleotide* (FAD). For each of the following reactions of these compounds, identify first what is oxidized and what is reduced and then the oxidizing agent and the reducing agent.

(a)
$$CH_3\overset{OH\ \ \ O}{\overset{|\ \ \ \ \|}{CH-COH}} + NAD^+ \longrightarrow$$

$$CH_3\overset{O\ \ \ O}{\overset{\|\ \ \ \|}{C-COH}} + NAD\!:\!H + H^+$$

(b) $NAD\!:\!H\ +\ FAD\!:\ +\ H^+ \longrightarrow$

$$NAD^+\ +\ F\overset{\displaystyle H}{\underset{\displaystyle}{A\overset{\cdot\cdot}{D}\!:\!H}}$$

*__11.13.__ (a) Explain why chlorine bleaches should never be mixed with acidic substances such as toilet-bowl cleaners.

(b) Explain why hanging laundry in the sun to dry is a fairly effective way to disinfect it.

(c) Explain why 3% hydrogen peroxide often effervesces (gives off a gas) when it is poured from a bottle or used as a mouthwash.

12 Nuclear Chemistry and Radioactivity

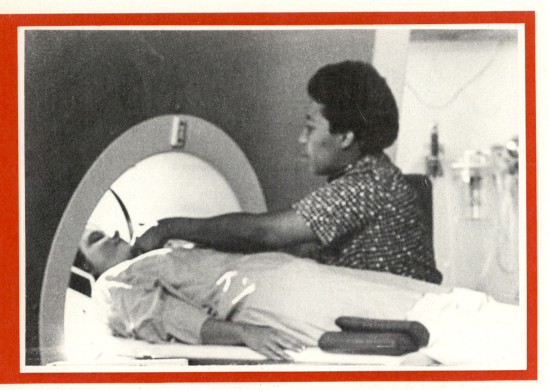

A patient in a CAT scanner. (© Richard Wood 1981/Taurus Photos.) A computerized axial tomography (CAT) scanner allows a radiologist to obtain a cross-sectional view of any part of a patient's body. The scanner rotates around the patient and takes hundreds of x-ray pictures, which are analyzed by a computer. The computer output, a detailed composite picture called a scan, can reveal tumors and many other types of abnormalities. This chapter describes radiation and the many ways that it is used.

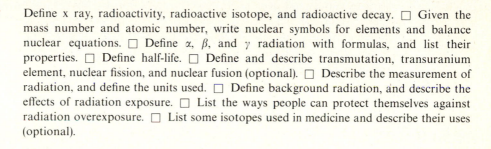

Objectives   Define x ray, radioactivity, radioactive isotope, and radioactive decay. ☐ Given the mass number and atomic number, write nuclear symbols for elements and balance nuclear equations. ☐ Define α, β, and γ radiation with formulas, and list their properties. ☐ Define half-life. ☐ Define and describe transmutation, transuranium element, nuclear fission, and nuclear fusion (optional). ☐ Describe the measurement of radiation, and define the units used. ☐ Define background radiation, and describe the effects of radiation exposure. ☐ List the ways people can protect themselves against radiation overexposure. ☐ List some isotopes used in medicine and describe their uses (optional).

In 1895, the German physicist Wilhelm Roentgen was conducting an experiment in which he was generating electrons in an evacuated glass tube. Roentgen observed that a nearby photographic plate, which was protected from light, was darkened by an emanation from the glass tube. He also observed that the plate darkened even when shielded from the tube by a sheet of metal. Roentgen correctly deduced that the darkening of the plate was caused by some new type of invisible rays. Because he did not know the nature of these rays, he named them **x rays**. Today, we know that x rays are a form of electromagnetic radiation similar to rays of visible light but much more energetic. Roentgen's x rays were produced by the collision of the electrons in the evacuated tube with the tube's glass walls.

One year after Roentgen's discovery of x rays, a French physicist, Henri Becquerel, stored some photographic film in a drawer where he also had stored a uranium compound. Imagine his surprise to find that the carefully wrapped film had darkened. Investigation by Becquerel showed that all uranium compounds affect film this way. Becquerel correctly reasoned that uranium compounds emit invisible radiation similar to x rays and that this natural radiation had caused the film to darken. Even today, special film badges are worn by workers who might be exposed to radiation. Darkening of the film is used as a measure of the amount of radiation to which the wearer of the badge has been exposed.

After Becquerel's discovery, Marie Curie and her husband Pierre Curie undertook to investigate and isolate **radioactive elements**—that is, elements that undergo changes in their nuclei and emit radiation. The Curies discovered polonium (named after Marie Curie's native country, Poland), then radium (so named because of its very high level of radioactivity). The Curies' dedication is attested to by the fact that to get 0.1 g of radium, they had to start with a ton of ore. Becquerel and the Curies received a joint Nobel prize in 1903 for their discoveries, but unfortunately only after Pierre Curie had been killed in an accident.

Today, radioactivity and nuclear chemistry affect all of us. Many cities obtain their electrical power from nuclear plants. Nuclear weaponry is always a controversial topic. In medicine, radioactive substances are used to monitor various glands and organs, to detect tumors, and to destroy cancerous cells. In this chapter, we will briefly survey the origins of radioactivity, types of radioactive emissions, nuclear fission, and nuclear fusion. We will also discuss how radioactivity is detected and measured. Finally, we will consider the effects of radioactivity on living systems and its uses in medicine.

12.1   TERMS USED IN NUCLEAR CHEMISTRY

Recall from Section 3.3 that *isotopes* of an element are forms of the element in which the nuclei contain different numbers of neutrons. Most elements and their naturally occurring isotopes contain stable nuclei; however, some isotopes (both natural and artificial) contain unstable nuclei that undergo spontaneous decomposition.

When nuclei decompose, one element may be converted to another, although this is not always the case. Decomposing nuclei emit radiation. The unstable isotopes are therefore referred to as *radioactive isotopes*, *radioisotopes*, or sometimes *radionuclides*. The decomposition of a radioactive isotope is called *radioactive decay*. The study of radioactive decay and other changes in the nuclei of atoms is called **nuclear chemistry**.

Because nuclear chemistry involves changes in the nuclei of atoms, we use a different type of symbolism from that used for ordinary chemical reactions in which the nuclei remain unchanged. When we write the symbol for an element involved in a nuclear reaction, we include the *mass number* (the number of protons plus the number of neutrons) at the top left of the atomic symbol, and we include the *atomic number* (the number of protons) at the lower left.

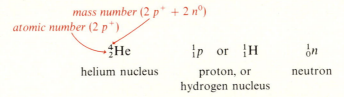

$$^{4}_{2}\text{He} \qquad ^{1}_{1}p \;\text{ or }\; ^{1}_{1}\text{H} \qquad ^{1}_{0}n$$

helium nucleus   proton, or   neutron
                     hydrogen nucleus

The mass number is often included in the names of elements encountered in nuclear chemistry. For example, tritium ($^{3}_{1}\text{H}$) is often called hydrogen-3.

Example   Describe in words the meanings of (a) $^{131}_{53}\text{I}$ and (b) $^{14}_{6}\text{C}$.

Solution   (a) An atom of iodine (atomic number 53) with a mass number of 131 amu (atomic mass units).
(b) An atom of carbon (atomic number 6) with a mass number of 14 amu.

Problem 12.1. Write the symbols for (a) uranium-235 (atomic number 92) and (b) chlorine-37 (atomic number 17).

12.2 WHY NUCLEI DECAY

The forces that hold a nucleus together are not entirely understood. We do know that there are other particles besides neutrons and protons in the nucleus, and we believe that they aid in binding neutrons and protons together.

The nuclei of the natural, stable isotopes of most elements of lower atomic mass contain equal numbers of neutrons and protons. The nucleus of carbon-12, for example, contains six protons and six neutrons; that of nitrogen-14 contains seven protons and seven neutrons. The nuclei of the natural, stable isotopes of the elements of higher atomic mass contain a greater proportion of neutrons than do the lighter-weight elements. For example, the nucleus of bromine-79 contains 35 protons and 44 neutrons.

If the ratio of neutrons to protons is changed dramatically from the normal, the result is a radioactive isotope that undergoes decay to yield a more stable isotope. Carbon-13 with six protons and seven neutrons is stable, but carbon-14 with six protons and eight neutrons is an unstable isotope.

$$^{12}_{6}C \qquad\qquad ^{13}_{6}C \qquad\qquad ^{14}_{6}C$$

carbon-12   carbon-13   carbon-14
abundant and stable   *stable*   *radioactive*

Another factor that affects the stability of a nucleus is the *size of the nucleus*. Every known isotope of every element with an atomic number greater than 82 is radioactive. In other words, lead is stable, but bismuth and the heavier elements are unstable and undergo radioactive decay.

12.3 TYPES OF RADIOACTIVE EMISSIONS

Alpha Radiation

There are three main types of radiation that arise from radioactive decay. The first is the emission of **alpha** (α) **particles**. An alpha particle is a helium nucleus ($^{4}_{2}He$), a positively charged particle containing two protons and two neutrons. Alpha particles travel at about one-tenth the speed of light. However, because of their relatively large size and their positive charge, they are not very penetrating and cannot get past the molecules in the first layer of human skin or through a piece of cardboard. Because they are not very penetrating, alpha particles are relatively harmless to humans unless the alpha-particle emitter is actually within the body. The radioactive isotope uranium-238 is an example of an alpha-particle emitter.

uranium-238 decay (first step):

an alpha (α) particle

$$^{238}_{92}\text{U} \longrightarrow ^{234}_{90}\text{Th} + ^{4}_{2}\text{He}$$

uranium-238   thorium-234   helium-4

The preceding equation for the first step in the decay of a uranium-238 nucleus illustrates how we write nuclear equations. The nuclear particles, or *nucleons*, on the left-hand side of the equation must balance with the nucleons on the right-hand side. In the preceding equation, you can see that the mass numbers on the right (234 + 4) equal the one on the left (238). Also, the number of protons on the right (90 + 2) equals the number on the left (92).

Problem 12.2.   Balance the following equations by filling in the missing numbers and symbols:

(a) $^{14}_{7}\text{N} + ^{1}_{0}n \longrightarrow ^{?}_{?}\text{C} + ^{1}_{1}\text{H}$

(b) $^{14}_{7}\text{N} + ^{1}_{1}\text{H} \longrightarrow ? + ^{4}_{2}\text{He}$

Beta Radiation

A second type of radiation is the emission of **beta (β) particles**, high-energy electrons that travel at an exceedingly high speed, up to nine-tenths the speed of light (over 100,000 miles/second). Because of their high speed and small size, beta particles are more penetrating than alpha particles and can penetrate a slight distance into the skin. For this reason, beta radiation can cause burns. An example of a nuclear reaction that yields beta particles follows:

a beta (β) particle

$$^{14}_{6}\text{C} \longrightarrow ^{14}_{7}\text{N} + ^{0}_{-1}e$$

carbon-14   nitrogen-14   electron

The electron that is emitted in this reaction comes from the nucleus itself, not from electronic shells around the nucleus. An atomic nucleus does not actually contain electrons; however, a neutron in an unstable nucleus can undergo decay to yield a proton, which remains in the nucleus, and an electron, which is expelled. Because a proton is created, the atomic number of a nucleus that has undergone decay in this manner is raised by 1. Thus, carbon-14 in the preceding reaction is converted to nitrogen-14, a different element with the same mass.

how a neutron decays:

$$^{1}_{0}n \longrightarrow ^{1}_{1}p + ^{0}_{-1}e$$

neutron   proton   electron

Figure 12.1.   The relative penetrating powers of alpha, beta, gamma, and x radiation (not to scale).

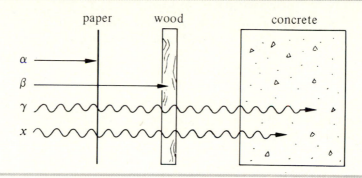

Gamma Radiation

A third type of radiation emission is the **gamma ray** (γ). Like x rays, gamma rays are electromagnetic waves; they have no mass or electrical charge. Their speed is the speed of light. Gamma rays are extremely penetrating and can pass through several feet of concrete or dirt. These rays may or may not accompany an emission of alpha or beta particles. Since gamma rays have no mass or charge, their emission does not alter the mass number or atomic number of an atom. See Table 12.1 for a summary of the types of radiation and Figure 12.1 for a depiction of their penetrating powers.

Table 12.1.   Types of Radiation from Nuclear Decay

Radiation	Symbol	Mass number	Charge	Speed
alpha (α)	^4_2He	4	$+2$	0.1 speed of light
beta (β)	$^{\ 0}_{-1}e$	0	-1	0.9 speed of light
gamma (γ)	—	0	0	speed of light

12.4   HALF-LIFE OF RADIOACTIVE DECAY

The **half-life** of a radioactive isotope is the time required for one-half of a given sample to decay to another element or isotope. The most common naturally occurring isotope of uranium is uranium-238, which goes through a series of radioactive-decay steps to yield a stable isotope of lead. The half-life of uranium-238 is 4.5 billion years. After 4.5 billion years, an initial 1.0-g sample of pure uranium-238 would contain only 0.50 g of uranium-238. In another 4.5 billion years, half of the 0.50 g of uranium-238 would decay, leaving only 0.25 g.

decay rate of uranium-238:

$$1.0 \text{ g } {}^{238}_{92}\text{U} \xrightarrow{4.5 \times 10^9 \text{ yr}} 0.50 \text{ g } {}^{238}_{92}\text{U} \xrightarrow{4.5 \times 10^9 \text{ yr}} 0.25 \text{ g } {}^{238}_{92}\text{U}$$

Not all radioactive isotopes have such long half-lives. For example, radium-226 has a half-life of 1590 years, while iodine-131 has a half-life of only eight days.

The rate of radioactive decay is always the same for different samples of the same isotope, regardless of its chemical form. The element iodine-131 ($^{131}_{53}\text{I}_2$) undergoes decay at the same rate as the iodide-131 anion in $\text{Na}^+ \: {}^{131}_{53}\text{I}^-$. The half-life of a radioactive isotope is also unaffected by temperature changes. There is no known method by which we can change the half-life of a radioactive isotope. The long half-lives of radioactive waste products from today's nuclear power plants make storage a problem. Radioactive isotopes with short half-lives are often used in medicine so as to decrease human exposure. Table 12.2 lists a few of the radioactive isotopes used in medicine, their half-lives, and the types of radiation they emit. We will discuss the uses of these isotopes later in this chapter.

Table 12.2. *Some Radioactive Isotopes Used in Medicine*

Isotope[a]	Symbol	Half-life	Radiation
iodine-131	$^{131}_{53}\text{I}$	8 days	β and γ
technetium-99^m	$^{99m}_{43}\text{Tc}$	6 hours	γ
phosphorus-32	$^{32}_{15}\text{P}$	14.3 days	β
cobalt-60	$^{60}_{27}\text{Co}$	5.3 years	β and γ

[a] The superscript *m* in technetium-99^m stands for metastable.

12.5 NUCLEAR REACTIONS IN SCIENCE AND TECHNOLOGY (optional)

Artificial Transmutations

In 1919, Ernst Rutherford bombarded nitrogen gas with alpha particles emitted from radium and obtained small amounts of oxygen-17 and hydrogen. Rutherford was the first to effect the artificial **transmutation of an element** (the changing of one element into another). Since Rutherford's experiment, hundreds of transmutations have been made.

the first artificial transmutation:

$$^{14}_{7}\text{N} \quad + \quad ^{4}_{2}\text{He} \quad \longrightarrow \quad ^{17}_{8}\text{O} \quad + \quad ^{1}_{1}\text{H}$$

nitrogen-14   alpha particle   oxygen-17   hydrogen-1

Until 1940, uranium was the heaviest element known. It is apparently the heaviest *natural* element. In 1940, McMillan and Abelson synthesized the first *transuranium element*, an element with an atomic mass greater than that of uranium, by bombarding uranium-238 with high-energy particles. In recent years, scientists have created a number of transuranium elements and also many artificial isotopes of naturally occurring elements. Many radioactive isotopes used in medicine are artificial isotopes not found in nature; iodine-131 and technetium-99^m are examples.

Nuclear Fission

In the late 1930s, Otto Hahn in Germany and others elsewhere noticed that when uranium was bombarded with neutrons, small amounts of barium and other elements from the middle of the periodic table were produced in the target samples. Was the bombardment actually causing uranium atoms to split in two? This, indeed, was the case. It was later shown that the isotope that was breaking up, or undergoing **nuclear fission**, was uranium-235, which occurs as less than 1% of naturally occurring uranium (mainly uranium-238).

the nuclear fission of uranium-235:

$$^{235}_{92}U + ^1_0n \longrightarrow ^{236}_{92}U \longrightarrow ^{141}_{56}Ba + ^{92}_{36}Kr + 3\ ^1_0n + \gamma + \text{energy}$$

With the discovery of nuclear fission came the dawn of the age of atomic energy. A few years after this discovery, the United States became involved in World War II and initiated the most fantastic problem-solving project the world had ever seen—the development of the atomic bomb.

When uranium-235 captures one neutron, it undergoes fission and produces *three* neutrons. When small quantities of uranium-235 undergo fission, these newly produced neutrons escape from the sample into the surroundings. When a large enough mass of uranium-235 is present in a small enough volume, the neutrons are contained and collide with other nuclei, causing more fissions and producing more neutrons. These neutrons, in turn, cause an even greater number of fission reactions. (See Figure 12.2.) A self-sustaining reaction such as the fission of uranium-235 is called a **chain reaction**.

When uranium-235 undergoes fission, tremendous amounts of energy are released. The energy arises from the conversion of a small amount of the uranium's mass to energy. The amount of energy generated can be calculated by Albert Einstein's famous equation $E = mc^2$, where E is the energy in ergs (1 erg = 2.4×10^{-8} cal), m is the mass in grams, and c is the speed of light (30,000,000,000 cm/sec). From Einstein's equation we can calculate the amount of energy formed from a given mass of matter. How much energy would be formed if 1.0 g of matter were completely converted to energy? Almost 20 billion kcal. This 1.0 g of matter could heat hundreds of homes during a long, hard winter.

Today's nuclear reactors use chain reactions of radioactive isotopes such as uranium-235. The chain reaction is controlled with rods that absorb excess neutrons and with moderating substances that decrease the speed of the neutrons. The energy produced by the controlled fission is trapped as heat to operate steam-powered electrical generators.

Figure 12.2. *Each fusion of a uranium-235 nucleus requires one neutron and produces three neutrons.*

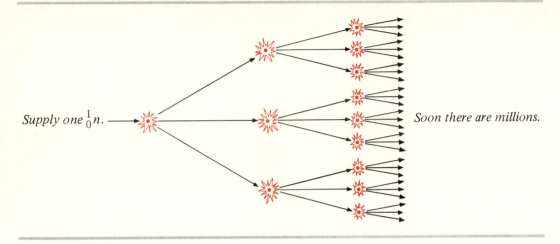

Supply one $_0^1 n$. ⟶ *Soon there are millions.*

Nuclear Fusion

Yet another source of energy may be found in **nuclear fusion reactions**, or **thermonuclear reactions**. *Fission* means "breaking apart"; *fusion* means "putting together." Nuclear fusion is the putting together of small atoms to make larger ones. The energy of the sun comes from the fusion of hydrogen atoms into helium atoms. By this reaction, the sun converts 5 million tons of matter into energy each second. In the production of only 1.0 mole (4.0 g) of helium from hydrogen, 620 million kcal of energy is produced.

a nuclear fusion reaction:

$$4\ _1^1 H \longrightarrow\ _2^4 He + 2\ _1^0 e + energy$$

The hydrogen bomb is an example of uncontrolled nuclear fusion. At the present time, controlled nuclear fusion is not a practical source of electricity because of the extremely high temperatures (greater than 1,000,000°C) needed to sustain fusion and because of the problems involved in physically containing such high-temperature reactions. ■

12.6 DETECTION AND MEASUREMENT OF RADIATION

Radiation Counters

Radiation can be detected by a *Geiger counter*, or *Geiger-Müller counter*, which contains a gas under low pressure in a glass tube. The gas molecules absorb energy from radiation entering the tube and become ionized. The ionized gas can conduct a pulse of electric current, which causes the clicking noise associated with Geiger counters.

Scintillation counters are more sensitive than Geiger counters. A scintillation counter contains a substance that emits tiny flashes of light (scintillations) when it is struck by radiation. These counters are used in *scanners*, which can detect the location and amount of radiation being given off by a radioactive substance in some portion of the human body. The scanner moves back and forth across the portion of the body being scanned, detecting the radiation being given off by each small area. This information is fed to a receiver/recorder, which produces a picture, or *scan*, of the portion of the body under investigation. Computer analysis of the data helps produce very detailed scans. One use of a scan is to detect malignant tumors, which, in many cases, absorb radioactive substances more rapidly than does normal tissue.

Measuring the Amount of Radiation

Because we use radioactive materials in the laboratory, in medicine, and in industry, we need units for measuring quantities of radiation. By necessity, several types of measurements, and therefore several units of measure, have been devised. We will briefly look at some of these.

Rate of Radioactive Decay.   The rate at which a radioactive substance decays is used to measure how radioactive a sample is. This rate is reported in **curies** (Ci), where 1.0 Ci is defined as 3.7×10^{10} nuclear disintegrations per second (the number of disintegrations per second produced by 1.0 g of radium). In medicine, the rate of radioactive decay is often expressed in *millicuries* (mCi) or *microcuries* (μCi). (Note that 1 mCi = 1000 μCi.)

$$1.0 \text{ curie (Ci)} = 3.7 \times 10^{10} \text{ disintegrations/sec}$$

$$1.0 \text{ millicurie (mCi)} = 10^{-3} \text{ Ci} = 3.7 \times 10^7 \text{ disintegrations/sec}$$

$$1.0 \text{ microcurie } (\mu\text{Ci}) = 10^{-6} \text{ Ci} = 3.7 \times 10^4 \text{ disintegrations/sec}$$

Example   Up to 30 μCi of radioactive sodium iodide is used for diagnostic purposes, while up to 30 mCi may be used for therapeutic treatment. Convert 30 mCi to μCi.

Solution
$$1 \text{ mCi} = 1000 \ \mu\text{Ci}$$

$$30 \times 1 \text{ mCi} = 30 \times 1000 \ \mu\text{Ci}$$

$$30 \text{ mCi} = 30,000 \ \mu\text{Ci}$$

Problem 12.3.   Make the following conversions:
(a) 0.50 mCi to μCi   (b) 0.036 Ci to mCi

Exposure to Radiation.   In biological or medical work, besides knowing how fast a radioactive sample decays, we need to know the intensity of a given exposure to radiation. The unit of exposure is called the **roentgen** (R), named after Wilhelm Roentgen, the discoverer of x rays. By definition, the roentgen applies only to the very penetrating x rays and gamma rays and is the amount of radiation that produces 2.1×10^9 units of ionic electrical charge in 1.0 cc of dry air at standard temperature and pressure.

From a practical standpoint, the number of roentgens of exposure can be used to determine what symptoms of radiation poisoning a person may develop. An exposure of 600 R, for example, is very likely to be fatal.

Actual Absorption of Radiation.   A measure of exposure to radiation does not tell us how much radiation is actually absorbed by a system. As a unit of absorption we use the **rad**, which stands for *radiation, absorbed dose*. (The rad unit applies to any type of radiation.)

1.0 rad = 100 ergs of energy absorbed by 1.0 g of tissue

Although 100 ergs is only 1/420,000,000 kcal, a dose of only 200 rads of gamma radiation will lead to symptoms of radiation poisoning. Statistically, about half of the recipients would die from a dose of 500 rads. We say that the LD_{50} (lethal dose, 50%) for humans is about 500 rads.

A specific quantity of rads from different types of radiation produces different effects on tissue. Even though alpha particles are not as penetrating as other forms of radiation, their actual absorption (from an internal source, for example) causes more damage than absorption of an equivalent dose of beta, gamma, or x rays. Therefore, a unit relating the effects of absorbed doses is necessary. This unit is the **rem**. Because very small doses of radiation are generally used in medicine, the *millirem* (mrem) is also encountered; 1.0 mrem = 0.001 rem.

Table 12.3.   Units Used in Measuring Radiation

Unit	Type of measurement	Definition
1.0 curie (Ci)	emission	3.7×10^{10} disintegrations/sec
1.0 roentgen (R)	exposure	quantity of x or γ radiation that produces 2.1×10^9 units of ionic charge in 1.0 cc of air
1.0 rad	absorbed dose	quantity of radiation (any type) equivalent to 100 ergs of energy per gram of tissue
1.0 rem	absorbed dose	quantity of radiation (any type) producing the same effects as 1.0 rad of therapeutic x rays

1.0 rem = absorbed dose that produces the same effect
as 1.0 rad of therapeutic x radiation

The units used for measuring radiation are summarized in Table 12.3.

Problem 12.4.   Make the following conversions:
(a) 150 mrems to rems   (b) 3 rems to mrems

12.7   THE BIOLOGICAL EFFECTS OF RADIATION EXPOSURE

Many natural radioactive isotopes are present in the air, in the soil, and in rocks. Other radioactive isotopes are formed in the upper atmosphere by the action of cosmic rays, which are subnuclear particles from outer space. Collectively, this radiation is called **background radiation**, the natural radiation to which the inhabitants of earth are constantly exposed. It would be difficult to avoid this background radiation.

Although exposure to normal background radiation seems to cause little harm, excessive exposure to radiation can cause illness and death. It is likely that many of the early workers in the field of nuclear chemistry died as a result of radiation poisoning. The symptoms of radiation sickness are nausea and a drop in the numbers of red and white blood cells. There may be a loss of appetite, a general feeling of illness, diarrhea, hair loss, and skin disorders. A cancer patient undergoing radiation treatment may experience some of these symptoms and may not be able to fight off infections such as pneumonia in the way a normal person could. Of course, a large dose of radiation or prolonged exposure may itself cause death, without the complications of secondary infections.

Radiation causes physiological damage partly because of its *ionizing properties*. Radiation can cause molecules to break apart into ions and free radicals (particles with unpaired electrons). For example,

$$H-\overset{..}{\underset{..}{O}}H \xrightarrow{\text{radiation}} H^+ \;+\; {}^-:\overset{..}{\underset{..}{O}}H \quad \textit{ions}$$

$$H-\overset{..}{\underset{..}{O}}H \xrightarrow{\text{radiation}} H\cdot \;+\; \cdot\overset{..}{\underset{..}{O}}H \quad \textit{free radicals}$$

unpaired e$^-$

Once formed, these unnatural ions and radicals can enter into chemical reactions that are foreign to the cells, thereby causing the destruction of biomolecules and eventually the death of the cells. In addition, radiation can damage the DNA molecules that constitute genes. This damage can result in cancer or hereditary disorders.

Rapidly dividing cells are more susceptible to radiation damage than are other cells. This is why radiation can be used to destroy cancer cells, which

Table 12.4. *Estimated Amounts of Radiation Received Yearly by an Average Member of the General Population in the United States*

Source	Dose equivalent, in mrems
natural background	100–150
medical and dental tests and treatment[a]	20–100
fallout from atmospheric nuclear bomb tests	1–5
other (radioactive pollutants, occupational exposure, television sets, smoke detectors, etc.)	3–5
total	124–260

[a] A typical diagnostic chest x ray has a dose equivalent of 25–30 mrems.

undergo more rapid division than do normal cells. This is also why pregnant women and children should avoid unnecessary exposure to radiation.

The National Council on Radiation Protection and Measurement recommends that an average person receive no more than 170 mrems of whole-body exposure per year, not including exposure to background radiation. (Greater exposures to small portions of the body for diagnostic or therapeutic purposes are allowable, as are higher whole-body exposures in some instances.) Table 12.4 lists the estimated exposure ranges of the average person living in the United States. As you can see from the table, the greatest quantities of radiation to which we are exposed are either of natural environmental origin or of medical-dental origin.

Protection from Overexposure

Gamma rays and x rays are very penetrating and thus present the greatest hazard in a radiology laboratory or other place where they are used. Two principal types of radiation hazards are encountered by medical and paramedical personnel: *external sources of radiation* (generally x rays or gamma rays) used to irradiate a patient and *internal sources of radiation* implanted in or applied to a portion of a patient's body.

A patient who has been subjected to an external source of radiation does not emit radioactivity after the treatment. However, an implanted radioactive needle or bead that emits gamma rays can be a source of harmful radiation to other people. Furthermore, the urine and other body secretions of a patient who has received a radioactive isotope orally or intravenously may be radioactive. These hazards are considered negligible when a patient has received small amounts of radioactive material for diagnostic purposes, but special precautions must be taken when one is dealing with patients undergoing therapy with greater amounts of radioactive substances.

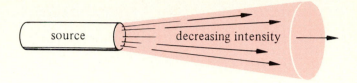

Figure 12.3.   *The intensity of radiation decreases with increasing distance from the source because the radiation becomes more diffuse.*

The three principal types of protection against exposure to radiation are *time*, *shielding*, and *distance*.

Time.   The amount of radiation exposure is proportional to the time of exposure. Nurses are asked to spend minimal time in a room with a patient emitting gamma radiation. Many radioactive isotopes used in medicine have short half-lives. A patient being treated with such an isotope will emit progressively less radiation with time. For example, a dose of a compound containing technetium-99^m (half-life, six hours) will lose half of its radioactivity in six hours and three-fourths of it in twelve hours.

Shielding.   Lead is one of the densest metals and is an effective shield against radiation. Hence, lead aprons and lead walls are commonly used to protect humans against radiation exposure.

Distance.   Another way of protecting oneself against radiation is by maximizing the distance from the source of radiation. The intensity, or actual quantity, of radiation drops off rapidly as the distance from the source is increased, because the radiation spreads out and becomes more diffuse. Mathematically, the intensity of radiation (I) varies inversely with the square of the distance (d) from the source:

$$I = \frac{1}{d^2} \times C$$

This means that at twice the distance, the radiation exposure is one-fourth the original value. (See Figure 12.3.)

To compare the intensities of radiation at two different distances (d_1 and d_2) from the source, we use the following equation:

$$\frac{I_1}{I_2} = \frac{d_2{}^2}{d_1{}^2}$$

Example   If a person standing 2 ft from a radiation source absorbs 50 mrems of radiation, how many mrems would the person absorb in the same amount of time from a distance of 8 ft?

Solution   List the data:

$$I_1 = 50 \text{ mrems} \quad d_1 = 2 \text{ ft}$$

$$I_2 = ? \qquad\qquad d_2 = 8 \text{ ft}$$

Set up the equation:

$$\frac{50 \text{ mrems}}{I_2} = \frac{(8 \text{ ft})^2}{(2 \text{ ft})^2}$$

Rearrange and solve:

$$I_2 = 50 \text{ mrems} \times \frac{(2 \text{ ft})^2}{(8 \text{ ft})^2}$$

$$= 50 \text{ mrems} \times \frac{4 \text{ ft}^2}{64 \text{ ft}^2}$$

$$= 3 \text{ mrems}$$

Problem 12.5.   Using the data in the preceding example, determine how many mrems of radiation would be absorbed from a 12-ft distance over the same period of time.

12.8   MEDICAL USES OF RADIOACTIVE ISOTOPES (optional)

Radiation is used in medicine in several ways. In *medical diagnosis*, a very small amount of a compound containing a radioactive isotope is administered. Such a compound exhibits the same chemistry as its nonradioactive counterpart, but its presence in certain parts of the body can be detected. Gamma emitters are generally more valuable than alpha or beta emitters because gamma radiation is more easily detected. (Alpha radiation and, to an extent, beta radiation cannot penetrate the body tissues.) Radioisotopes with short half-lives are used to minimize the patient's exposure.

By contrast, in *medical therapy*, such as cancer treatment, larger amounts of radioactive materials are used to destroy cells. The radiation may be introduced internally (orally, intravenously, or as needles or capsules), applied externally as a mold, or supplied by an outside source such as an x-ray machine or a cobalt "bomb." An alpha emitter may be the isotope of choice for an internal source, because alpha radiation is destructive but is blocked by surrounding tissue and thus remains localized. An isotope with a long half-life may be used in therapy involving an outside source of radiation or a temporary implant. Let us consider just a few of the radioisotopes that are of medical value in order to see *how* they are used.

Iodine-131. Iodine-131 is a beta- and gamma-radiation emitter with a half-life of eight days. This isotope can be administered orally for determining the activity of the liver, kidneys, or other organs. Because iodide ions concentrate in the thyroid gland, this isotope is especially useful in diagnoses of thyroid problems.

The thyroid gland secretes hormones that regulate the metabolism, or utilization of nutrients, of the cells of the body. If the thyroid gland is over-active (*hyperthyroidism*), the metabolism is higher than normal and the person may be restless and underweight. If the thyroid gland is underactive (*hypothyroidism*), the person may be lethargic and overweight.

The activity of the thyroid gland determines the rate at which it takes up iodide ions. When a small dose of radioactive sodium iodide is given to a patient, the uptake of iodine-131 can be measured with a scanner. A normal thyroid accumulates 12% of the iodine-131 in a few hours. (Most of the rest of the iodine-131 is eliminated in the urine.) A greater uptake indicates hyperthyroidism, while a lesser uptake indicates hypothyroidism. In larger doses, radioactive sodium iodide can be used in treatment of hyperthyroidism to destroy some of the cells of the thyroid gland.

In detection of other problems, the radioactive iodine atom may be incorporated into other molecules. For example, liver function can be measured by incorporating iodine-131 into a compound that is ordinarily removed from the body by way of the liver, such as the dye rose bengal. In this case, the iodine-131 is used as a label or tracer for the compound so that it can be detected.

Technetium-99^m. Technetium-99^m is an artificial radioactive isotope that is said to be *metastable*. This isotope emits only gamma radiation and becomes technetium-99. The half-life of technetium-99^m is six hours. Incorporated into such compounds as sodium pertechnetate (a soluble salt) or serum albumins (blood proteins), technetium-99^m is used to test thyroid function and to scan the liver, brain, placenta, lung, bone marrow, and kidneys.

Cobalt-60. Cobalt-60, which has a half-life of 5.3 years and gives off both beta and gamma radiation, is used in large machines that aim the gamma radiation toward a cancerous tumor in the patient's body. (The beta particles are removed prior to tumor irradiation by passing the radiation through a thin piece of aluminum.) The patient is usually turned during the treatment so that other parts of the body receive less radiation than the tumor does. Cobalt-60 is also sometimes implanted as needles or capsules in a tumor.

Other Isotopes. Carbon-14, oxygen-15, and certain other radioisotopes emit *positrons* (positively charged electrons, e^+). Positron emission can be scanned in a process called *positron emission transverse tomography* (PETT). This procedure is used in detecting brain disorders and tracing the body's use of glucose.

Photo 12.1. *A brain scan. The bright spot at the left indicates a tumor, which has absorbed a greater amount of radioactive material than has the normal brain tissue. (Photograph by Martin M. Rotker/Taurus Photos.)*

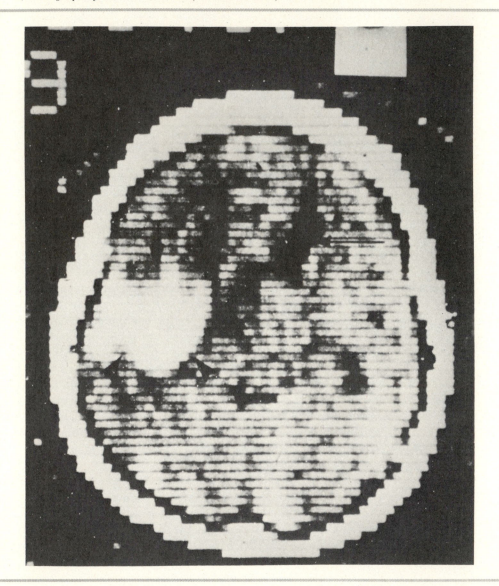

Phosphorus-32 (a beta emitter with a half-life of 14.3 days) as a phosphate salt can be used to treat leukemia, a cancer of the bone marrow, because phosphates become concentrated in the bones, which are primarily calcium phosphate.

SUMMARY

Nuclear chemistry is the study of changes in the nuclei of atoms. Two causes of nuclear instability are an imbalance in the numbers of protons and neutrons and too large a nucleus. Decomposition of radioactive isotopes, called *radioactive decay*, is accompanied by *alpha radiation* ($_2^4$He), *beta radiation* ($_{-1}^0 e$), and *gamma radiation*. Gamma rays are far more penetrating than alpha or beta radiation and are somewhat more penetrating than x rays. *Half-life* is the time required for one-half of the mass of a given sample to undergo radioactive decay.

Bombarding some elements with high-energy nuclear or subnuclear particles can lead to *transmutation of the element* (change to a different element) or *nuclear fission* (breakup of a nucleus into smaller nuclei). *Nuclear fusion* is the combination of two or more nuclei to yield a larger nucleus.

Radiation is detected by *Geiger counters* and *scintillation counters*; the latter are used in *scanners*. The units used for quantitative measurements of radiation emission and absorption are summarized in Table 12.3.

Radiation damage to tissues and biomolecules is caused by the ionizing properties of radiation; rapidly dividing cells are more susceptible than other cells. Lead metal can be used as a radiation shield. The intensity of radiation varies inversely with the square of the distance from the radiation source.

Radioactive isotopes used for diagnostic purposes are generally gamma emitters with short half-lives; examples are iodine-131 and technetium-99^m. Other isotopes have therapeutic uses, such as destroying cancer cells.

KEY TERMS

x ray
radioactive decay
nuclear chemistry
alpha particle
beta particle
gamma ray

half-life
*transmutation
*nuclear fission
*chain reaction
*nuclear fusion
 radiation counter

curie
roentgen
rad
rem
background radiation
radiation sickness

STUDY PROBLEMS

12.6. Which of the following would you expect to be a stable isotope? Explain.
(a) carbon-14
(b) nitrogen-14
(c) sodium-24
(d) oxygen-18

12.7. Define the following terms:
(a) $_{19}^{40}$K   (b) $_{-1}^0 e$
(c) $_0^1 n$

12.8. Match the following terms. More than one answer may be correct.
(a) alpha particle
(b) beta particle
(c) gamma ray

(1) electromagnetic waves
(2) mass of 4 amu
(3) mass of 0 amu (or approximately so)
(4) helium nucleus
(5) electron

12.9.   Compare the penetration abilities of alpha, beta, and gamma radiation.

12.10.   Complete and balance the following equations:

(a) $^{214}_{83}Bi \longrightarrow ^{214}_{84}Po + $ _____

(b) $^{226}_{88}Ra \longrightarrow $ _____ $ + ^{4}_{2}He$

12.11.   Write a balanced nuclear equation for each of the following reactions:
(a) loss of a beta particle by $^{115}_{48}Cd$
(b) loss of an alpha particle by $^{220}_{86}Rn$

12.12.   The half-life for radon-222 is four days. If you start with 10.0 g of radon, how much will be left in twelve days?

*12.13.   Tell whether each of the following reactions is an example of nuclear fission or fusion:

(a) $^{3}_{1}H + ^{2}_{1}H \longrightarrow ^{4}_{2}He + ^{1}_{0}n + $ energy

(b) $^{235}_{92}U + ^{1}_{0}n \longrightarrow$

$^{139}_{54}Xe + ^{94}_{38}Sr + 3\,^{1}_{0}n + $ energy

12.14.   The radioactivity of a sample is found to be 500 disintegrations per second. What is the rate of decay in microcuries?

12.15.   Make the following conversions:
(a) 10 mCi to μCi   (b) 200 mrems to rems

12.16.   (a) List the ways a person can be protected against unnecessary exposure to radiation.

(b) How much radiation will be absorbed by a person 6 ft away from a source if the absorption is 20 mrems at a distance of 1 ft?

*12.17.   Which of the following isotopes is generally used for diagnosis only and which for therapy only? Explain.
(a) cobalt-60   (b) technetium-99^{m}

*12.18.   Explain why iodine-131 is frequently used for diagnosis and treatment of thyroid problems, while phosphorus-32 is used in the treatment of bone problems.

PART **2** **Organic Chemistry**

13 Introduction to Organic Chemistry

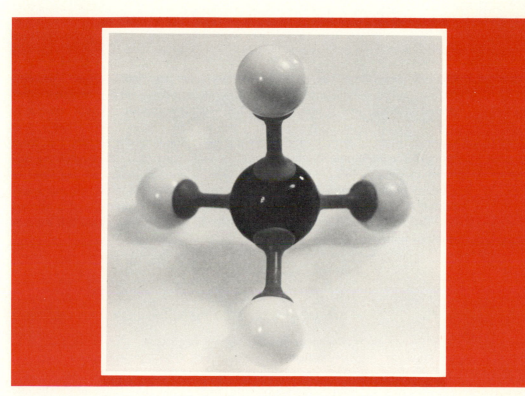

Molecular model of methane (CH$_4$). The shapes of molecules are important in organic chemistry and biochemistry, and molecular models help us visualize these shapes. This photograph shows a ball-and-stick model. In this model, the black ball represents a carbon atom, and each light-colored ball represents a hydrogen atom.

Objectives   Define organic chemistry. ☐ Compare the properties of organic compounds and inorganic salts. ☐ Write formulas for organic compounds containing single, double, and triple bonds. ☐ Write formulas for structural isomers. ☐ Predict bond angles from the types of bonds in a structure. ☐ Define conformation and write formulas in different projections. ☐ Write polygon formulas for cyclic compounds. ☐ Identify some common functional groups.

Organic chemistry is the *chemistry of compounds containing carbon*. The term organic comes to us from an early misconception concerning organic compounds. At one time, it was believed that organic compounds could come only from living, or *organic*, systems. This belief, called the *vital force theory*, survived until chemists proved that organic compounds could be prepared from inorganic compounds in the laboratory. An experiment performed by the German chemist Friedrich Wöhler in 1828 is generally cited as the dawn of modern organic chemistry. Wöhler was able to show that the organic compound urea (a waste product from the breakdown of proteins, and a normal constituent of urine) could be prepared by heating the inorganic salt ammonium cyanate.

Wöhler's experiment:

$$NH_4{}^+ CNO^- \xrightarrow{\text{heat}} H_2N-\overset{\displaystyle O}{\overset{\|}{C}}-NH_2$$

ammonium   urea
cyanate

What is so unique about carbon compounds that they should be the subject of a major division in the study of chemistry? The answer lies in the ability of carbon atoms to form stable covalent bonds with other carbon atoms. Elemental carbon itself is found as graphite and diamonds; coal and soot are also primarily elemental carbon. In compounds, carbon atoms can be bonded together in molecules to form short chains, long chains, branched chains, small rings, large rings, and connected rings. In addition, carbon can form stable covalent bonds with other elements, notably hydrogen, oxygen, and nitrogen. The number of ways these elements can be combined is beyond comprehension! Consequently, the field of organic chemistry includes the study of such diverse substances as natural gas and petroleum, plastics, drugs, vitamins, soaps and detergents, solvents, varnishes, natural and synthetic fabrics, natural and synthetic rubber, and a host of other compounds.

some of the compounds of carbon and hydrogen:

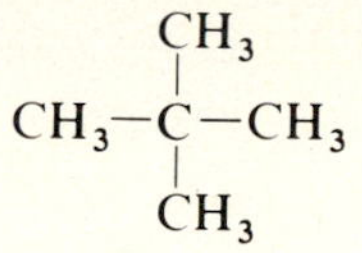

$$CH_3-\underset{\underset{\displaystyle CH_3}{|}}{\overset{\overset{\displaystyle CH_3}{|}}{C}}-CH_3$$

dimethylpropane

$$CH_3CH_2CH_2CH_2CH_3$$

pentane

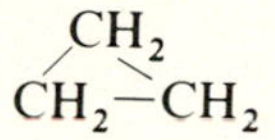

$$\underset{\displaystyle CH_2-CH_2}{\overset{\displaystyle CH_2}{}}$$

cyclopropane

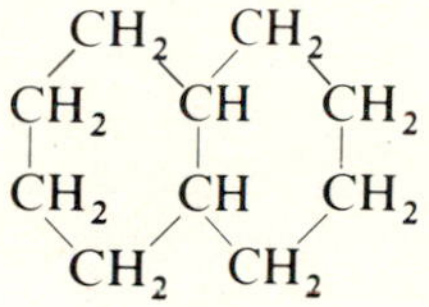

decalin

Living systems are composed primarily of organic compounds and water. Biochemistry, in many respects, is organic chemistry applied to living systems. For this reason, we must survey the structures and properties of some carbon compounds before we can discuss the chemistry of living systems. In this introduction to organic chemistry, we will compare the properties of organic compounds with those of inorganic compounds, we will briefly review covalent bonding, and we will introduce the types of formulas that are used to represent organic compounds. Finally, we will discuss structural isomers, the shapes of organic molecules, and some types of organic compounds.

13.1   COMPARISON OF THE PROPERTIES OF COVALENT AND IONIC COMPOUNDS

Most organic compounds are composed of covalently bonded molecules. Because the attractive forces between covalent molecules are weaker than the attractive forces between ions, the properties of organic compounds are quite different from the properties of ionic compounds. Let us review very briefly the principal differences between covalent and ionic compounds.

Melting Points and Boiling Points.   Most ionic compounds are high-melting salts. While high-formula-weight organic compounds may be solids at room temperature, their melting points are usually much lower than those of inorganic salts. Low-formula-weight organic compounds are generally gases or liquids at room temperature.

$$CH_3CH_2CH_2CH_3 \qquad CH_3CH_2OH \qquad C_{27}H_{46}O$$

butane
(boiling point, $-0.5°C$)
*used in cigarette
lighters*

ethanol
(boiling point, $78.5°C$)
*used as a solvent
and in beverages*

cholesterol
(melting point, $148.5°C$)
*implicated in
hardening of the
arteries*

Water Solubility. Although small polar organic molecules such as CH_3CH_2OH can dissolve in water, most organic compounds are relatively nonpolar and are water insoluble. These compounds are more likely to dissolve in nonpolar solvents, such as hexane, or slightly polar solvents, such as diethyl ether—like dissolves like.

typical solvents for nonpolar organic compounds:

$$CH_3CH_2CH_2CH_2CH_2CH_3 \qquad\qquad CH_3CH_2OCH_2CH_3$$
hexane   diethyl ether

In the human digestive tract, digestive enzymes help break down large, water-insoluble organic molecules (proteins, carbohydrates, and fats) into small, water-soluble molecules that can be absorbed through the intestinal walls into the bloodstream.

Aqueous solutions of water-soluble organic compounds do not conduct an electric current unless the compound contains an ionic bond, such as in sodium acetate ($Na^+ \ ^-O_2CCH_3$), or is weakly ionized, such as acetic acid (CH_3CO_2H). Most organic compounds are nonelectrolytes.

Rate of Reaction.   Because most organic molecules do not ionize when they are in solution, covalent bonds must be broken before a chemical reaction can occur. Ionic reactions are generally fast reactions because it is easy for two ions to combine. By contrast, reactions of covalent compounds are generally quite slow.

Combustion.   Ionic compounds, with rare exceptions, do not burn. Most organic compounds are flammable. Methane, gasoline, and ethanol are examples of flammable organic substances.

Table 13.1. *Comparisons of the Properties of Inorganic Compounds and Organic Compounds*

Inorganic ionic compounds	Organic compounds
generally do not contain carbon[a]	contain carbon
are ionic	are rarely ionic
generally have high melting points	have relatively low melting points
generally have high boiling points	have relatively low boiling points
are sometimes water soluble; are rarely soluble in nonpolar liquids	are rarely water soluble; are generally soluble in nonpolar liquids
are electrolytes	are generally nonelectrolytes
undergo fast reactions in solution	generally undergo slow reactions in solution
are generally nonflammable	are generally flammable

[a] Some exceptions are carbon dioxide (CO_2), carbonates ($CO_3{}^{2-}$), cyanides (CN^-), and cyanates (CNO^-).

$$CH_4 \quad + 2\,O_2 \xrightarrow{\text{spark}} 2\,H_2O + CO_2$$

methane
in natural gas

$$2\,C_8H_{18} \quad + 25\,O_2 \xrightarrow{\text{spark}} 18\,H_2O + 16\,CO_2$$

octane
in gasoline

$$CH_3CH_2OH + 3\,O_2 \xrightarrow{\text{spark}} 3\,H_2O + 2\,CO_2$$

ethanol

Table 13.1 compares the properties of ionic compounds and organic compounds.

13.2   A SHORT REVIEW OF COVALENT BONDING

Since the chemistry of carbon compounds is the chemistry of covalently bonded compounds, we will summarize some of the important features of covalent bonds here. We also suggest a review of Chapter 4, which covers covalent bonds in more detail.

1. Atoms held together by covalent bonds are not usually free to dissociate in solution, as are atoms joined by ionic bonds.
2. Atoms joined by covalent bonds are held at fixed distances from each other and at fixed angles in relation to each other. (We will discuss the shapes of organic molecules in Section 13.5.)
3. Atoms in covalent molecules can be joined by single, double, or triple bonds.

only single bonds:

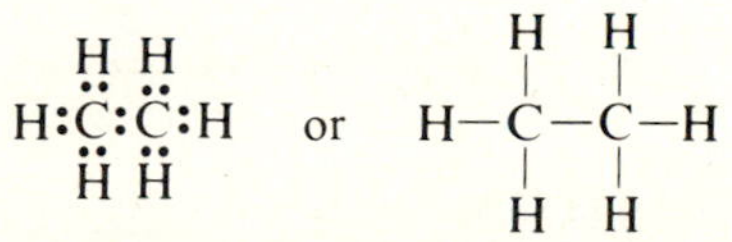

Ethane (CH_3CH_3) contains seven single bonds.

one double bond:

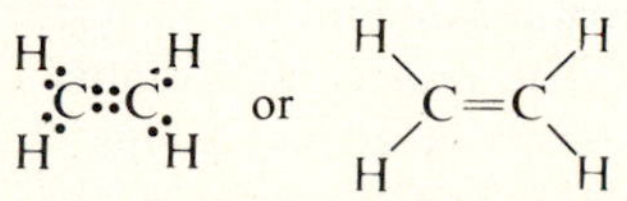

Ethylene ($CH_2{=}CH_2$) contains four single bonds ($C{-}H$)
and one double bond ($C{=}C$).

one triple bond:

$$H\text{:}C\text{:}\text{:}\text{:}C\text{:}H \quad \text{or} \quad H-C\equiv C-H$$

Acetylene ($CH\equiv CH$) contains two single bonds ($C-H$)
and one triple bond ($C\equiv C$).

Table 4.2 (Section 4.8) lists the number of covalent bonds that atoms common in organic chemistry can form. Remember that *carbon always forms 4 covalent bonds and oxygen always forms 2 covalent bonds.*

13.3   ORGANIC FORMULAS AND ISOMERISM

A molecular formula, as you will recall from Section 4.11, tells how many of each type of atom are in a molecule. Although molecular formulas, such as Na_2SO_4 or H_2SO_4, are used extensively in inorganic chemistry, we find that molecular formulas are not very useful in organic chemistry, where we need to know the *molecular structure*, or the order in which atoms are attached in a molecule. Therefore, *expanded structural formulas*, showing each atom and bond, or *condensed structural formulas* are used in organic chemistry.

formulas for butane:

$$
\begin{array}{ccc}
C_4H_{10} & \text{or} &
\begin{array}{c}
\quad H \;\; H \;\; H \;\; H \\
\quad | \quad | \quad | \quad | \\
H-C-C-C-C-H \\
\quad | \quad | \quad | \quad | \\
\quad H \;\; H \;\; H \;\; H
\end{array}
& \text{or} & CH_3CH_2CH_2CH_3
\end{array}
$$

molecular   *expanded structural*   *condensed structural*

formulas for ethanol:

$$
\begin{array}{ccc}
C_2H_6O &
\begin{array}{c}
\quad H \;\; H \\
\quad | \quad | \\
H-C-C-O-H \\
\quad | \quad | \\
\quad H \;\; H
\end{array}
& CH_3CH_2OH
\end{array}
$$

molecular   *expanded structural*   *condensed structural*

Structural Isomerism

The order of attachment of atoms in a molecule is important in organic chemistry because one molecular formula can represent two or more different compounds. The molecular formula C_4H_{10} can represent butane (with four carbons attached in a continuous chain), as we have just shown, or it can represent methylpropane (with a branched chain), a different compound with different physical and chemical properties. For example, butane boils at $-0.5°C$ while methylpropane boils at $-12°C$, even though both compounds have the molecular formula C_4H_{10}.

formulas for methylpropane:

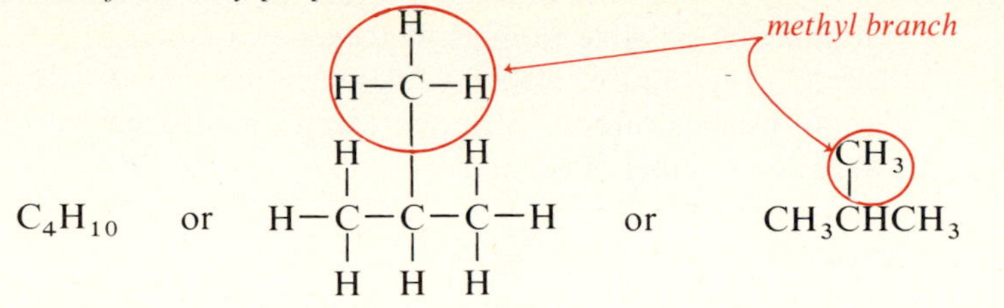

C_4H_{10} or   expanded structural   or   condensed structural

methyl branch

molecular   expanded structural   condensed structural

Similarly, the molecular formula C_2H_6O can represent ethanol, a liquid at room temperature, or dimethyl ether, a gas.

O joins a C and an H   *O joins two C's*

CH_3CH_2OH $H_3C-O-CH_3$

ethanol   dimethyl ether
(boiling point, $78.5°C$)   (boiling point, $-23.6°C$)

Compounds that have the same molecular formula but differ in how their atoms are bonded together are called structural isomers of each other. Butane and methylpropane are structural isomers, as are ethanol and dimethyl ether.

Structural isomers are a group of two or more compounds with the same molecular formula but different structures, or orders of attachment of their atoms.

Other examples of structural isomers follow:

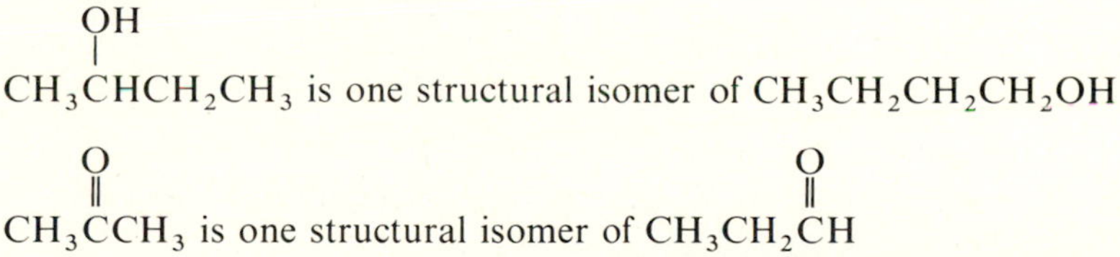

$$\overset{\overset{\displaystyle OH}{|}}{CH_3CHCH_2CH_3} \text{ is one structural isomer of } CH_3CH_2CH_2CH_2OH$$

$$\overset{\overset{\displaystyle O}{\parallel}}{CH_3CCH_3} \text{ is one structural isomer of } \overset{\overset{\displaystyle O}{\parallel}}{CH_3CH_2CH}$$

The number of isomers for a given molecular formula is dependent upon the number of atoms in the molecule. In a compound of only carbon and hydrogen with one, two, or three carbon atoms (CH_4, CH_3CH_3, and $CH_3CH_2CH_3$), there is only one possible structure. As we have seen, the formula C_4H_{10} can represent two structural isomers. The formula C_5H_{12} represents three structural isomers.

the three structural isomers of C_5H_{12}:

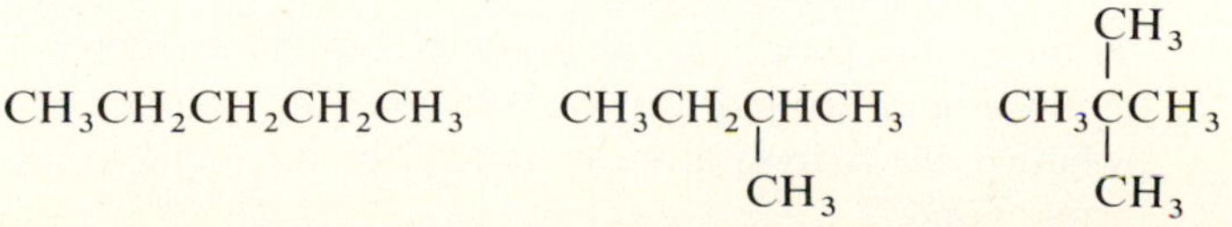

$$CH_3CH_2CH_2CH_2CH_3 \qquad \overset{\displaystyle CH_3CH_2CHCH_3}{\underset{\displaystyle CH_3}{|}} \qquad \overset{\overset{\displaystyle CH_3}{|}}{\underset{\overset{\displaystyle |}{CH_3}}{CH_3CCH_3}}$$

pentane   methylbutane   dimethylpropane

As the number of carbon atoms in a molecule increases, we find that the number of possible isomers increases at a fantastic rate. There are 75 isomers for $C_{10}H_{22}$ and 62,500,000,000,000 isomers for $C_{40}H_{82}$. The presence of an atom besides carbon, such as oxygen or nitrogen, in a molecule increases the possible number of isomers.

Example   Write the formula for a structural isomer for each of the following compounds. (Remember to consider the number of covalent bonds that each atom forms: C = 4, H = 1, O = 2. Also note that Cl forms *one* covalent bond in organic compounds.)

(a) $CH_3CH_2-O-CH_3$ (b) $CH_3CH_2CH_2Cl$

Solution   (a) Two structural isomers would be

$$CH_3CH_2CH_2-OH \quad \text{and} \quad CH_3\overset{\overset{\displaystyle OH}{|}}{C}HCH_3$$

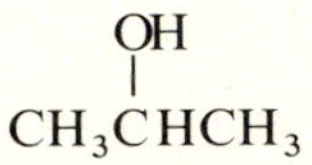

In the first case, an $-OH$ group is attached to an end carbon; in the second case, the $-OH$ group is attached to the central carbon. $CH_3-O-CH_2CH_3$ is *not* an isomer of (a) because the atoms are still attached in the same order; see Section 13.5.

(b) The only structural isomer of this compound has the Cl atom attached to the central carbon instead of to the end carbon:

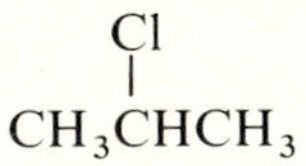

$$CH_3\overset{\overset{\displaystyle Cl}{|}}{C}HCH_3$$

Problem 13.1.   Write the formula for a structural isomer of each of the following compounds. There may be more than one correct answer.

(a) 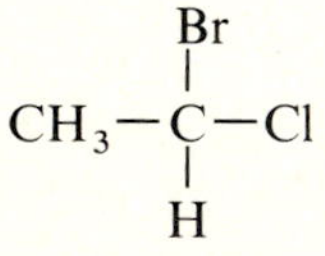 $CH_3-\overset{\overset{\displaystyle Br}{|}}{\underset{\underset{\displaystyle H}{|}}{C}}-Cl$ (b) 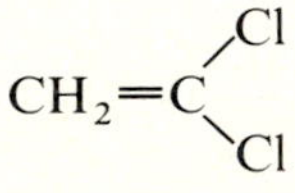 $CH_2{=}C{\overset{\displaystyle \diagup Cl}{\diagdown Cl}}$

13.4   BOND ANGLES IN CARBON COMPOUNDS

Some of the properties of organic compounds depend on their shapes. The bond angles around any carbon atom in an organic compound depend on whether the carbon atom is bonded by all single bonds, a double bond, or a triple bond.

Figure 13.1.   *Carbon forms four single bonds with bond angles of approximately 109.5°.*

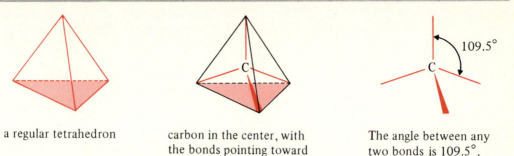

a regular tetrahedron   carbon in the center, with   The angle between any
                          the bonds pointing toward   two bonds is 109.5°.
                          the corners

1. *Four single bonds.* When carbon is bonded by four single bonds, as in methane (CH_4) or ethane (CH_3CH_3), the bonds from the carbon atom point toward the corners of a regular tetrahedron, as shown in Figure 13.1. All four bonds are identical (or nearly so), and the bond angle between any two of the four bonds is 109.5° or close to that value. Other factors, such as attractions or repulsions between atoms, have an effect on bond angles and can change them from the ideal 109.5°.
2. *A double bond.* When a carbon atom is bonded by one double bond and two single bonds, the atoms so bonded lie in a plane with bond angles of 120°, as shown in Figure 13.2.
3. *A triple bond.* When carbon is bonded by a triple bond and a single bond, the atoms lie in a line with bond angles of 180°.

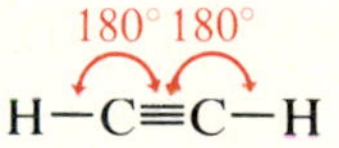

$$\text{H}-\text{C}\equiv\text{C}-\text{H}$$
180° 180°

a linear molecule

Most organic molecules are more complex than those we have used as examples. One molecule may have a variety of types of bonds and, therefore, different bond angles, depending on which part of the molecule is being con-

Figure 13.2.   *Carbon forms a double bond and two single bonds with bond angles of approximately 120°.*

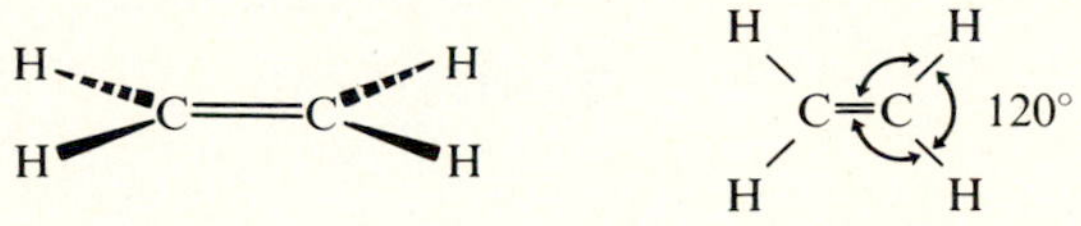

side view: The atoms of ethylene ($CH_2{=}CH_2$) lie in a plane.

top view: The bond angles around each double-bonded carbon are approximately 120°.

Figure 13.3.   *The bond angles in propylene.*

These C-C and C-H bond angles are approximately 120°.

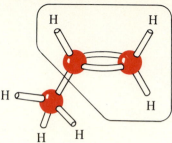

$$H_3C \diagup \underset{H}{\overset{H}{C}} = \underset{H}{\overset{H}{C}}$$

These C-H bond angles are approximately 109.5°.

sidered. In the example in Figure 13.3, a ball-and-stick model is used to emphasize the shape of the molecule. Each unlabeled ball represents a carbon atom. For clarity, we have used the symbol H for hydrogen atoms.

Problem 13.2.   What is the approximate value of each bond angle in the following compound?

$$\overset{O}{\overset{\|}{HCC}} \equiv CH$$

13.5   CONFORMATIONS AND PROJECTIONS OF MOLECULES

The structure of butane may be drawn on paper in a number of ways. Are these different-looking structures isomers? No, they are different projections of the same compound. In each of the three projections, the four carbon atoms are connected to each other in a continuous chain. The order of attachment of the atoms is not changed. The projections only appear different on paper.

three ways of writing butane (each structure represents the same compound):

$$CH_3CH_2CH_2CH_3 \qquad \begin{array}{c} CH_2CH_3 \\ | \\ CH_3CH_2 \end{array} \qquad \begin{array}{c} CH_3CH_2 \\ | \\ CH_2 \\ | \\ CH_3 \end{array}$$

The two ball-and-stick models of butane in Figure 13.4 show how the **conformation**, or shape, of butane can change by rotation of atoms around single

Figure 13.4. *Ball-and-stick models of two conformations of butane* ($CH_3CH_2CH_2CH_3$) *showing how the overall shape of a molecule can change even though the bond angles remain the same.*

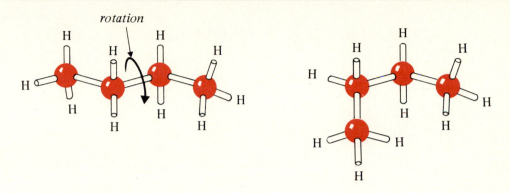

bonds. (Note that although this molecule can twist and bend, the bond angles and bond distances remain the same.) The butane molecule can assume a variety of shapes. Therefore, the point to consider in determining whether or not one structure is a structural isomer of another is the *order in which the atoms are joined*. If a bond must be broken to convert one structure to the other, the structures are isomers. If we can convert one paper structure to the other merely by rotating the structure around bonds or moving it as a whole, the structures are not isomers.

Example   Do the following structural formulas represent the same compound or a pair of structural isomers?

$$\underset{\text{ClCHCH}_3}{\overset{\text{Cl}}{|}} \qquad \underset{\text{CH}_3\text{CH}}{\overset{\text{Cl}}{|}}\atop{\underset{\text{Cl}}{|}}$$

Solution   The order of attachment of the atoms is the same. In each of the two formulas, one carbon is bonded to two chlorine atoms and one hydrogen atom, and the other carbon is bonded to three hydrogen atoms. The two formulas represent the same compound.

Problem 13.3. Three of the following formulas represent the same compound, while one represents a structural isomer. Which is the isomer?

(a) $\underset{\text{OH}}{\overset{}{\underset{|}{\text{CH}_3\text{CHCH}_3}}}$ (b) $\underset{\text{OH}}{\overset{\text{OH}}{\overset{|}{\text{CH}_3\text{CHCH}_3}}}$ (c) $\underset{\text{CH}_3}{\overset{}{\underset{|}{\text{CH}_3\text{CHOH}}}}$ (d) $CH_3CH_2CH_2OH$

Problem 13.4. Which of the following pairs represent a pair of structural isomers?

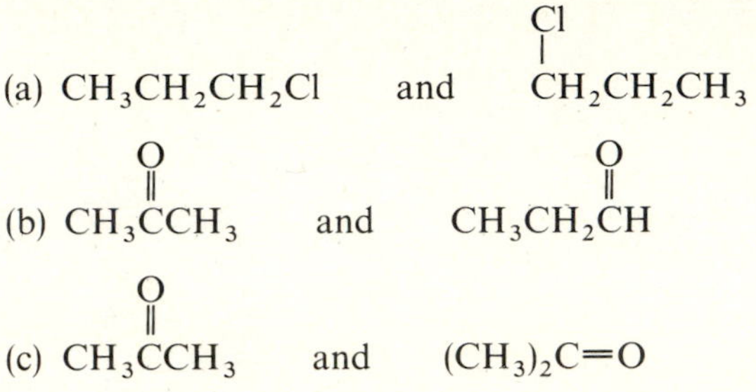

$$\text{Cl}$$
$$\mid$$
(a) $CH_3CH_2CH_2Cl$ and $CH_2CH_2CH_3$

$$\quad\quad\quad O \quad\quad\quad\quad\quad\quad\quad\quad O$$
$$\quad\quad\quad \parallel \quad\quad\quad\quad\quad\quad\quad\quad \parallel$$
(b) CH_3CCH_3 and CH_3CH_2CH

$$\quad\quad\quad O$$
$$\quad\quad\quad \parallel$$
(c) CH_3CCH_3 and $(CH_3)_2C{=}O$

13.6   CYCLIC COMPOUNDS

Carbon atoms can be joined in rings as well as in continuous chains and branched chains. A compound with carbon atoms joined in one or more rings is called a **cyclic compound**.

two cyclic compounds:

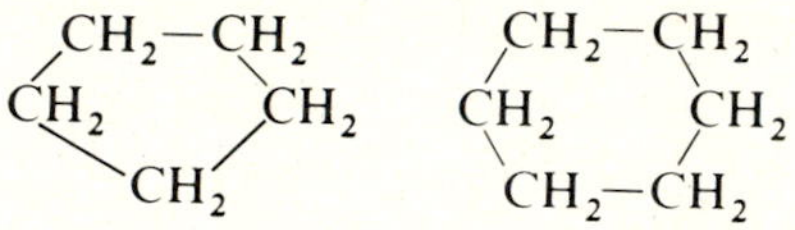

cyclopentane   cyclohexane

When we write the formula for a cyclic compound, we generally do not write out the atomic symbols. Instead, we simply draw a polygon. Thus the formula for cyclohexane becomes a hexagon, with each corner of the hexagon understood to represent a carbon atom with its hydrogens. The sides of the hexagon represent the covalent bonds between carbon atoms.

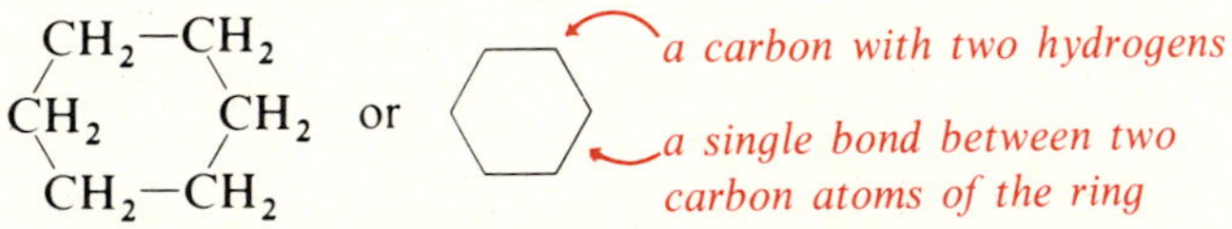

a carbon with two hydrogens

a single bond between two carbon atoms of the ring

This type of shorthand formula may also be used when there are other groups attached to a ring carbon. (Note that the number of hydrogens on a ring carbon with a substituent is decreased, so each carbon atom still has four bonds.)

formulas for cyclic compounds:

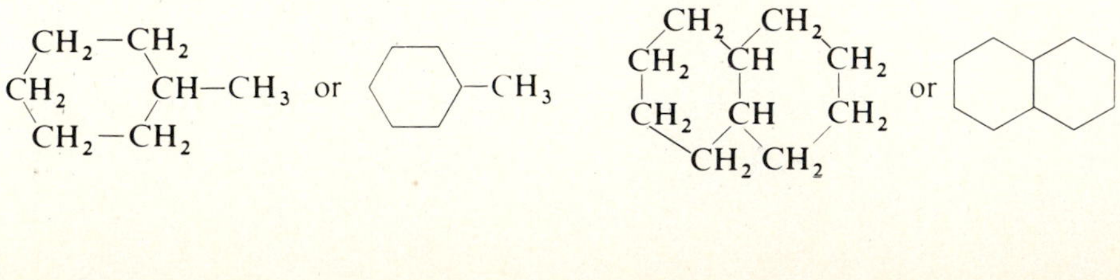

Rings containing four or more ring carbons are not actually flat, as the polygon formulas imply; they are puckered. For example, the puckering of a cyclohexane ring results in the bond angles being close to the optimal 109.5° for each carbon that has four single bonds.

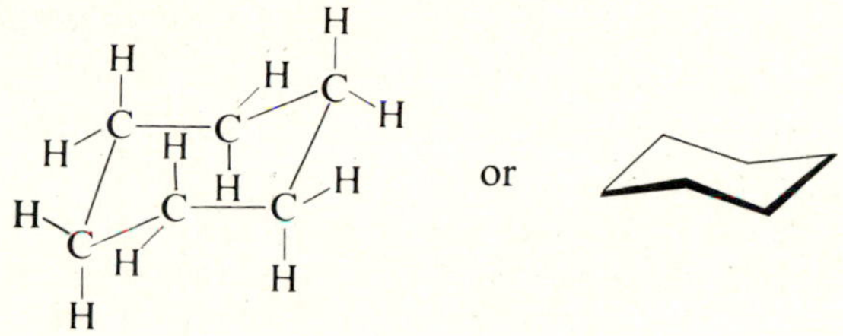

"chair form" of cyclohexane

Even though these rings exist in puckered conformations, we usually ignore the puckering when writing out the formulas for these compounds, just as we ignore the shape of a butane molecule when we write $CH_3CH_2CH_2CH_3$.

Problem 13.5.   Ignoring the actual shapes of cyclic molecules, rewrite the following formulas to show the C and H atoms:

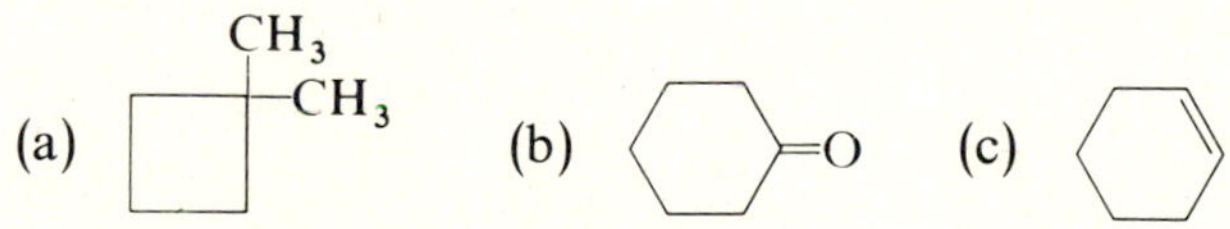

(a)   (b)   (c)

Problem 13.6.   Which of the following pairs of cyclic compounds represent pairs of structural isomers? Explain your answer.

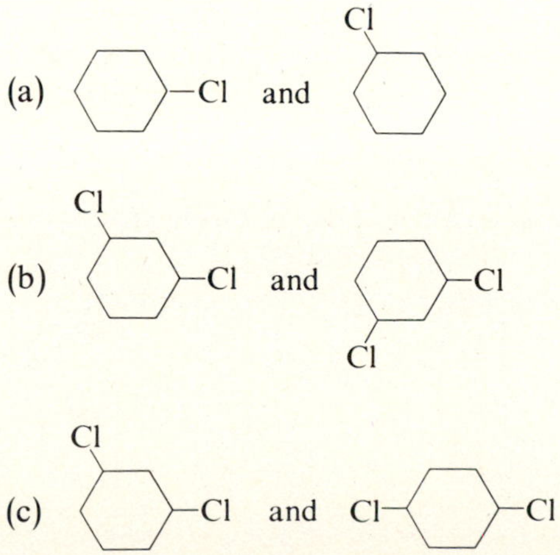

(a)

(b)

(c)

13.7   FUNCTIONAL GROUPS

Carbon-carbon and carbon-hydrogen single bonds are relatively nonpolar. For this reason, compounds containing only C-C and C-H single bonds are relatively nonreactive in chemical reactions other than combustion. On the other hand, a carbon-carbon double bond, a triple bond, or an electronegative atom such as O or N in an organic molecule is a site of reactivity in the molecule. A group of atoms containing a double bond, a triple bond, or an electronegative atom is called a **functional group**. Different compounds containing the same functional group undergo similar chemical reactions. For example, these compounds contain the $-OH$ functional group and behave similarly in chemical reactions:

$$CH_3OH \qquad CH_3CH_2CH_2OH \qquad \langle \rangle -OH$$

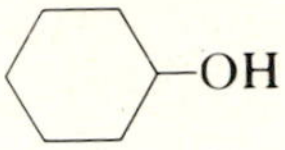

methanol   1-propanol   cyclohexanol

Because all compounds with an $-OH$ group, for example, behave similarly, we sometimes use R to represent the portion of a molecule that contains only carbon and hydrogen joined by single bonds. Using this technique, we can generalize the formula for an alcohol (a compound with an $-OH$ group) as ROH.

$R-$ means CH_3-, $\langle \rangle -$, or any other singly bonded carbon-hydrogen group. Therefore,

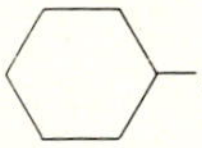

$$ROH \text{ can mean } CH_3-OH, \langle \rangle -OH, \text{ etc.}$$

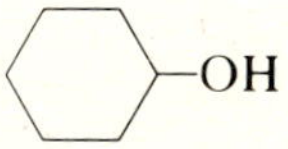

$$\overset{Cl}{\underset{|}{}}$$
$$RCl \text{ can mean } CH_3CH_2Cl, CH_3CHCH_3, \text{ etc.}$$

Table 13.2 lists some functional groups and examples of structures in which they are found. You will encounter other functional groups in later chapters.

Example   Write the general R formula for each of the following groups of compounds:

(a) CH_3CO_2H $CH_3CH_2CH_2CO_2H$ $\langle \rangle -CO_2H$

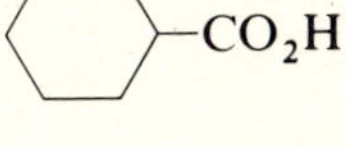

(b) $CH_3CH_2NH_2$ CH_3NH_2 $\langle \rangle_{NH_2}$

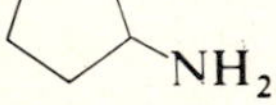

Solution   (a) RCO_2H, (b) RNH_2.

Table 13.2. *Some Common Functional Groups*

Name of group	Compound class	Structure	Specific example
double bond	alkene	$\diagdown C=C \diagup$	$CH_3CH=CH_2$
triple bond	alkyne	$-C\equiv C-$	$CH_3C\equiv CH$
hydroxyl	alcohol	$-OH$	$CH_3CH_2CH_2OH$
amino	amine	$-NH_2$	$CH_3CH_2CH_2NH_2$
carboxyl[a]	carboxylic acid	$-\overset{\overset{\displaystyle O}{\|\|}}{C}-OH$	$CH_3CH_2\overset{\overset{\displaystyle O}{\|\|}}{C}OH$

[a] The carboxyl group is also written $-COOH$ or $-CO_2H$. Therefore, the formula for the carboxylic acid shown can also be written CH_3CH_2COOH or $CH_3CH_2CO_2H$.

Problem 13.7. Write a general formula for each of the following groups of compounds:

(a) $CH_3C\equiv CCH_3$ $CH_3CH_2C\equiv CCH_3$ $CH_3CH_2C\equiv CCH_2CH_3$

(b) $CH_3\overset{\overset{\displaystyle O}{\|\|}}{C}CH_3$ $CH_3CH_2\overset{\overset{\displaystyle O}{\|\|}}{C}CH_3$ $CH_3CH_2\overset{\overset{\displaystyle O}{\|\|}}{C}CH_2CH_3$

Problem 13.8. Referring to Table 13.2, circle and name the functional groups in the following formulas:

(a) $CH_3CH_2CH_2\overset{\overset{\displaystyle O}{\|\|}}{C}OH$

(b) $CH_3\overset{\overset{\displaystyle NH_2}{\|}}{C}HCH_3$

(c)

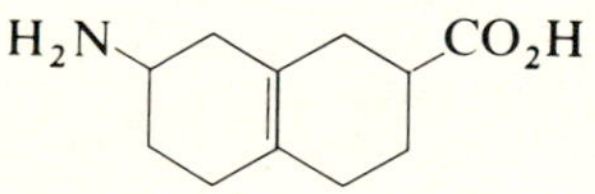

(d)

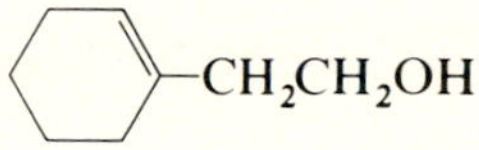

Problem 13.9. The following structural formulas represent compounds with functional groups different from those shown in Table 13.2. Circle each functional group.

(a) CH_3CH_2Cl

(b) $CH_3CH_2OCH_2CH_3$

(c) $CH_3\overset{\overset{\displaystyle CH_3}{\|}}{N}CH_3$

(d)

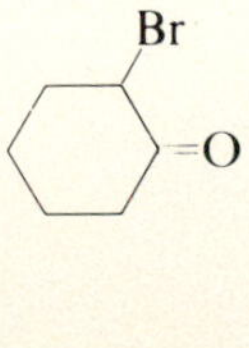

SUMMARY

Organic chemistry is the chemistry of the compounds of carbon. Organic compounds are almost all covalent compounds; they are usually relatively low melting and low boiling, water insoluble (but soluble in less polar solvents), and flammable.

Carbon can form single, double, or triple bonds, but always forms a total of four bonds. The usual formulas used for organic compounds are *structural formulas*.

structural formulas for cyclopropane, an anesthetic gas:

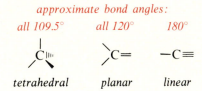

expanded   *condensed*

Structural isomers are two or more compounds with the same molecular formula but different *structures* (order of attachment of atoms): CH_3OCH_3 and CH_3CH_2OH both have the molecular formula C_2H_6O, and thus are examples of structural isomers.

The bond angles around a carbon atom depend on whether the carbon is involved in four single bonds, one double bond and two single bonds, or one triple bond and one single bond.

approximate bond angles:

all 109.5°   *all 120°*   *180°*

tetrahedral   *planar*   *linear*

Because groups can rotate around single bonds, most organic molecules can assume a variety of *conformations*, or shapes. *Cyclic compounds* may have puckered molecules, even though we usually represent them as flat polygons.

A *functional group*, such as a double bond, a triple bond, or a group containing an electronegative atom, is a site of reactivity in an organic molecule. The symbol R— can be used to represent a group of singly bonded C and H atoms, such as CH_3— or CH_3CH_2—.

KEY TERMS

single covalent bond   conformation   alcohol
double bond   cyclic compound   amine

triple bond   functional group   carboxylic acid
structural isomer   alkene
bond angle   alkyne

STUDY PROBLEMS

13.10. Predict the following properties of propane, $CH_3CH_2CH_3$:
(a) physical state at room temperature (solid, liquid, or gas)
(b) solubility in water
(c) solubility in hexane (Table 14.1, Section 14.1)
(d) combustibility
(e) ability to conduct an electric current

13.11. In stable compounds, carbon always forms how many bonds?

13.12. Write Lewis (dot) formulas for the following compounds:
(a) CH_3CH_3 (b) $CH_3CH=CH_2$
(c) CH_3OH

13.13. Determine whether each of the following is a molecular formula, an expanded structural formula, or a condensed structural formula, and rewrite it in the two forms not shown.

(a) $CH_3\underset{\underset{OH}{|}}{C}HCH_3$

(b) 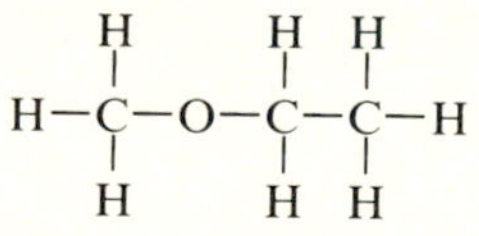
$$H-\underset{\underset{H}{|}}{\overset{\overset{H}{|}}{C}}-O-\underset{\underset{H}{|}}{\overset{\overset{H}{|}}{C}}-\underset{\underset{H}{|}}{\overset{\overset{H}{|}}{C}}-H$$

(c) $CH_3CH_2CH_2OH$ (d) C_3H_8O

13.14. Draw one structural isomer for each of the following compounds. (In each case there may be more than one possible isomer.)
(a) $CH_3CH_2CH_2OH$ (b) $CH_3CH=CHCH_3$

13.15. Which of the following pairs are structural isomers and which are different representations of the same compound?

(a) $HOCH_2\underset{\underset{OH}{|}}{C}H_2$ $CH_3\underset{\underset{OH}{|}}{\overset{\overset{OH}{|}}{C}}H$

(b) 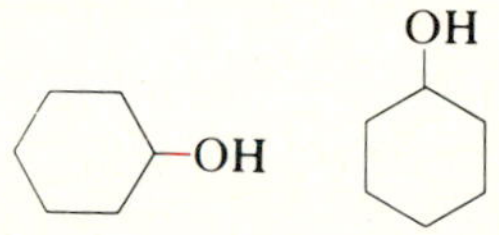 $-OH$

(c) CH_3CH_2OH $HOCH_2CH_3$

13.16. Rewrite the following formula for acetone and indicate (with numbers) the approximate bond angles:

$$CH_3\overset{\overset{O}{\|}}{C}CH_3$$

13.17. Draw the structures of the following compounds, showing each C and each H:

(a) 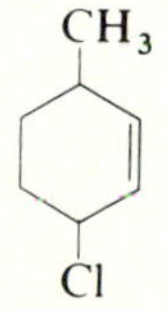   (b)

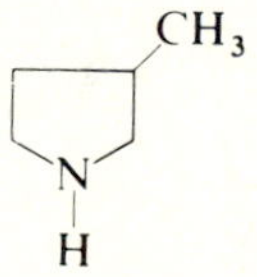

13.18. Draw the following structures as polygon formulas:

(a) 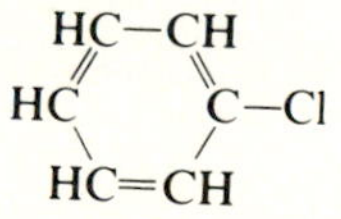 $HC\overset{\overset{HC=CH}{}}{\underset{\underset{HC=CH}{}}{}}C-Cl$ (b) 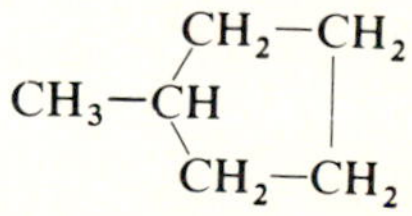 $CH_3-\overset{\overset{CH_2-CH_2}{|}}{\underset{\underset{CH_2-CH_2}{|}}{C}}H$

13.19. Write a formula for a specific example of each of the following general compound classes:

(a) $RCH=CHR$ (b) $R\overset{\overset{O}{\|}}{C}H$ (c) R_2NH

13.20. Referring to Table 13.2, circle and name each functional group, then name each compound class for the following compounds:

(a) $CH_3\underset{\underset{OH}{|}}{C}HCH_3$ (b) $CH_3\underset{\underset{OH}{|}}{C}HCO_2H$

14 Saturated Hydrocarbons

Occupied France in the early 1940s: A driver stokes a fuel converter with charcoal to power his car. (© SPADEM, Paris/ VAGA, New York, 1983.) The world's petroleum supplies are limited. Before they are depleted, scientists and engineers must develop alternative sources, such as coal and wood, for fuel and chemicals. During the German occupation of France in World War II, some French citizens responded to gasoline shortages by attaching fuel converters to their automobiles. These devices produced a gaseous fuel from the burning of wood or charcoal.

Objectives   Define and give examples of saturated and unsaturated hydrocarbons, alkanes, and cycloalkanes. □ Name the first ten continuous-chain alkanes and cycloalkanes, name the common alkyl groups, and name the branched alkanes. □ List some physical properties of alkanes and describe their uses. □ Write equations for the combustion and halogenation of alkanes, name the halogenated alkanes and describe their uses. □ Describe the processes of refining petroleum and increasing the octane number of gasoline (optional).

Hydrocarbons are compounds that contain carbon and hydrogen and no other elements. Hydrocarbons, in turn, can be classified as saturated or unsaturated. A *saturated* hydrocarbon is one that contains only single bonds and no double or triple bonds. Saturated hydrocarbons that have carbon atoms attached in chains are called *alkanes*, or sometimes *paraffins*. (Paraffin wax is a mixture of high-formula-weight alkanes.) If the carbon atoms of a saturated hydrocarbon are attached in one or more rings, the compound is called a *cycloalkane*.

typical saturated hydrocarbons:

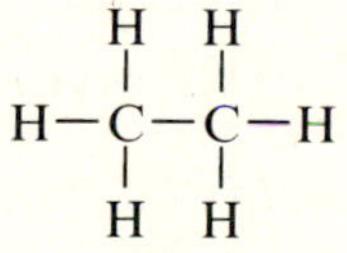

ethane
an alkane

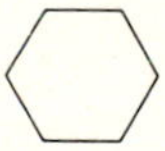

cyclohexane
a cycloalkane

Compounds that contain one or more double or triple bonds are said to be *unsaturated*. An unsaturated hydrocarbon has a functional group—that is, the double bond or the triple bond.

typical unsaturated hydrocarbons:

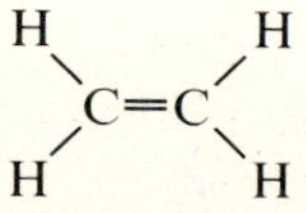

ethylene
an alkene

acetylene
an alkyne

The term *saturated* is derived from the idea that a compound is saturated with hydrogen; that is, a saturated hydrocarbon will not react with hydrogen. An unsaturated hydrocarbon can react with hydrogen to become saturated.

$$\underset{\text{a saturated hydrocarbon}}{H-\overset{\overset{H}{|}}{\underset{\underset{H}{|}}{C}}-\overset{\overset{H}{|}}{\underset{\underset{H}{|}}{C}}-H} \quad + H_2 \xrightarrow{\text{catalyst}} \text{no reaction}$$

$$\underset{\text{an unsaturated hydrocarbon}}{\overset{H}{\underset{H}{>}}C=C\overset{H}{\underset{H}{<}}} \quad + H_2 \xrightarrow{\text{catalyst}} \underset{\text{a saturated hydrocarbon}}{H-\overset{\overset{H}{|}}{\underset{\underset{H}{|}}{C}}-\overset{\overset{H}{|}}{\underset{\underset{H}{|}}{C}}-H}$$

In this chapter, we will discuss the alkanes and cycloalkanes. The unsaturated hydrocarbons will be discussed in Chapter 15. The alkanes are used as the basis for naming organic compounds; therefore, we will cover the nomenclature of these compounds in some detail. We will discuss the physical and chemical properties of alkanes and cycloalkanes, petroleum as a source of alkanes, and the uses of some alkanes. We will also show how organohalogen compounds can be obtained from alkanes, and we will mention some useful members of that class of compounds.

14.1   NAMING ALKANES AND CYCLOALKANES

In the nineteenth century, before chemists had learned the structures of many compounds, a new organic compound was named after its source, after some chemical or physical property, or even after a friend or relative. Such names are called **trivial names**, or **common names**.

$$\underset{\substack{\text{urea} \\ \textit{found in urine}}}{H_2N-\overset{\overset{O}{\|}}{C}-NH_2} \qquad \qquad \underset{\substack{\text{barbituric acid} \\ \textit{from Barbara}}}{\text{barbituric acid structure}}$$

In the chemical and medical sciences, trivial names are still used, as are more formal, or *systematic*, names. To introduce the basic principles of organic nomenclature, we will use systematic names. Then, throughout the remainder of the book, we will generally choose the most commonly used name, whether it is trivial or systematic.

Before 1900, chemists realized that some system of nomenclature should be devised to give the thousands of organic compounds names based on molecular structure. The nomenclature system that was developed is now known as the **IUPAC system**, after the International Union of Pure and Applied Chemistry, the group responsible for the continual upgrading and revising of the system. Although chemists still use many trivial names, the IUPAC system forms the foundation for almost all organic nomenclature.*

The IUPAC system is based on the premise that each organic compound has a basic skeletal structure, called the **parent**, to which hydrocarbon branches and functional groups have been added. In the following formula, the branch and functional groups are circled. The parent structure is in a box:

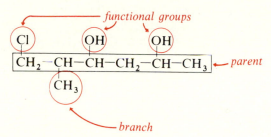

Continuous-Chain Alkanes and Cycloalkanes

The names of the continuous-chain alkanes are used as parent names for most simple organic compounds. The names for the first ten continuous-chain alkanes are given in Table 14.1.

Table 14.1. *Names and Structures of the First Ten Continuous-Chain Alkanes*

Name	Number of carbon atoms in chain	Structure
methane	1	CH_4
ethane	2	CH_3CH_3
propane	3	$CH_3CH_2CH_3$
butane	4	$CH_3CH_2CH_2CH_3$ or $CH_3(CH_2)_2CH_3$
pentane	5	$CH_3CH_2CH_2CH_2CH_3$ or $CH_3(CH_2)_3CH_3$
hexane	6	$CH_3(CH_2)_4CH_3$
heptane	7	$CH_3(CH_2)_5CH_3$
octane	8	$CH_3(CH_2)_6CH_3$
nonane	9	$CH_3(CH_2)_7CH_3$
decane	10	$CH_3(CH_2)_8CH_3$

* The *generic name* of a drug or other compound used in the medical sciences is not necessarily the same as either the IUPAC name or the trivial name used in organic chemistry. The generic name is a type of trivial name not legally owned by any individual or corporation. A *proprietary name*, on the other hand, is a legally owned name; only the owner of the proprietary name can market a compound under this name.

The names of the continuous-chain alkanes are composed of two parts. The first part of the name tells us the number of carbon atoms in the chain; the second part tells us that the compound is saturated. For example, the name *pentane* starts with *pent-*, indicating the presence of five carbon atoms, and ends with *-ane*, indicating that the compound is a saturated hydrocarbon, or an alk*ane*. (The names of unsaturated hydrocarbons have different endings.) Sometimes the prefix *n-* for *normal* is used to emphasize that an alkane has a continuous chain and no branches. Pentane can thus also be called *n-*pentane.

The names of the *cycloalkanes* are taken directly from the names of the continuous-chain parents, with the prefix *cyclo-* added. The number of carbon atoms in the ring determines the parent name. Cyclohexane is a cycloalkane with six carbons in a ring. Cyclooctane is a cycloalkane with eight carbons in a ring.

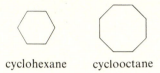

cyclohexane   cyclooctane

Problem 14.1.   Name the following cycloalkanes:

(a) △   (b) ☐   (c) ⬠

Branched Alkanes

The parent of a branched-chain alkane is the *longest continuous chain*. The branches are designated by prefixes to the parent name. When naming a branched-chain alkane, *first find and name the longest continuous chain of carbon atoms, and then number that chain starting at the end closer to the branch*.

Let us illustrate these rules with a specific alkane:

$$CH_3CH_2CH_2\underset{\overset{|}{CH_3}}{C}HCH_3$$

This structure has a total of six carbons, but only five of the carbons atoms are in one continuous chain. The structure is named as a pentane.

Now the five-carbon chain must be numbered. We start at the end closer to the branch:

$$\overset{5}{C}-\overset{4}{C}-\overset{3}{C}-\overset{2}{\underset{\overset{|}{C}}{C}}-\overset{1}{C} \quad not \quad \overset{1}{C}-\overset{2}{C}-\overset{3}{C}-\overset{4}{\underset{\overset{|}{C}}{C}}-\overset{5}{C}$$

The one-carbon branch at carbon number 2 is called a methyl group (*meth-*, indicating one carbon; *-yl*, indicating that it is attached to something

Table 14.2. *Names of Continuous-Chain Alkyl Groups*

Name	Structure
methyl	CH_3-
ethyl	CH_3CH_2-
propyl, or *n*-propyl	$CH_3CH_2CH_2-$
butyl, or *n*-butyl	$CH_3(CH_2)_2CH_2-$
pentyl, or *n*-pentyl	$CH_3(CH_2)_3CH_2-$
hexyl, or *n*-hexyl	$CH_3(CH_2)_4CH_2-$

else). The complete name of this compound is 2-methylpentane. The 2 designates the position of the branch; *methyl* tells what group is at that position; and *pentane* is the parent alkane. The position number is separated by a hyphen from the rest of the name; methylpentane is one word. If there is only one possibility for the position of attachment, the prefix number is usually omitted.

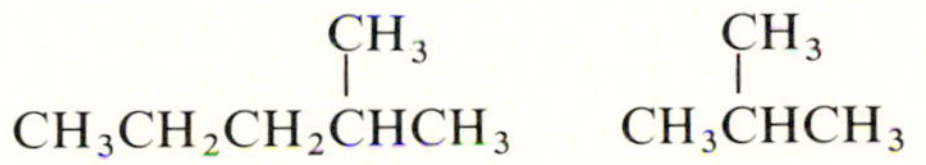

$$CH_3CH_2CH_2\overset{\overset{\displaystyle CH_3}{|}}{C}HCH_3 \qquad CH_3\overset{\overset{\displaystyle CH_3}{|}}{C}HCH_3$$

2-methylpentane   methylpropane

Alkyl Groups. The names of the commonly encountered continuous-chain branches are listed in Table 14.2. In each case, the name of the group is taken directly from the name of the continuous-chain parent alkane, with the *-ane* ending of the alkane replaced by *-yl*. These groups are referred to as **alkyl groups**. As we mentioned in Chapter 13, an alkyl group can be represented as R when we want to generalize a formula.

Alkyl substituents may have branching within themselves. A few branched alkyl groups are so common that they have specific names. The commonly used prefixes for these branched alkyl groups are *iso-*, *secondary-* (abbreviated. *sec-*), and *tertiary-* (abbreviated *t-* or *tert-*). (See Table 14.3.)

Problem 14.2. Write condensed structural formulas for the following compounds:
(a) 4-*n*-propyloctane   (b) 4-isopropyloctane
(c) *t*-butylcyclobutane

Compounds with More Than One Substituent. When a compound contains two or more *different* substituents, their prefixes are listed alphabetically. For example:

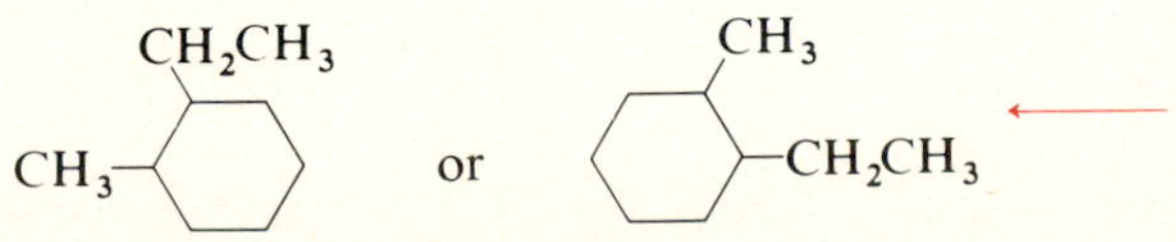

Start numbering at the first substituent, regardless of how the structure is drawn.

1-ethyl-2-methylcyclohexane

Table 14.3.   *Names of Common Branched Alkyl Groups*

Name	Structure	Example

isopropyl

$$CH_3CH-$$ with CH_3 above

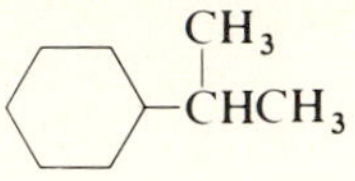

isopropylcyclohexane

isobutyl

$$CH_3CHCH_2-$$ with CH_3 above

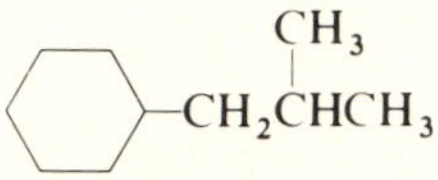

isobutylcyclohexane

sec-butyl

$$CH_3CH_2CH-$$ with CH_3 above

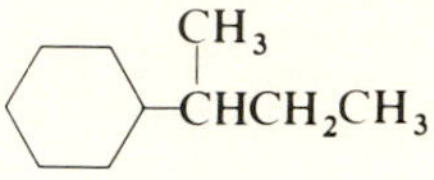

sec-butylcyclohexane

t-butyl
   or *tert*-butyl

$$CH_3-C- $$ with CH_3 above and CH_3 below

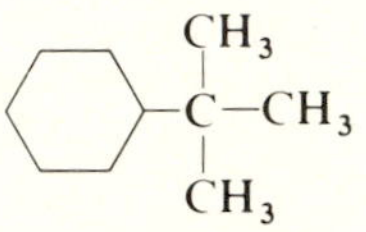

t-butylcyclohexane

When a compound contains two or more of the *same* substituent, the prefix is preceded by *di-*, *tri-*, and so forth. This additional prefix tells us the number of times that a substituent is found in the molecule. For example, 2,3-dimethylbutane is a four-carbon continuous chain (*butane*) that has two (*di*) methyl groups located at carbons 2 and 3, respectively. The prefixes indicating number are located in Table 4.4, Section 4.11. In alphabetizing prefix names, we ignore the portion of the prefix that designates number. For example, dimethyl is alphabetized as methyl, and triethyl as ethyl.

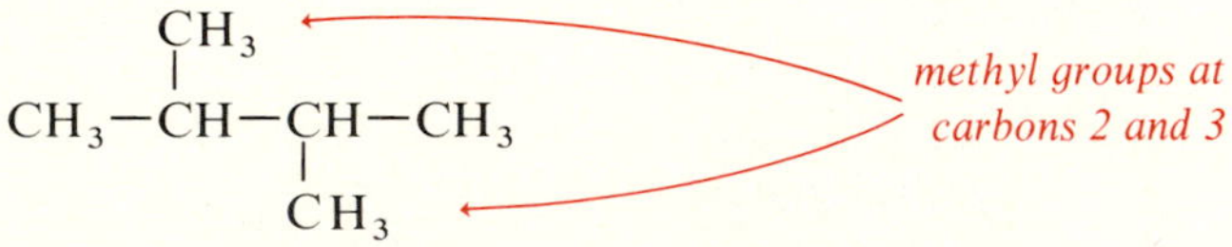

$$CH_3-CH-CH-CH_3$$ with CH_3 above the second carbon and CH_3 below the third carbon

methyl groups at carbons 2 and 3

2,3-dimethylbutane

In summary, when naming a complex branched alkane, proceed by the following steps:

1. Find the longest continuous chain, which may or may not be written in a straight line.
2. Number this chain, starting at the end that will give the smaller prefix numbers.

3. Name the parent.
4. Add the prefixes alphabetically, grouping like prefixes together.

Example   Write the IUPAC name for the following compound:

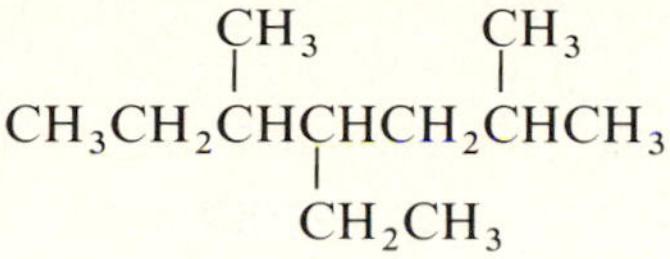

$$\underset{\displaystyle \overset{|}{CH_2CH_3}}{CH_3CH_2CHCHCH_2CHCH_3}$$

with CH_3 branches above.

Solution   (1) Find the longest continuous chain.
(2) Number this chain from the end that will give the lower prefix numbers.

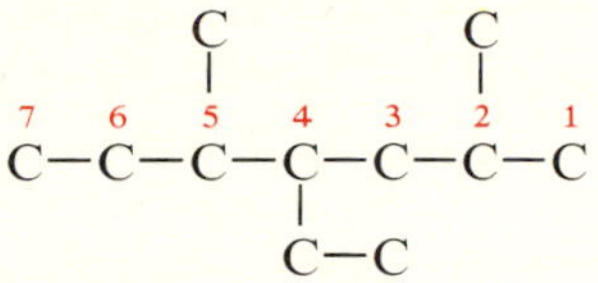

(3) Name the parent: heptane.
(4) Add the branches (ethyl before methyl), grouping like branches together:

4-ethyl-2,5-dimethylheptane

Example   Write the structure for 1-isopropyl-3-methylcyclopentane.

Solution   (1) Write the structure of the parent (cyclopentane) and number the carbon atoms. (In this case, it does not matter where we start numbering or whether we number clockwise or counterclockwise.)

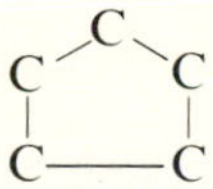

(2) Insert the branches (1-isopropyl and 3-methyl).

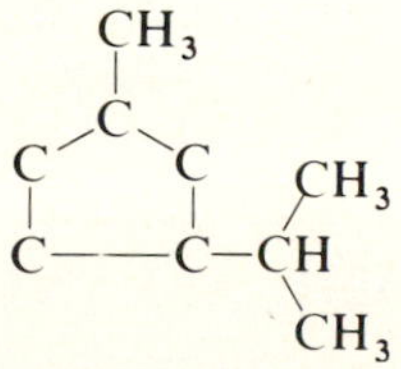

(3) Insert the hydrogens so that each carbon has four bonds.

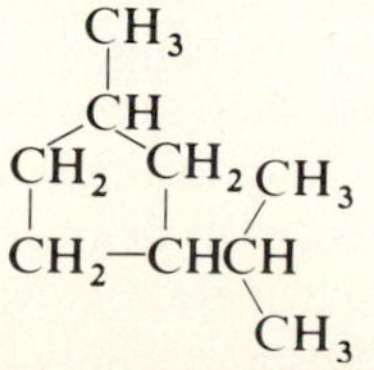

or

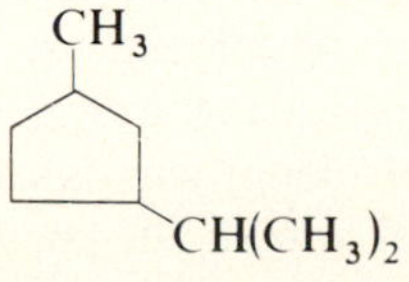

Problem 14.3.   Name the following structures:

(a) 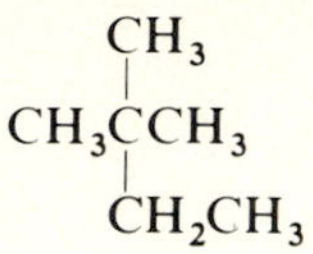
$$CH_3CCH_3$$
with CH_3 above and CH_2CH_3 below the central carbon

(b) 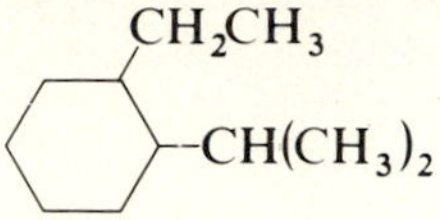 a cyclohexane ring with CH_2CH_3 and $-CH(CH_3)_2$ substituents

Problem 14.4.   Write formulas for the following compounds:
(a) 1-*n*-butyl-2-*n*-propylcyclopentane
(b) 2,2,3,3-tetramethylhexane

14.2   PHYSICAL PROPERTIES OF THE ALKANES

Because alkanes and cycloalkanes are relatively nonpolar, the attractions between molecules are quite weak. The alkanes containing 1–4 carbons are gases at room temperature, while the alkanes with 5–17 carbons are liquids. (Table 14.4 lists the boiling points of a few alkanes.) Alkanes that contain more than 17 carbons are solids. It is generally true that as formula weights in a series of covalent compounds increase, melting points and boiling points also increase.

Table 14.4.   *Boiling Points of Some Alkanes*

Formula	Bp (°C)	Formula	Bp (°C)
CH_4	-162	$CH_3(CH_2)_2CH_3$	-0.5
CH_3CH_3	-88.5	$CH_3(CH_2)_3CH_3$	36
$CH_3CH_2CH_3$	-42	$CH_3(CH_2)_4CH_3$	69

Because the saturated hydrocarbons are nonpolar, they are all insoluble in water (which is very polar), but are soluble in nonpolar solvents such as other hydrocarbons. All the alkanes are less dense than water. For example, gasoline and motor oil (primarily mixtures of alkanes) float on water.

14.3   REACTIONS OF THE ALKANES

Alkanes do not contain a functional group. For the most part, alkanes are chemically unreactive and relatively nontoxic. They do not react with acids, bases, or other common laboratory reagents. Because alkanes are immiscible with water and are nonreactive, they are widely used as protective coatings. Lubricating oils, for example, help protect automobile parts from rusting. Petroleum jelly is used to protect skin from moisture. Paraffin wax is used to seal homemade jams and jellies. Mineral oil, which can coat the intestinal wall, is used as a laxative. All these substances are mixtures of alkanes.

Alkanes react with the halogens (which are strong oxidizing agents) in the presence of ultraviolet (uv) light. They will also burn if they are ignited. The following two equations show the halogenation and the combustion of methane.

$$CH_4 + Cl_2 \xrightarrow{\text{uv light}} CH_3Cl + CH_2Cl_2 + CHCl_3 + CCl_4 + HCl \text{ (not balanced)}$$

$$CH_4 + 2\,O_2 \xrightarrow{\text{spark}} CO_2 + 2\,H_2O + energy$$

Halogenation

A reaction in which a halogen atom such as a chlorine atom (Cl) displaces a hydrogen atom is called a **halogenation reaction**. The products of the reaction of a hydrocarbon with a halogen such as chlorine gas are called *halogenated hydrocarbons*, or *organohalogen compounds*.

a halogenation reaction:

$$CH_3CH_3 + Cl_2 \xrightarrow{\text{uv light}} CH_3CH_2Cl + HCl + \text{other halogenated products}$$

ethane $\qquad\qquad\qquad$ *a halogenated hydrocarbon*

The naming of halogenated alkanes is similar to that of alkanes. In the IUPAC system, the halogen atom is named as a prefix: *fluoro-* (F), *chloro-* (Cl), *bromo-* (Br), or *iodo-* (I).

$$CH_3CH_2Cl$$

chloroethane
(*trivial name:* ethyl chloride)
a topical anesthetic

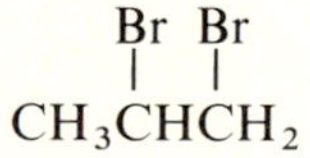

$$\overset{\displaystyle Br\ Br}{\underset{\displaystyle |\ \ |}{CH_3CHCH_2}}$$

1,2-dibromopropane

Organohalogen compounds find use in industry as solvents. Because these compounds contain a functional group, the electronegative halogen atom, they are also useful as reagents for preparing other organic compounds. Many synthetic organohalogen compounds are toxic and cause liver damage. Some of these compounds are suspected carcinogens (cancer-causing agents). People who work with common halogen-containing solvents such as chloroform or carbon tetrachloride should take precautions to avoid excessive inhalation or contact.

three common solvents:

$$CHCl_3 \qquad\qquad CCl_4$$

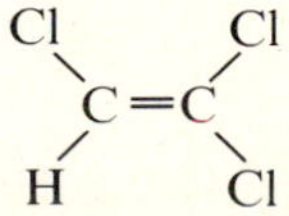

IUPAC name: trichloromethane	tetrachloromethane	trichloroethene
trivial name: chloroform	carbon tetrachloride	trichloroethylene

At one time chloroform was used as an anesthetic because it is fast-acting and nonflammable; however, its toxicity precludes its use as an anesthetic today. Halothane (Section 6.2) is a nontoxic organohalogen compound now used as an anesthetic. Some organohalogen compounds, such as tetra-chloroethylene ($Cl_2C\!=\!CCl_2$), find medical and veterinary use in the treatment of intestinal worms. Ethyl chloride (CH_3CH_2Cl) is used as an external topical anesthetic. Many insecticides (DDT, dieldrin, aldrin, etc.) are organohalogen compounds. Because organohalogen compounds are not readily broken down by the action of air, water, or microorganisms, these insecticides are long acting, or "persistent."

Naturally occurring organohalogen compounds are rather rare. One example is thyroxine, a hormone of the thyroid gland.

$$HO-\overset{\displaystyle I}{\underset{\displaystyle I}{\bigcirc}}-O-\overset{\displaystyle I}{\underset{\displaystyle I}{\bigcirc}}-CH_2\overset{\displaystyle }{\underset{\displaystyle NH_2}{CH}}CO_2H$$

thyroxine

Problem 14.5. Complete the following equation leading to chloromethane, showing the products as Lewis (dot) formulas.

$$H\!:\!\overset{\displaystyle H}{\underset{\displaystyle H}{\overset{\displaystyle ..}{C}}}\!:\!H \quad + \quad :\!\overset{..}{\underset{..}{Cl}}\!:\!\overset{..}{\underset{..}{Cl}}\!: \quad \xrightarrow{\text{uv light}}$$

methane   chlorine

Problem 14.6. First complete the following equations, showing the mono-chlorinated (one Cl substituted) products. Then name the products.

(a) $H_3C-\overset{\displaystyle CH_3}{\underset{\displaystyle CH_3}{\overset{\displaystyle |}{\underset{\displaystyle |}{C}}}}-CH_3 + Cl_2 \xrightarrow{\text{uv light}}$

(b) $\bigcirc + Cl_2 \xrightarrow{\text{uv light}}$

Problem 14.7. Any or all hydrogen atoms in an alkane can be displaced by chlorine atoms. (a) List all the organic products that can be obtained by the chlorination of ethane, CH_3CH_3. (b) Name these products.

Combustion

The combustion of organic compounds was discussed in Section 11.4. The combustion of hydrocarbons is of particular interest because most of the energy of modern society comes from the combustion of natural gas (primarily methane), liquefied petroleum gas (primarily propane), gasoline, and furnace oil. Equations for the combustion of methane and octane are shown in Section 13.1.

Problem 14.8.   (a) Write an equation for the complete combustion of propane ($CH_3CH_2CH_3$). (b) What product(s) would you expect from burning propane in an inadequate supply of oxygen? (*Hint:* See Section 11.4.)

■ **14.4   PETROLEUM (optional)**

Our modern society runs on energy, and we obtain much of this energy from petroleum fuels. Petroleum is also the raw material for over 90% of the organic chemicals produced in the United States. These chemicals include the compounds used to make plastics, drugs, synthetic fibers, and detergents.

Petroleum has been formed by the anaerobic decay (decay without air) of organic materials originally buried in sediment but now found in underground deposits. Crude oil, or petroleum, is a complex mixture of organic compounds: straight-chain and branched-chain hydrocarbons; saturated and unsaturated hydrocarbons; cyclic hydrocarbons; and sulfur-, nitrogen-, and oxygen-containing organic compounds.

The conversion of crude oil to usable products is called **refining**. The first step in refining crude oil is *straight-run distillation*: distillation of the crude oil into fractions of different boiling ranges. The principal fractions obtained from this distillation are gases, gasoline, kerosene, gas oil, and residue. The residue is distilled under reduced pressure to yield lubricating oils and petroleum waxes. The residue from the vacuum distillation is used primarily as asphalt.

Gasoline is a mixture of hydrocarbons boiling in the 40–200°C range. Because of the great demand for this product, some higher-boiling petroleum compounds are subjected to *cracking*: heating in the presence of a catalyst to break the molecules into smaller, more useful molecules.

a cracking reaction:

$$CH_3(CH_2)_8CH_3 \xrightarrow[\text{catalyst}]{\text{heat, pressure}} CH_3(CH_2)_6CH_3 + CH_2{=}CH_2$$

decane   octane   ethylene

can be used in   *can be used to*
gasoline   *synthesize*
other compounds

The quality of a gasoline is rated by an **octane number**, which is based on an empirical scale. Heptane, a continuous-chain alkane, has very poor burning characteristics in gasoline engines and has been assigned an octane number of 0. At one time, the branched-chain alkane 2,2,4-trimethylpentane (trivial name: isooctane) was the best antiknock fuel known. This compound was assigned an octane number of 100. A gasoline that has an octane number of 90 has the same burning characteristics as a mixture of 90% isooctane and 10% heptane.

$$CH_3(CH_2)_5CH_3$$

heptane
octane number 0

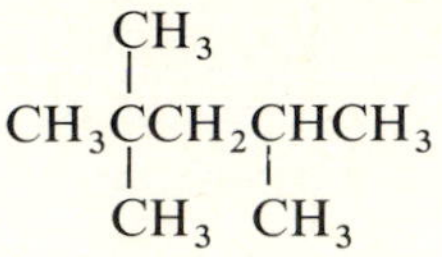

$$CH_3\overset{\overset{\displaystyle CH_3}{|}}{\underset{\underset{\displaystyle CH_3}{|}}{C}}CH_2\overset{}{\underset{\underset{\displaystyle CH_3}{|}}{C}}HCH_3$$

2,2,4-trimethylpentane (isooctane)
octane number 100

Straight-run gasolines have octane numbers of about 25 to 80. Octane numbers are increased to the desired values (around 90) by the addition of *cracking products*, which contain a greater percentage of branched alkanes than does straight-run gasoline, and *additives*, such as benzene, ethanol, and tetraethyllead.

three gasoline additives:

benzene
octane number 106

$$CH_3CH_2OH$$

ethanol
octane number 106
(used in gasohol)

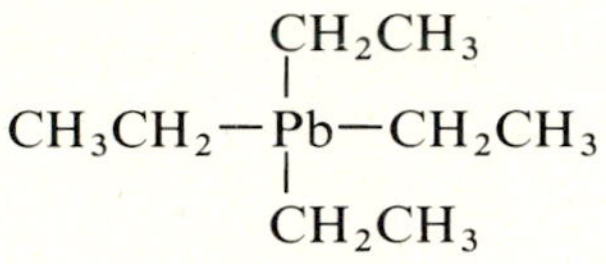

$$CH_3CH_2-\overset{\overset{\displaystyle CH_2CH_3}{|}}{\underset{\underset{\displaystyle CH_2CH_3}{|}}{Pb}}-CH_2CH_3$$

tetraethyllead

SUMMARY

Hydrocarbons, compounds containing only C and H, may be *saturated* (contain only single bonds) or *unsaturated* (contain one or more double or triple bonds). Saturated hydrocarbons are called *alkanes* or *cycloalkanes*.

The names of the first ten continuous-chain alkanes are listed in Table 14.1. Cycloalkane names are taken from these names, with the prefix *cyclo-* added. The steps in naming a branched alkane are listed in Section 14.1.

Alkanes are low melting, low boiling, water insoluble, and less dense than water. Alkanes can undergo *halogenation* in the presence of a halogen and ultraviolet light and can undergo *combustion*. Otherwise, alkanes are quite inert.

Petroleum, primarily a mixture of alkanes, is *refined* by *straight-run distillation* and other procedures, such as *cracking*, to yield gasoline and other fuels, as well as raw materials for the chemical industry.

KEY TERMS

hydrocarbon
saturated compound
unsaturated compound
alkane
cycloalkane

IUPAC nomenclature system
parent name
alkyl group
halogenation reaction
*petroleum

*refining
*cracking reaction
*octane number

STUDY PROBLEMS

14.9. Which of the following compounds is a saturated hydrocarbon? Which is unsaturated?

(a) 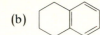 (b)

14.10. Circle the functional groups and branches, then draw boxes around the parent skeletons:

(a)
$$CH_3{-}\overset{\displaystyle CH_3}{\underset{\displaystyle CH_3}{C}}{-}OH$$

(b) $CH_3{-}$⬡$\overset{O}{-}COH$

14.11. From memory, write the structures and names of the first ten continuous-chain alkanes.

14.12. Name the following cycloalkanes:

(a) (b) ⬡

14.13. For each of the following carbon skeletons, give the number of carbon atoms in the longest continuous chain:

(a)
```
    C—C
    |   |
C—C   C—C
    |   |
    C—C C—C
```
(b)
```
C—C—C C
    |   |
C—C—C—C
    |
    C—C–C
```

14.14. Name the following compounds:

(a)
$$CH_3\overset{\displaystyle CH_3}{\underset{\displaystyle CH_3}{C}}{-}\underset{\displaystyle CH_3}{C}HCH_2CH_3$$

(b) $CH_3{-}$⬡${-}\underset{\displaystyle CH_3}{C}HCH_3$

(c)
$$CH_3\underset{\displaystyle \underset{\displaystyle CH_3}{CHCH_3}}{C}HCH_2CH_2CH_3$$

14.15. Draw a cyclopentane ring with each of the following substituents:
(a) isopropyl   (b) isobutyl
(c) *t*-butyl   (d) *sec*-butyl

14.16. Which of the following phrases describe alkanes?
(a) heavier than water
(b) unreactive toward most other chemicals
(c) insoluble in water
(d) soluble in hydrocarbon solvents

14.17. What is the physical state of each of the following compounds at room temperature and normal pressures?
(a) ethane   (b) butane   (c) hexane

14.18. Complete and balance the following equations. If no reaction occurs, write "no reaction."

(a) $CH_4 + H_2SO_4 \longrightarrow$

(b) ⬡ $+ H_2 \xrightarrow{\text{catalyst}}$

(c) $CHCl_3 + Cl_2 \xrightarrow{\text{uv light}}$

(d) $CH_3CH_2CH_2CH_3 + O_2 \xrightarrow{\text{spark}}$

(e) ⬡ $+ H_2 \xrightarrow{\text{catalyst}}$

14.19. Name the following compounds:

(a) $CH_3CH_2CH_2I$ (b) $\overset{\displaystyle Cl}{\underset{}{C}}H_2\overset{\displaystyle Cl}{\underset{}{C}}HCH_2Cl$

14.20.   Write formulas for the following compounds:

(a) 1,1-difluorocyclopentane

(b) 1-bromo-2,2-dichloroethane

*14.21.   In Section 14.4, we show an equation for a cracking reaction of decane. Write an equation for the cracking reaction of decane that leads to $CH_3CH{=}CH_2$ as one product.

*14.22.   A gasoline has the same burning characteristics as a mixture of 65% isooctane and 35% heptane. What is the octane number of the gasoline?

*14.23.   Unleaded gasoline is more expensive to produce than leaded gasoline. Suggest a reason for this.

15 Unsaturated Hydrocarbons

Girl with Silly Putty. (From Chemical Week. *Photo by J. Bruce Haag.) Silly Putty® is an example of a synthetic polymer, a manufactured substance composed of extremely long-chain molecules. Silly Putty is amusing because it can be slowly drawn out like taffy, yet can be shattered when struck. This particular polymer is an organosilicon compound called a silicone:*

$$\{-O-\underset{\underset{R}{|}}{\overset{\overset{R}{|}}{Si}}-O-\underset{\underset{R}{|}}{\overset{\overset{R}{|}}{Si}}-O-\underset{\underset{R}{|}}{\overset{\overset{R}{|}}{Si}}-O-\underset{\underset{R}{|}}{\overset{\overset{R}{|}}{Si}}-O-\}$$

Silicones are generally quite inert and find use as greases, as protective coatings, and even as implants in human tissue. In Chapter 15, we will introduce some common synthetic polymers made from unsaturated organic compounds.

Objectives   Define and give examples of alkenes and alkynes. ☐ Explain why carbon-carbon double bonds and triple bonds are considered functional groups. ☐ Name alkenes and alkynes. ☐ Define *cis,trans* isomers. ☐ Describe the role of *cis,trans* isomerism in vision (optional). ☐ Write equations for hydrogenation, hydration, and polymerization of alkenes. ☐ List some synthetic polymers and their uses. ☐ Draw two types of polygon formulas for benzene, and explain why benzene does not contain three double bonds. ☐ Define aromatic compound and heterocyclic compound; list some examples. ☐ Name substituted benzenes. ☐ Write an equation for an aromatic substitution reaction.

An **alkene** (sometimes called an *olefin*) is a hydrocarbon containing a carbon-carbon double bond, while an **alkyne** (sometimes called an *acetylene*) contains a carbon-carbon triple bond. Recall from Chapter 14 that these compounds are called *unsaturated compounds* because they are not saturated with hydrogen.

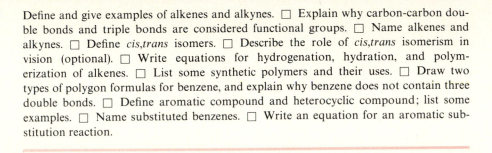

an alkene   *an alkyne*

In this chapter, we will discuss the naming and the physical properties of alkenes and alkynes. We will also introduce a second type of isomerism, called *cis,trans* isomerism. Then we will discuss some reactions of alkenes, including the formation of *polymers* such as polyethylene. Finally, we will briefly discuss another class of unsaturated compounds called *aromatic compounds*.

15.1 ALKENES AND ALKYNES

Unsaturation in an organic molecule is a site of reactivity in the molecule; therefore, a double bond or a triple bond is a functional group. Compounds containing carbon-carbon double bonds are found so universally in nature and are used so widely in industrial processes that it would be impossible to list all the useful and interesting compounds that contain this functional group. Throughout this chapter and the remainder of the book you will find examples of compounds with double bonds found in plants and animals—for example, vegetable oils and steroids (Chapter 19). The sex attractant of the housefly is a biologically active hydrocarbon containing a double bond as the only functional group.

$$CH_3(CH_2)_{12}CH{=}CH(CH_2)_7CH_3$$

sex attractant of the housefly

Figure 15.1.   *The second bond of a double bond and the second and third bonds of a triple bond arise from side-to-side overlap of p orbitals, so that the electrons are farther from the nuclei than are the electrons in the first bond.*

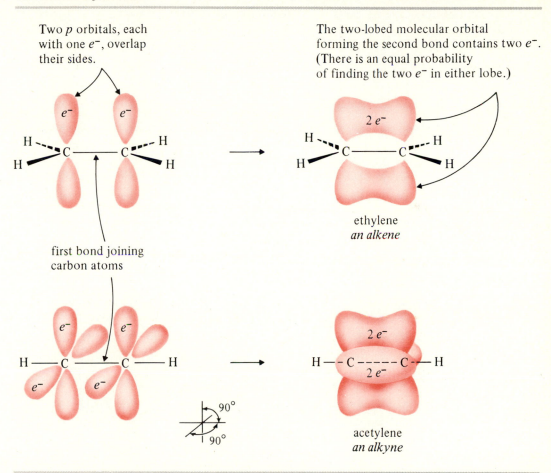

Two *p* orbitals, each with one *e⁻*, overlap their sides.

first bond joining carbon atoms

The two-lobed molecular orbital forming the second bond contains two *e⁻*. (There is an equal probability of finding the two *e⁻* in either lobe.)

ethylene
an alkene

acetylene
an alkyne

Structure of Alkenes and Alkynes

Recall that the carbon-carbon double bond in an alkene results from the sharing of *two* pairs of electrons and that the atoms involved in a double bond lie in a plane with bond angles of approximately 120°. An alkyne's triple bond is formed by the sharing of *three* pairs of electrons, and the atoms involved lie in a straight line. The bond angles of these classes of compounds are shown in Section 13.4.

A double or triple bond is a site of reactivity in an organic molecule because the second and third bonds are not the same as the first bond. The first bond of a double or triple bond is similar to the C—C bond in an alkane and is not very reactive. The electrons in the second and third bonds are farther from the carbon nuclei than are the electrons in the first bond. They are more loosely held and are relatively more exposed. The electrons in the second and third bonds are more easily attacked by incoming reagents. For example, strong acids react with alkenes because the H^+ of the acid is attracted to the negative charge of the exposed electrons. Figure 15.1 shows double and triple carbon-carbon bonds with orbitals.

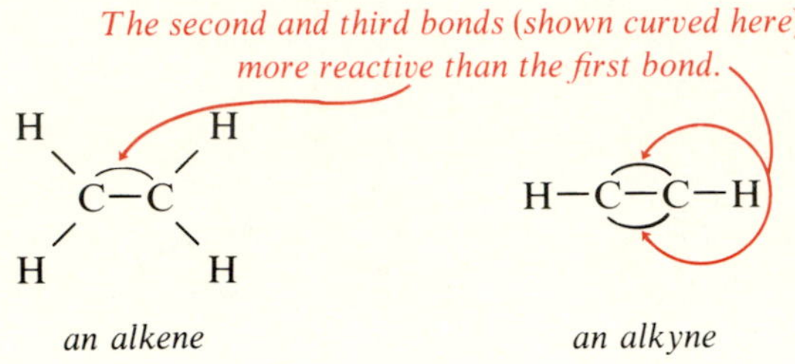

The second and third bonds (shown curved here) are more reactive than the first bond.

an alkene   *an alkyne*

Physical Properties of Alkenes and Alkynes

Like alkanes, alkenes and alkynes are composed of relatively nonpolar molecules. The physical properties (but not the chemical properties) of alkenes and alkynes are quite similar to those of their alkane counterparts. Alkenes and alkynes are relatively low melting and low boiling (depending on their formula weights) and are insoluble in water. Like alkanes, these compounds are less dense than water. Table 15.1 lists the boiling points of a few important alkenes and alkynes. Note that these low-formula-weight alkenes and alkynes are all gases at room temperature. Also note that the boiling points increase with increasing formula weight, just as they do for alkanes.

Table 15.1.   *Names and Boiling Points of Some Alkenes and Alkynes*

Name[a]	Formula	Bp (°C)
alkenes:		
ethene	$CH_2{=}CH_2$	-102
(ethylene)		
propene	$CH_3CH{=}CH_2$	-48
(propylene)		
1-butene	$CH_3CH_2CH{=}CH_2$	-7
(1-butylene)		
alkynes:		
ethyne	$CH{\equiv}CH$	-84
(acetylene)		
propyne	$CH_3C{\equiv}CH$	-23
(methylacetylene)		

[a] Trivial names are in parentheses.

15.2 NAMING ALKENES AND ALKYNES

The IUPAC nomenclature of the alkenes and alkynes is a direct extension of the nomenclature of the alkanes. In an alkene, the ending *-ane* of the alkane name is changed to *-ene*. Ethene is the IUPAC name for the two-carbon compound containing a double bond. (The trivial name for ethene is ethylene.) An alkyne is named by changing the *-ane* ending of the alkane name to *-yne*. Ethyne is the IUPAC name for the two-carbon compound with a triple bond. (Its trivial name is acetylene.)

CH_3CH_3	$CH_2{=}CH_2$	$CH{\equiv}CH$
ethane	ethene	ethyne
an alkane	*an alkene*	*an alkyne*
$CH_3CH_2CH_3$	$CH_3CH{=}CH_2$	$CH_3C{\equiv}CH$
propane	propene	propyne
an alkane	*an alkene*	*an alkyne*

In compounds with carbon chains longer than three carbon atoms, there is more than one possibility for the position of the double or triple bond. For example, in a four-carbon chain, a double bond could be located between carbons 1 and 2 or between carbons 2 and 3. Because these two structures contain atoms with different orders of attachment, they are structural isomers.

two structural isomers of a four-carbon alkene:

$CH_3CH_2CH{=}CH_2$	$CH_3CH{=}CHCH_3$
1-butene	2-butene
double bond between carbons 1 and 2	*double bond between carbons 2 and 3*

In order to distinguish between the possible structural isomers, a number preceding the name is used to tell where the double or triple bond is located in the parent skeleton. The following rules are followed in naming such an alkene or alkyne:

1. Find the longest chain *containing the double or triple bond*. This chain is the parent and is named after the corresponding alkane with the ending changed to *-ene* or *-yne*.
2. Number the chain beginning at the end *closer to the double or triple bond*.
3. Use the single position number where the *double or triple bond begins* as the prefix number just before the alkene or alkyne name.

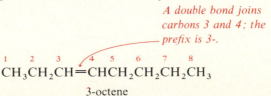

A double bond joins carbons 3 and 4; the prefix is 3-.

$$\overset{1}{C}H_3\overset{2}{C}H_2\overset{3}{C}H{=}\overset{4}{C}H\overset{5}{C}H_2\overset{6}{C}H_2\overset{7}{C}H_2\overset{8}{C}H_3$$

3-octene

4. Add any branches or other prefix substituents with their appropriate prefix numbers.

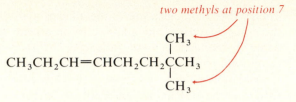

two methyls at position 7

$$CH_3CH_2CH=CHCH_2CH_2CCH_3$$

7,7-dimethyl-3-octene

5. In **dienes**, compounds containing two double bonds, the ending *-adiene* is used.

Double bonds begin at carbons 1 and 3.

$$\overset{1}{C}H_2=\overset{2}{C}H\overset{3}{C}H=\overset{4}{C}H_2$$

1,3-butadiene

Example   Name the following alkene:

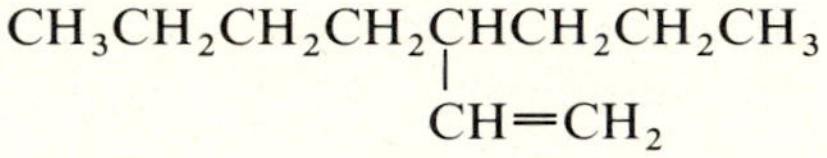

$$CH_3CH_2CH_2CH_2CHCH_2CH_2CH_3$$
$$CH=CH_2$$

Solution   (1) Find the longest continuous chain that contains the double bond.

(2) Number the chain, starting at the end closer to the double bond. The parent is *heptene*.

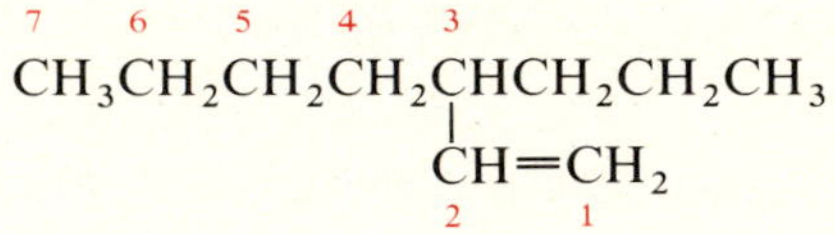

$$\overset{7}{C}H_3\overset{6}{C}H_2\overset{5}{C}H_2\overset{4}{C}H_2\overset{3}{C}HCH_2CH_2CH_3$$
$$\overset{2}{C}H=\overset{1}{C}H_2$$

(3) Insert the number of the position where the double bond begins.

The double bond starts at carbon 1.   1-heptene   *There are seven carbons in the chain.*

(4) Add the name and number for the branch—a three-carbon, or propyl, branch in this case:

3-propyl-1-heptene

Problem 15.1.   Name the following compounds:

(a) 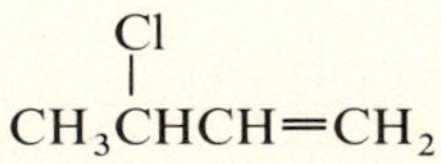 $CH_3\overset{Cl}{CH}CH=CH_2$ (b)

Problem 15.2.   Write formulas for the following compounds:
(a) 1,2-dibromo-1-heptene
(b) 2,3-dimethyl-1,3-cyclopentadiene

15.3   *CIS,TRANS* ISOMERISM

In Chapter 13, we stated that atoms and groups are free to rotate about single bonds. Atoms joined by double bonds are *not free to rotate* because rotation would break the second bond.

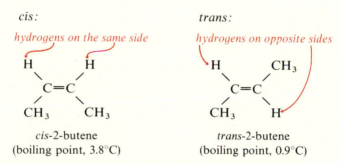

Because there is no rotation about double bonds, the portion of a molecule containing a double bond is held rigidly. The result of this rigidity is that there are *two* 2-butenes, $CH_3CH=CHCH_3$.

cis: *trans:*

hydrogens on the same side   *hydrogens on opposite sides*

cis-2-butene   trans-2-butene
(boiling point, 3.8°C)   (boiling point, 0.9°C)

Because the groups adjacent to the double bond are held rigidly, the pair of compounds cannot be interconverted under normal circumstances. Thus, the two compounds are isomers of each other. Because the order of attachment of the atoms is the same, the two compounds are not structural isomers. This pair of isomers is called a pair of **cis,trans isomers**. *Cis-* means *on the same side*; in *cis*-2-butene, the H atoms are on the same side of the double bond. *Trans-* means *on opposite sides*; in *trans*-2-butene, the H atoms are on opposite sides of the double bond. Note how the *cis* and *trans* prefixes are incorporated into the names of the two 2-butenes under the preceding formulas.

Not all alkenes are capable of *cis,trans* isomerism. In order to have *cis,trans* isomers, the two carbons of the double bond must each have two different atoms or groups attached. The following two structures are the same compound, and not *cis,trans* isomers, because one carbon of the double bond has two atoms (hydrogen) that are the same.

not cis,trans isomers:

$$\underset{CH_3}{\overset{Cl}{\diagdown}}C=C\underset{H}{\overset{H}{\diagup}} \qquad \underset{Cl}{\overset{CH_3}{\diagdown}}C=C\underset{H}{\overset{H}{\diagup}}$$

Substituted cyclic compounds are the other class of organic compounds that can have *cis,trans* isomers. In a cycloalkane, such as cyclopropane, the ring carbons cannot rotate about their single bonds without breaking the ring. Therefore, when two groups are attached to two different ring carbon atoms, they can be on the *same* side of the ring (*cis*) or on *opposite* sides of the ring (*trans*), as shown in Figure 15.3. In this figure, we also show one way to represent substituents as being on the same side or opposite sides of a ring.

Problem 15.3.   Name each of the following compounds, including a *cis* or *trans* designation:

(a) $\underset{H}{\overset{Cl}{\diagdown}}C=C\underset{H}{\overset{Cl}{\diagup}}$

(b) $\overset{H}{\diagdown}\underset{Cl}{\diagup}\overset{CH_2CH_3}{\diagup}$ C ‖ C (structure)

(c) (cyclohexane ring with Cl, Cl, H)

Problem 15.4.   Draw the structures of the following compounds:
(a) *cis*-3-hexene
(b) *trans*-1-ethyl-2-methylcyclopentane

Figure 15.2.   *Models showing the difference between the cis and trans isomers of 2-butene.*

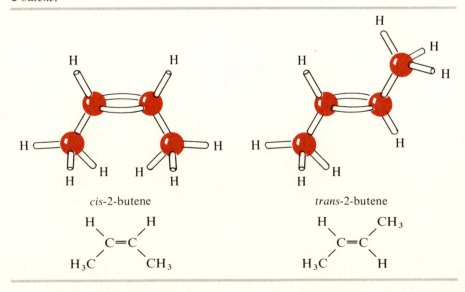

cis-2-butene

trans-2-butene

$$\underset{H_3C}{\overset{H}{\diagdown}}C=C\underset{CH_3}{\overset{H}{\diagup}} \qquad \underset{H_3C}{\overset{H}{\diagdown}}C=C\underset{H}{\overset{CH_3}{\diagup}}$$

Figure 15.3.   *Cis,trans isomers of 1,2-dimethylcyclopropane.*

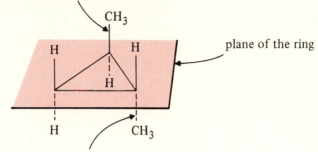

The solid bond indicates
a group above the plane of the ring.

plane of the ring

The dashed bond indicates
a group below the plane of the ring.

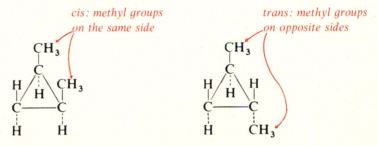

cis-1,2-dimethylcyclopropane   *trans*-1,2-dimethylcyclopropane

■ *Cis, Trans* Isomerism in Vision (optional)

Cis,trans isomerism plays an important role in many biochemical processes, including reception of light by the retina of the eye. The retina contains a compound called *rhodopsin*, or *visual purple*, which is responsible for black-and-white vision and for vision in dim light.

Rhodopsin consists of a relatively small molecule called 11-*cis*-retinal that just fits in a pocket of a larger protein molecule called opsin. When light strikes the 11-*cis*-retinal molecule in rhodopsin, the 11-*cis* double bond is converted to a *trans* double bond. This conversion changes the shape of the retinal molecule so that it can no longer fit in the opsin pocket. The all-*trans*-retinal is therefore released from the opsin and, in the process, triggers a nerve impulse that results in our being able to see.*

---

* Retinal is synthesized in the eye from vitamin A. One of the symptoms of vitamin A deficiency is night blindness, which results from a lack of rhodopsin.

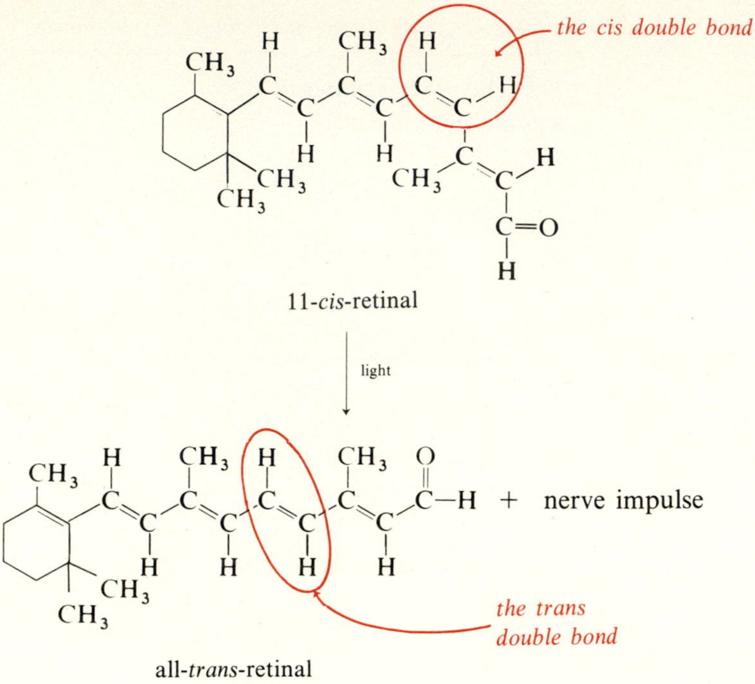

11-*cis*-retinal

light

all-*trans*-retinal   +   nerve impulse

the cis double bond

the trans double bond

15.4   REACTIONS OF ALKENES

The most characteristic reaction of an alkene is one in which a reagent adds to the double bond. In the reaction, one of the bonds of the double bond is broken and two new single bonds are formed. Such a reaction is a **combination reaction**, or **addition reaction**. Alkynes undergo similar addition reactions but are far less important in biological processes; therefore, we will not discuss the reactions of alkynes.

Hydrogenation

Hydrogen gas (H_2) can add across a double bond of an alkene. The addition of hydrogen to a compound is called **hydrogenation**. The hydrogenation of an alkene is an oxidation-reduction reaction in which the alkene gains two H atoms and is reduced to an alkane, while H_2 is oxidized.

a hydrogenation reaction:

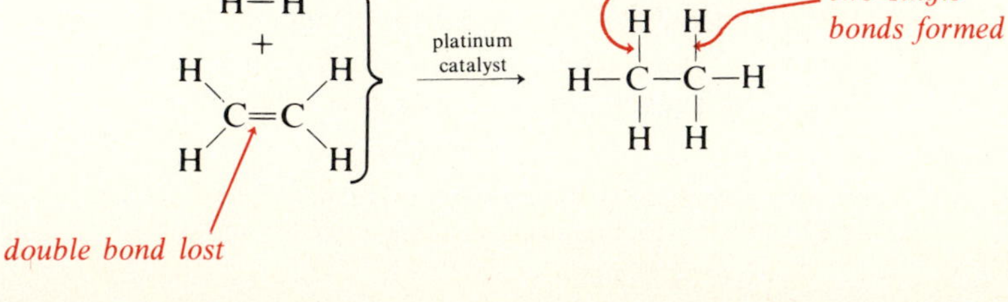

two single bonds formed

double bond lost

Hydrogenation reactions are general for alkenes:

$$CH_3CH\!=\!CHCH_3 \quad + \quad H_2 \quad \xrightarrow{\text{Pt}} \quad CH_3CH_2CH_2CH_3$$

2-butene
(*cis* or *trans*)

butane

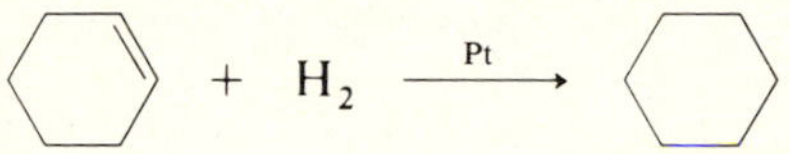

cyclohexene   cyclohexane

A hydrogenation reaction does not take place without the aid of a catalyst. Platinum metal (Pt) is a very effective laboratory catalyst for the reaction of alkenes with hydrogen. The catalyst functions by adsorbing the hydrogen gas onto its surface. The association between H_2 and the metal catalyst weakens the H-H bond and lowers the energy of activation necessary for the reaction with the alkene. (See Section 5.3 for a discussion of catalysts.)

Margarine is made by hydrogenation of some of the double bonds in vegetable oils such as corn oil or soybean oil. This process is called *partial hydrogenation*, or *hardening*. In the commercial production of peanut butter, peanut oil is hardened by a similar process.

$$
\begin{array}{l}
\overset{\displaystyle O}{\overset{\|}{}} \\
CH_2OC(CH_2)_7CH\!=\!CHCH_2CH\!=\!CHCH_2CH\!=\!CHCH_2CH_3 \\
| \qquad\;\; O \\
\qquad\;\;\, \| \\
CHOC(CH_2)_7CH\!=\!CHCH_2CH\!=\!CHCH_2CH\!=\!CHCH_2CH_3 \\
| \qquad\;\; O \\
\qquad\;\;\, \| \\
CH_2OC(CH_2)_7CH\!=\!CH(CH_2)_7CH_3
\end{array}
$$

A typical vegetable oil is polyunsaturated,
or contains many carbon-carbon double bonds.

$$\xrightarrow[\text{catalyst}]{6\,H_2}$$

$$
\begin{array}{l}
\overset{\displaystyle O}{\overset{\|}{}} \\
CH_2OC(CH_2)_{16}CH_3 \\
| \qquad\;\; O \\
\qquad\;\;\, \| \\
CHOC(CH_2)_7CH\!=\!CH(CH_2)_7CH_3 \\
| \qquad\;\; O \\
\qquad\;\;\, \| \\
CH_2OC(CH_2)_{16}CH_3
\end{array}
$$

A typical solid fat contains few
carbon-carbon double bonds.

Hydration

In the presence of an acidic catalyst, such as H_2SO_4, water adds across the double bond of an alkene to yield an alcohol (ROH). The addition of water to a compound is called **hydration**. In the hydration of an alkene, an H from H—OH adds to one carbon of the double bond, and —OH adds to the other carbon.

a hydration reaction:

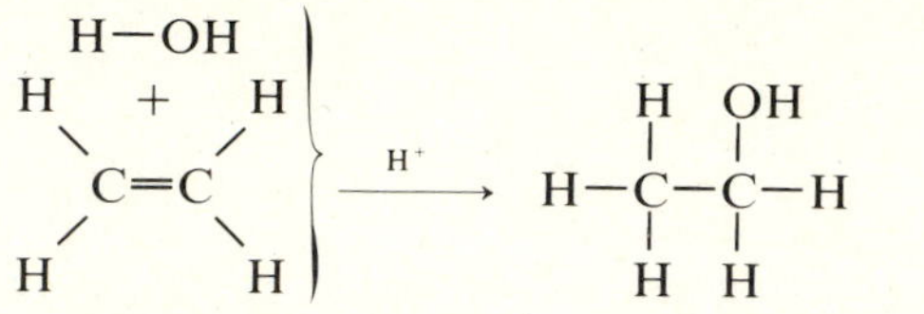

$$
\begin{matrix}
\text{H—OH} \\
\underset{\text{ethylene}}{\underset{\displaystyle H}{\overset{\displaystyle H}{\diagdown}} C = C \underset{\displaystyle H}{\overset{\displaystyle H}{\diagup}}}
\end{matrix}
\xrightarrow{\;H^+\;}
\underset{\text{ethanol, } CH_3CH_2OH}{H - \underset{H}{\overset{H}{C}} - \underset{H}{\overset{OH}{C}} - H}
$$

Industrial ethanol (not beverage ethanol) and isopropyl alcohol (rubbing alcohol) are made this way, from alkenes obtained by petroleum cracking reactions. The addition of water to a double bond is also a common biochemical reaction; in these cases, enzymes (not acids) are the catalysts.

Example   Write the equation for the reaction of propene (propylene) with water to yield isopropyl alcohol:

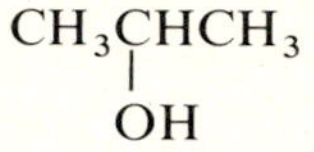

$$CH_3\underset{\displaystyle OH}{\overset{\displaystyle |}{C}}HCH_3$$

Solution

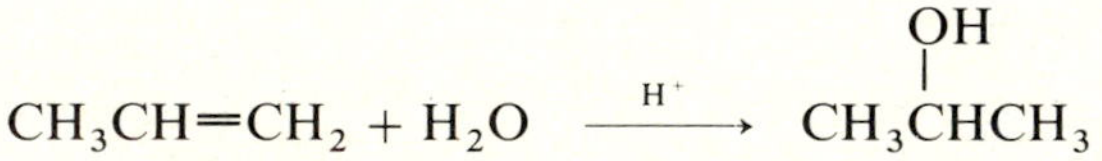

$$
\underset{\text{propylene}}{CH_3CH = CH_2} + H_2O \xrightarrow{\;H^+\;} \underset{\substack{\text{2-propanol} \\ \text{(isopropyl alcohol)}}}{CH_3\underset{\displaystyle OH}{\overset{\displaystyle |}{C}}HCH_3}
$$

Problem 15.5.   Complete the following equations:

(a) $CH_3CH = CH_2 + H_2 \xrightarrow{\;Pt\;}$

(b) $CH_3CH = CHCH_3 + H_2O \xrightarrow{\;H^+\;}$

(c) 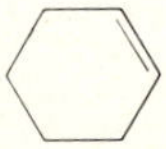 ⬡ $+ H_2O \xrightarrow{\;H^+\;}$

Polymerization

The word *polymer* is derived from the Greek *poly* (many) and *mer* (units). A **polymer** is a giant molecule that contains many repeating units. Polymeric molecules form nerve cells, skin, hair, muscles, and chromosomes. Plant fibers are also composed of polymers. We will discuss some of these natural polymers in Chapters 18, 20, and 26.

Synthetic polymers, such as plastics, synthetic fibers, synthetic rubber, and films like Saran wrap, are tremendously useful. We sit on polymer-covered chairs and write on polymer-coated table tops. Synthetic polymers are used to make fibers for the rugs on our polymer floors and for the clothing we wear. In this section, we will consider polymers formed from alkenes, a group of polymers that includes polyethylene, polypropylene, acrylic fibers, and synthetic rubber.

The compound used to make a polymer is called the *monomer* (*mono-*, one). The monomer used to make polyethylene is ethylene. Most polymers are named after their monomer, with the prefix *poly-* added. Under the influence of a catalyst, two ethylene molecules combine to form a *dimer* (*di-*, two). Then another molecule of ethylene is added to form the *trimer* (*tri-*, three). The process continues until a giant molecule has been formed from thousands of ethylene molecules—this, of course, is the polymer. A **polymerization reaction** is the overall reaction in which monomers are combined to form the polymer.

a polymerization reaction:

many molecules of $CH_2{=}CH_2$ $\xrightarrow{\text{catalyst}}$

ethylene
a monomer

the repeating unit

$$\{-CH_2CH_2-CH_2CH_2-\overbrace{CH_2CH_2}-\}$$

portion of a polyethylene molecule

To simplify writing a polymerization equation, chemists use the letter x or n to designate a very large number. The end groups of a polymer are not usually written out. Thus, the polymerization of ethylene may be represented by the following abbreviated equation:

$$x\ CH_2{=}CH_2 \xrightarrow{\text{catalyst}} {+}CH_2CH_2{+}_x$$

ethylene \qquad\qquad polyethylene

The properties of polymers can be changed by using alkenes with different substituents. While polyethylene is suitable for plastic squeeze bottles and for plastic bags to use in the kitchen and to protect clothes coming home from the cleaners, polypropylene is more suitable for indoor-outdoor carpeting. Examples of some common polymerization reactions follow:

$$x\ CH_2{=}\underset{\underset{Cl}{|}}{CH} \xrightarrow{\text{catalyst}} \left(\underset{\underset{Cl}{|}}{CH_2CH}\right)_x$$

vinyl chloride \qquad\qquad polyvinyl chloride (PVC)
a carcinogen \qquad\qquad *used for vinyl floor*
covering, piping,
phonograph records, and
garbage bags

$$x \ CH_2{=}CH \xrightarrow{\text{catalyst}} \left(CH_2CH \right)_x$$
$$\quad\quad | \quad\quad\quad\quad\quad\quad\quad\quad | $$
$$\quad\quad CH_3 \quad\quad\quad\quad\quad\quad CH_3$$

propylene   polypropylene

*used for piping and
carpeting*

Example   Polytetrafluoroethylene (commonly called Teflon) is used as a heat- and corrosion-resistant material for gaskets and valves. Cookware is sometimes coated with this polymer. From its name, deduce the structure of the polymer, and write the equation for its formation from its monomer.

Solution

$$x \ CF_2{=}CF_2 \xrightarrow{\text{catalyst}} \left(CF_2CF_2 \right)_x$$

tetrafluoroethylene   polytetrafluoroethylene

Problem 15.6.   The synthetic fiber Orlon and the plastic polystyrene (used in such items as toothbrush handles and styrofoam) are made from the following monomers, respectively. Write the formulas for the polymers.

$$\quad\quad\quad\quad CN$$
$$\quad\quad\quad\quad |$$
(a) $CH_2{=}CH$ (b) $CH_2{=}CH$

15.5   BENZENE

Alkenes and alkynes are two of the important classes of unsaturated hydrocarbons. A third class is the *aromatic hydrocarbons*. In this section, we will consider *benzene*, the simplest aromatic hydrocarbon. In the next section, we will briefly discuss other aromatic compounds.

Benzene (C_6H_6) is a toxic, volatile, water-insoluble liquid that is the starting material for a variety of important compounds. Like most other commercially important organic compounds, benzene is synthesized from alkanes found in petroleum.

Until the late 1970s, benzene was widely used commercially as a solvent for other nonpolar organic compounds. However, because continued inhalation of benzene vapors can cause a drop in blood-cell count and because benzene is considered somewhat carcinogenic, the use of benzene as a solvent has declined in recent years. Many substituted benzene compounds are also toxic, but many are biologically indispensable. For example, some of the amino acids that make up proteins are substituted benzenes.

Figure 15.4. *The bonding in benzene, showing how the six p electrons are shared equally by the six carbon atoms.*

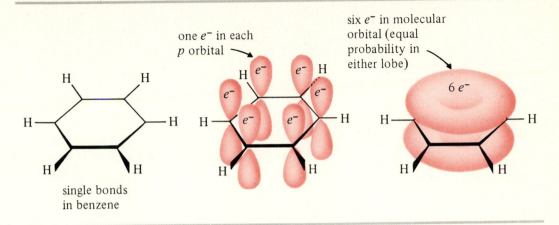

A molecule of benzene contains six carbon atoms joined to form a ring. Each carbon atom is also bonded to one hydrogen atom. Contrast the structure of benzene to that of cyclohexane, in which each carbon atom is bonded to *two* hydrogen atoms.

$$CH-CH$$
$$CH \qquad CH$$
$$CH=CH$$

benzene
a total of six hydrogens

$$CH_2-CH_2$$
$$CH_2 \qquad CH_2$$
$$CH_2-CH_2$$

cyclohexane
a total of twelve hydrogens

With three double bonds, benzene might be expected to behave like an alkene; however, this is not the case. Benzene does *not* undergo typical alkene addition reactions. The reason is that the six "double-bond" electrons in benzene do not actually form three double bonds. Instead, these six electrons are shared *equally* by all six ring carbon atoms. Each carbon-carbon bond in benzene, therefore, is halfway between a single and a double bond. (Figure 15.4 shows the bonding in benzene with an orbital picture.)

Although the structure of benzene is often represented as a hexagon containing three double bonds, keep in mind that benzene does not truly have any double bonds. In this text, we will generally use a circle in the hexagon instead of three double bonds to represent the sharing of these six electrons.

three ways of representing benzene:

Table 15.2. Some Important Substituted Benzenes

Name	Formula	Use
toluene	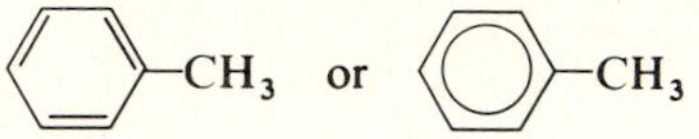 ⟨ring⟩—CH$_3$ or ⟨ring⟩—CH$_3$	solvent and reagent
phenol (carbolic acid)	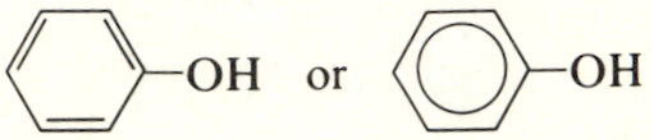 ⟨ring⟩—OH or ⟨ring⟩—OH	toxic antiseptic, can be absorbed through the skin
aniline	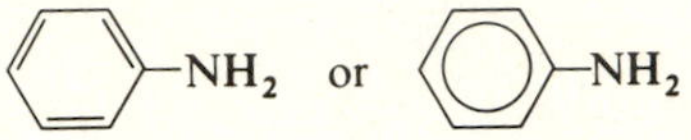 ⟨ring⟩—NH$_2$ or ⟨ring⟩—NH$_2$	toxic, used in the manufacture of some dyes and drugs
benzoic acid	⟨ring⟩—$\overset{\text{O}}{\overset{\|}{\text{C}}}$OH or ⟨ring⟩—$\overset{\text{O}}{\overset{\|}{\text{C}}}$OH	reagent and used in some skin ointments
p-aminobenzoic acid (PABA)	H$_2$N—⟨ring⟩—$\overset{\text{O}}{\overset{\|}{\text{C}}}$OH or H$_2$N—⟨ring⟩—$\overset{\text{O}}{\overset{\|}{\text{C}}}$OH	used in some sunscreens; part of the structure of the class of vitamins called *folic acids*

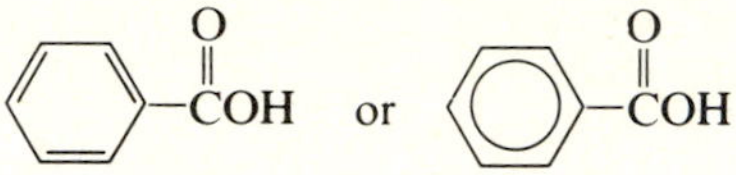

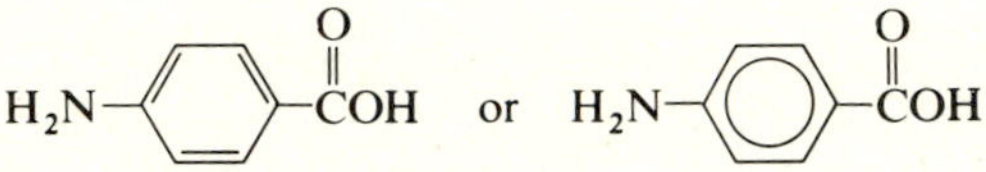

Benzene is one member of a class of compounds called **aromatic compounds**, cyclic compounds that can be represented with alternate single and double bonds but that do not behave like alkenes. Some important aromatic compounds that are substituted benzenes are shown in Table 15.2. In Section 15.6, we will present a few other types of aromatic compounds.

a substituted benzene:

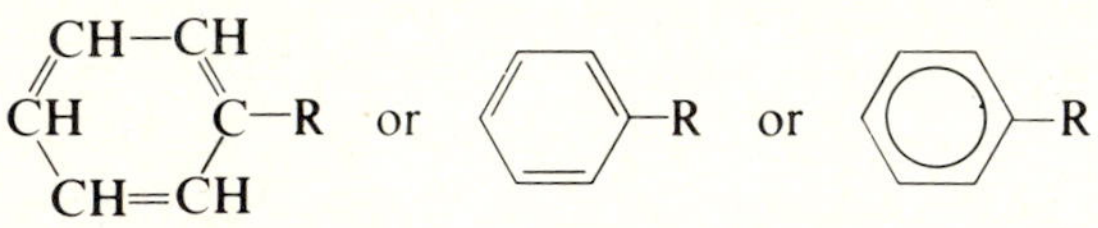

CH—CH
CH   C—R or ⟨ring⟩—R or ⟨ring⟩—R
CH=CH

Naming Substituted Benzenes

Substituted benzene compounds are named somewhat differently from substituted alkanes or alkenes. Several of the substituted benzenes have specific names, as shown in Table 15.2.

Some functional groups and alkyl groups attached to the benzene ring are designated by prefixes, such as *bromo-* or *isopropyl-*.

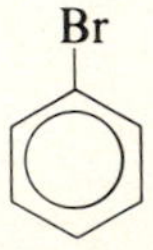

Br
⟨ring⟩

CH$_3$CHCH$_3$
⟨ring⟩

bromobenzene 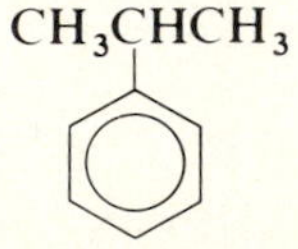   isopropylbenzene

When a benzene ring has two substituents, there is the possibility of three structural isomers:

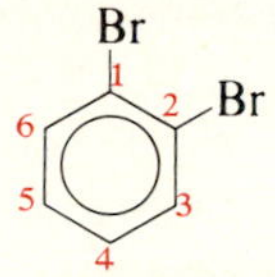

o-dibromobenzene	m-dibromobenzene	p-dibromobenzene
o- means 1,2	*m- means 1,3*	*p- means 1,4*

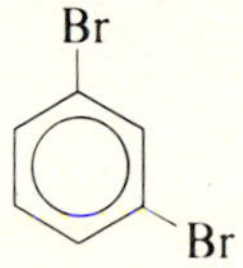

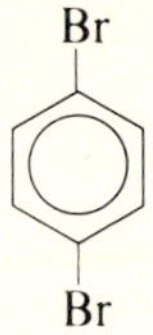

These three isomers are distinguished from one another by the prefixes *ortho-*, *meta-*, and *para-*, abbreviated *o-*, *m-*, and *p-*. *Ortho-* means that the two groups are on adjacent carbon atoms on the benzene ring, or in the 1,2-positions. *Meta-* means the substituted carbons are separated by one carbon atom, or that the groups are in the 1,3-positions. *Para-* means that the two groups are in the 1,4-positions. (The *o-*, *m-*, and *p-* designations are reserved solely for disubstituted benzenes. The substituted cyclohexanes are named with numbers, never with *o*, *m*, or *p*). Using these prefixes, we can designate the positions of two substituents on a benzene ring as we have done for the dibromobenzenes and for *p*-aminobenzoic acid (Table 15.2).

If benzene is attached to a molecule larger than itself or if the rest of the structure is more important than the benzene ring, the benzene is named by the prefix *phenyl-*. Any aromatic attachment, such as a phenyl group, can be called an **aryl group** to differentiate it from alkyl groups.

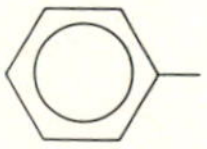

phenyl group	2-phenylheptane
an aryl group	

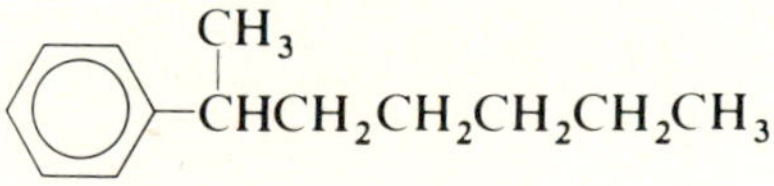

Problem 15.7.   Name the following compounds. (More than one answer may be correct.)

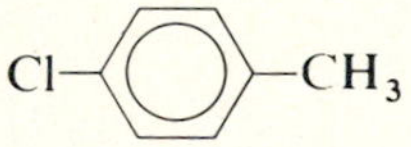

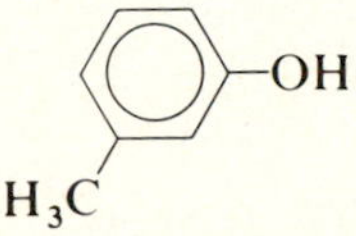

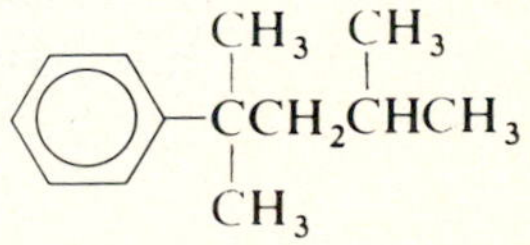

Problem 15.8.   Write formulas for (a) *m*-dimethylbenzene and (b) *p*-chlorobenzoic acid.

Aromatic Substitution Reactions

Although aromatic rings tend to be unreactive in alkene addition reactions, there is one important class of reactions that aromatic rings undergo. These reactions, called **aromatic substitution reactions**, involve substitution of another atom or group for a hydrogen atom attached to the ring.

One example of an aromatic substitution reaction is the *bromination* of benzene. When benzene is treated with bromine in the presence of an iron(III) bromide catalyst, a bromine atom replaces a hydrogen atom on the benzene ring. The products of this bromination reaction are bromobenzene and hydrogen bromide. Other substituted benzenes are synthesized by similar reactions.

an aromatic substitution reaction:

| benzene | bromine | | bromobenzene | hydrogen bromide |

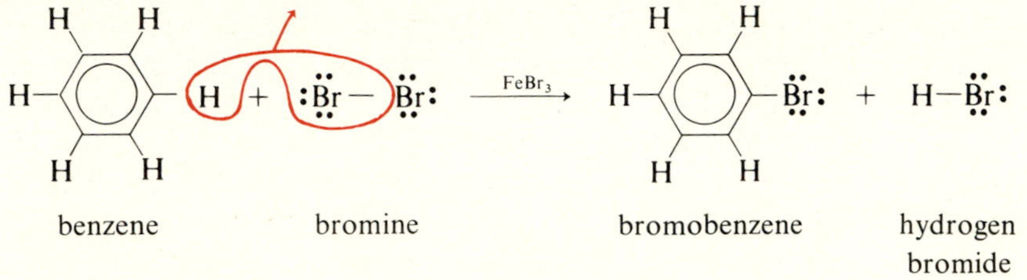

or simply

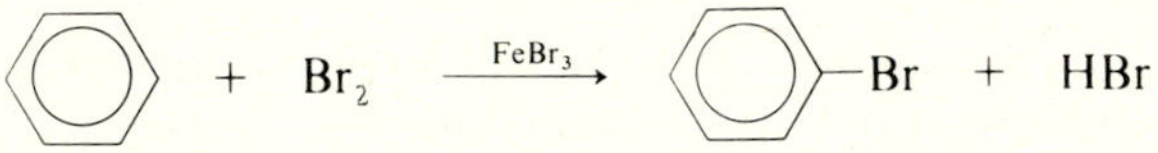

Some aromatic substitution reactions are accomplished enzymatically in animal organisms. The amino acid phenylalanine from ingested proteins is normally converted to tyrosine before it is broken down, or metabolized. The substitution reaction is catalyzed by the enzyme *phenylalanine hydroxylase.*

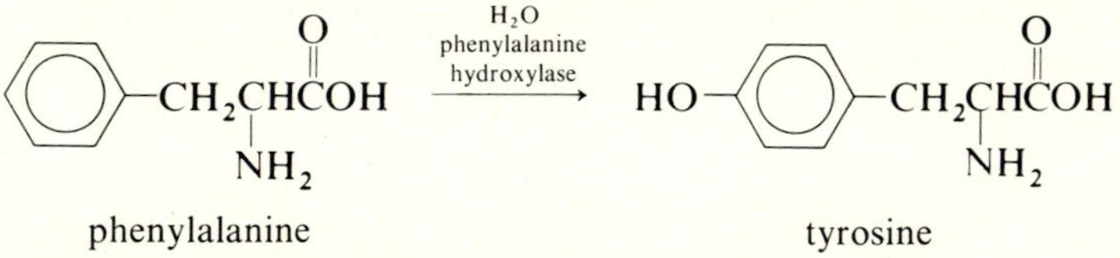

phenylalanine   tyrosine

One out of ten thousand humans lacks the enzyme that catalyzes this conversion to tyrosine. In these individuals, phenylalanine is converted to phenylpyruvic acid, which contains a ketone group ($R_2C{=}O$) adjacent to the carboxyl group ($-CO_2H$).

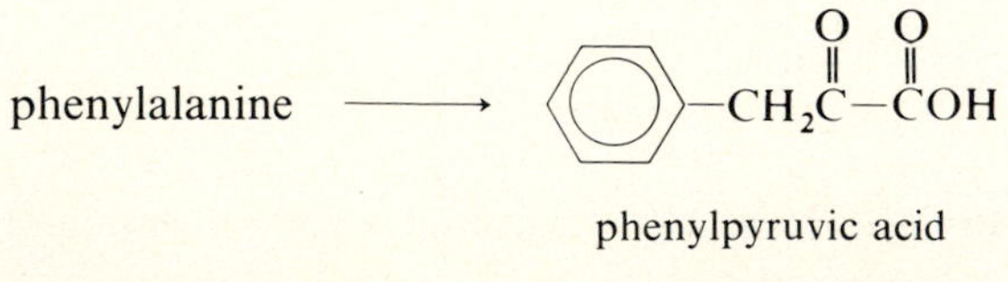

phenylpyruvic acid

The phenylpyruvic acid is not broken down further; instead, it builds up in the bloodstream and spills over into the urine. This condition is known as **phenyl-ketonuria**, meaning "a phenyl ketone in the urine," often abbreviated PKU. The presence of phenylpyruvic acid in the blood prevents normal development of the brain, resulting in mental retardation. PKU can be detected by a blood test a few days after a baby's birth and is controlled by limiting phenylalanine in the diet.

15.6   OTHER AROMATIC COMPOUNDS

As we have mentioned, benzene and substituted benzenes are only one class of aromatic compounds. Many other cyclic compounds with formulas that can be written with alternate single and double bonds are also aromatic compounds and do not undergo alkene addition reactions. For example, two or more benzenoid (benzene-like) rings can be joined together as in naphthalene, a toxic solid that has been used in mothballs and is widely used in the dye industry.

formulas for naphthalene:

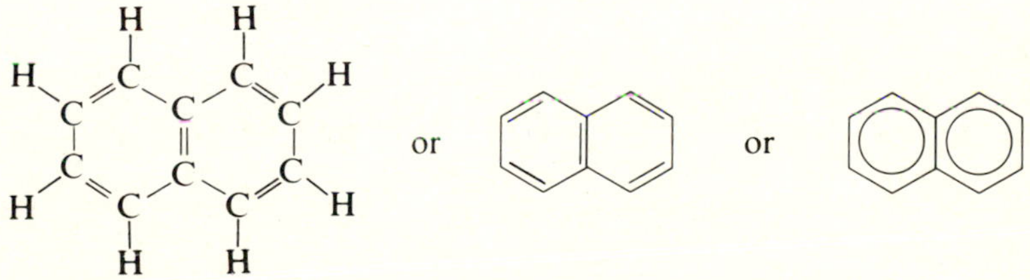

Many aromatic compounds contain one or more nitrogen atoms in place of carbon atoms in the ring. Pyridine, a toxic, pungent-smelling liquid, is an example:

formulas for pyridine:

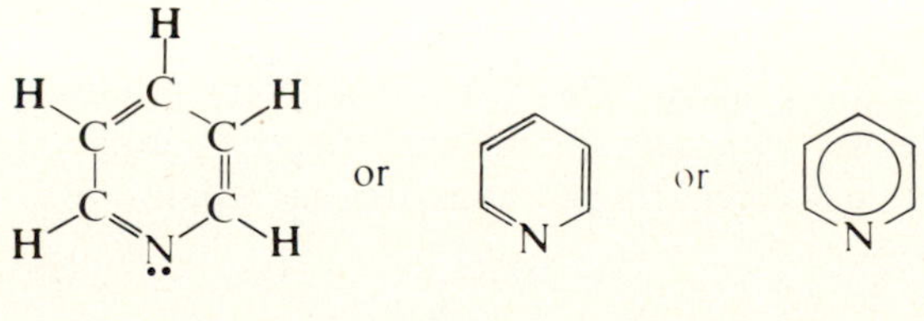

A cyclic compound, such as pyridine, containing carbon atoms and one or more other atoms in the ring is called a **heterocyclic compound** (from the Greek *hetero*, meaning "different"). Table 15.3 lists a few aromatic heterocyclic compounds that are of biological importance. Note that two of the B vitamins and nicotine contain pyridine rings.

Table 15.3.   *Some Important Aromatic Heterocyclic Compounds*

Name	Formula	Comments
nicotinic acid (niacin)	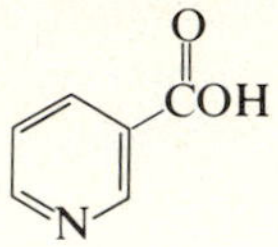	a B vitamin (sometimes called vitamin B_3 or vitamin PP) essential for the cells' oxidation of nutrients
pyridoxine	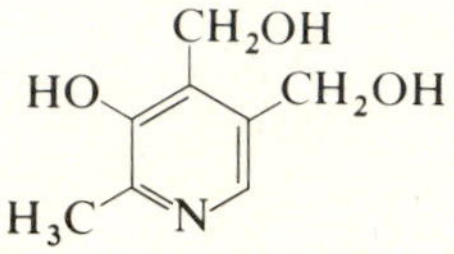	vitamin B_6, essential for the cells' metabolism of amino acids from proteins
nicotine	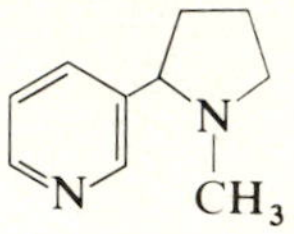	a stimulant of the central nervous system, found in tobacco; also used as an insecticide
purine	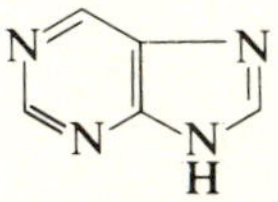	substituted purines form part of the structure of nucleic acids

Problem 15.9.   Draw the structure of purine (Table 15.3), showing each C, H, and N atom. Also show any unshared valence electrons as dots.

SUMMARY

Alkenes and *alkynes* are hydrocarbons containing one double bond and one triple bond, respectively. The second and third bonds in a double or triple bond are more reactive than the first bond.

An alkene or alkyne is named by changing the ending of the parent alkane's name to *-ene* or *-yne*, respectively. Dienes (two double bonds) are named with the ending *-adiene*. A prefix number is used to designate the position at which a double or triple bond begins.

An alkene that has two different groups on each double-bonded carbon exists as a pair of *cis,trans* isomers. A cyclic compound with substituents on two different ring carbons also has a pair of *cis,trans* isomers.

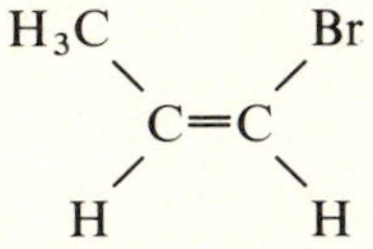

cis-1-bromopropene   trans-1-bromopropene
(*H's on same side*)   (*H's on opposite sides*)

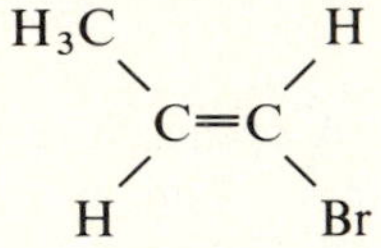

Alkenes and alkynes are nonpolar, low melting, low boiling, and water insoluble. Alkenes undergo *addition reactions* and *polymerization.*

$$R_2C{=}CR_2 \quad + \quad H_2 \quad \xrightarrow{\text{Pt}} \quad R_2\overset{\overset{\displaystyle H}{|}}{C}{-}\overset{\overset{\displaystyle H}{|}}{C}R_2$$

$$R_2C{=}CR_2 \quad + \quad H_2O \quad \xrightarrow{\text{H}^+} \quad R_2\overset{\overset{\displaystyle H}{|}}{C}{-}\overset{\overset{\displaystyle OH}{|}}{C}R_2$$

$$x\ R_2C{=}CR_2 \quad \xrightarrow{\text{catalyst}} \quad \left(\!\!\begin{array}{c} \overset{\displaystyle R}{|}\ \ \overset{\displaystyle R}{|} \\ {-}C{-}C{-} \\ \underset{\displaystyle R}{|}\ \ \underset{\displaystyle R}{|} \end{array}\!\!\right)_{\!x}$$

the monomer   *the polymer*

Benzene (C_6H_6) is an *aromatic compound,* a cyclic compound that can be represented with alternate single and double bonds, but that does not behave like an alkene. Substituents on a benzene ring may be named with prefixes, although some substituted benzenes have specific names (Table 15.2). A disubstituted benzene is named by the *o-, m-, p-* system, discussed in Section 15.5.

Other aromatic compounds include naphthalene and *aromatic heterocyclic compounds,* compounds that contain aromatic rings with at least one ring atom different from carbon.

KEY TERMS

alkene	hydrogenation reaction	benzene ring
alkyne	hydration reaction	*o-, m-, p-* nomenclature system
diene	polymer	aromatic substitution reaction
cis,trans isomerism	polymerization reaction	heterocyclic compound
addition reaction	aromatic compound	

STUDY PROBLEMS

15.10.   Name the following hydrocarbons:

(a) $CH_3\overset{\overset{\displaystyle }{|}}{C}HCH{=}CH_2$
    $\overset{\displaystyle |}{CH_3}$

(b) $(CH_3)_2CHC{\equiv}CH$

(c)   〈benzene ring〉$-CH_2\overset{\overset{\displaystyle }{}}{C}HCH{=}CHCH_3$
    $\overset{\displaystyle |}{CH(CH_3)_2}$

15.11.   Draw the condensed structural formula of each of the following compounds:

(a) 2,4-dimethyl-3-hexene
(b) 4-methyl-1-cyclohexene
(c) *cis*-1-bromo-1-butene
(d) *trans*-1-bromo-1-butene
(e) 4-ethyl-4-methyl-2-heptyne

15.12. Which of the following compounds have *cis,trans* isomers?

(a) $CH_3C\equiv CCH_3$

(b) $Cl_2C=CHCl$

(c) $BrCH=CHBr$

(d) $CH_3CH_2CH=CHCH_2CH_2Cl$

15.13. Write the structures for the *cis,trans* isomers in Problem 15.12. Name each one.
15.14. Draw formulas for all possible *cis,trans* isomers of 1,2,3-cyclohexanetriol (OH groups at positions 1, 2, and 3).
15.15. What would be the product of the complete hydrogenation of each of the following? [In (c), the $-CO_2H$ group is unaffected.]

(a) 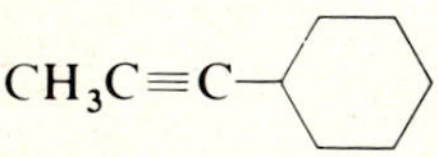 $CH_3C\equiv C-$

(b) $CH_2=CHCH_2CH_2CH=CH_2$

(c) $CH_3(CH_2)_7CH=CH(CH_2)_7CO_2H$

15.16. 1-Butene is treated with water and a trace of H_2SO_4 as a catalyst. What two alcohols would you expect as products?
15.17. One of the steps in the biological conversion of glucose to CO_2, H_2O, and energy is the conversion of fumaric acid to malic acid. What is the structure of malic acid?

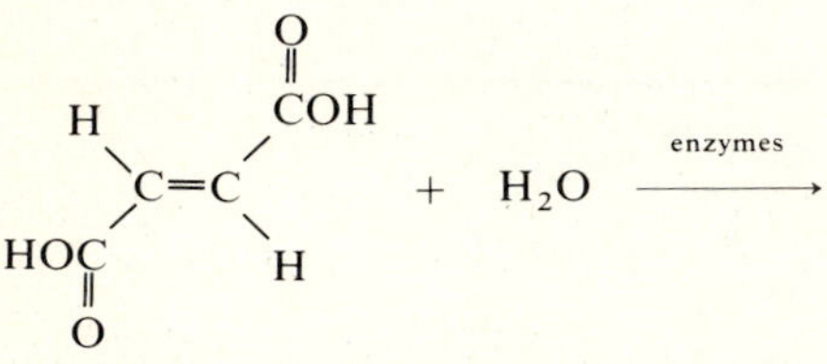

fumaric acid

malic acid

15.18. Predict the structure of the polymer formed from each of the following monomers:
(a) 1-butene   (b) trichloroethylene
15.19. What is the monomer for each of the following polymers?

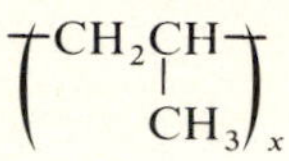

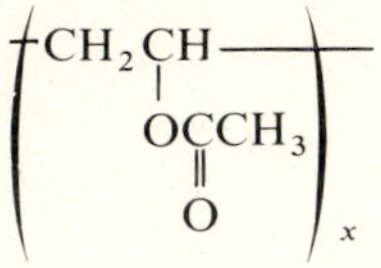

15.20. Name the following substituted benzenes. More than one answer may be correct.

(a)

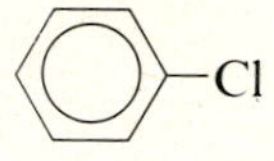

(b)

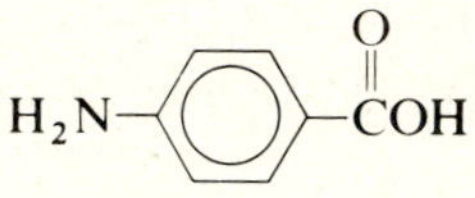

(c)

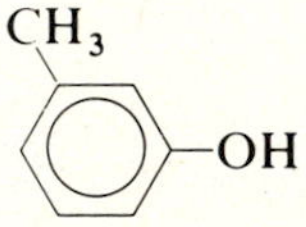

(d)

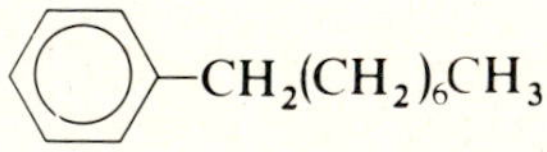

15.21. Complete the equation for the following aromatic monosubstitution reaction:

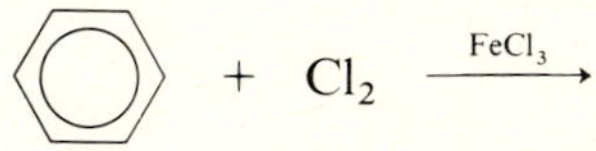

15.22. (a) Referring to Table 15.3, Section 15.6, draw the structure of nicotinic acid, showing all unshared electrons as dots.
(b) Recalling what you learned about acids and bases in Chapter 10, predict the product of the reaction of nicotinic acid with aqueous NaOH.
(c) Predict the product of its reaction with aqueous HCl.
15.23. In nucleic acid chemistry, substituted purines (Table 15.3) are called *bases*. Suggest a reason for this.

16 Alcohols, Amines, and Related Compounds

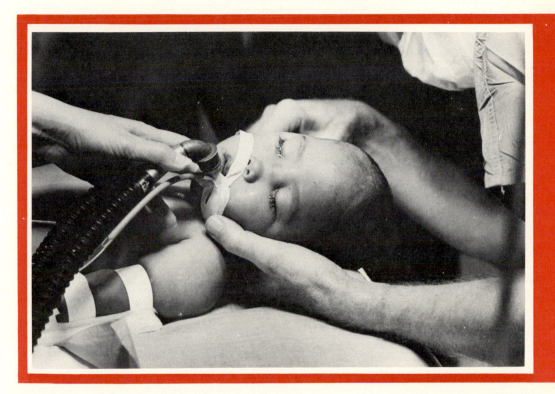

A child receiving an inhalation anesthetic. (© Jeffrey Grosscup 1978.) Ethers are compounds with the general formula ROR'. Most low-formula-weight ethers are volatile liquids, and many of these ethers have been used as inhalation anesthetics. Except for their combustibility, alkyl ethers are almost as unreactive as alkanes and are thus generally physiologically safe. For example, the LD_{50} (lethal dose, 50% kill) of diethyl ether is 2.2 grams per kilogram of body weight when fed to rats.

Objectives Define and give examples of alcohols, ethers, and phenols. □ Name and classify alcohols. □ Compare the physical properties of alcohols with those of hydrocarbons. □ Write equations for the dehydration and oxidation of alcohols. □ Relate the structures of thiols and sulfides to those of oxygen compounds; write equations for the formation of a disulfide and a sulfoxide (optional). □ Name and classify amines. □ Show how amines form hydrogen bonds. □ Write equations to illustrate the basicity of amines. □ Define alkaloid. □ Write formulas for quaternary ammonium salts and describe their properties. □ Illustrate complex-ion formation and chelation.

Over 90% of the weight of the human body arises from compounds of carbon, hydrogen, and oxygen. Nitrogen is the fourth most common element in the body. A glance through the biochemical chapters toward the end of this book will reveal that these four elements alone form carbohydrates, fats, and most proteins. For this reason, we will consider the organic compounds containing oxygen and nitrogen in some detail.

In this chapter, we will discuss organic compounds containing carbon-oxygen single bonds, such as *alcohols* (ROH), *ethers* (ROR), and related *sulfur-containing compounds*. Then we will present some chemistry of compounds containing carbon-nitrogen single bonds—the *amines* (RNH_2, R_2NH, and R_3N) and some of their relatives. In Chapter 17, we will consider compounds containing carbon-oxygen double bonds.

16.1 ALCOHOLS, ETHERS, AND PHENOLS

Compounds containing carbon with a single bond to oxygen are similar in structure to water.

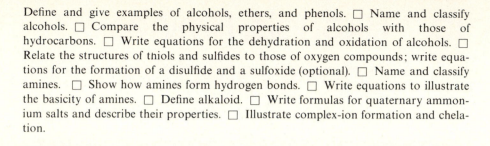

water   methanol   diethyl ether
               an alcohol   *an ether*

An **alcohol** (ROH) is a compound that contains one alkyl group replacing one hydrogen of water. The functional group of an alcohol is the polar *hydroxyl group* (—OH), not to be confused with the hydroxide ion. Unlike inorganic hydroxides, alcohols do not undergo ionization to yield hydroxide ions because the —OH group is firmly bonded to a carbon. In biological systems, hydroxyl groups are frequently found in compounds containing other functional groups as well. For example, inspect the structures of the sugars in Figure 18.1 or of the prostaglandins (Section 19.4).

Photo 16.1. *Demonstration of anesthesia to the world by William T. G. Morton (holding ether inhaler) while Dr. John C. Warren operates at the Massachusetts General Hospital. (Photograph courtesy of MGH News and Archives.)*

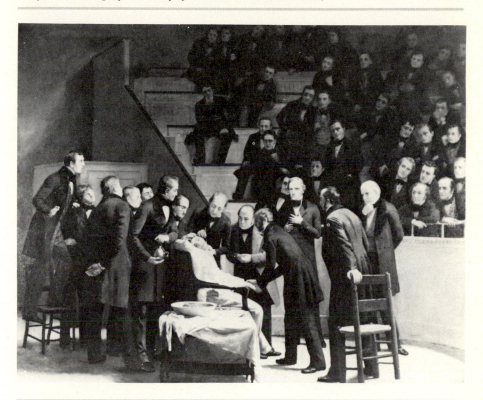

An **ether** (ROR) has two alkyl groups (or aryl groups) in place of the two hydrogens of water. The most commonly encountered ether is diethyl ether (also called *ethyl ether* or simply *ether*). Diethyl ether is a volatile liquid (bp 34.5°C) that is used as a surgical anesthetic and as a laboratory solvent. We will not discuss the chemistry of ethers because they are not very reactive. However ethers, like hydrocarbons, burn readily in the air if ignited. Diethyl ether, because of its high volatility, is especially dangerous in this regard.

$$CH_3CH_2OCH_2CH_3 + 6\ O_2 \xrightarrow{\text{spark}} 4\ CO_2 + 5\ H_2O + \text{energy}$$
diethyl ether

A **phenol** is similar to an alcohol except that it has an aryl group, instead of an alkyl group, attached to the oxygen.

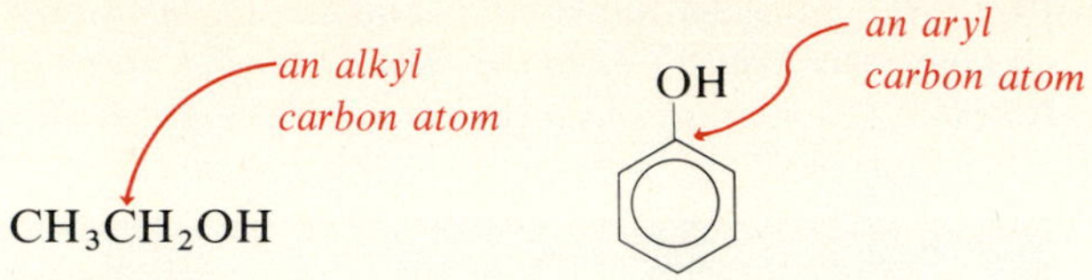

an alkyl carbon atom

an aryl carbon atom

OH

CH$_3$CH$_2$OH

ethanol
an alcohol

phenol
a phenol

Example   Classify each of the following compounds as an alcohol, ether, or phenol, or as containing a combination of these functional groups:

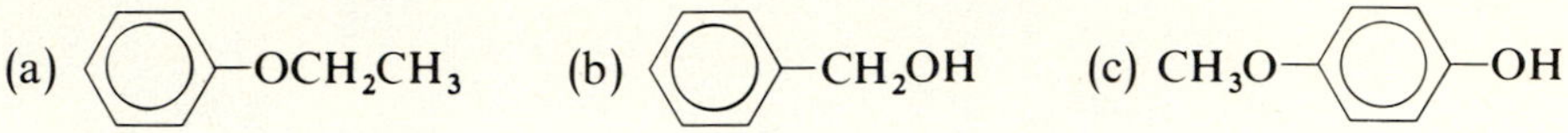

(a) ⬡—OCH$_2$CH$_3$ (b) ⬡—CH$_2$OH   (c) CH$_3$O—⬡—OH

Solution   (a) is an ether.
(b) is an alcohol because the OH is attached to the CH$_2$ carbon, not the carbon of an aromatic ring.
(c) contains both an ether group (CH$_3$O—) *and* a phenol group.

Problem 16.1.   Circle and name (as a hydroxyl group, for example) the functional groups in the following formulas:

(a) CH$_3$CHCH$_3$ (with OH on central carbon)

(b) CH$_3$OCH$_2$CH$_2$CH$_2$OH

(c) 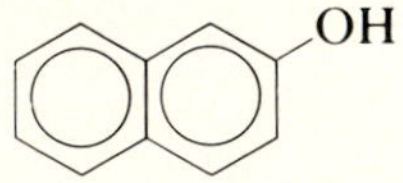 [naphthalene with OH]

(d) CH$_2$=CH—O—CH=CH$_2$

Naming and Classifying Alcohols

Alcohols may be named by the IUPAC system. The name of the alkane is used as the parent name. The final *-e* is dropped from the alkane name, and the ending *-ol* is added. Prefix numbers are used to show the position of the hydroxyl group, if necessary.

CH$_3$CH$_2$CH$_2$CH$_3$ CH$_3$CH$_2$CH$_2$CH$_2$OH   CH$_3$CHCH$_2$CH$_3$ (with OH on second carbon)

butane   1-butanol   2-butanol

Simple alcohols are often also named by using the name of the alkyl group that is attached to the hydroxyl group and adding the word *alcohol*. Table 16.1 lists the IUPAC names and trivial names of a few important alcohols, along with those of one diol (*two* OH groups) and one triol (*three* OH groups).

Table 16.1.   *Some Important Alcohols*

IUPAC name	Trivial name	Formula	Bp (°C)	Comments
methanol	methyl alcohol	CH_3OH	65	also called wood alcohol because it can be obtained from wood; toxic (about one tablespoon can cause blindness; two tablespoons, death)
ethanol	ethyl alcohol (alcohol)	CH_3CH_2OH	78.5	also called grain alcohol because it can be obtained by fermentation of grain; a depressant; also used as a rubbing alcohol
1-propanol	*n*-propyl alcohol	$CH_3CH_2CH_2OH$	97	toxic
2-propanol	isopropyl alcohol	$(CH_3)_2CHOH$	82	70% aqueous solution, called rubbing alcohol, is used as an antiseptic and coolant; toxic
1,2-ethanediol	ethylene glycol	$\underset{CH_2-CH_2}{\overset{OH\quad OH}{\mid\qquad\mid}}$	197	used in automobile radiator antifreeze; toxic
1,2,3-propanetriol	glycerol (glycerin)	$\underset{CH_2-CH-CH_2}{\overset{OH\quad OH\quad OH}{\mid\qquad\mid\qquad\mid}}$	290	can be obtained from animal fats and vegetable oils

Alcohols are classified as *primary*, *secondary*, or *tertiary*, depending on the number of alkyl groups (such as CH_3- or CH_3CH_2-) attached directly to the "head" carbon, the carbon with the OH group. Alcohols of these three types exhibit somewhat different behavior in chemical reactions.

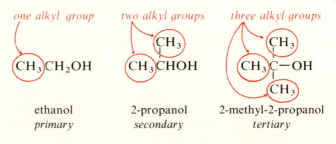

ethanol	2-propanol	2-methyl-2-propanol
primary	*secondary*	*tertiary*

Example   Name the following alcohol by the IUPAC system, and classify it as primary, secondary, or tertiary:

$$\underset{\underset{CH_3}{\mid}}{\overset{\overset{CH_3}{\mid}}{CH_3CH_2CH_2COH}}$$

Solution   2-Methyl-2-pentanol, a tertiary alcohol. The three alkyl groups are two CH_3- groups and one $CH_3CH_2CH_2-$ group.

Problem 16.2.   Name the following alcohols, and classify each as primary, secondary, or tertiary. In (c) and (d), you may want to write out the formulas showing each ring C and H.

(a) $(CH_3)_2CHCH_2OH$

(b) 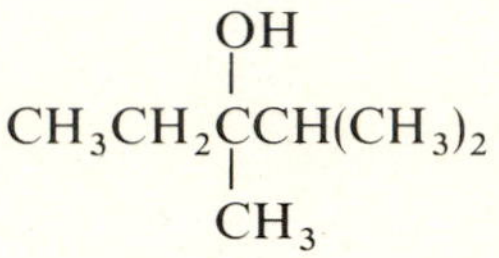 $CH_3CH_2\overset{\overset{\displaystyle OH}{|}}{\underset{\underset{\displaystyle CH_3}{|}}{C}}CH(CH_3)_2$

(c) [cyclohexane ring]—OH

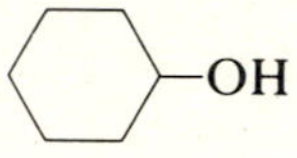

(d) 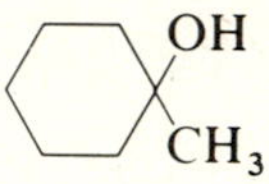 [cyclohexane ring with OH and CH_3]

Physical Properties of Alcohols and Phenols

Like water, alcohols and phenols are capable of *hydrogen bonding*. An alcohol or phenol molecule can form a hydrogen bond with another alcohol molecule or with a water molecule. The lower-formula-weight alcohols (up to four carbons) are therefore water soluble. The sugars (Chapter 18) are of high formula weight but contain several hydroxyl groups. These compounds, too, are generally water soluble.

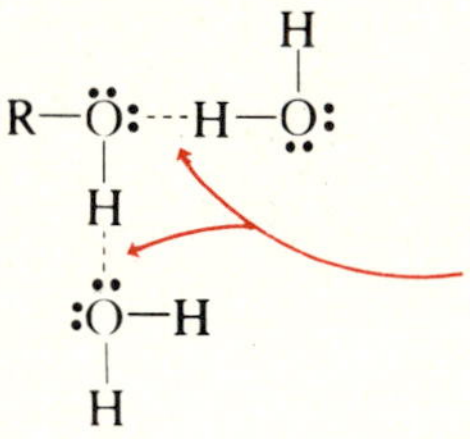

$$R-\overset{\overset{\displaystyle H}{|}}{\underset{\underset{\displaystyle H}{|}}{\ddot{O}}}\cdots H-\ddot{\underset{\underset{\displaystyle :\ddot{O}-H}{|}}{O}}:$$

hydrogen bonds between an alcohol and water

Alcohols containing a greater proportion of hydrocarbon character, such as $CH_3(CH_2)_3CH_2OH$, are insoluble in water. These alcohols are soluble in other alcohols, ethers, and other less-polar solvents because the solvent's hydrocarbon portion is attracted to the hydrocarbon chain of the solute alcohol.

Hydrogen bonding also affects the boiling points of alcohols. Unlike the hydrocarbons, low-formula-weight alcohols are liquid at room temperature because of the strong attractions between alcohol molecules. Even methanol (CH_3OH), the smallest alcohol, is a liquid at room temperature. Table 16.1 lists the boiling points of a few important alcohols.

Problem 16.3.   Suggest reasons why diethyl ether is about as soluble in water as $CH_3CH_2CH_2CH_2OH$ (about 8 g/100 mL), even though the boiling point of diethyl ether (34.5°C) is much lower than that of this four-carbon alcohol (117°C).

Problem 16.4.   Which one of the following compounds would be the most soluble in a relatively nonpolar solvent such as hexane? Explain your answer.

$$\begin{matrix} OH & OH \\ | & | \end{matrix}$$
(a) CH_2-CH_2 (b) CH_3CH_2OH (c) $CH_3CH_2CH_2CH_2OH$

16.2   REACTIONS OF ALCOHOLS

Dehydration

Heating an alcohol with a strong acid can cause it to lose water and form an alkene. A reaction in which water is lost from a compound is called a **dehydration reaction**. Concentrated sulfuric acid, a good dehydrating agent, is usually the acid used in this reaction. Phenols do not undergo dehydration because the aromatic ring is already unsaturated.

a dehydration reaction:

$$\begin{matrix} H & OH \\ | & | \end{matrix}$$
$$CH_2-CH_2 + H_2SO_4 \xrightarrow{\text{180°C}} CH_2=CH_2 + H_2SO_4 \cdot H_2O$$
ethanol   ethylene

generalized:

$$\begin{matrix} H & OH \\ | & | \end{matrix}$$
$$R_2C-CR_2 + H_2SO_4 \xrightarrow{\text{heat}} R_2C=CR_2 + H_2SO_4 \cdot H_2O$$

Note that this reaction is the reverse of the hydration of an alkene. Hydration of alkenes and dehydration of alcohols are both common reactions in biological systems, where they are enzyme catalyzed. In the laboratory, the concentration of acid and the temperature determine the direction in which the reaction proceeds.

$$\begin{matrix} OH \\ | \end{matrix}$$
$$CH_3CHCH_3 \underset{\text{H}_2\text{O (H}_2\text{SO}_4\text{ catalyst)}}{\overset{\text{conc. H}_2\text{SO}_4, 100 C}{\rightleftharpoons}} CH_3CH=CH_2 + H_2O$$
2-propanol   propene
(isopropyl alcohol)   (propylene)

Problem 16.5.   Write an equation for the dehydration of cyclohexanol.

Oxidation

Like most other organic compounds, alcohols burn when ignited. Flaming shish kebab and cherries jubilee depend on the ignition of the vapors of ethanol from warmed brandy or another beverage with a high alcohol content.

$$CH_3CH_2OH + 3\ O_2 \xrightarrow{\text{spark}} 2\ CO_2 + 3\ H_2O + \text{energy}$$

ethanol

The term *proof*, referring to the alcohol content of a beverage, is related to the combustion of ethanol. In days of old, a sample of a shipment of rum or other alcoholic beverage was poured on some gunpowder and ignited. If the gunpowder burned after the alcohol had burned away, it was *proof* that the rum was not watered down. The number given as proof is twice the percent by volume of alcohol—for example, a whiskey that is 80 proof is 40% alcohol by volume.

Because alcoholic beverages are taxed in most countries of the world, nontaxable ethanol for laboratory or industrial purposes is often *denatured* by the addition of a small quantity of toxic material such as methanol or benzene, so that it is unfit to drink. The denaturing agent selected depends on the use for which the alcohol is intended.

Recall from Chapter 11 that oxidation of organic compounds is the *addition of oxygen* or the *removal of hydrogen*. Using these criteria, we can assign relative states of oxidation to a series of organic compounds:

$$CH_3CH_3 \xrightarrow{+O} CH_3CH_2OH \xrightarrow{-2H} CH_3\overset{\displaystyle O}{\overset{\|}{C}H} \xrightarrow{+O} CH_3\overset{\displaystyle O}{\overset{\|}{C}OH} \xrightarrow{\text{4 more steps}} 2\ CO_2$$

alkane   alcohol   aldehyde   carboxylic
acid

increasing oxidation state of C

It is difficult to obtain alcohols from the direct oxidation of alkanes. Instead, alcohols are generally obtained from the hydration of alkenes. However, alcohols themselves are easily oxidized. In the body, enzymes can catalyze the oxidation of alcohols. In the laboratory, a hot aqueous solution of potassium permanganate ($KMnO_4$) is a commonly used oxidizing agent. The product of an alcohol oxidation is usually either a ketone or a carboxylic acid. Which product is formed depends on the number of hydrogen atoms on the hydroxyl group's carbon atom. (Tertiary alcohols and simple phenols, with no H on their hydroxyl carbons, cannot be oxidized in this fashion.)

$$\underset{\textit{a primary alcohol}}{\overset{\textit{two H's}}{RCH_2OH}} \xrightarrow{-2H} \underset{\substack{\textit{an aldehyde} \\ \textit{(oxidized further)}}}{R\overset{\displaystyle O}{\overset{\|}{C}H}} \xrightarrow{+O} \underset{\textit{a carboxylic acid}}{R\overset{\displaystyle O}{\overset{\|}{C}OH}}$$

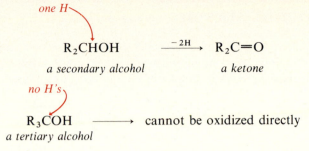

specific examples of alcohol oxidation:

$$CH_3CH_2CH_2OH \xrightarrow[\substack{\text{or other} \\ \text{oxidizing agent}}]{\text{hot KMnO}_4 \text{ solution}} CH_3CH_2\overset{\displaystyle O}{\overset{\|}{C}}H \longrightarrow CH_3CH_2\overset{\displaystyle O}{\overset{\|}{C}}OH$$

1-propanol   *an aldehyde*   propanoic acid
a primary alcohol   *(oxidized further)*   *a carboxylic acid*

$$CH_3\overset{\displaystyle OH}{\overset{|}{C}}HCH_3 \xrightarrow[\substack{\text{or other} \\ \text{oxidizing agent}}]{\text{hot KMnO}_4 \text{ solution}} CH_3\overset{\displaystyle O}{\overset{\|}{C}}CH_3$$

2-propanol   acetone
a secondary alcohol   *a ketone*

The oxidation of a primary alcohol, as shown in the preceding equations, proceeds by way of an *aldehyde*. Because aldehydes are readily oxidized to carboxylic acids, the final product of such a laboratory oxidation is usually a carboxylic acid. Under the milder oxidizing conditions of living systems, however, an aldehyde may be the actual product.

Ethanol does not occur naturally in the body. When ingested, ethanol is oxidized enzymatically to acetaldehyde, then to acetate ions, which can be converted to fats and other compounds or can be used for energy. Methanol is poisonous because its oxidation results in formaldehyde, which is toxic.

$$CH_3CH_2OH \xrightarrow{\text{enzyme}} CH_3\overset{\displaystyle O}{\overset{\|}{C}}H \xrightarrow{\text{enzyme}} CH_3\overset{\displaystyle O}{\overset{\|}{C}}O^- \begin{array}{l} \xrightarrow{\substack{\text{many} \\ \text{steps}}} CO_2 + H_2O + \text{energy} \\ \\ \xrightarrow[\substack{\text{many} \\ \text{steps}}]{} \text{fats, etc.} \end{array}$$

ethanol   acetaldehyde   acetate
              ion

$$CH_3OH \xrightarrow{\text{enzyme}} \overset{\displaystyle O}{\overset{\|}{H}}CH$$

methanol   formaldehyde
          toxic

Problem 16.6.   Predict the organic products of the oxidation of the following alcohols and diol with a hot $KMnO_4$ solution:

(a) $CH_3CH_2\overset{\displaystyle OH}{\overset{|}{C}}HCH_3$ (b)   (c) $\overset{\displaystyle OH}{\overset{|}{C}}H_2CH_2\overset{\displaystyle OH}{\overset{|}{C}}H_2$

■ 16.3 THIOLS, SULFIDES, AND DISULFIDES (optional)

Like oxygen, sulfur contains six valence electrons and can form covalent compounds by bonding with two other atoms. Many sulfur compounds, such as **thiols** (RSH) and **sulfides** (RSR), are directly analogous to oxygen compounds.

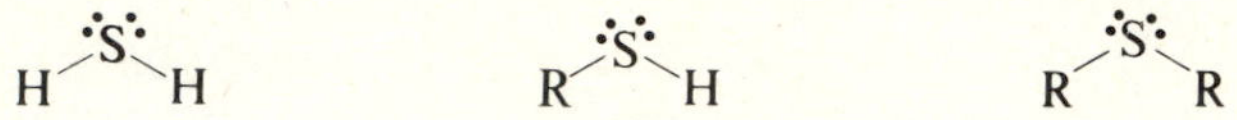

hydrogen sulfide (*analogous to H_2O*) — *a thiol, or mercaptan* (*analogous to ROH*) — *a sulfide* (*analogous to ROR*)

Sulfur compounds of low formula weight have distinctive odors. Cabbage being cooked, skunks, and garlic all owe their odors, in part, to simple organic sulfur compounds. The human nose can detect the odor of a volatile thiol when its concentration in air is as low as 0.02 part per billion.

Thiols undergo some reactions similar to reactions of alcohols. (One is discussed in Section 17.8.) Thiols also exhibit some different chemical reactivity. For example, they are readily oxidized to **disulfides** (R—S—S—R). The amino acid cysteine, which can be obtained from proteins, can form cross-links between protein chains by this type of oxidation. The positions of these cross-links determine whether human hair is straight or curly. (Permanent waves involve changing the positions of the cross-links between protein molecules in hair.)

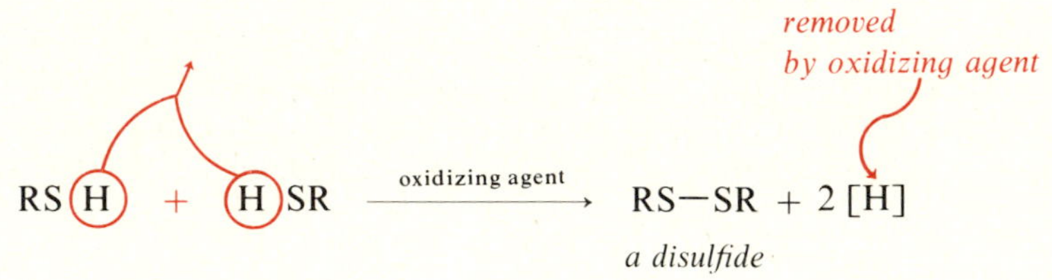
removed
by oxidizing agent
RSH + HSR
oxidizing agent
RS—SR + 2 [H]
a disulfide

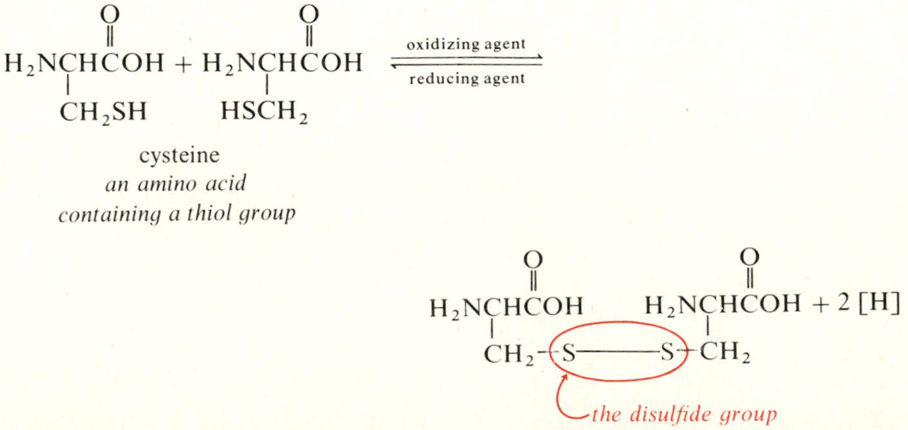
oxidizing agent
reducing agent
cysteine
an amino acid
containing a thiol group
+ 2 [H]
the disulfide group

Sulfides can be oxidized to **sulfoxides** ($R_2S\!=\!O$) because a sulfur atom, with an unfilled third electron shell, can accommodate more than eight electrons in this shell. One important sulfoxide is dimethyl sulfoxide (DMSO), used in veterinary medicine and sometimes in human medicine. One useful property of DMSO is that it is very readily absorbed by the skin and can carry other drugs into the bloodstream. However, it can also carry undesirable compounds and toxins into the bloodstream, so DMSO must be used with care.

$$CH_3SCH_3 \ + 30\% \ H_2O_2 \ \xrightarrow[25°C]{H^+} \ CH_3\overset{\overset{\displaystyle O}{\|}}{S}CH_3 \ + H_2O \quad \blacksquare$$

dimethyl sulfide   hydrogen   dimethyl sulfoxide
                    peroxide   (DMSO)

16.4 AMINES

Life as we know it depends on nitrogen. Some less hospitable planets of our solar system have atmospheres composed primarily of methane and ammonia. Speculation about whether amino acids, the components of living proteins, could arise from these compounds has led to research along these lines. Indeed, amino acids can be formed when a discharge of electricity is passed through such an atmosphere. In addition, nonterrestrial amino acids, along with other organic compounds, have been isolated from carbonaceous meteorites that have fallen to the earth.

Nitrogen forms many types of compounds with carbon. Among these are the **amines**, compounds in which nitrogen is attached to alkyl or aryl groups by single bonds. Table 16.2 lists a few representative examples.

Table 16.2. *Names and Boiling Points of Some Amines*

Name	Formula	Bp (°C)
methylamine	CH_3NH_2	−6
dimethylamine	$(CH_3)_2NH$	8
trimethylamine	$(CH_3)_3N$	3
piperidine	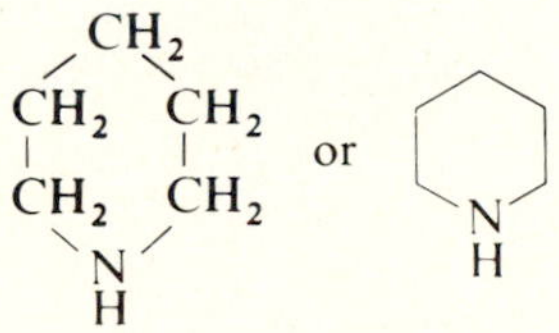	106
aniline	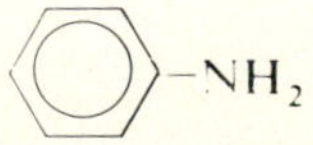	184

Amines are related to ammonia in their structures and properties. In ammonia, the N forms three single bonds to H atoms and has a pair of unshared valence electrons. In an amine, the N forms three single bonds to one or more alkyl or aryl groups and H atoms. The N of an amine also has a pair of unshared valence electrons. The functional group of an amine ($-NH_2$, $-NHR$, or $-NR_2$) is called an **amino group.**

unshared electrons

$$H-\overset{\cdot\cdot}{\underset{|}{N}}-H \qquad CH_3CH_2-\overset{\cdot\cdot}{\underset{|}{N}}-CH_3$$
$$\underset{\text{ammonia}}{H} \qquad \underset{\text{ethyldimethylamine}}{CH_3}$$

The nitrogen in an amine can be part of a ring system. Piperidine, shown in Table 16.2, is an example of a nonaromatic heterocyclic compound of this type.

Naming and Classifying Amines

The most common way to name simple amines is to list alphabetically the alkyl or aryl groups that are attached to the nitrogen and then add the ending -*amine*. Methylamine, dimethylamine, and trimethylamine are examples. If necessary, a number is used to designate the position of the amino group. Note that amine names are run together as one word.

$$CH_3CH_2CH_2NH_2 \qquad \overset{\displaystyle NH_2}{\underset{\displaystyle CH_3CHCH_3}{|}} \qquad \text{(cyclohexyl)}-NHCH_3$$

1-propylamine   2-propylamine   cyclohexylmethylamine
or *n*-propylamine   or isopropylamine

Amines may be classified as *primary* (one alkyl or aryl substituent on the N), *secondary* (two substituents), or *tertiary* (three substituents).

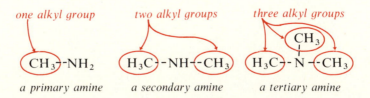

a primary amine   *a secondary amine*   *a tertiary amine*

Problem 16.7.   Name the following amines:

(a) $CH_3CH_2NHCH_2CH_3$ (b) $CH_3CH_2NHCH_3$

Problem 16.8.   Classify the following amines as primary, secondary, or tertiary:

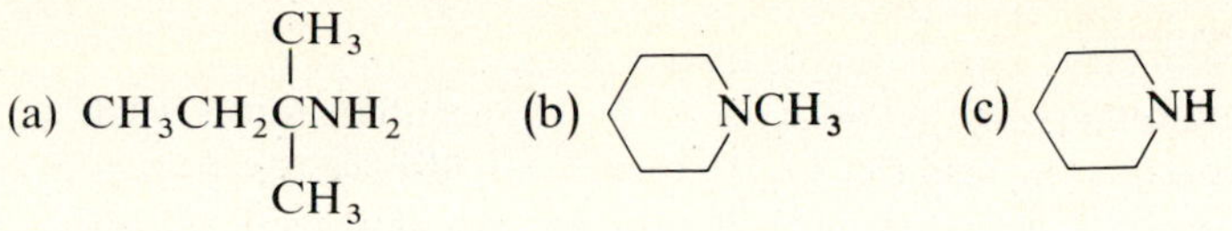

(a) $CH_3CH_2\overset{\displaystyle CH_3}{\underset{\displaystyle CH_3}{C}}NH_2$ (b) $\bigcirc NCH_3$ (c) $\bigcirc NH$

Physical Properties of Amines

Amines that have low formula weights are gases or low-boiling liquids and often have objectionable odors. For example, trimethylamine smells like dead salt-water salmon and piperidine smells like dead fresh-water fish. Two diamines, putrescine and cadaverine, contribute to the odor of decaying flesh. These compounds arise from the bacterial degradation of proteins.

$$H_2NCH_2CH_2CH_2CH_2CH_2NH_2 \qquad H_2NCH_2CH_2CH_2CH_2NH_2$$

1,5-pentanediamine   1,4-butanediamine
(cadaverine)   (putrescine)

Like water, amines are capable of forming hydrogen bonds. Because of this ability, low-formula-weight amines (up to about six carbon atoms) are water soluble.

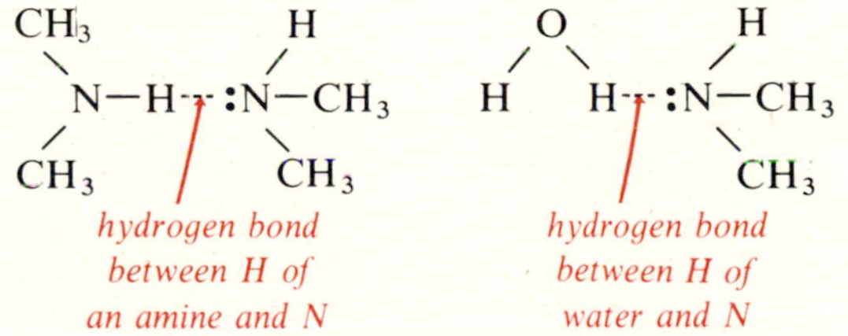

$$\underset{CH_3}{\overset{CH_3}{N}}-H\cdots:\underset{CH_3}{\overset{H}{N}}-CH_3 \qquad \overset{O}{\underset{H}{\diagup}}\quad H\cdots:\underset{CH_3}{\overset{H}{N}}-CH_3$$

hydrogen bond   *hydrogen bond*
between H of   *between H of*
an amine and N   *water and N*

16.5   BASICITY OF AMINES

Like ammonia, amines are bases. In both ammonia and amines, the unshared pair of electrons on the nitrogen can be donated to form a bond with a hydrogen ion.

$$R_3N: \quad + \quad H^+ \quad \longrightarrow \quad R_3\overset{+}{N}-H$$

Also like ammonia, amines are *weak* bases. In water solution, less than about 1% of the amine molecules react with water to form ions. The exact percentage depends on the structure of the amine.

partial ionization in water:

$$CH_3\ddot{N}H_2 + H-\ddot{O}H \;\rightleftharpoons\; CH_3\overset{H}{\underset{+}{N}}H_2 + :\ddot{O}H$$

methylamine
a weak base

Amine Salts

If an amine is treated with an acid stronger than water, such as HCl, H_2SO_4, or a carboxylic acid (RCO_2H), then the acid-base reaction goes to completion. The product of an amine with an acid is an **amine salt** ($R_3NH^+ A^-$, where A^- is an anion). Amine salts, like most inorganic salts, are high-melting, odorless solids.

salt formation:

$$\ddot{N}H_3 + HCl \longrightarrow \overset{+}{N}H_4 \quad Cl^-$$

ammonium chloride
a salt

$$CH_3\ddot{N}H_2 + HCl \longrightarrow CH_3\overset{+}{N}H_3 \quad Cl^-$$

methylammonium chloride
(methylamine hydrochloride)
an amine salt

Problem 16.9.   Write the equation for the reaction of acetic acid (CH_3CO_2H) and methylamine.

Many naturally occurring amines are biologically active, and some are used as drugs. (See Figures 16.1 and 16.2.) These amines, usually of high formula weight and water insoluble, may be made soluble and extracted from plant material by an acidic solution. Because they react with acids, these naturally occurring nitrogen compounds are called **alkaloids**, meaning "like a base." The soluble salts of amine-containing drugs are also used when it is desired to administer an insoluble drug in solution.

If an amine salt is treated with any base stronger than the amine, the base pulls H^+ from the amine cation, and the amine is regenerated.

Figure 16.1.   Some alkaloids: physiologically active nitrogen compounds obtained from plants.

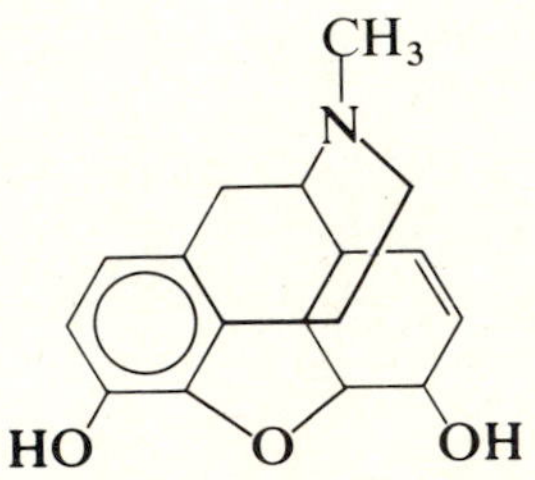

morphine
an addictive pain reliever

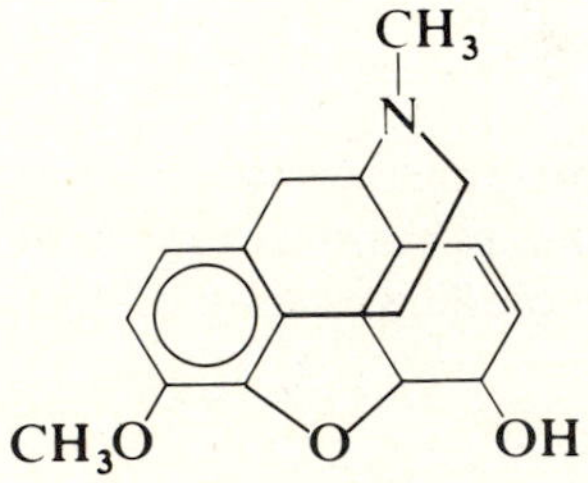

codeine
*an addictive pain reliever
and cough suppressant*

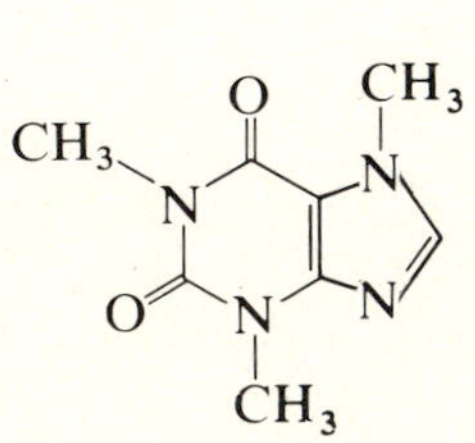

caffeine
*a stimulant to the
central nervous system*

Figure 16.2. *Some physiologically active amines. Note that these are all related in structure to β-phenylethylamine:* ◯—$CH_2CH_2NH_2$.

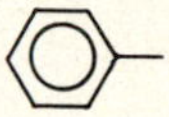

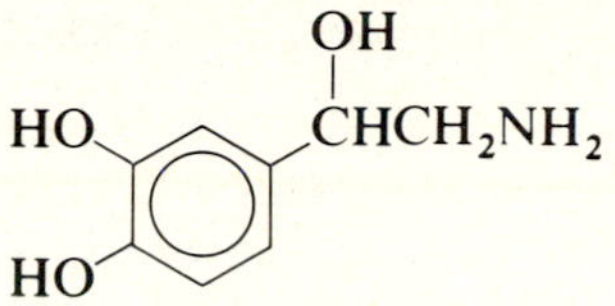

norepinephrine
involved in nerve impulses

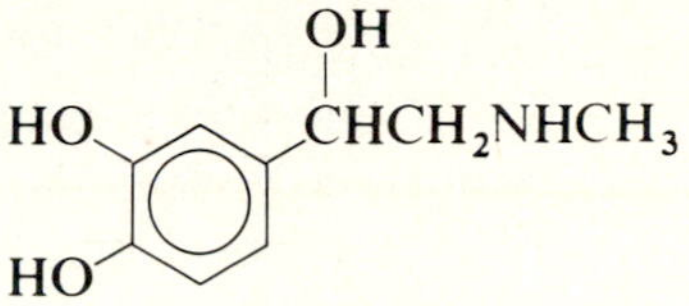

epinephrine
(adrenaline)
stimulates the sympathetic
("fight or flight") nervous system

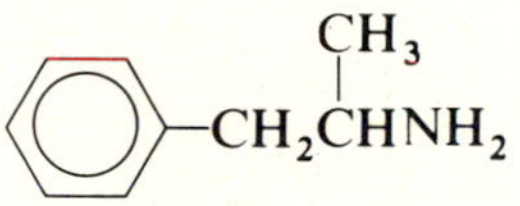

amphetamine
(Benzedrine)
a synthetic stimulant

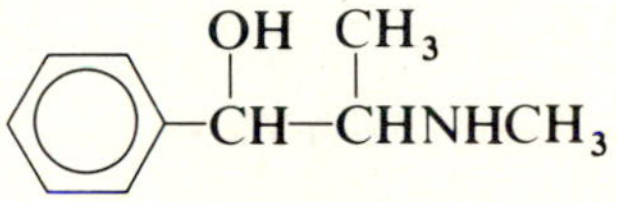

ephedrine
a decongestant and
mild stimulant

regeneration of the amine:

$$CH_3\overset{+}{N}H_2 \;+\; {}^-OH \longrightarrow CH_3\overset{..}{N}H_2 \;+\; H_2O$$

cation of   base stronger   *the amine*
amine salt   than CH_3NH_2

Problem 16.10.   Glycine ($H_2NCH_2\overset{\overset{\textstyle O}{\|}}{C}OH$) is an amino acid that can be obtained from proteins. Although glycine has a low formula weight, its melting point (262°C) is quite high. Suggest a reason for this physical property.

Problem 16.11.   Predict the product if the amine salt $(CH_3)_3NH^+\ Cl^-$ is treated with aqueous $Na^+\ {}^-OH$.

16.6   QUATERNARY AMMONIUM SALTS

The unshared electrons of the nitrogen of a tertiary amine (R_3N) can be used to form a covalent bond with a *fourth* alkyl or aryl group to yield a salt called a **quaternary ammonium salt**.

one or more H's

four R groups;
no H's

$$R_3\overset{+}{N}H \; Cl^-$$

$$R_4\overset{+}{N} \; Cl^-$$

an amine salt   *a quaternary ammonium salt*

Compounds containing quaternary ammonium groups are important in biological systems. Choline chloride and acetylcholine chloride are examples of quaternary ammonium salts that are involved in transmitting nerve impulses across synapses (gaps between nerve cells).

$$HOCH_2CH_2-\overset{\overset{\displaystyle CH_3}{|}}{\underset{\underset{\displaystyle CH_3}{|}}{\overset{+}{N}}}-CH_3 \; Cl^-$$

$$CH_3\overset{\overset{\displaystyle O}{\|}}{C}OCH_2CH_2-\overset{\overset{\displaystyle CH_3}{|}}{\underset{\underset{\displaystyle CH_3}{|}}{\overset{+}{N}}}-CH_3 \; Cl^-$$

choline chloride   acetylcholine chloride

If a quaternary ammonium salt contains one or more large hydrocarbon groups, it can act as an *emulsifying agent* or a *detergent*: the ionic end is attracted to water, while grease and oils are attracted to the hydrocarbon group. Such compounds are also an important component of cell membranes, where the hydrocarbon groups help form a barrier to water and ions. (See Section 22.2.) Some quaternary ammonium salts have antibacterial properties as well as detergent activity and are used to store sterilized surgical instruments, to disinfect diapers, and so forth.

Problem 16.12.   Complete the following equations. If no reaction occurs, write "no reaction."

(a) $HOCH_2CH_2\overset{+}{N}H_3 \; Cl^- + \; ^-OH \longrightarrow$

(b) $HOCH_2CH_2\overset{+}{N}(CH_3)_3 \; Cl^- + \; ^-OH \longrightarrow$

16.7   COMPLEX COMPOUNDS AND CHELATES

The unshared electrons of nitrogen in ammonia, an amine, or a heterocyclic compound can be donated to a metal ion as well as to a hydrogen ion to yield a **complex ion** or a **complex compound**, an ion or compound in which a covalent molecule is bonded to or strongly associated with the metal ion.

complex ion formation:

$$H-\overset{\overset{\displaystyle \cdot\cdot}{}}{\underset{\underset{\displaystyle H}{|}}{N}}-H + Ag^+ + H-\overset{\overset{\displaystyle \cdot\cdot}{}}{\underset{\underset{\displaystyle H}{|}}{N}}-H \longrightarrow$$

$$[H_3N \rightarrow Ag \leftarrow NH_3]^+ \quad \text{or} \quad Ag(NH_3)_2{}^+$$

the silver-ammonia complex ion

An interesting type of complex can be formed when two or more nitrogen atoms are in the same molecule. In the following example, two molecules of ethylenediamine ($H_2NCH_2CH_2NH_2$) have added to a copper(II) ion:

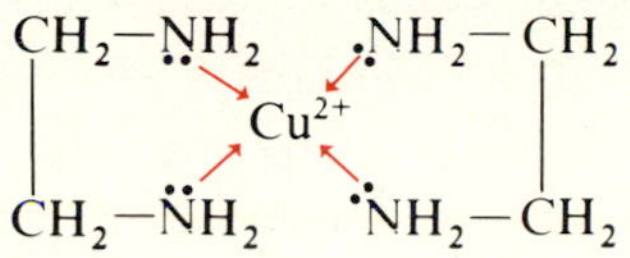

$$CH_2-\overset{..}{N}H_2 \quad \overset{.}{N}H_2-CH_2$$
$$| \qquad \searrow Cu^{2+} \swarrow \qquad |$$
$$CH_2-\overset{..}{N}H_2 \quad \overset{..}{N}H_2-CH_2$$

Figure 16.3. *Two important naturally occurring chelates.*

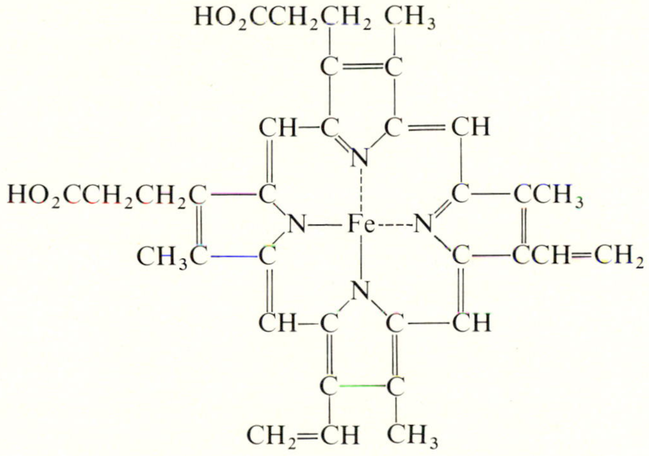

heme

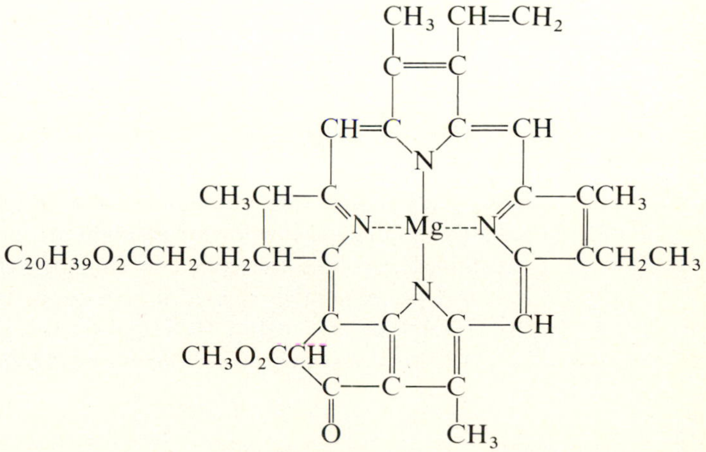

chlorophyll a

This type of complex ion or compound is called a **chelate**, from the Greek word *kela* (crab's claw)—the imagery is that of a cation being caught in a pincers-like claw. Ethylenediaminetetraacetic acid (EDTA), an organic compound with several atoms that have unshared electrons, is a very effective chelating agent and is often used as an antidote in heavy-metal poisoning.

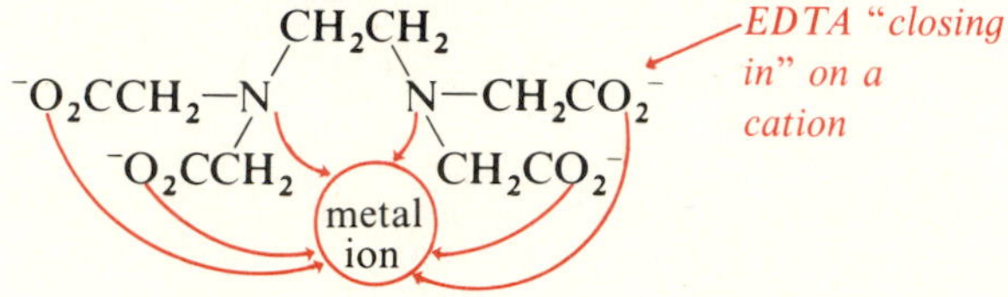

EDTA "closing in" on a cation

Many indispensable natural compounds are chelates—for example, heme (the oxygen-carrier portion of hemoglobin) and chlorophyll (see Figure 16.3). Note the similarities in the organic portions of heme and chlorophyll. The ring system, called the *porphyrin* ring system, is ideally structured for the chelation of metal ions.

Problem 16.13.   Write the equation for the reaction you would expect to occur when an aqueous solution of silver nitrate ($Ag^+ NO_3^-$) is treated with $H_2NCH_2CH_2NH_2$. Show the electrons as dots on the nitrogen atoms in your equation.

SUMMARY

Alcohols, *phenols*, and *ethers* are structurally related to water.

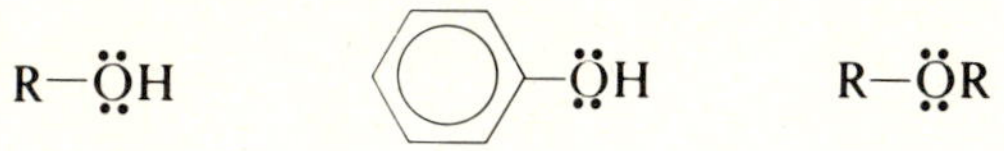

an alcohol   *a phenol*   *an ether*

Alcohols, which contain a *hydroxyl group* (—OH), can undergo a variety of reactions, but ethers are relatively unreactive. Because alcohols undergo hydrogen bonding, they are high boiling compared to hydrocarbons of similar formula weight, and low-formula-weight alcohols are water soluble.

Simple alcohols are named by changing the *-e* of the parent alkane name to *-ol*. A prefix number is used if necessary. Alcohols are classified as *primary* (RCH_2OH), *secondary* (R_2CHOH), or *tertiary* (R_3COH).

Alcohols can undergo *dehydration* and *oxidation*.

dehydration:

$$\underset{\substack{|\\H}}{R_2C}-\underset{\substack{|\\OH}}{CR_2} \xrightarrow[\text{heat}]{H_2SO_4} R_2C{=}CR_2 + H_2O$$

oxidation:

$$RCH_2OH \xrightarrow[\text{heat}]{KMnO_4} RC\overset{\overset{\displaystyle O}{\|}}{H} \longrightarrow RC\overset{\overset{\displaystyle O}{\|}}{O}H$$

$$R_2CHOH \xrightarrow[\text{heat}]{KMnO_4} R_2C{=}O$$

Thiols (RSH) are sulfur analogs of alcohols, while *sulfides* (RSR) are sulfur analogs of ethers. Thiols can be coupled to yield *disulfides* (RSSR) or oxidized to *sulfoxides* ($R_2S{=}O$).

Amines are structurally related to ammonia and may be classified as *primary* (RNH_2), *secondary* (R_2NH), or *tertiary* (R_3N). If the nitrogen is bonded to four alkyl or aryl groups, the compound is a *quaternary ammonium salt* ($R_4N^+ A^-$).

Amines (but not quaternary ammonium salts) undergo hydrogen bonding. They are also basic and yield *amine salts* when treated with an acid. (The amine can be regenerated by treatment with a strong base.)

$$R_3N{:} \;+\; H^+ \;+\; Cl^- \;\longrightarrow\; R_3\overset{+}{N}H \; Cl^-$$

The unshared electrons of an amine nitrogen can be donated to a metal cation to yield a *complex ion*, a *complex compound*, or a *chelate*.

KEY TERMS

alcohol	*sulfide	quaternary ammonium salt
ether	*disulfide	complex ion or compound
phenol	*sulfoxide	chelate
dehydration reaction	amine	
*thiol	alkaloid	

STUDY PROBLEMS

16.14. Identify each of the following compounds as an ether, an alcohol, or a phenol:

(a) 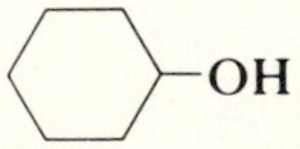 [cyclohexane]–OH   (b) 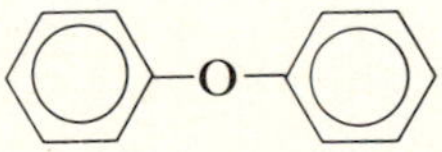 [benzene]–O–[benzene]

(c) 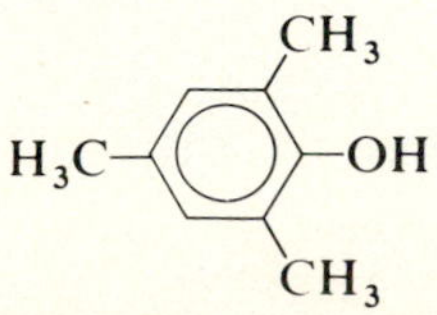 H_3C–[benzene with two CH_3]–OH

(d) 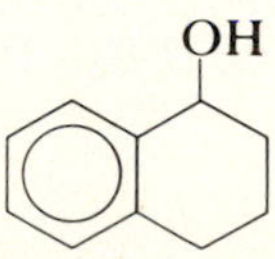 [naphthalene-like bicyclic structure]–OH

16.15. Name the following alcohols first by the IUPAC system and then as alkyl alcohols. Classify each as primary, secondary, or tertiary.

(a) $(CH_3)_2CHOH$ (b) $CH_3CH_2CH_2CH_2OH$

(c) 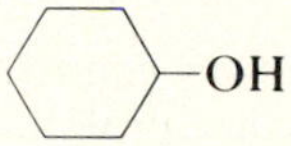 [cyclohexane]–OH   (d) $(CH_3)_3COH$

16.16. Which alcohols can be (a) used as a rubbing alcohol? (b) safely ingested?

16.17.   Show hydrogen bonds between the following molecules:

(a) cyclohexanol and water

(b) two molecules of 1-propanol

16.18.   Write equations for the dehydration of the following alcohols. If more than one alkene can be formed, show all possibilities.

(a) $CH_3CH_2CH_2OH$ (b) $CH_3\overset{\text{OH}}{\underset{|}{C}}HCH_2CH_3$

16.19.   State whether each of the following biochemical conversions is an oxidation or a reduction:

(a) $HO_2CCH_2\overset{\text{OH}}{\underset{|}{C}}HCO_2H \longrightarrow$
malic acid

$HO_2CCH_2\overset{\text{O}}{\overset{\|}{C}}CO_2H$
oxaloacetic acid

(b) $HO_2CCH_2CH_2CO_2H \longrightarrow$
succinic acid

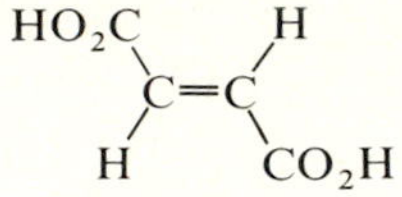

$\underset{H}{\overset{HO_2C}{}}C=C\underset{CO_2H}{\overset{H}{}}$
fumaric acid

16.20.   What is the organic product of each of the following reactions?

(a) $CH_3CH_2CH_2CH_2OH$ +
hot $KMnO_4$ solution

(b) $CH_3\overset{\text{OH}}{\underset{|}{C}}HCO_2H$ + hot $KMnO_4$ solution
lactic acid

*__16.21.__   Identify the compound class of each of the following compounds:

(a) CH_3CH_2SH (b) $CH_3CH_2S-SCH_3$

(c) $CH_3CH_2SCH_3$ (d) $CH_3CH_2\overset{\text{O}}{\overset{\|}{S}}CH_2CH_3$

16.22.   Name each of the following amines, and classify each as primary, secondary, or tertiary:

(a) $CH_3CH_2NH_2$ (b) $(CH_3)_2CHNH_2$

(c) $(CH_3)_3CNH_2$ (d) $(CH_3)_3CNHCH_3$

16.23.   Show the hydrogen bonding between two molecules of cyclohexylamine.

16.24.   Complete the following equations:

(a) $(CH_3)_3CNH_2$ + HCl $\longrightarrow$

(b) 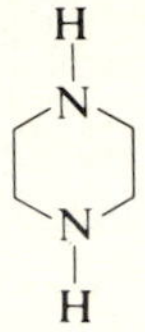   + $2 CH_3CO_2H \longrightarrow$

16.25.   The formula for *dextromethorphan*, a nonaddictive cough suppressant, follows. (Note its close structural relationship to the addictive cough suppressant codeine, Figure 16.1.) Write an equation that shows how this water-insoluble drug could be made water soluble.

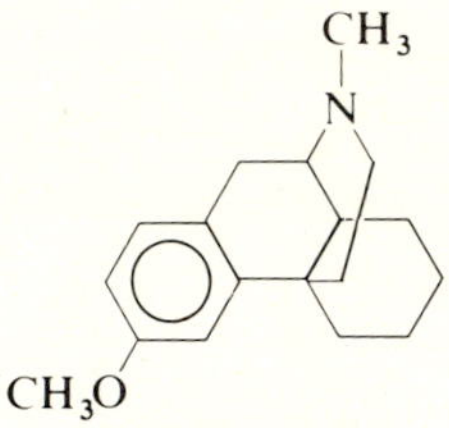

dextromethorphan

16.26.   Which of the following compounds are alkaloids? Explain your answer.

(a)

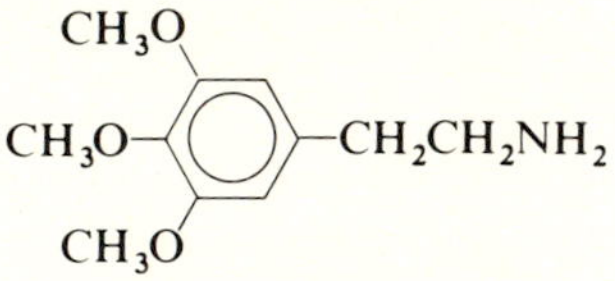

mescaline
*a hallucinogen found in the
peyote cactus*

(b) 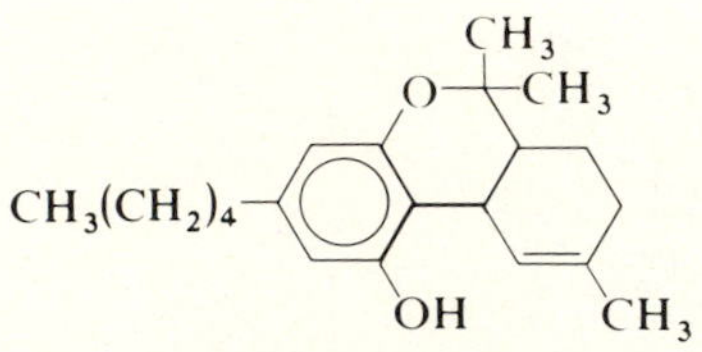 $CH_3(CH_2)_4-$

tetrahydrocannibinol
a depressant in marijuana

(c)

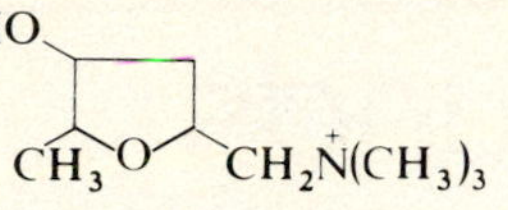

muscarine cation
*found in the poisonous
mushroom* Amanita muscaria

(d) nicotine (Table 15.3, Section 15.6)

16.27. Which of the following compounds is a quaternary ammonium cation?

(a)

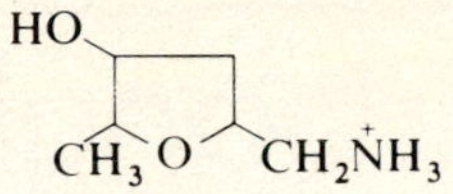

(b) muscarine (Problem 16.26)

16.28. Which of the following compounds could chelate a metal ion? Explain your answer.

(a) 1,3-propanediamine

(b) amphetamine (Figure 16.2)

17 Carbonyl Compounds

A male tsetse fly attempting to breed a cork treated with a synthetic sex-stimulant pheromone. (Courtesy of David A. Carlson, U.S. Department of Agriculture.) An insect pheromone is an organic compound secreted by an insect to communicate with other insects of the same species. Small amounts of a pheromone can be extraordinarily effective. A female gypsy moth may carry only 0.00000001 gram of a sex-attractant pheromone, but she can attract millions of males from miles away. Today scientists are studying how pheromones can be used to control insects such as the gypsy moth caterpillar, which defoliates trees, and the African tsetse fly, which is a vector of sleeping sickness.

Many insect pheromones are compounds containing carbonyl groups (C=O), a type of compound discussed in Chapter 17.

Objectives   List names and structures of carbonyl compound classes. ☐ Draw structures showing hydrogen bonds to carbonyl groups. ☐ Name aldehydes, ketones, carboxylic acids, esters, and amides; describe their physical properties; write equations for their principal reactions. ☐ Write formulas for a polyester, a thioester, and an inorganic ester (optional).

The **carbonyl group** (C=O) is a part of many functional groups found in proteins, fats, carbohydrates, and other biological compounds. Table 17.1 contains a summary of the important carbonyl-containing functional groups and their compound classes. In this chapter, we will discuss each of these compound classes in turn, including their nomenclature and chemistry, and we will survey a few important individual compounds in each class.

Table 17.1. *Some Classes of Carbonyl Compounds*

Name of class	Functional group	General formula[a]	Example
aldehyde	$-\overset{\displaystyle O}{\overset{\|}{C}}-H$	$\overset{\displaystyle O}{\overset{\|}{R}CH}$	$CH_3\overset{\displaystyle O}{\overset{\|}{C}}H$
ketone	$-\overset{\displaystyle O}{\overset{\|}{C}}-$	$R\overset{\displaystyle O}{\overset{\|}{C}}R'$	$CH_3\overset{\displaystyle O}{\overset{\|}{C}}CH_3$
carboxylic acid	$-\overset{\displaystyle O}{\overset{\|}{C}}-OH$	$R\overset{\displaystyle O}{\overset{\|}{C}}OH$	$CH_3\overset{\displaystyle O}{\overset{\|}{C}}OH$
ester	$-\overset{\displaystyle O}{\overset{\|}{C}}-O-$	$R\overset{\displaystyle O}{\overset{\|}{C}}OR'$	$CH_3\overset{\displaystyle O}{\overset{\|}{C}}OCH_2CH_3$
amide	$-\overset{\displaystyle O}{\overset{\|}{C}}-N\overset{/}{_\backslash}$	$R\overset{\displaystyle O}{\overset{\|}{C}}NR'_2$	$CH_3\overset{\displaystyle O}{\overset{\|}{C}}NH_2$

[a] In each case, R may be alkyl or aryl. In formulas showing R', the R' groups may be the same as or different from R.

17.1 THE CARBONYL GROUP

The chemistry of carbonyl compounds results partly from the bonding in the carbonyl group. The carbon atom and the oxygen atom in the carbonyl group are joined by a double bond. Unlike a carbon-carbon double bond, the carbon-oxygen double bond is *polar*. Because the oxygen atom is more electronegative than the carbon atom, the oxygen atom carries a partial negative charge, while the carbon atom carries a partial positive charge. Another important feature of the carbonyl group is that the oxygen atom has two pairs of unshared valence electrons.

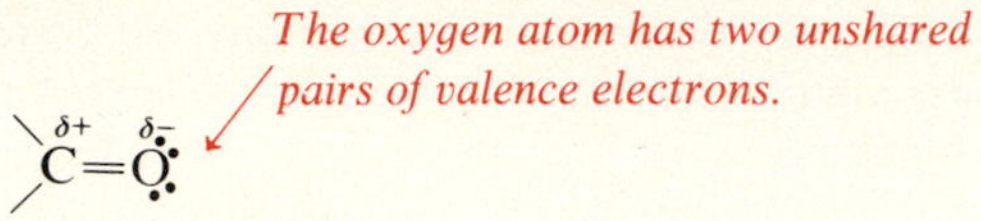

The oxygen atom has two unshared pairs of valence electrons.

$$\ce{\backslash C = O}$$

The carbonyl group is polar.

Because they contain oxygen atoms with unshared electrons, carbonyl compounds undergo hydrogen bonding with compounds that contain NH or OH bonds. You will learn in Chapter 20 that the shapes of protein molecules are one of their distinguishing features and a requirement for their biological activity. For example, the shapes of proteins in muscle fibers are different from those of proteins in hemoglobin. Protein molecules are held in their distinctive shapes primarily because of hydrogen bonding between their carbonyl groups and NH groups, either within the same molecule or in a neighboring protein molecule.

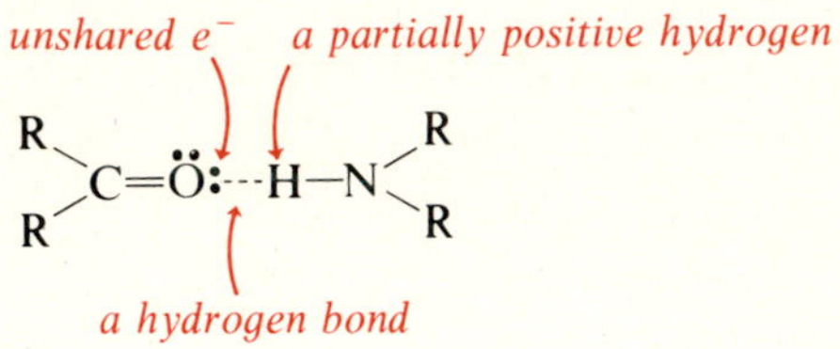

unshared e⁻   a partially positive hydrogen

$$\ce{R2C=O\bond{...}H-NR2}$$

a hydrogen bond

Example   Refer to Table 17.1 (and Chapter 16), then name the functional groups in the following compounds:

(a) HCCH₂CCH₂COH   (b)

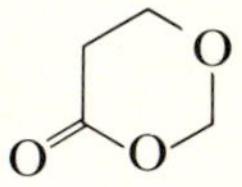

Solution

(a) 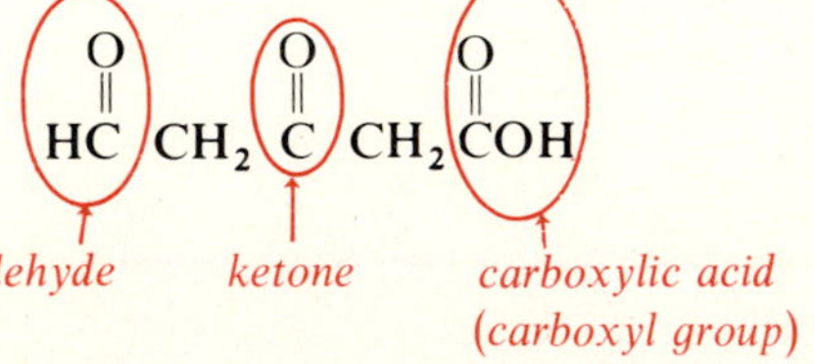   aldehyde   ketone   carboxylic acid (carboxyl group)

(b) 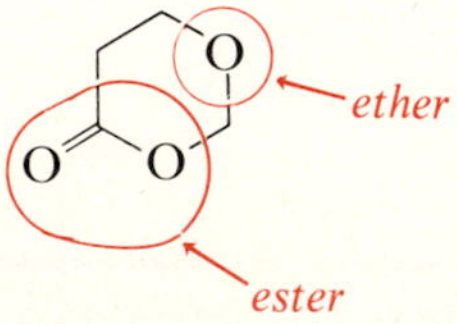   ether   ester

Example   Draw structures that show a hydrogen bond between a water molecule and an acetone molecule, (CH₃)₂C=O.

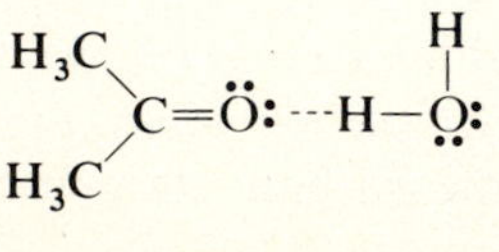

Solution   The hydrogen atom in water forms hydrogen bonds with the unshared pair of electrons of the carbonyl oxygen. (The C-H bond is nonpolar; therefore, the hydrogens of the —CH$_3$ groups do not participate in hydrogen bonding.)

Problem 17.1.   Circle and name the functional groups in each of the following formulas:

(a)

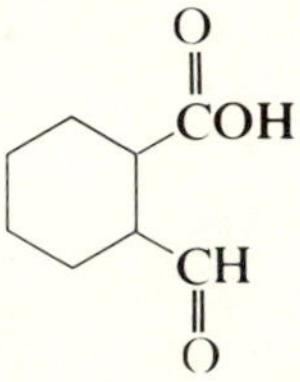

(b) 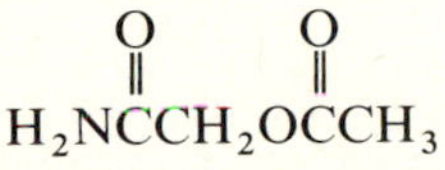 H$_2$NCCH$_2$OCCH$_3$

Problem 17.2.   Draw all the hydrogen bonds that can be formed between two acetic acid molecules (CH$_3$CO$_2$H).

17.2   ALDEHYDES AND KETONES

Compounds in which the carbonyl carbon is bonded to at least one hydrogen atom are called **aldehydes**. Compounds in which the carbonyl carbon is bonded to two carbon atoms are called **ketones**. In either type of compound, the organic group can be alkyl or aryl.

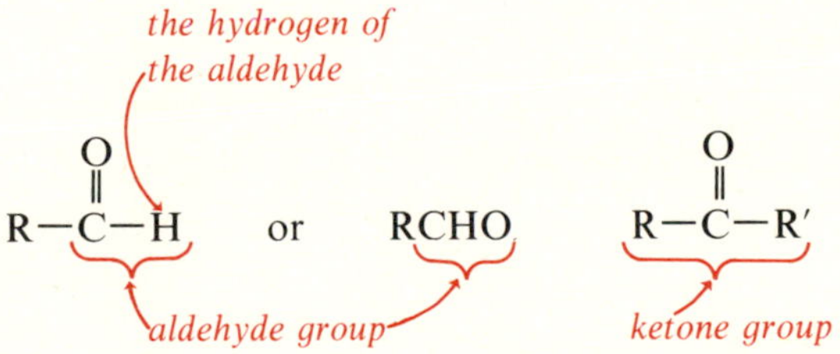

the hydrogen of the aldehyde

$$R—C—H \quad or \quad RCHO \qquad R—C—R'$$

aldehyde group   *ketone group*

Figure 17.1.   Model of acetone, a compound containing a carbonyl group.

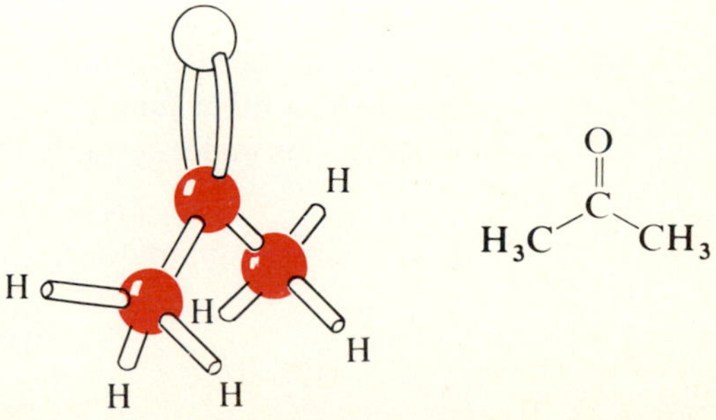

Table 17.2.   *Names and Boiling Points of Some Aldehydes and Ketones*

Formula	IUPAC name	Trivial name	Bp (°C)
aldehydes:			
HCHO	methanal	formaldehyde	−21
CH_3CHO	ethanal	acetaldehyde	20
CH_3CH_2CHO	propanal	propionaldehyde	49
$CH_3CH_2CH_2CHO$	butanal	butyraldehyde	76
ketones:			
$CH_3\overset{\overset{O}{\|\|}}{C}CH_3$	propanone	acetone	56
$CH_3\overset{\overset{O}{\|\|}}{C}CH_2CH_3$	butanone	methyl ethyl ketone	80

Simple aldehyde and ketone molecules cannot form hydrogen bonds with other aldehyde or ketone molecules because they do not contain a hydrogen atom bonded to a nitrogen or oxygen atom. However, they can form hydrogen bonds with water; simple aldehydes and ketones containing less than five carbon atoms are water soluble. The polarity of the carbonyl group in aldehydes and ketones results in moderately strong attractions between molecules; their boiling points are between those of analogous hydrocarbons (weak intermolecular attractions) and alcohols (strong intermolecular attractions); see Table 17.2.

Naming Aldehydes and Ketones

In the IUPAC system, aldehydes are named by dropping the final *-e* from the name of the parent alkane and adding *-al*.

$$CH_3CH_3 \qquad CH_3\overset{\overset{O}{\|\|}}{C}H \qquad CH_3CH_2CH_3 \qquad CH_3CH_2\overset{\overset{O}{\|\|}}{C}H$$

ethane   ethanal   propane   propanal

The carbonyl carbon in an aldehyde must be at the end of a carbon chain because it is bonded to a hydrogen atom. In the numbering of the parent chain, the carbonyl carbon atom of an aldehyde is considered the first carbon, regardless of other functional groups. We do not use the number in the name, however; it is understood that the aldehyde group is at position 1.

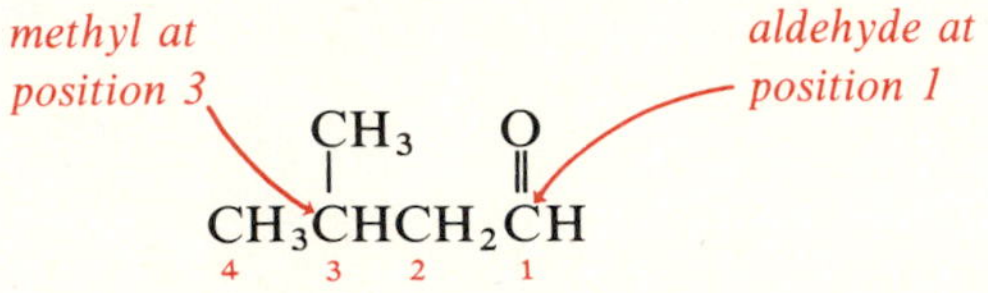

methyl at position 3   *aldehyde at position 1*

$$\underset{4}{CH_3}\underset{3}{\overset{\overset{CH_3}{\|}}{C}H}\underset{2}{CH_2}\underset{1}{\overset{\overset{O}{\|\|}}{C}H}$$

3-methylbutanal

Ketones are named in the IUPAC system with the ending *-one* replacing the *-e* of the parent alkane name. If there is more than one possible position in which a ketone group could appear in a molecule, a prefix number is used in the name to designate the position.

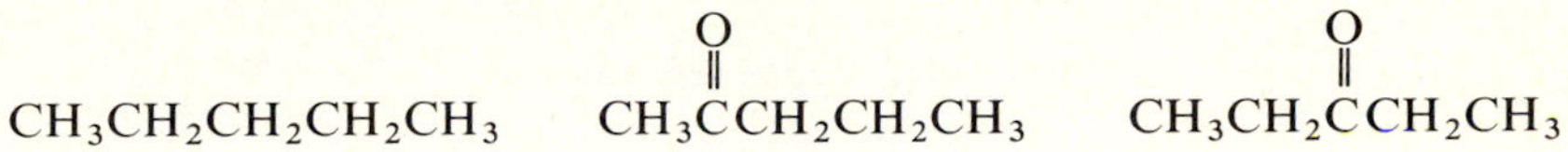

$$CH_3CH_2CH_2CH_2CH_3 \qquad CH_3\overset{\overset{\displaystyle O}{\|}}{C}CH_2CH_2CH_3 \qquad CH_3CH_2\overset{\overset{\displaystyle O}{\|}}{C}CH_2CH_3$$

pentane   2-pentanone   3-pentanone

Many small aldehydes and ketones are commonly referred to by their trivial names. A few of these aldehydes and ketones, along with their trivial names and boiling points, are listed in Table 17.2.

Example   Write the structure for 3-propyl-1-cyclopentanone.

Solution   (1) Draw and number the parent skeleton.

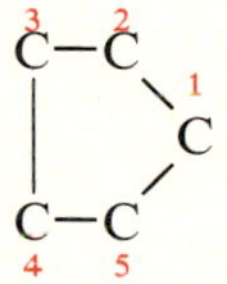

(2) Add functional groups and branches: ketone at position 1, propyl at position 3.

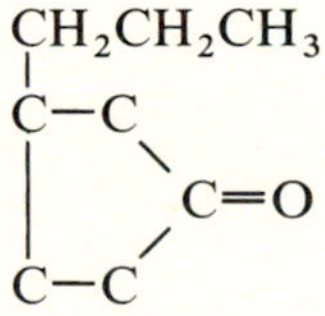

(3) Add hydrogens.

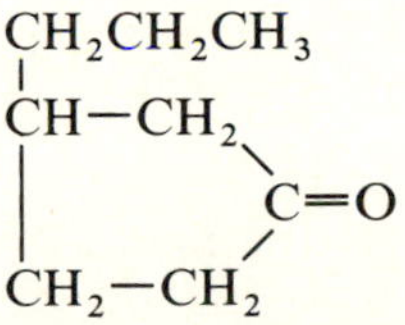

Problem 17.3.   Name the following compounds:

(a) 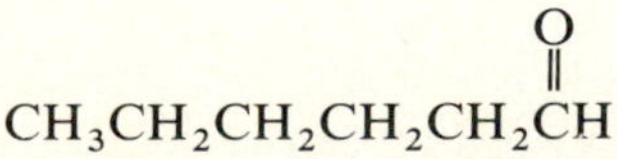 $CH_3CH_2CH_2CH_2CH_2\overset{\overset{\displaystyle O}{\|}}{C}H$ (b) 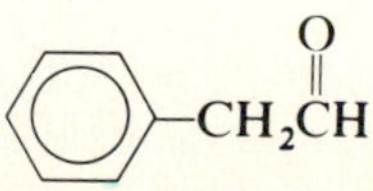 $-CH_2\overset{\overset{\displaystyle O}{\|}}{C}H$

$$\text{(c)} \quad \underset{\underset{Cl}{|}}{CH_3CHC}\overset{\overset{O}{\|}}{C}CH_2CH(CH_3)_2$$

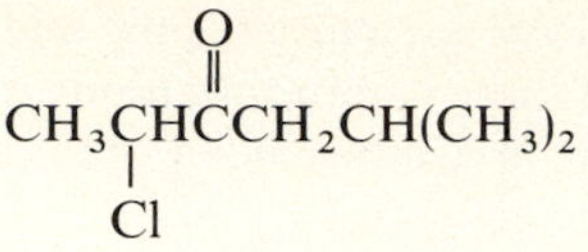

Some Important Aldehydes and Ketones

Formaldehyde ($H_2C{=}O$) is an irritating gas formed in smog. An aqueous solution of formaldehyde in water, called *formalin*, is a common preservative for biological specimens.

Formaldehyde reacts with many other compounds to yield polymers. Formica is one example of such a polymer. Formaldehyde is also used in plywood and particle-board glues and in urea-formaldehyde foam insulation. Formaldehyde itself can polymerize. The formaldehyde polymer, called *paraformaldehyde*, is a solid that regenerates gaseous formaldehyde when it is heated. This polymer provides a convenient method for shipping and storing formaldehyde as a solid, rather than as a gas. Paraformaldehyde is a sedative when taken internally and is occasionally used for that purpose in medical practice.

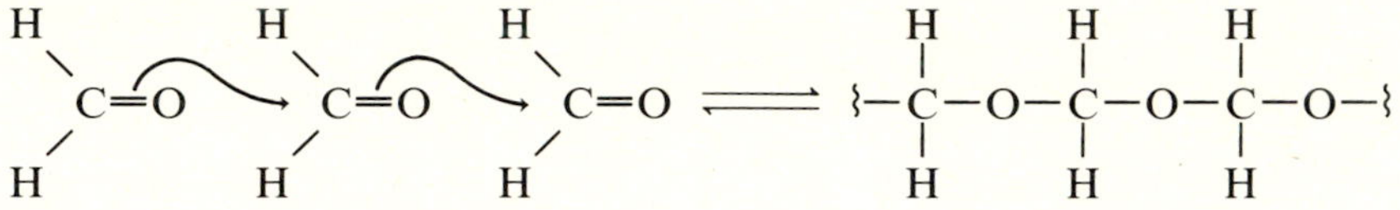

formaldehyde

*portion of a
paraformaldehyde polymer*

Acetone $\left(CH_3\overset{\overset{O}{\|}}{C}CH_3 \right)$ is a rather sweet-smelling ketone that is used as a solvent in industry and in the laboratory. It is useful because it is relatively inexpensive and is miscible with water as well as with organic solvents. Acetone is also one of the products of metabolism of fats and thus is found in small amounts in the blood. Persons suffering from *diabetes mellitus* often have such excessive amounts of this ketone in their bloodstream that its odor can be detected on the breath and in the urine, conditions called *ketone breath* and *ketonuria*.

17.3 REACTIONS OF ALDEHYDES AND KETONES

Reduction

In the presence of hydrogen gas and a suitable catalyst, such as finely divided nickel, the double bond of a carbonyl group can be reduced. The hydrogen gas adds across the carbonyl double bond in a reaction that is similar to the hydrogenation of alkenes. In living systems, the H atoms can be added to the C=O group enzymatically.

$$
\underset{\substack{\text{acetaldehyde}\\ \textit{an aldehyde}}}{\overset{\displaystyle O \atop \displaystyle \|}{CH_3CH}} + H_2 \xrightarrow[\text{catalyst}]{\text{heat, pressure}} \underset{\text{ethanol}}{\overset{\displaystyle OH \atop \displaystyle |}{CH_3CH_2}}
$$

reduced to alcohols

$$
\underset{\substack{\text{acetone}\\ \textit{a ketone}}}{\overset{\displaystyle O \atop \displaystyle \|}{CH_3CCH_3}} + H_2 \xrightarrow[\text{catalyst}]{\text{heat, pressure}} \underset{\text{2-propanol}}{\overset{\displaystyle OH \atop \displaystyle |}{CH_3CHCH_3}}
$$

Oxidation

Under mild oxidizing conditions, aldehydes are easily oxidized to carboxylic acids. Ketones resist oxidation. Compare the reactions to the oxidation scheme and reactions in Section 16.2.

$$
\underset{\text{acetaldehyde}}{\overset{\displaystyle O \atop \displaystyle \|}{CH_3CH}} \xrightarrow{\overset{\text{mild}}{\text{oxidizing agent}}} \underset{\text{acetic acid}}{\overset{\displaystyle O \atop \displaystyle \|}{CH_3COH}}
$$

$$
\underset{\text{acetone}}{\overset{\displaystyle O \atop \displaystyle \|}{CH_3CCH_3}} \xrightarrow{\overset{\text{mild}}{\text{oxidizing agent}}} \text{No reaction}
$$

generalized:

$$
\underset{\textit{an aldehyde}}{RCHO} \xrightarrow{\overset{\text{mild}}{\text{oxidizing agent}}} \underset{\textit{a carboxylic acid}}{RCO_2H}
$$

Many chemical tests used to distinguish aldehydes from other functional groups are based on the oxidation of the aldehyde group. **Tollens reagent**, an alkaline solution of the silver-ammonia complex ion, is used in one such test. An aldehyde, but not a simple ketone, reacts with the silver ion (Ag^+) of Tollens reagent to form the anion of the carboxylic acid and silver metal, which plates out on the container as a silver mirror. A similar reaction is used in the manufacture of silver mirrors.

$$
\underset{\text{acetaldehyde}}{\overset{\displaystyle O \atop \displaystyle \|}{CH_3CH}} \xrightarrow[\text{(Tollens reagent)}]{Ag^+} \underset{\text{acetate ion}}{\overset{\displaystyle O \atop \displaystyle \|}{CH_3CO^-}} + \underset{\text{silver metal}}{Ag\downarrow}
$$

Two other aldehyde tests, the *Fehling test* and the *Benedict test*, are based on the ability of aldehydes to reduce Cu^{2+} to Cu^+, which is detected as a brick-red precipitate of Cu_2O.

Many sugars contain aldehyde groups and react with these test reagents. Because they can reduce Ag^+ or Cu^{2+}, these sugars are called **reducing sugars**. In the past, these reactions were used as clinical tests to detect glucose in the urine, a symptom of diabetes. Today, test papers containing a specific catalyst for the oxidation of glucose, the enzyme *glucose oxidase*, are used.

Problem 17.4.   Write the structures of the organic products in the following equations:

(a) $\underset{\substack{\|\\ \text{O}}}{CH_3CH_2CH} + H_2 \xrightarrow[\text{catalyst}]{\text{heat, pressure}}$

(b) $\underset{\substack{\|\\ \text{O}}}{CH_2{=}CHCH} + \text{excess } H_2 \xrightarrow[\text{catalyst}]{\text{heat, pressure}}$

(c) $\underset{\substack{\|\\ \text{O}}}{CH_3CH_2CH} \xrightarrow{Cu^{2+}}$

Addition of Alcohols

Because of the polarity of the carbonyl group, alcohols can add across the carbonyl double bond of an aldehyde and some ketones in the presence of a trace of acid. The addition product of an alcohol and an aldehyde is called a **hemiacetal**. A hemiacetal contains an —OH group and an —OR group bonded to the same carbon atom.

hemiacetal formation:

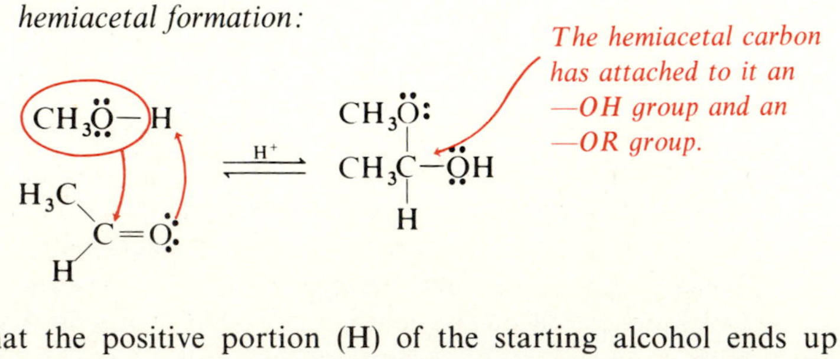

The hemiacetal carbon has attached to it an —OH group and an —OR group.

Note that the positive portion (H) of the starting alcohol ends up on the partially negative carbonyl oxygen and that the negative portion (RO) of the alcohol ends up on the partially positive carbonyl carbon: opposite charges attract. Also note that the alcohol is cleaved at the O—H bond, not at the C—O bond as in dehydration.

generalized:

$$
\underset{\delta-}{RO}\!\!-\!\!\underset{\delta+}{H}
\qquad
\underset{\text{alcohol and aldehyde}}{\overset{\displaystyle R}{\underset{H}{\diagdown}}C\!\!=\!\!\underset{\delta-}{\underset{\delta+}{O}}}
\quad\underset{H^+}{\rightleftharpoons}\quad
\underset{\text{a hemiacetal}}{\overset{RO\ \ H}{\underset{H}{RC\!\!-\!\!O}}}
$$

alcohol and aldehyde   *a hemiacetal*

A second molecule of an alcohol can undergo reaction with the hemi-acetal to give a product with *two* —OR groups on the same carbon. This product is called an **acetal**.

acetal formation:

The acetal carbon has two —OR groups.

$$
\underset{\text{a hemiacetal}}{\overset{OH}{\underset{OCH_3}{CH_3CH}}}
\;+\;
\underset{\text{an alcohol}}{CH_3O\,H}
\;\underset{H^+}{\rightleftharpoons}\;
\underset{\text{an acetal}}{\overset{OCH_3}{\underset{OCH_3}{CH_3CH}}}
\;+\; HOH
$$

a hemiacetal   *an alcohol*   *an acetal*

The formations of hemiacetals and acetals are *reversible* reactions. In most cases, the aldehyde is favored over the hemiacetal or acetal in the equilibrium mixture. Thus, in an acidic solution of an aldehyde and an alcohol, we would expect to find primarily the aldehyde and the alcohol, and only small amounts of hemiacetal and acetal.

favored

$$
\underset{\text{an aldehyde}}{\overset{O}{\overset{\|}{RCH}}}
\;+\;
\underset{\text{an alcohol}}{R'OH}
\;\underset{H^+}{\xleftarrow{\hspace{1cm}}}\;
\underset{\text{a hemiacetal}}{\overset{OH}{\underset{OR'}{RCH}}}
\;\underset{\substack{R'OH\\H^+}}{\xleftarrow{\hspace{1cm}}}\;
\underset{\text{an acetal}}{\overset{OR'}{\underset{OR'}{RCH}}}
\;+\; H_2O
$$

an aldehyde   *an alcohol*   *a hemiacetal*   *an acetal*

Hemiacetals and acetals are very important in the chemistry of sugars. Sugars form *cyclic hemiacetals*, which, unlike ordinary hemiacetals, are quite stable. Cyclic hemiacetals are favored in the equilibrium with the aldehyde. Sugars also form acetals readily. Starch and cellulose are polymers formed from sugar units linked together by acetal bonds. We will discuss these reactions and structures in greater detail in Chapter 18.

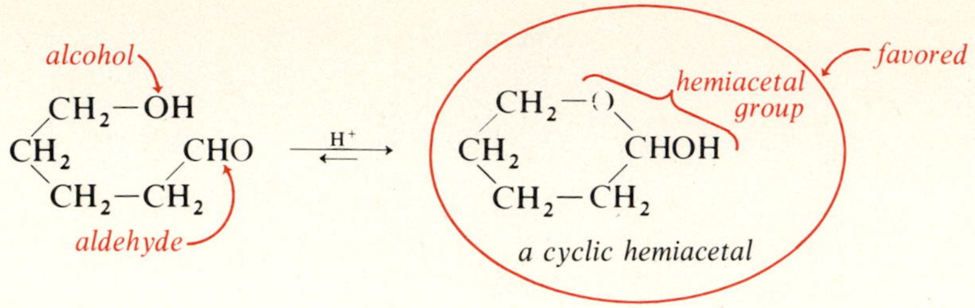

a cyclic hemiacetal

Example   Write formulas for the hemiacetal and acetal formed from formaldehyde and methanol.

Solution

the hemiacetal   *the acetal*

In writing equations for *hemiacetal formation*, determine the product from the *addition* of ROH to C=O, remembering that the product contains OH and OR on the original carbonyl carbon.

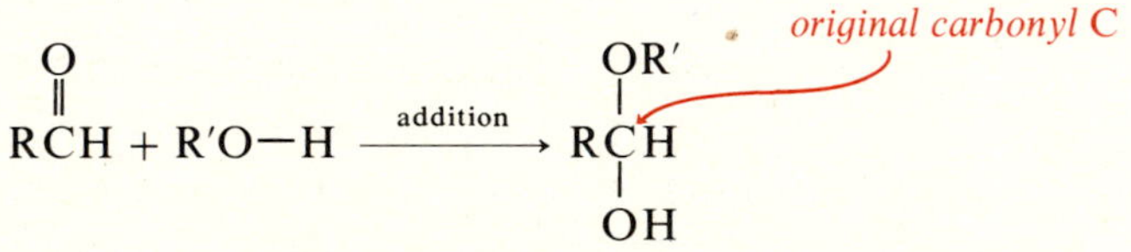

In writing equations for *acetal formation*, replace the hemiacetal OH group with OR′ from the alcohol—a *substitution reaction*.

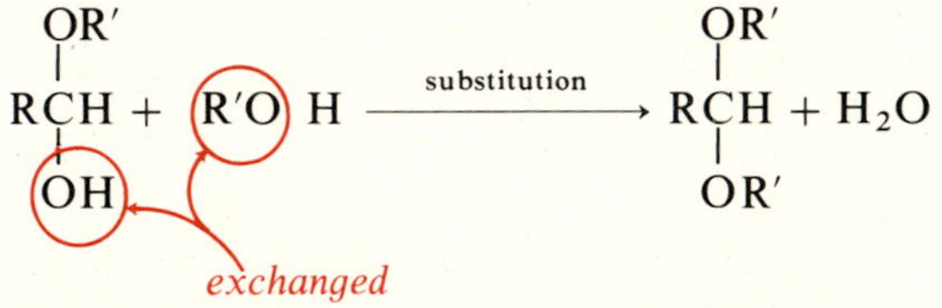

Problem 17.5. Show the hemiacetal and acetal products. (Remember, cyclic acetals are favored.)

$$\text{(a)} \quad CH_3CH_2\overset{\displaystyle O}{\overset{\|}{C}}H + CH_3CH_2OH \xrightarrow{\;H^+\;}$$

$$\text{(b)} \quad CH_3\overset{\displaystyle O}{\overset{\|}{C}}H + HOCH_2CH_2OH \xrightarrow{\;H^+\;}$$

17.4 CARBOXYLIC ACIDS

A **carboxylic acid** is a compound that contains a hydroxyl group bonded to the carbon of a carbonyl group. The functional group of the carboxylic acid is called a **carboxyl group**.

three ways to represent a carboxyl group:

$$-\overset{\overset{\displaystyle O}{\|}}{C}OH \quad \text{or} \quad -COOH \quad \text{or} \quad -CO_2H$$

We have encountered acetic acid (CH_3CO_2H) many times already. Commercially, acetic acid is made by oxidizing ethanol or acetaldehyde. Dilute solutions (vinegar) can be made by an oxidative fermentation of ethanol, which, in turn, can be produced from glucose in a fermentation process with yeast. Acetic acid inhibits microbial growth, and vinegar has been used since ancient times as a preservative for meat, fish, and vegetables.

an enzyme-catalyzed fermentation sequence:

$$glucose \longrightarrow CH_3CH_2OH \longrightarrow CH_3\overset{\overset{\displaystyle O}{\|}}{C}H \longrightarrow CH_3\overset{\overset{\displaystyle O}{\|}}{C}OH$$

ethanol $\qquad$ acetaldehyde $\qquad$ acetic acid

Such diverse compounds as the analgesic aspirin and the weed killer 2,4-D contain carboxyl groups. Many other compounds containing this functional group will be encountered in the biochemical chapters of this book.

acetylsalicylic acid $\qquad$ 2,4-dichlorophenoxyacetic acid (2,4-D)
(aspirin) $\qquad$ *an herbicide*

Carboxylic acids contain a hydrogen atom bonded to an oxygen atom and thus can undergo hydrogen bonding with water or with other carboxylic acid molecules. The carboxylic acids of up to four carbon atoms are water soluble and fairly high boiling for their formula weights. (See Table 17.3.)

Naming Carboxylic Acids

The carboxylic acids that contain fewer than ten carbons have strong, and often very unpleasant, odors. For example, the strong smell of vinegar is due to acetic acid. Some of the common names of the carboxylic acids are based on their odors or sources. The carboxylic acid with four carbon atoms is called butyric acid because it is the odorous component of rancid butter. The six-carbon carboxylic acid, caproic acid, is found in the skin secretions of goats

Table 17.3.   *Names and Boiling Points of Some Carboxylic Acids*

Formula	IUPAC name	Trivial name	Latin origin	Bp (°C)
$\overset{\text{O}}{\overset{\|}{\text{HCOH}}}$	methanoic acid	formic acid	*formica*, ants	100
$\overset{\text{O}}{\overset{\|}{\text{CH}_3\text{COH}}}$	ethanoic acid	acetic acid	*acetum*, vinegar	118
$\overset{\text{O}}{\overset{\|}{\text{CH}_3\text{CH}_2\text{CH}_2\text{COH}}}$	butanoic acid	butyric acid	*butyrum*, butter	163
$\overset{\text{O}}{\overset{\|}{\text{CH}_3(\text{CH}_2)_4\text{COH}}}$	hexanoic acid	caproic acid	*caper*, goat	205

(*caper* in Latin) and has an unforgettable, permeating odor. Formic acid was at one time obtained by the distillation of red ants (*formica*).

In the IUPAC nomenclature of carboxylic acids, the number of carbon atoms in the longest chain containing the carboxyl group determines the parent. Because the carboxyl group must be at the end of the chain, the carbonyl carbon is numbered as carbon 1. (As with aldehydes, the number 1 is omitted from the name.) The ending of the IUPAC name for a carboxylic acid is -*oic acid*.

Example Give the IUPAC name for the following compound:

$$\underset{\text{CH}_3\text{CHCH}_2\text{CHCO}_2\text{H}}{\overset{\overset{\text{CH}_3}{|}\qquad\overset{\text{Br}}{|}}{}}$$

Solution (1) Number the chain.

$$\underset{5\quad\ 4\quad\ 3\quad\ 2\ \ 1}{\underset{\text{CH}_3\text{CHCH}_2\text{CHCO}_2\text{H}}{\overset{\overset{\text{CH}_3}{|}\qquad\overset{\text{Br}}{|}}{}}}$$

(2) Name the parent:

pentanoic acid

(3) Add the branches and the functional groups:

2-bromo-4-methylpentanoic acid

Problem 17.6.   Name the following carboxylic acids:

(a) $\text{ClCH}_2\text{CO}_2\text{H}$ (b) ⬡—$\text{CH}_2\text{CH}_2\text{CH}_2\text{CO}_2\text{H}$

17.5   REACTIONS OF CARBOXYLIC ACIDS

Acidity

Carboxylic acids are very weak acids compared to HCl or H_2SO_4, but are stronger acids than most other classes of organic compounds. The water-soluble carboxylic acids, such as acetic acid, undergo partial ionization in water.

$$\underset{\text{acetic acid}}{CH_3\overset{\displaystyle O}{\overset{\|}{C}}OH} \underset{H_2O}{\longleftarrow\!\!\longrightarrow} \underset{\text{acetate ion}}{CH_3\overset{\displaystyle O}{\overset{\|}{C}}O^- + H^+}$$

Carboxylic acids react with bases such as NaOH or $NaHCO_3$ to form **carboxylate salts**. Most sodium and potassium carboxylates are soluble in water (even when the carboxylic acid itself is not). High-formula-weight sodium carboxylates, such as $C_{15}H_{31}CO_2^-\ Na^+$, which are soaps, are only sparingly soluble.

reaction with base:

$$\underset{\text{acetic acid}}{CH_3\overset{\displaystyle O}{\overset{\|}{C}}OH} + Na^+OH^- \xrightarrow{H_2O} \underset{\substack{\text{sodium acetate}\\ \textit{a salt}}}{CH_3\overset{\displaystyle O}{\overset{\|}{C}}O^-\ Na^+} + H_2O$$

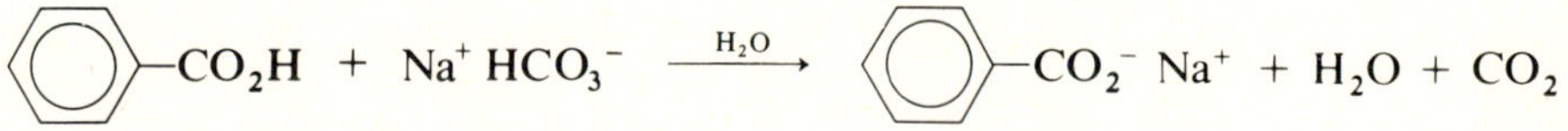

$$\underset{\text{benzoic acid}}{\bigcirc\!\!\!\!-CO_2H} + Na^+HCO_3^- \xrightarrow{H_2O} \underset{\substack{\text{sodium benzoate}\\ \textit{a salt}}}{\bigcirc\!\!\!\!-CO_2^-\ Na^+} + H_2O + CO_2$$

When a carboxylate ion is treated with a strong mineral acid, such as aqueous HCl, the carboxylic acid is regnerated.

regeneration of the carboxylic acid:

$$CH_3\overset{\displaystyle O}{\overset{\|}{C}}O^-\ Na^+ + H^+Cl^- \longrightarrow CH_3\overset{\displaystyle O}{\overset{\|}{C}}OH + Na^+Cl^-$$

Many carboxylic acids found in biological systems are *polyfunctional*—that is, they contain other functional groups. When proteins are digested, for example, they are broken down to amino acids, which can undergo an internal acid-base reaction to yield salt-like, water-soluble substances called *dipolar ions.*

*The amino acid alanine
exists as a dipolar ion.*

$$\underset{\text{CH}_3\text{CHCO}_2(\text{H})}{\overset{\overset{\text{NH}_2}{|}}{}} \longrightarrow \underset{\text{CH}_3\text{CHCO}_2{}^-}{\overset{\overset{\overset{+}{\text{NH}_3}}{|}}{}}$$

Decarboxylation

Biological energy arises from the cellular oxidation of nutrients. In the production of energy, the nutrients are converted to CO_2 and H_2O. The carbon dioxide formed comes from **decarboxylation reactions**, reactions in which CO_2 is lost from a carboxylic acid. In most decarboxylations (laboratory or enzymatic), the original carboxylic acid contains a second carbonyl group in the 2 or 3 position:

$$\overset{\overset{O}{\|}\ \overset{O}{\|}}{\{-C-COH} \quad \text{or} \quad \overset{\overset{O}{\|}\quad\overset{O}{\|}}{\{-CCH_2COH}$$

The following equation shows how acetone can be formed from acetoacetic acid, a reaction that can occur by heating or enzymatically.

$$\underset{\text{acetoacetic acid}}{\text{CH}_3\overset{\overset{O}{\|}}{\text{C}}\text{CH}_2(\text{CO}_2)\text{H}} \xrightarrow[\text{decarboxylase enzyme}]{\text{heat or}} \underset{\text{acetone}}{\text{CH}_3\overset{\overset{O}{\|}}{\text{C}}\text{CH}_3 + \text{CO}_2}$$

Reaction with Alcohols

Carboxylic acids undergo reaction with alcohols in the presence of an acidic catalyst to yield **esters**. These reactions are called *esterification reactions*.

an esterification reaction:

$$\underset{\text{acetic acid}}{\text{CH}_3\overset{\overset{O}{\|}}{\text{C}}(\text{OH})} + \underset{\text{ethanol}}{(\text{H})\text{OCH}_2\text{CH}_3} \underset{\text{heat}}{\overset{\text{H}^+}{\rightleftharpoons}} \underset{\substack{\text{ethyl acetate}\\ \textit{an ester}}}{\text{CH}_3\overset{\overset{O}{\|}}{\text{C}}\text{OCH}_2\text{CH}_3 + \text{H}_2\text{O}}$$

Esterification in acidic solution is a reversible reaction. In the laboratory, an excess of the alcohol is usually used to drive the equilibrium to the right (Le Châtelier's principle). The reverse of the esterification reaction is a **hydrolysis reaction**, which can be used to prepare a carboxylic acid from its ester. (*Hydrolysis* means "cleavage with water.") In a hydrolysis reaction, an excess of water is used to drive the equilibrium to the carboxylic acid side.

esterification:

$$\underset{excess}{RCOH + R'OH} \ \underset{heat}{\overset{H^+}{\rightleftarrows}} \ RCOR' + H_2O$$

ester hydrolysis (the reverse reaction of esterification):

$$\underset{excess}{RCOR' + H_2O} \ \underset{heat}{\overset{H^+}{\rightleftarrows}} \ RCOH + R'OH$$

Example   What are the esterification products of benzoic acid and methanol?

Solution

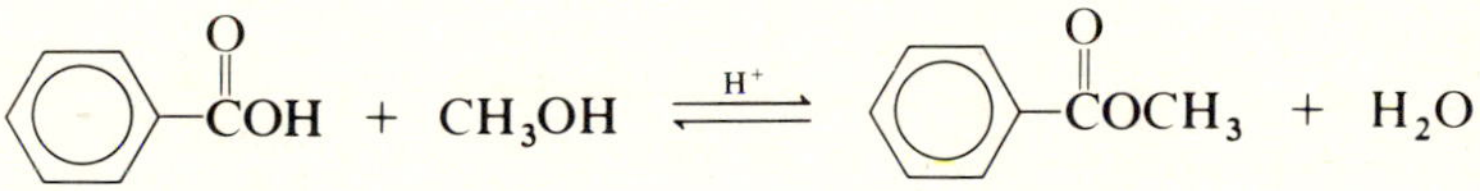

benzoic acid   methanol   methyl benzoate
an ester

In solving a problem involving *esterification*, first write the structures of the carboxylic acid and the alcohol; then determine what is left after H_2O is removed from the two.

Similarly, in solving a hydrolysis problem, remove R′O— from the ester and H from H_2O to determine the products.

Problem 17.7.   Predict the *esterification* products:
(a)   pentanoic acid + 1-butanol ⟶
(b)   acetic acid + 2-propanol ⟶

Problem 17.8.   Predict the *hydrolysis* products:

(a) $CH_3CH_2COCH_2CH_2CH_3 + H_2O \overset{H^+}{\longrightarrow}$

(b)

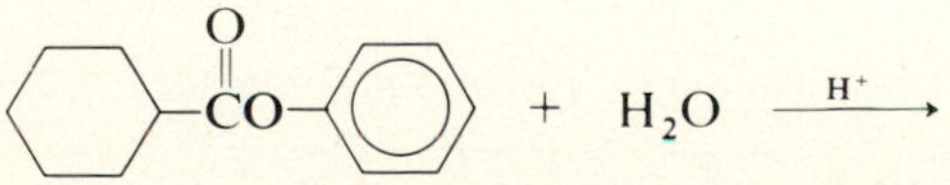

17.6   ESTERS OF CARBOXYLIC ACIDS

Esterification reactions lead us to that exceedingly important class of compounds, the **esters of carboxylic acids**. An ester is a compound that contains a $-CO_2R$ grouping, a carbonyl group bonded to an $-OR$ group.

three ways to represent an ester group:

$$\overset{\overset{\displaystyle O}{\|}}{-C}OR \quad \text{or} \quad -COOR \quad \text{or} \quad -CO_2R$$

Aspirin (Section 17.4) is an example of one important ester. Two other examples follow:

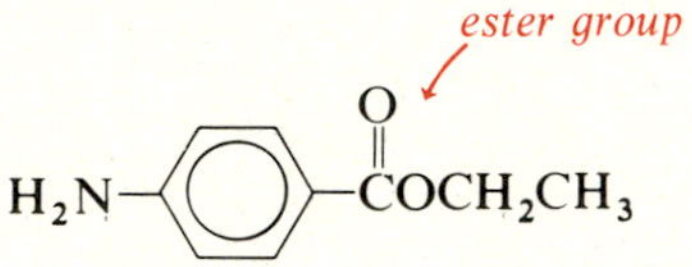

ethyl *p*-aminobenzoate
(benzocaine)
a local anesthetic

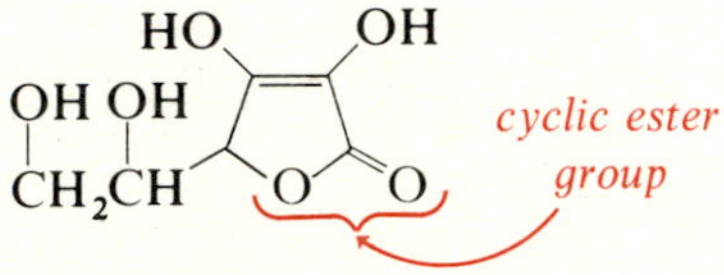

ascorbic acid
(vitamin C)

Table 17.4 lists a few simple esters, along with some of their properties. Note that esters have lower boiling points than do carboxylic acids of similar formula weight. Their boiling points are lower because esters contain no OH or NH groups and thus cannot undergo hydrogen bonding in the pure liquid state. Also, most esters are insoluble in water. Note from Table 17.4 that volatile esters, unlike carboxylic acids, often have sweet, pleasant, fruity odors.

Table 17.4.   *Names and Properties of Some Esters*

Formula	IUPAC name	Trivial name	Bp (°C)	Odor
$CH_3\overset{\overset{\displaystyle O}{\|}}{C}OCH_2CH_3$	ethyl ethanoate	ethyl acetate	77	pleasant
$CH_3\overset{\overset{\displaystyle O}{\|}}{C}OCH_2CH_2CH_3$	*n*-propyl ethanoate	*n*-propyl acetate	102	like pears
$CH_3CH_2CH_2\overset{\overset{\displaystyle O}{\|}}{C}OCH_2CH_3$	ethyl butanoate	ethyl butyrate	121	like pineapple
$CH_3CH_2\overset{\overset{\displaystyle O}{\|}}{C}OCH_2\overset{\overset{\displaystyle CH_3}{\|}}{C}HCH_3$	isobutyl propanoate	isobutyl propionate	137	like rum

Naming Esters

In the IUPAC name of an ester, the alkyl or aryl group that has replaced the H^+ of the carboxylic acid precedes the name as a separate word. The ending *-ic acid* of the parent carboxylic acid is then changed to *-ate*.

$$
\begin{matrix}
& O \\
& \parallel \\
CH_3CH_2C&OH
\end{matrix}
\qquad
\begin{matrix}
& O \\
& \parallel \\
CH_3CH_2C&OCH_3
\end{matrix}
$$

propanoic acid   methyl propanoate
a carboxylic acid   *an ester*

Note that the first word of the ester name is taken from the alcohol's name and is the name of the alkyl or aryl group that is *attached to the oxygen*. The second word is taken from the carboxylic acid's name and refers to the portion of the molecule that *contains the carbonyl group*.

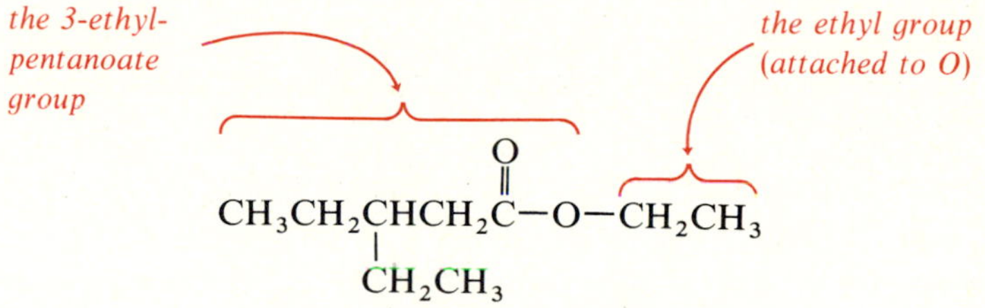

the 3-ethyl-
pentanoate
group

the ethyl group
(attached to O)

$$
\begin{matrix}
& & & & O \\
& & & & \parallel \\
CH_3CH_2C&HCH_2&C&-O&-CH_2CH_3 \\
& \mid \\
& CH_2CH_3
\end{matrix}
$$

ethyl 3-ethylpentanoate

Example   Name the following ester:

$$
\begin{matrix}
& & O \\
& & \parallel \\
CH_3CH_2O&C&CH_2CH_2CH_3
\end{matrix}
$$

Solution   Note that the formula is written in the reverse of the usual fashion. The parent carboxylic acid is the portion of the molecule that contains the C=O group.

from ethanol

from the four-carbon acid:
butanoic acid

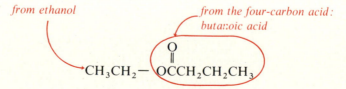

The name of this ester is thus ethyl butanoate.

Problem 17.9.   Write formulas for (a) cyclohexyl propanoate and (b) 2-chloro-*n*-propyl butanoate.

Problem 17.10.   Name the following esters by the IUPAC system. In (b) you may want to refer to Chapter 15.

(a) $CH_3COCH_2CH_2CHCH_3$, commonly called isoamyl acetate or banana oil, also emitted by some species of bees to communicate danger to other bees.

(b) 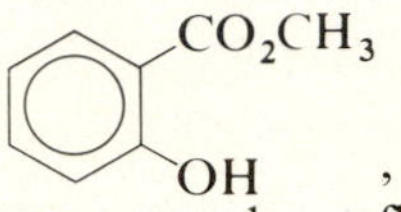 , commonly called methyl salicylate or oil of winter-green, used as a flavoring and as an irritant in liniments.

17.7   REACTIONS OF ESTERS

We mentioned in Section 17.5 that acidic hydrolysis of esters is a reversible reaction and gives carboxylic acids and alcohols as products. Esters may also be hydrolyzed by a water solution of a *strong base* to yield the salt of a carboxylic acid. This reaction, called a **saponification reaction**, is not reversible.

saponification in alkaline solution:

$$CH_3COCH_3 + Na^+ OH^- \xrightarrow[\text{heat}]{H_2O} CH_3CO^- Na^+ + CH_3OH$$

methyl acetate   sodium acetate   methanol
an ester   *a salt*   *an alcohol*

The carboxylic acid can be obtained by acidification of the solution following the saponification.

step 1 (saponification):

$$RCO_2R' + OH^- \xrightarrow[\text{heat}]{H_2O} RCO_2^- + R'OH$$

an ester   *the anion of a*
                                    carboxylic acid

step 2 (acidification):

$$RCO_2^- + H^+ \xrightarrow{H_2O} RCO_2H$$

the anion   *a carboxylic acid*

The term *saponification* comes from the word *soap*. A soap is the sodium salt of a long-chain carboxylic acid, called a *fatty acid*. In nature, these acids are found in fats and oils as esters of the triol glycerol. Tristearin, a fat found in cattle, is the ester formed from one molecule of glycerol and three molecules of the fatty acid stearic acid. Saponification of tristearin with aqueous sodium hydroxide yields a mixture of sodium stearate (a soap) and glycerol. Recall from Section 8.3 that a soap is effective in removing grease and soil because it has an ionic end (attracted to water) and a large hydrocarbon end (attracted to oils and other hydrocarbon material).

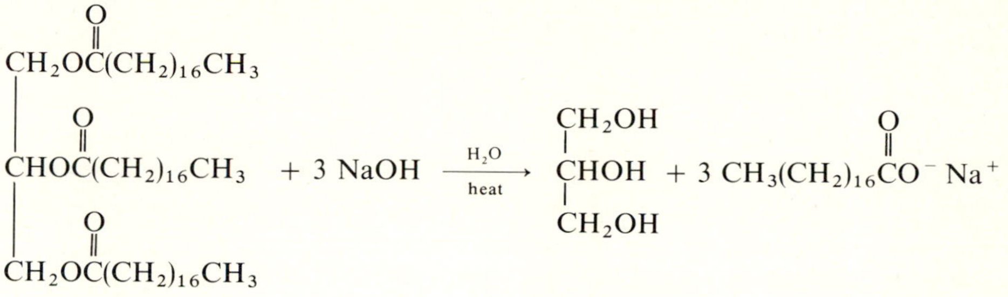

$$\begin{array}{c} \underset{\text{tristearin}}{\underset{a\ fat}{\left. \begin{array}{l} \overset{\text{O}}{\overset{\|}{\text{CH}_2\text{OC(CH}_2)_{16}\text{CH}_3}} \\[4pt] \overset{\text{O}}{\overset{\|}{\text{CHOC(CH}_2)_{16}\text{CH}_3}} \\[4pt] \overset{\text{O}}{\overset{\|}{\text{CH}_2\text{OC(CH}_2)_{16}\text{CH}_3}} \end{array}\right.}} + 3\ \text{NaOH} \xrightarrow[\text{heat}]{\text{H}_2\text{O}} \underset{\substack{\text{glycerol}\\ a\ triol}}{\begin{array}{l}\text{CH}_2\text{OH}\\ \text{CHOH}\\ \text{CH}_2\text{OH}\end{array}} + 3\ \underset{\substack{\text{sodium stearate}\\ a\ soap}}{\text{CH}_3(\text{CH}_2)_{16}\overset{\text{O}}{\overset{\|}{\text{C}}}\text{O}^-\,\text{Na}^+} \end{array}$$

Example   Write equations for (a) the acidic hydrolysis and (b) the saponification of *n*-propyl benzoate.

Solution

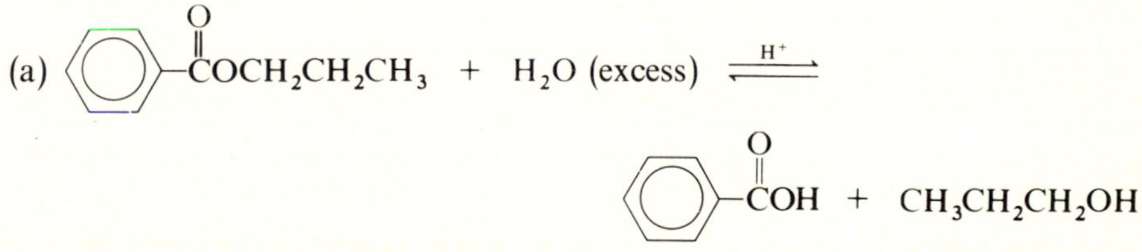

(a)   C₆H₅—$\overset{\text{O}}{\overset{\|}{\text{C}}}$OCH₂CH₂CH₃ + H₂O (excess) $\underset{}{\overset{\text{H}^+}{\rightleftharpoons}}$

C₆H₅—$\overset{\text{O}}{\overset{\|}{\text{C}}}$OH + CH₃CH₂CH₂OH

benzoic acid   1-propanol

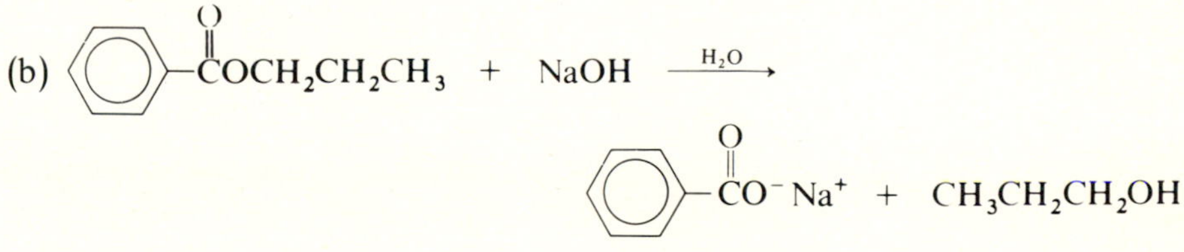

(b)   C₆H₅—$\overset{\text{O}}{\overset{\|}{\text{C}}}$OCH₂CH₂CH₃ + NaOH $\xrightarrow{\text{H}_2\text{O}}$

C₆H₅—$\overset{\text{O}}{\overset{\|}{\text{C}}}$O⁻ Na⁺ + CH₃CH₂CH₂OH

sodium benzoate   1-propanol

In the saponification reaction, as in the acidic hydrolysis reaction, the —OR′ portion of the ester yields the alcohol, and the carbonyl portion yields the carboxylate.

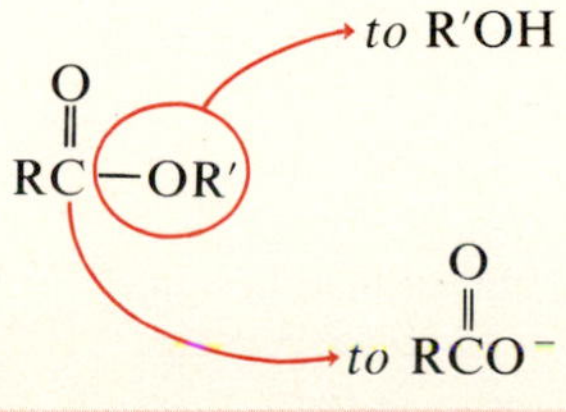

$$\text{R}\overset{\text{O}}{\overset{\|}{\text{C}}}-\text{OR}'$$

to R′OH

to RCO⁻

Problem 17.11. Predict the saponification products of the following esters with excess aqueous NaOH.

(a) isobutyl propanoate (Table 17.4)

(b) diethyl oxalate, $CH_3CH_2OC\overset{O}{\underset{\|}{C}}-\overset{O}{\underset{\|}{C}}OCH_2CH_3$

Problem 17.12. Write the formulas and names for the carboxylic acids that will be obtained when the products in Problem 17.11 are acidified. [*Hint:* In (b), the name of the acid can be deduced from the name of the ester.]

■ **17.8   SOME COMPOUNDS RELATED TO ESTERS (optional)**

Polyesters are polymers with repeating units bonded together by ester linkages. Dacron is a polyester used as a textile fiber.

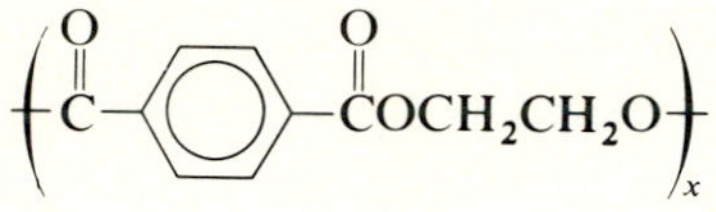

Dacron
a polyester

A **thioester** is an ester that has a sulfur atom in place of the —OR′ oxygen. A thioester is formed from a carboxylic acid and a thiol (R′SH) instead of an alcohol.

$$\overset{O}{\underset{\|}{RCOR'}} \qquad \overset{O}{\underset{\|}{RCSR'}}$$

an ester   *a thioester*

Acetylcoenzyme A is a biologically important thioester formed by the enzymatic esterification of acetate ions and the thiol coenzyme A (see Figure 17.2). Acetylcoenzyme A plays a key role in the metabolic cycles of living systems. Its acetyl group can be used in the biological synthesis of carbohydrates, fats, cholesterol, and other compounds. The acetyl group can also be converted to CO_2, H_2O, and energy when an organism needs energy.

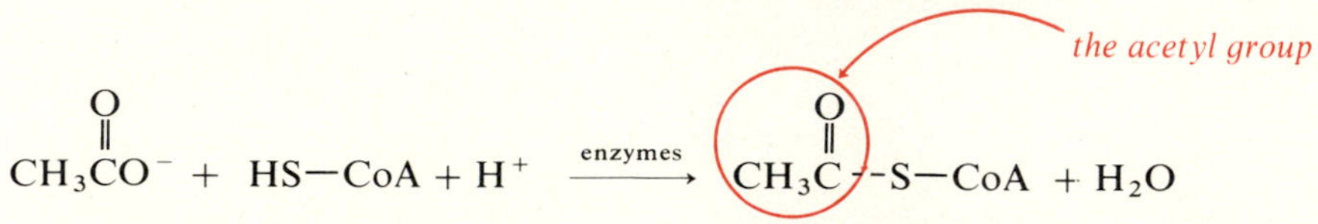

the acetyl group

$$CH_3\overset{O}{\underset{\|}{C}}O^- + HS-CoA + H^+ \xrightarrow{\text{enzymes}} CH_3\overset{O}{\underset{\|}{C}}-S-CoA + H_2O$$

acetate ion   abbreviated formula   abbreviated formula
for coenzyme A,   for acetylcoenzyme A,
a thiol   a thioester

Figure 17.2.   *The structure of coenzyme A (abbreviated CoA or HSCoA).*

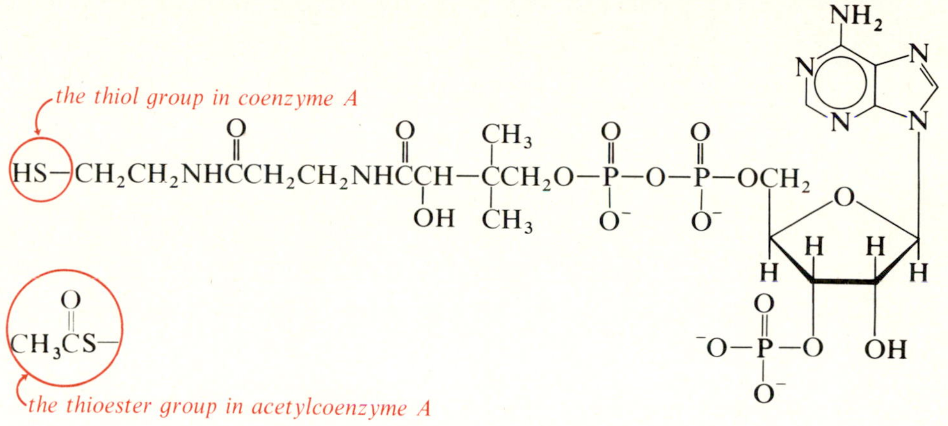

the thiol group in coenzyme A

$$\text{HS—CH}_2\text{CH}_2\text{NHCCH}_2\text{CH}_2\text{NHCCH—CCH}_2\text{O—P—O—P—OCH}_2$$

the thioester group in acetylcoenzyme A

Inorganic esters are esters of alcohols with *inorganic acids*. For example, nitrate esters can be prepared from nitric acid and alcohols.

$$\text{RO}\,\boxed{\text{H}} \;+\; \boxed{\text{HO}}\,\text{NO}_2 \;\longrightarrow\; \text{RONO}_2 \;+\; \text{H}_2\text{O}$$

an alcohol   nitric acid   *an alkyl nitrate*

Many nitrate esters are powerful explosives. Nitroglycerin, the trinitrate ester of glycerol, is just one example of this type of compound. The nitrate esters are also strong smooth-muscle relaxants and vasodilators (agents that can relax the stomach and heart muscles and dilate blood vessels). They are used clinically to lower blood pressure and to treat the heart condition *angina pectoris*. Nitroglycerin and pentaerythritol tetranitrate (PETN) are commonly used for these purposes.

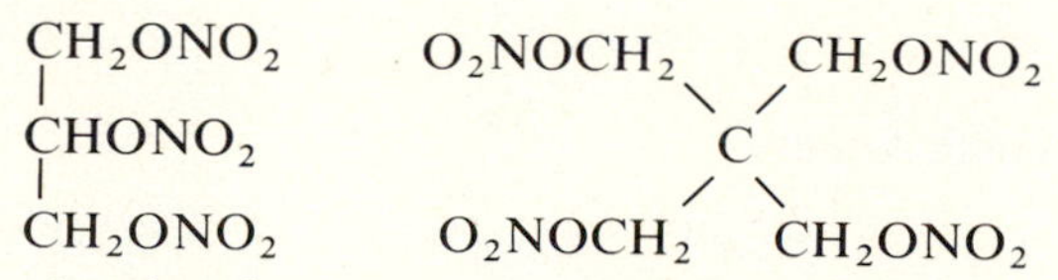

$$\begin{array}{l}
\text{CH}_2\text{ONO}_2 \\
\mid \\
\text{CHONO}_2 \\
\mid \\
\text{CH}_2\text{ONO}_2
\end{array}$$

nitroglycerin

$$\begin{array}{c}
\text{O}_2\text{NOCH}_2 \quad\diagdown\quad\diagup\quad \text{CH}_2\text{ONO}_2 \\
\text{C} \\
\text{O}_2\text{NOCH}_2 \quad\diagup\quad\diagdown\quad \text{CH}_2\text{ONO}_2
\end{array}$$

pentaerythritol tetranitrate

In Section 22.3, you will learn that biological energy is stored in inorganic *phosphate esters*. Note that coenzyme A (Figure 17.2) is also a phosphate ester.

17.9 AMIDES

An **amide** is a compound that contains a nitrogen atom attached to a carbonyl group.

general formulas for amides:

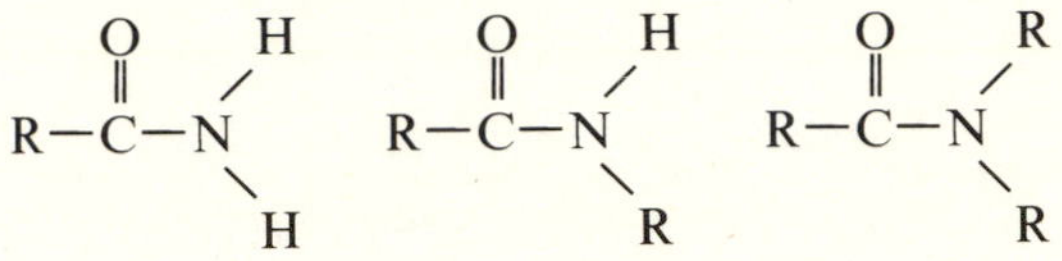

Proteins are natural polyamides, while nylon is a synthetic polyamide.

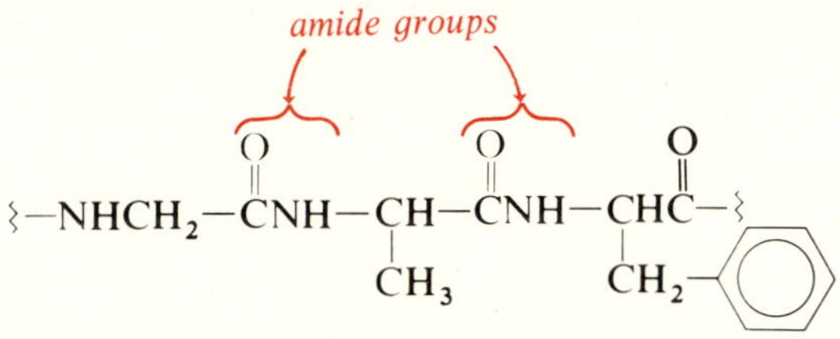

portion of a protein molecule

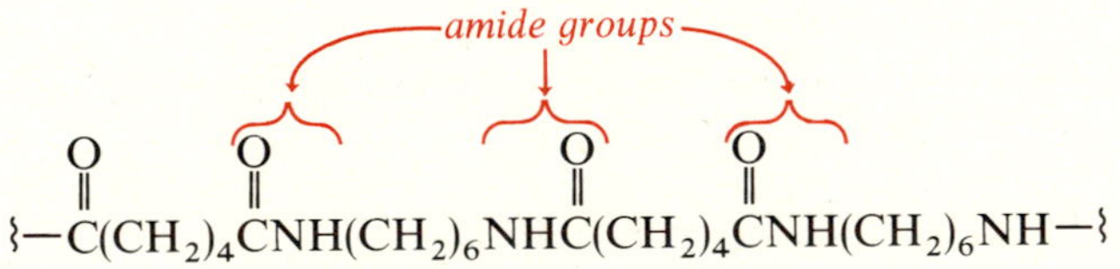

portion of a nylon molecule

The barbiturates, which are used as sedatives, are examples of *cyclic amides.*

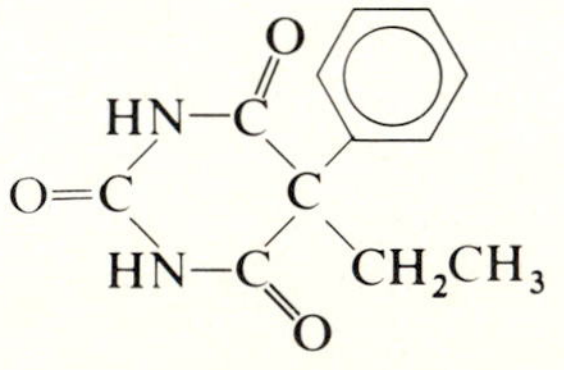

phenobarbital 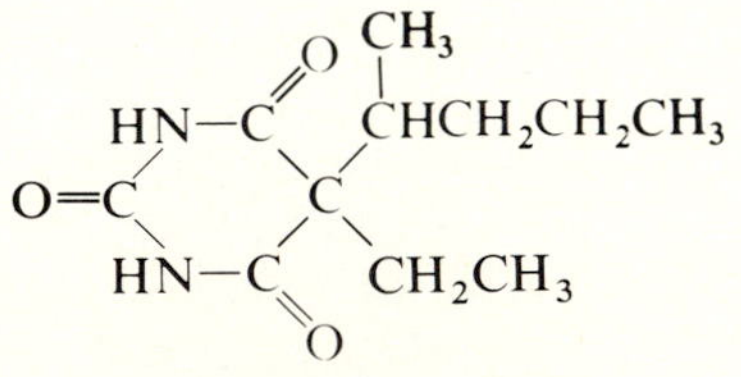   Nembutal

Naming Amides

An amide is named by dropping the *-ic acid* or *-oic acid* ending of the carboxylic acid name and adding the ending *-amide.*

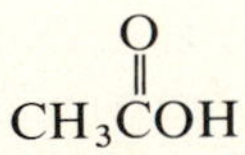

$$\underset{\text{acetic acid}}{CH_3\overset{\overset{O}{\|}}{C}OH} \qquad \underset{\text{acetamide}}{CH_3\overset{\overset{O}{\|}}{C}NH_2}$$

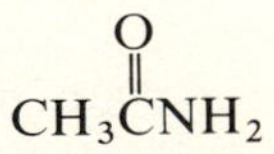

If an alkyl or aryl group is attached to the nitrogen, the group is designated in the name by *N*- (a capital *N* for nitrogen), followed by the name of the group.

methyl group on N

$$\text{C}_6\text{H}_5-\overset{\overset{\displaystyle O}{\|}}{\text{C}}\text{NHCH}_3$$

N-methylbenzamide

Problem 17.13. Write formulas for (a) propanamide and (b) *N,N*-dimethylopropanamide.

Hydrolysis of Amides

Amides, like esters, can be hydrolyzed. Acidic hydrolysis yields a carboxylic acid and an ammonium salt (or amine salt).

$$\underset{\text{acetamide}}{\overset{\overset{\displaystyle O}{\|}}{\text{CH}_3\text{C}}-\text{NH}_2} + \text{H}_2\text{O} + \text{H}^+\,\text{Cl}^- \xrightarrow{\text{heat}} \underset{\text{acetic acid}}{\overset{\overset{\displaystyle O}{\|}}{\text{CH}_3\text{C}}\text{OH}} + \underset{\textit{a salt}}{\overset{+}{\text{N}}\text{H}_4\,\text{Cl}^-}$$

Example   Write the equation for the acidic hydrolysis of *N*-methylbenzamide, shown previously in this section.

Solution

$$\overset{\overset{\displaystyle O}{\|}}{\text{C}_6\text{H}_5-\text{C}}-\text{N}\overset{\text{H}}{\underset{\text{CH}_3}{\big<}} \;+\; \text{H}_2\text{O} \;+\; \text{H}^+\,\text{Cl}^- \xrightarrow{\text{heat}}$$

$$\underset{\text{benzoic acid}}{\overset{\overset{\displaystyle O}{\|}}{\text{C}_6\text{H}_5-\text{C}}\text{OH}} \;+\; \underset{\substack{\text{methylammonium}\\\text{chloride}\\\textit{an amine salt}}}{\text{CH}_3\overset{+}{\text{N}}\text{H}_3\,\text{Cl}^-}$$

Note that, as in ester hydrolysis, the carbonyl portion of the molecule yields the carboxylic acid. The nitrogen part of the molecule yields ammonia or an amine, which reacts with the acid to yield a salt.

$$\overset{\overset{\displaystyle O}{\|}}{\text{RC}}-\text{N}\overset{\text{R}'}{\underset{\text{R}'}{\big<}}$$

$$\text{to}\;\; \text{H}-\overset{..}{\underset{\text{R}'}{\text{N}}}-\text{R}' \xrightarrow{\text{H}^+} \text{H}-\overset{\overset{\displaystyle \text{H}}{|}}{\underset{\text{R}'}{\overset{+}{\text{N}}}}-\text{R}'$$

$$\text{to}\;\; \overset{\overset{\displaystyle O}{\|}}{\text{RC}}\text{OH}$$

Problem 17.14. The protein molecule shown at the start of Section 17.9 is subjected to acidic hydrolysis. What are the structures of the organic products from this portion of the protein molecule? (*Hint:* Be careful to break the protein molecule between the carbonyl carbon and the N, as shown in the preceding example.)

SUMMARY

The *carbonyl group* $C{=}O$ is found in a variety of functional groups, summarized in Table 17.1.

The endings used in the IUPAC names are as follows:

aldehydes	-al
ketones	-one
carboxylic acids	-oic acid
esters	-oate
amides	-amide

Aldehydes (RCHO) and *ketones* ($R_2C{=}O$) can be reduced to alcohols by the addition of H_2. Aldehydes can be oxidized to carboxylic acids (in acid) or their anions (in base).

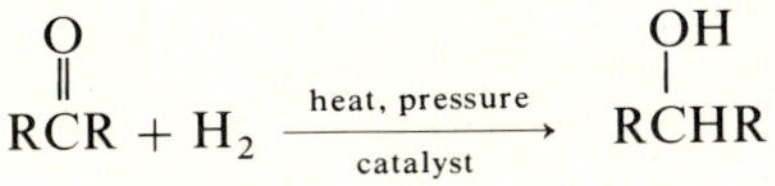

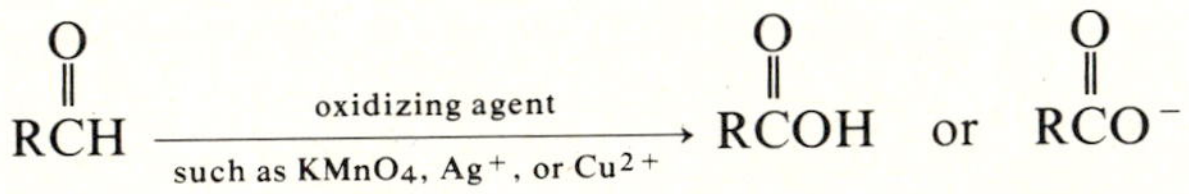

$$\underset{\text{RCR}}{\overset{\text{O}}{\|}} + H_2 \xrightarrow[\text{catalyst}]{\text{heat, pressure}} \underset{\text{RCHR}}{\overset{\text{OH}}{|}}$$

$$\underset{\text{RCH}}{\overset{\text{O}}{\|}} \xrightarrow[\text{such as KMnO}_4,\ \text{Ag}^+,\ \text{or Cu}^{2+}]{\text{oxidizing agent}} \underset{\text{RCOH}}{\overset{\text{O}}{\|}} \quad \text{or} \quad \underset{\text{RCO}^-}{\overset{\text{O}}{\|}}$$

An alcohol can add to an aldehyde to yield a *hemiacetal*. Further reaction yields the *acetal*.

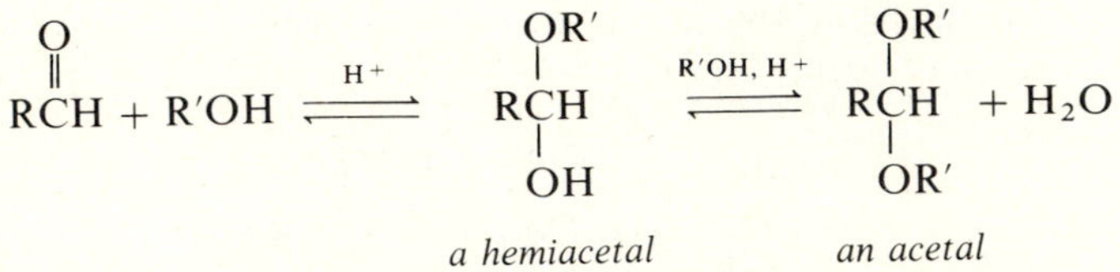

$$\underset{\text{RCH}}{\overset{\text{O}}{\|}} + R'OH \rightleftharpoons \underset{\underset{\text{OH}}{|}}{\overset{\overset{\text{OR}'}{|}}{\text{RCH}}} \underset{}{\overset{R'OH,\ H^+}{\rightleftharpoons}} \underset{\underset{\text{OR}'}{|}}{\overset{\overset{\text{OR}'}{|}}{\text{RCH}}} + H_2O$$

$$\qquad\qquad\qquad\qquad\quad \textit{a hemiacetal} \qquad\qquad \textit{an acetal}$$

Carboxylic acids (RCO_2H) are weak acids that yield carboxylate salts when treated with a base or esters when heated with an alcohol and an acidic catalyst. Some carboxylic acids can undergo decarboxylation.

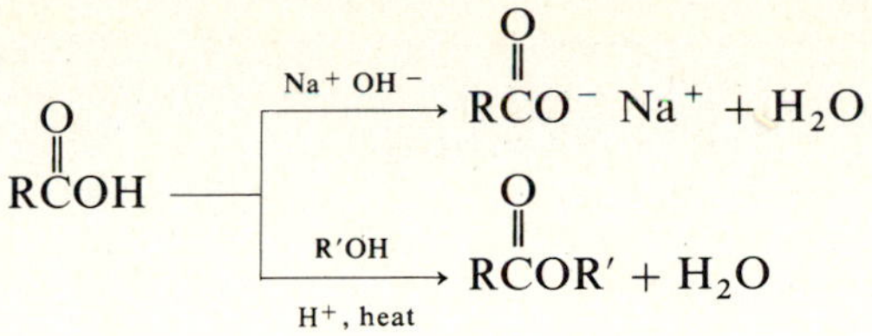

$$\text{RCOH} \begin{cases} \xrightarrow{\text{Na}^+ \text{OH}^-} \text{RCO}^- \text{Na}^+ + H_2O \\ \xrightarrow[\text{H}^+,\text{ heat}]{\text{R'OH}} \text{RCOR'} + H_2O \end{cases}$$

$$\text{RCCH}_2\text{COH} \xrightarrow{\text{heat or enzyme}} \text{RCCH}_3 + CO_2$$

Esters RCOR' and *amides* RCNR$'_2$ can be hydrolyzed to carboxylic acids. *Saponification* is the alkaline hydrolysis of an ester.

acidic hydrolysis:

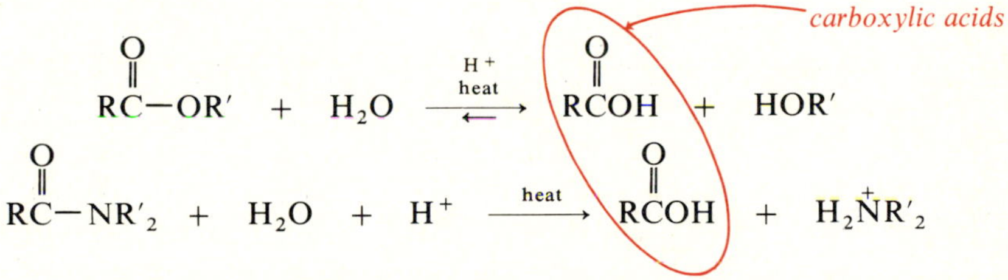

carboxylic acids

$$\text{RC}-\text{OR'} + H_2O \xrightleftharpoons[\text{heat}]{H^+} \text{RCOH} + \text{HOR'}$$

$$\text{RC}-\text{NR'}_2 + H_2O + H^+ \xrightarrow{\text{heat}} \text{RCOH} + H_2\overset{+}{N}R'_2$$

KEY TERMS

carbonyl group	carboxyl group	*polyester
aldehyde	carboxylate	*thioester
ketone	decarboxylation	*acetylcoenzyme A
Tollens reagent	ester	*inorganic ester
hemiacetal	esterification reaction	amide
acetal	hydrolysis reaction	
carboxylic acid	saponification reaction	

STUDY PROBLEMS

17.15. Label each of the following compounds by class (Table 17.1):

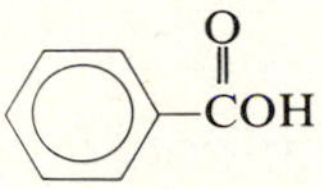

(a) 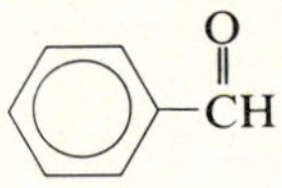 C$_6$H$_5$–CH (with =O)   (b) C$_6$H$_5$–COH (with =O)

(c) CH$_3$CH$_2$CH$_2$CNH$_2$ (d) HCOCH$_2$CH$_3$

(e) HCCH$_2$CH$_3$ (f) CH$_3$CCH$_2$CH$_3$

17.16. (a) Of the compound classes in Table 17.1, which can form hydrogen bonds with water?
(b) Which can form hydrogen bonds in their pure states?
(c) Using formulas, show a hydrogen bond between formaldehyde ($H_2C=O$) and water.

17.17. Name the following compounds using the IUPAC system:

(a) BrCH$_2$CH (with =O)   (b) CH$_3$CH$_2$CCH$_2$CH$_3$

(c) $CH_3\overset{O}{\overset{\|}{C}}HCCH_2Br$
         CH_3

(d) 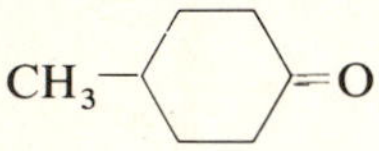 $CH_3-\langle\;\rangle=O$

17.18. Write formulas for (a) 1,3-dihydroxy-acetone and (b) 2,3-dihydroxypropanal. (The hydroxy group is —OH.)

17.19. Complete the following equations. If no reaction occurs, write "no reaction." In (a), the —CO₂H group does not react under these conditions. In (c), show only the organic product.

(a) $CH_3\overset{O}{\overset{\|}{C}}CO_2H$ + H_2 $\xrightarrow[\text{catalyst}]{\text{heat, pressure}}$
        pyruvic acid

(b) $\overset{O}{\overset{\|}{C}}H$
        $\overset{|}{C}HOH$ + H_2 $\xrightarrow[\text{catalyst}]{\text{heat, pressure}}$
        $\overset{|}{C}H_2OH$

        glyceraldehyde

(c) $CH_3\overset{O}{\overset{\|}{C}}CH_2\overset{O}{\overset{\|}{C}}H$ $\xrightarrow{\text{Tollens reagent}}$

17.20. Which of the following structures are hemiacetals? Which are acetals?

(a) $CH_3CH_2\overset{OCH_2CH_3}{\overset{|}{C}}HOCH_3$

(b) $CH_3CH_2\overset{OH}{\overset{|}{C}}HOCH_2CH_3$

(c)   ⟨ring⟩—OCH₃

(d) 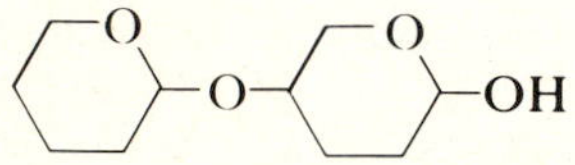   ⟨ring⟩—O—⟨ring⟩—OH

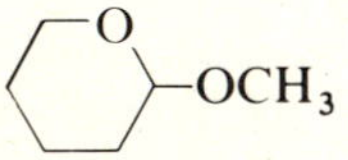

17.21. Write formulas for the hemiacetal and the acetal products in the following mixtures:

(a) $H\overset{O}{\overset{\|}{C}}H$ + CH_3CH_2OH $\rightleftharpoons$

(b) 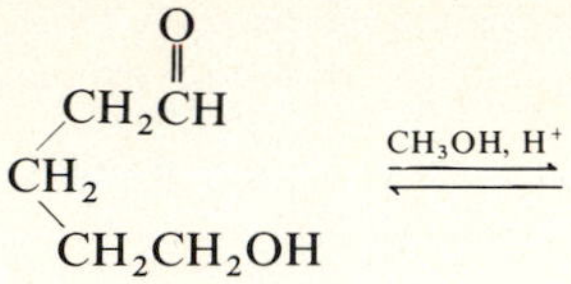 $CH_2\overset{\overset{\displaystyle CH_2\overset{O}{\overset{\|}{C}}H}{|}}{\underset{|}{}}CH_2CH_2OH$ $\xrightleftharpoons{CH_3OH,\ H^+}$

17.22. Write formulas for the following carboxylic acids, carboxylates, and esters:
(a) 2-hydroxypropanoic acid (lactic acid)
(b) sodium lactate
(c) methyl lactate
(d) 2-bromoethyl acetate
(e) 2-bromoacetic acid
(f) ethyl 2-bromoacetate

17.23. Write the equations for the reactions of aspirin (Section 17.4) with (a) cold aqueous NaHCO₃ and (b) hot aqueous HCl.

17.24. Write the equation for the reaction of acetoacetic acid (Section 17.5) with cold, aqueous NaOH. (Ketones do not react with bases under these conditions.)

17.25. In the metabolism sequence for glucose, CO₂ is lost from the following acids. Circle the CO₂ that would be lost. Then write the formulas for the decarboxylated products, assuming no other reactions take place. More than one answer may be correct.

(a) $HO_2CCH_2CH_2\overset{O}{\overset{\|}{C}}CO_2H$

        α-ketoglutaric acid

(b) $HO_2C\overset{O}{\overset{\|}{C}}-\overset{CO_2H}{\overset{|}{C}}HCH_2CO_2H$

        oxalosuccinic acid

17.26. Complete and balance the following equations:

(a) $CH_3CO_2H + CH_3OH$ $\xrightleftharpoons[]{\overset{H^+}{\text{heat}}}$

(b) $CH_3\overset{O}{\overset{\|}{C}}OCH_3 + H_2O$ $\xrightleftharpoons[]{\overset{H^+}{\text{heat}}}$

(c) $CH_3\overset{O}{\overset{\|}{C}}NHCH_3 + H_2O + H^+$ $\xrightarrow{\text{heat}}$

(d) $CH_3\overset{O}{\overset{\|}{C}}OCH_3 + Na^+\ {}^-OH$ $\xrightarrow[\text{heat}]{H_2O}$

(e) $CH_3\overset{O}{\underset{\|}{C}}SCH_3 + Na^+ \ ^-OH \xrightarrow[\text{heat}]{H_2O}$

17.27. Write equations to show how you would prepare (a) isoamyl acetate and (b) methyl salicylate from carboxylic acids and alcohols. (See Section 17.6, Problem 17.10, for the structures of these compounds.)

17.28. Show with an equation how you would convert methyl salicylate to salicylic acid.

*17.29. Write an equation to show how you would prepare PETN (Section 17.8).

17.30. Name the following amides:

(a) $CH_3CH_2\overset{O}{\underset{\|}{C}}NH_2$

(b)

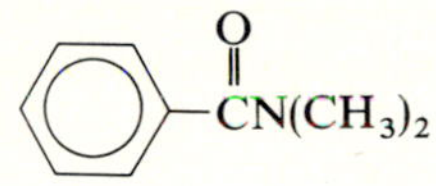

$\langle\bigcirc\rangle - \overset{O}{\underset{\|}{C}}N(CH_3)_2$

17.31. The product of acid hydrolysis of *nicotinamide*, one of the B vitamins, is *nicotinic acid*. What is the structure of nicotinic acid?

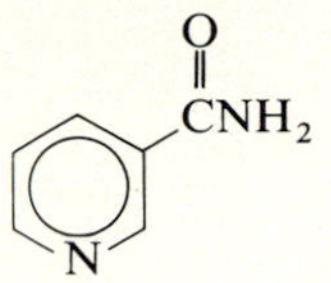

nicotinamide (niacinamide)

17.32. *Xylocaine* is used as a local anesthetic. What are the organic products of its acidic hydrolysis?

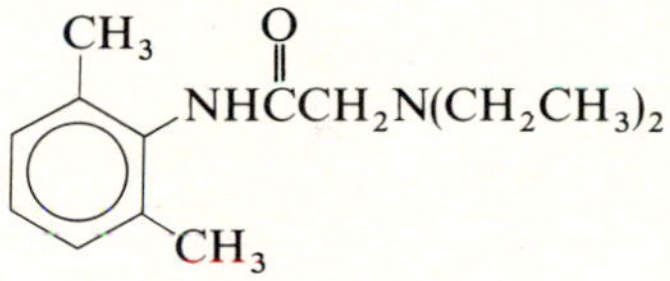

xylocaine (lidocaine)

PART 3 Important Biological Compounds

18 Carbohydrates

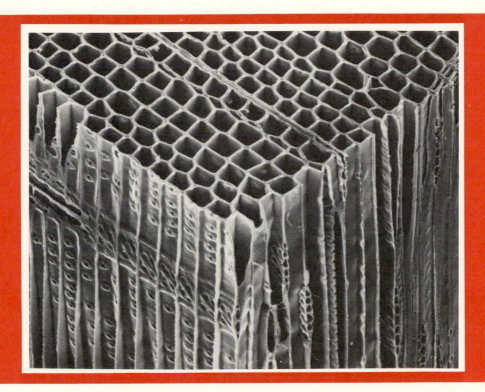

An electron micrograph showing the wood cells of a sugar pine. (Courtesy of Dr. Wilfred A. Côté, State University of New York College of Environmental Science and Forestry.) The walls of longitudinal wood cells, such as those in the photograph, consist primarily of cellulose, a type of polymeric carbohydrate. This picture is an example of an electron micrograph, an image obtained from an electron microscope. Electron microscopy allows magnification far greater than can be obtained with ordinary microscopy. These wood cells are magnified only about 200×, but magnifications of over 2000,000× are possible with an electron microscope. At high magnifications, such tiny objects as viruses and relatively large protein molecules can be visualized.

Write names and formulas for some common monosaccharides. ☐ Define and give examples of an aldose, a ketose, a triose, a tetrose, a pentose, and a hexose. ☐ Define stereoisomers. ☐ Differentiate between D- and L-monosaccharides. ☐ Define chiral molecule, show how one of these can be recognized, and define optical isomers (optional). ☐ Write equations to illustrate fermentation, oxidation, reduction, and cyclization of monosaccharides. ☐ Define and give examples of reducing sugars. ☐ Draw polygon formulas for α- and β-D-glucose. ☐ Write an equation to show the formation of an alkyl glycoside. ☐ Show why glycosides are nonreducing sugars. ☐ Compare the structures of maltose, lactose, and sucrose. ☐ Describe and differentiate the structures of amylose, amylopectin, cellulose, and glycogen. ☐ Define glycogenesis and glycogenolysis.

The three principal foodstuffs for humans are carbohydrates, fats, and proteins. In the coming chapters, we will discuss the structures and chemical properties of these classes of compounds, beginning with carbohydrates.

The name *carbohydrate* comes from the fact that analyses for the elements in these compounds often show molecular formulas that are multiples of CH_2O. For example, the molecular formula for glucose is $C_6H_{12}O_6$, or $(CH_2O)_6$. At one time, the carbohydrates were erroneously thought to be hydrates of carbon, or $C \cdot H_2O$. We know now that the carbohydrates are polyhydroxy aldehydes and polyhydroxy ketones, or are dimers or larger molecules formed from these aldehydes and ketones.

Carbohydrates are synthesized from CO_2 and H_2O by green plants, both on land and in the oceans, in a process called **photosynthesis**. Photosynthesis relies on chlorophyll to trap the sun's energy and on a series of enzymes that catalyze each step in the reaction sequence.

in green plants:

$$CO_2 + H_2O \xrightarrow[\text{many steps}]{\text{sunlight}} \text{carbohydrates} + O_2$$

Plant and animal cells use the simple sugar glucose as a cell nutrient. A plant stores excess glucose that it has synthesized in the form of *starch*, a polymer of glucose. In plants, *cellulose*, another polymer of glucose, serves as a structural component.

A balanced human diet includes carbohydrates: starch and sugars from fruits, vegetables, or grains. Carbohydrates undergo hydrolysis in the process of digestion to yield soluble sugars, primarily glucose, which are absorbed into the bloodstream. In animals, excess glucose can be polymerized to *glycogen* to be stored for later use.

In this chapter, we will look at the structures of only a few carbohydrates. We will begin with the simplest sugars, or monosaccharides (such as glucose); then we will turn to the disaccharides (maltose, lactose, and sucrose). Finally, we will discuss how glucose is bonded together to form three important polysaccharides: starch, cellulose, and glycogen.

18.1 MONOSACCHARIDES—THE SIMPLEST SUGARS

A **sugar** is any carbohydrate, such as glucose or sucrose, that is sweet tasting and water soluble. However, sweetness is not a characteristic unique to sugars. Saccharin and calcium cyclamate are examples of sweet-tasting compounds that are not sugars.

calcium cyclamate
500 times sweeter than sucrose

saccharin
30 times sweeter than sucrose

A **simple sugar**, or **monosaccharide**, is the smallest type of carbohydrate. Monosaccharides are the monomers, or building blocks, for the more complex carbohydrates.

Although a large number of monosaccharides are found in nature, we will limit our discussion to only five: glucose, galactose, fructose, ribose, and 2-deoxyribose. The structures of these monosaccharides are shown in Figure 18.1. Their molecular formulas and comments concerning their importance are listed in Table 18.1. We will stress glucose because it is the monomer for many of the larger carbohydrate structures. Alone or in polymers, glucose is probably the most abundant organic compound on the earth.

All monosaccharides are very similar in structure; each contains an aldehyde or ketone group and two or more hydroxyl groups. For example,

Table 18.1. *Some Important Monosaccharides*

Name	Molecular formula	Comments
ribose	$C_5H_{10}O_5$	found in vitamin B_{12}; forms the backbone for ribonucleic acids (RNA)
2-deoxyribose	$C_5H_{10}O_4$	forms the backbone for deoxyribonucleic acids (DNA)
glucose	$C_6H_{12}O_6$	also called *dextrose*, *blood sugar*, and *grape sugar*; provides energy for cellular processes; a structural unit of many disaccharides and polysaccharides
galactose	$C_6H_{12}O_6$	part of the structure of lactose; converted to glucose in cells
fructose	$C_6H_{12}O_6$	also called *levulose*; the sweetest sugar; found in honey, invert sugar, and fruit

Figure 18.1.   *Five important monosaccharides. (The prefix* D *is defined later in Section 18.2.)*

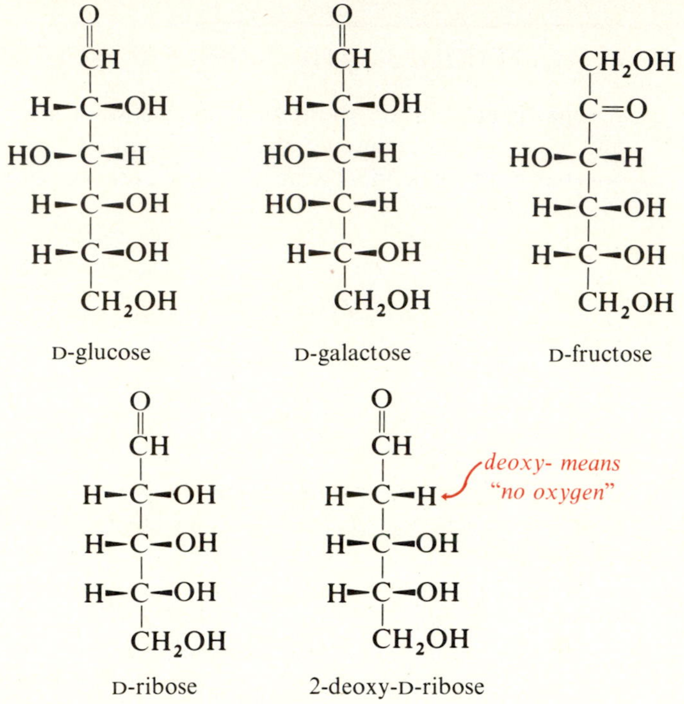

glucose, galactose, ribose, and 2-deoxyribose are polyhydroxy aldehydes. Fructose is a polyhydroxy ketone. Because of the large number of hydrogen-bonding hydroxyl groups in their structures, all of these monosaccharides are water soluble.

Monosaccharide names, as well as the names of some higher carbohydrates, end in *-ose*: for example, glucose or fructose. A monosaccharide can be classified according to its principal functional group or according to the number of carbons in its molecules. The ending -ose is also used in these classification names.

classification by functional group
   an aldehyde monosaccharide: an aldose
   a ketone monosaccharide: a ketose

classification by number of carbons
   three carbons: a triose
   four carbons: a tetrose
   five carbons: a pentose
   six carbons: a hexose

Often the terms are combined. Glucose and galactose both contain six carbon atoms and an aldehyde group; each is an *aldohexose*. Fructose contains six carbon atoms and a keto group; fructose is a *ketohexose*, sometimes called a *hexulose* (the internal *-ul-* referring to a ketone sugar).

Problem 18.1.   Referring to Figure 18.1, classify each of the following monosaccharides as an aldose or a ketose and by the number of carbon atoms each contains:

(a)   D-ribose

(b)   2-deoxy-D-ribose

(c) 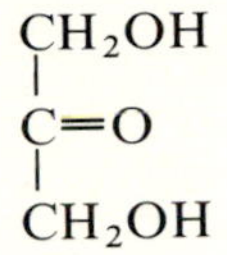
$$CH_2OH$$
$$|$$
$$C{=}O$$
$$|$$
$$CH_2OH$$
reductone
(dihydroxyacetone)

(d) 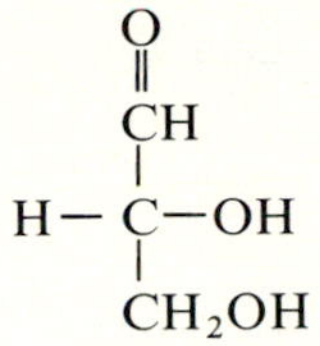
$$O$$
$$\|$$
$$CH$$
$$|$$
$$H{-}C{-}OH$$
$$|$$
$$CH_2OH$$
D-glyceraldehyde

18.2   FORMULAS AND ISOMERS OF MONOSACCHARIDES

Note how the formulas for the monosaccharides in Figure 18.1 are represented. The aldehyde or ketone group is placed at or near the top; then the hydroxyl groups are shown to the right or the left of the stretched-out carbon chain. Of course, the molecules are not actually stretched out in this fashion, but exist as flexible chains. The wedges in the formulas show bonds angled toward the viewer in the stretched-out chain.

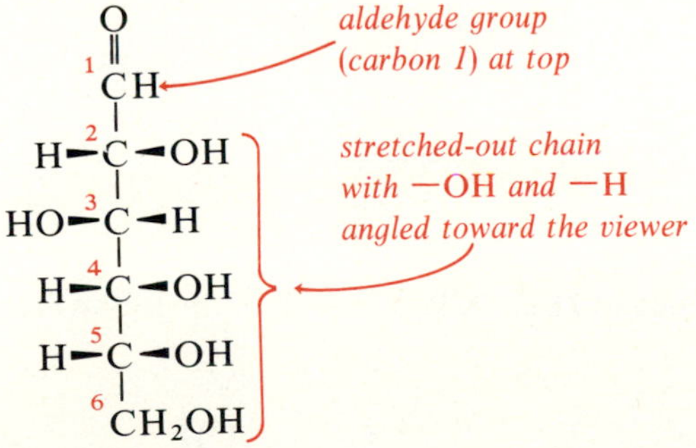

$$\begin{array}{l}
O \\
\| \\
{}^{1}CH \leftarrow \quad \text{\textit{aldehyde group}} \\
\qquad\qquad \text{\textit{(carbon 1) at top}} \\
H{-}{}^{2}C{-}OH \\
\qquad\qquad \text{\textit{stretched-out chain}} \\
HO{-}{}^{3}C{-}H \quad \text{\textit{with }}{-}OH\text{ \textit{and} }{-}H \\
\qquad\qquad \text{\textit{angled toward the viewer}} \\
H{-}{}^{4}C{-}OH \\
H{-}{}^{5}C{-}OH \\
{}^{6}CH_2OH
\end{array}$$

D-glucose

Monosaccharide molecules are represented in this way because the arrangement of groups around the carbon atoms in monosaccharides is important. In glucose, switching a hydroxyl group on carbon 2, 3, 4, or 5 from the right side of the chain to the left side of the chain, or vice versa, results in a different compound. Referring to Figure 18.1, you will see that galactose differs from glucose only in the position of the OH at carbon 4.

Glucose and galactose are *stereoisomers* of each other.

Stereoisomers are compounds with the same molecular formula and the same order of attachment of atoms, but with a different arrangement of their groups in space.

There are 16 stereoisomeric aldohexoses (each with its own name) arising from the right- or left-handed arrangement of the OH groups at carbons 2–5, but only D-glucose and D-galactose are important in human metabolism.

D- and L-Monosaccharides

In the names of monosaccharides, the letters D (for right) and L (for left) are used to tell the arrangement of the OH group on the carbon that is farthest from carbon 1 and that has four *different* groups attached. In the carbohydrates we are considering, this hydroxyl is bonded to the second-to-last carbon in the chain (carbon 5 in glucose). Almost all monosaccharides found in nature, either uncombined or as part of larger structures, are members of the D series, with the —OH in question shown to the right.

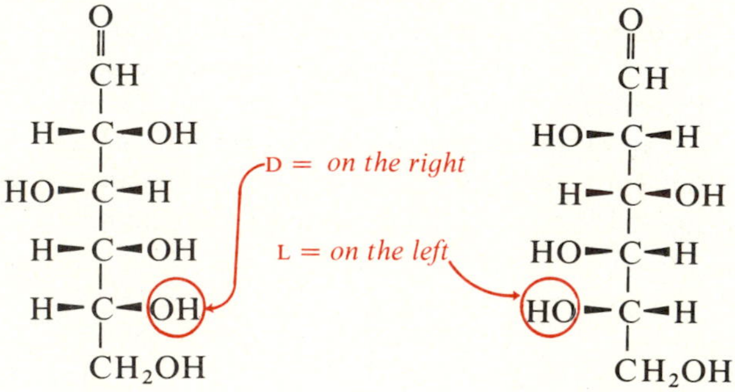

D-glucose
important in human metabolism

L-glucose
not important in human metabolism

Problem 18.2. Classify each of the following monosaccharides as D or L:

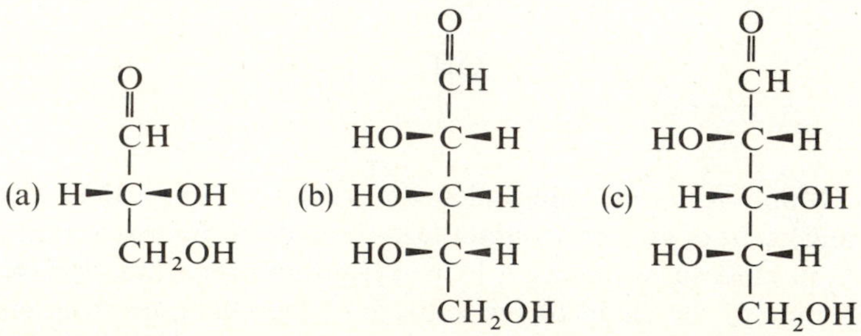

(a)

(b)

(c)

■ 18.3 OPTICAL ISOMERISM IN MONOSACCHARIDES (optional)

One interesting aspect of stereoisomerism in carbohydrates is that the right- or left-handed arrangement of groups around certain carbons leads to a special type of stereoisomerism called *optical isomerism.* Before we define optical isomers, let us consider some aspects of right- and left-handedness, using the human hand. At first glance, your two hands may appear identical, but in actuality they are mirror images of each other. Hold your left hand up to a mirror and you will see a reflection that looks like a right hand. (See Figure 18.2.)

Your two hands cannot be exactly superimposed, or merged finger-for-finger, on each other or on their mirror images. If you keep the backs of both your hands up and try to superimpose them, your thumbs will be on opposite sides. (See Figure 18.3.) An object that *cannot* be superimposed on its mirror image is called a *chiral* ("handed") object.

Gloves and shoes are other examples of objects that cannot be superimposed on their mirror images and thus are chiral. A plain cup or a bottle,

Figure 18.2. *Your two hands are mirror images.*

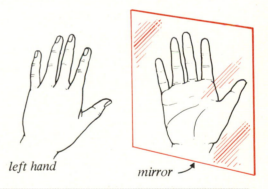

left hand

mirror

Figure 18.3. *A chiral object cannot be superimposed on its mirror image.*

Two left hands being superimposed.

A left hand cannot be superimposed on a right hand.

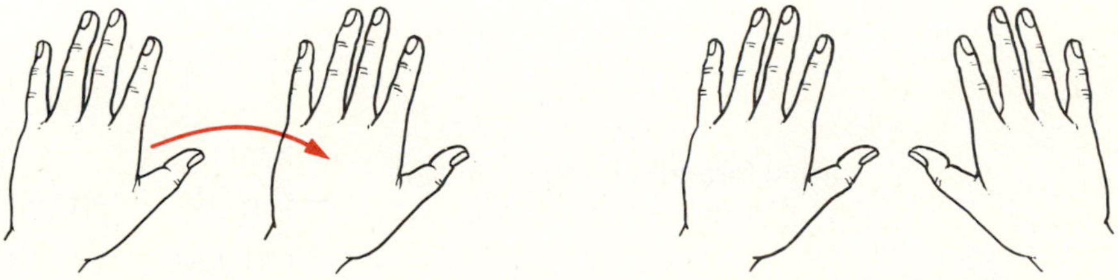

Figure 18.4.   *Inexpensive molecular models made from gumdrops and toothpicks may be used to illustrate chirality of molecules.*

two or more groups the same

four groups different

not chiral, superimposable: same compound

chiral, not superimposable: isomers

however, is different; each can be superimposed on its mirror image and is not chiral. All objects can be classified as being either chiral or not chiral.

Now, let us consider a molecule in which *one carbon atom has four different groups* bonded to it. This molecule, like the human hand, is chiral and cannot be superimposed group-for-group on its mirror image. No matter how

Figure 18.5.   *The triose glyceraldehyde exists as a pair of nonsuperimposable mirror-image isomers.*

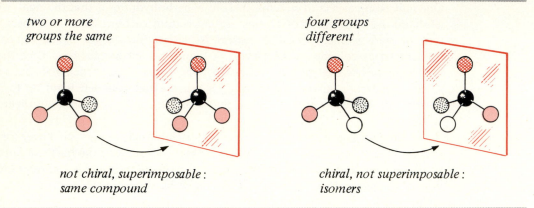

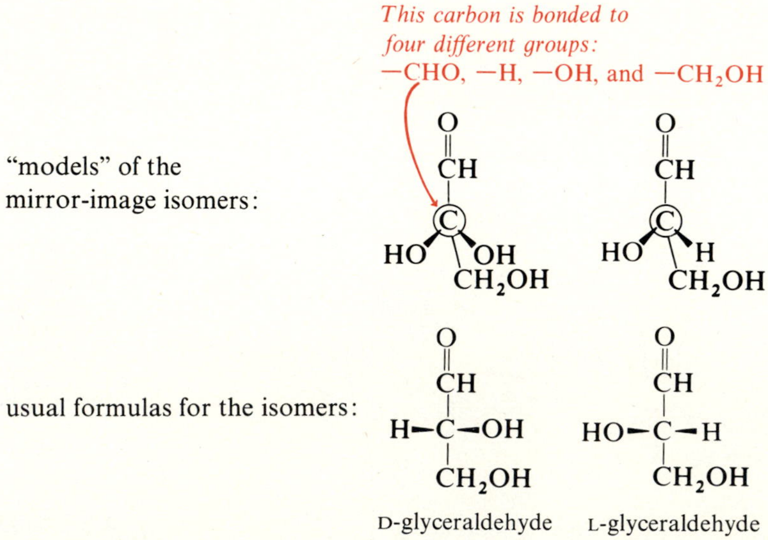

This carbon is bonded to four different groups:
—CHO, —H, —OH, and —CH_2OH

"models" of the mirror-image isomers:

usual formulas for the isomers:

D-glyceraldehyde   L-glyceraldehyde

the molecule is twisted and turned, two of the four groups will be in the wrong positions. Figure 18.4 shows two molecular models. The first is a molecule that is not chiral (its carbon does *not* have four different groups bonded to it), and the second is a chiral molecule (four different groups and not superimposable on its mirror image). You can see in the figure that the chiral molecule cannot be merged with its mirror image. By constructing your own models of such molecules, you can see the chirality better.

If a molecule cannot be superimposed on its mirror image, the molecule and its mirror image must be different compounds. A molecule and its non-superimposable mirror image are called **optical isomers**, **mirror-image isomers**, or **enantiomers**. Figure 18.5 shows the mirror-image isomers of the simplest sugar of all, the triose glyceraldehyde.

Example   Which of the following molecules are chiral? (*Hint:* Which carbon atoms have four different groups bonded to them?)

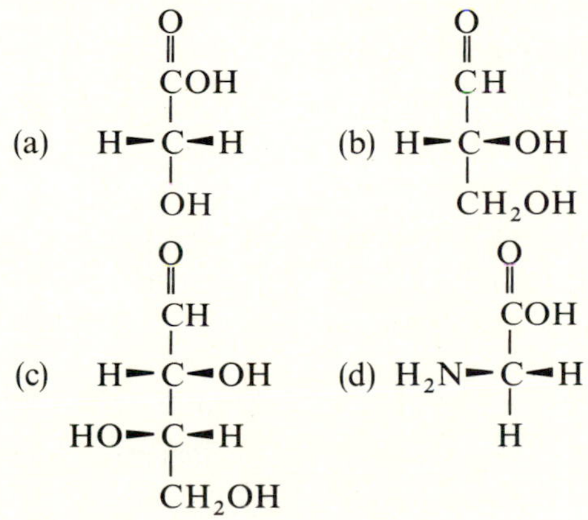

(a)
$$\begin{array}{c} O \\ \| \\ COH \\ | \\ H-C-H \\ | \\ OH \end{array}$$

(b)
$$\begin{array}{c} O \\ \| \\ CH \\ | \\ H-C-OH \\ | \\ CH_2OH \end{array}$$

(c)
$$\begin{array}{c} O \\ \| \\ CH \\ | \\ H-C-OH \\ | \\ HO-C-H \\ | \\ CH_2OH \end{array}$$

(d)
$$\begin{array}{c} O \\ \| \\ COH \\ | \\ H_2N-C-H \\ | \\ H \end{array}$$

Solution   Molecules (a) and (d) are *not* chiral because no carbon atom in these molecules has four different groups bonded to it. (In each case, carbon 2 is bonded to two H atoms.) Molecules (b) and (c) are both chiral. Note that the chiral molecule in (c) contains *two* carbons (carbons 2 and 3) with four different groups:

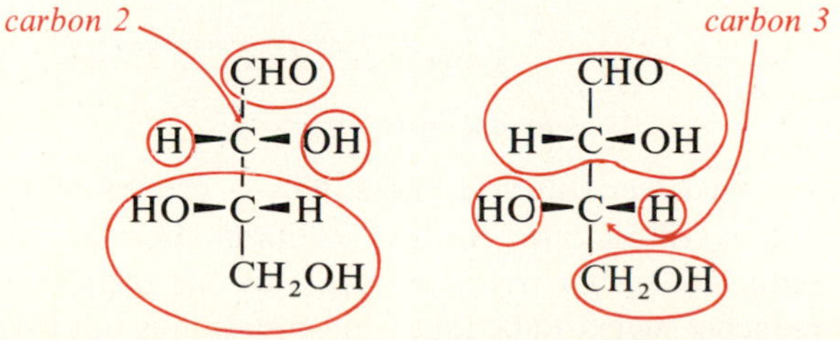

carbon 2
$$\begin{array}{c} CHO \\ | \\ H-C-OH \\ | \\ HO-C-H \\ | \\ CH_2OH \end{array}$$

carbon 3
$$\begin{array}{c} CHO \\ | \\ H-C-OH \\ | \\ HO-C-H \\ | \\ CH_2OH \end{array}$$

Because of structural similarities, pairs of mirror-image isomers have identical physical and chemical properties, such as melting point or reactivity toward an acid, with very few exceptions. In fact, the *only* chemical difference between D- and L-glyceraldehyde or any other pair of mirror-image isomers is that they exhibit *different chemical reactivities in chiral environments*—that is, these isomers behave differently when they undergo reaction with other chiral molecules or in the presence of a chiral catalyst. Almost all biomolecules, including enzymes, are chiral. Therefore, a pair of mirror-image isomers, which exhibit identical laboratory reactions, can undergo different reactions in the human body. L-Glucose, the mirror-image isomer of D-glucose (shown in Section 18.2), cannot enter into the biological reactions characteristic of D-glucose. ■

18.4   REACTIONS OF MONOSACCHARIDES

Four important reactions of the monosaccharides are fermentation, oxidation, reduction, and cyclization. We will present the first three in this section, and cyclization in Section 18.5.

Fermentation

In the presence of certain enzymes found in yeast, D-glucose or D-fructose is converted to ethanol and carbon dioxide. This enzyme-catalyzed reaction is used in the production of alcoholic beverages.

$$\text{D-glucose or D-fructose} \xrightarrow{\text{enzymes}} 2\ CH_3CH_2OH + 2\ CO_2\uparrow$$
$$\text{ethanol}$$

Oxidation

Recall from Chapter 17 that aldehydes are readily oxidized to carboxylate ions in alkaline solutions of mild oxidizing agents such as Ag^+ (in Tollens reagent) or Cu^{2+} (in Fehling or Benedict reagent).

$$\underset{\textit{an aldehyde}}{\overset{\overset{\displaystyle O}{\|}}{RCH}} \xrightarrow{\text{mild oxidizing agent}} \underset{\textit{a carboxylate ion}}{\overset{\overset{\displaystyle O}{\|}}{RCO^-}}$$

In water solution, the aldehyde groups of glucose, galactose, ribose, and 2-deoxyribose are readily oxidized. Because the monosaccharides act as reducing agents (reducing Ag^+ to Ag metal, for example), they are called **reducing sugars**. (The following equation is not balanced.)

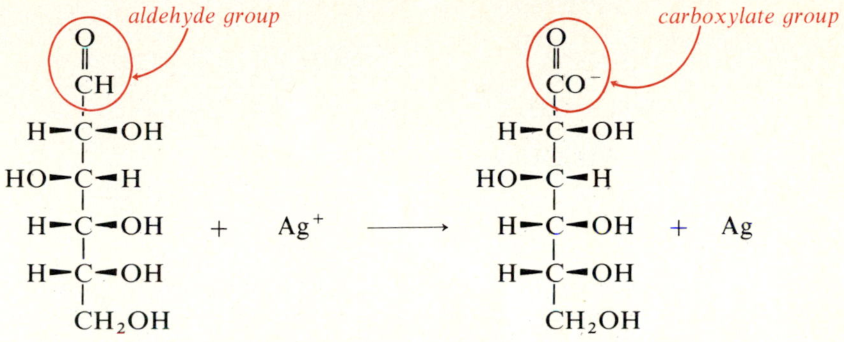

aldehyde group

carboxylate group

$$
\begin{array}{c}
\overset{O}{\underset{\|}{C}}H \\
H-C-OH \\
HO-C-H \\
H-C-OH \\
H-C-OH \\
CH_2OH
\end{array}
\quad + \quad Ag^+ \quad \longrightarrow \quad
\begin{array}{c}
\overset{O}{\underset{\|}{C}}O^- \\
H-C-OH \\
HO-C-H \\
H-C-OH \\
H-C-OH \\
CH_2OH
\end{array}
\quad + \quad Ag
$$

D-glucose   silver ion   D-gluconate ion   silver metal

*the oxidizing agent
in Tollens reagent*

Fructose, which is a ketose, does not contain an aldehyde group. However, fructose is also a reducing sugar because under the alkaline conditions of the test reagents, it can be converted to an aldehyde sugar. However, we will not discuss this reaction of fructose here.

The Tollens, Fehling, and Benedict tests can all be used to test for glucose in the urine, a symptom of diabetes. Other reducing sugars would also yield positive test results, but glucose is the only sugar usually found in urine.

Example   Write the equation for the reaction of D-galactose with the silver ion in Tollens reagent, as we have just done for D-glucose.

Solution   Write the structure of D-galactose (see Figure 18.1), and convert the aldehyde group to a carboxylate group.

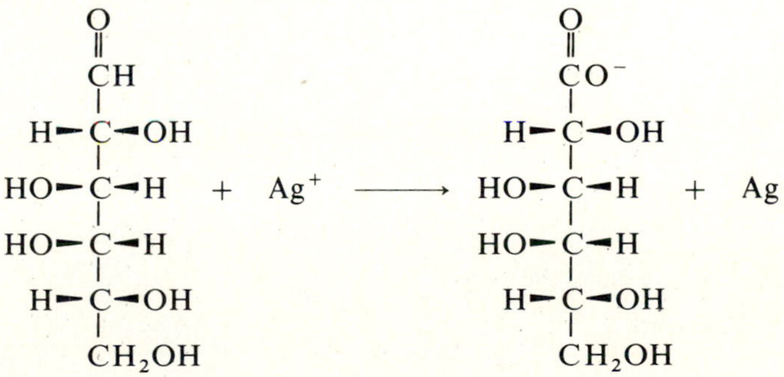

$$
\begin{array}{c}
\overset{O}{\underset{\|}{C}}H \\
H-C-OH \\
HO-C-H \\
HO-C-H \\
H-C-OH \\
CH_2OH
\end{array}
\quad + \quad Ag^+ \quad \longrightarrow \quad
\begin{array}{c}
\overset{O}{\underset{\|}{C}}O^- \\
H-C-OH \\
HO-C-H \\
HO-C-H \\
H-C-OH \\
CH_2OH
\end{array}
\quad + \quad Ag
$$

Problem 18.3. Which of the following compounds would give a positive Tollens test?

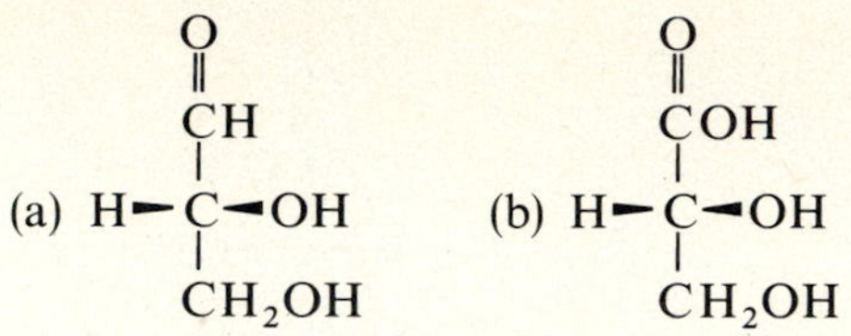

$$
\text{(a)} \quad
\begin{array}{c}
\overset{\displaystyle O}{\underset{\displaystyle |}{\overset{\displaystyle \|}{C}}}H \\
H\!-\!C\!-\!OH \\
| \\
CH_2OH
\end{array}
\qquad
\text{(b)} \quad
\begin{array}{c}
\overset{\displaystyle O}{\underset{\displaystyle |}{\overset{\displaystyle \|}{C}}}OH \\
H\!-\!C\!-\!OH \\
| \\
CH_2OH
\end{array}
$$

Reduction

The aldehyde or ketone group in a monosaccharide can be reduced to a hydroxyl group. The reduction product of glucose is called *glucitol* or *sorbitol*. Sorbitol is used as an artificial sweetener because it does not affect blood sugar levels to the same extent as does glucose or sucrose. (However, sorbitol is not calorie free as are saccharin and the cyclamates.) Sorbitol also occurs naturally in certain seaweeds and many fruits.

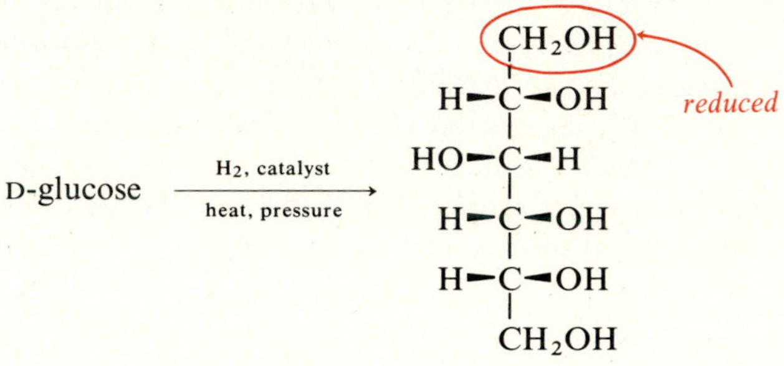

$$
\text{D-glucose} \quad \xrightarrow[\text{heat, pressure}]{H_2,\ \text{catalyst}} \quad
\begin{array}{c}
CH_2OH \qquad reduced \\
H\!-\!C\!-\!OH \\
HO\!-\!C\!-\!H \\
H\!-\!C\!-\!OH \\
H\!-\!C\!-\!OH \\
CH_2OH
\end{array}
$$

sorbitol

an artificial sweetener

Problem 18.4.   Write the formula for the organic product of the reaction of D-ribose (Figure 18.1) with H_2 plus a catalyst, heat, and pressure.

18.5   CYCLIZATION OF MONOSACCHARIDES

Aldehydes react with alcohols to form hemiacetals (Chapter 17).

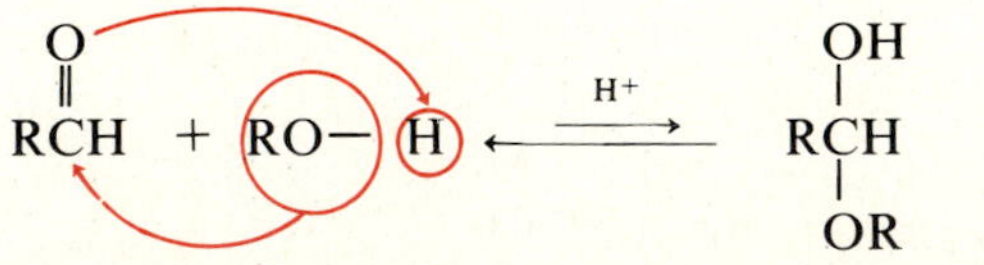

$$
\overset{\displaystyle O}{\underset{\displaystyle }{\overset{\displaystyle \|}{R}}}CH \ + \ RO\!-\!H \quad \underset{}{\overset{H^+}{\rightleftarrows}} \quad
\begin{array}{c}
OH \\
| \\
RCH \\
| \\
OR
\end{array}
$$

an aldehyde   an alcohol   a hemiacetal

If the aldehyde group and a hydroxyl group are positioned in the same molecule so that they can react to form either a five- or six-membered ring, a *cyclic hemiacetal* will be formed. Glucose contains an aldehyde group and hydroxyl groups that are correctly positioned in the molecule for cyclic-hemiacetal formation. Glucose preferentially forms the six-membered cyclic hemiacetal. In water, 99.8% of glucose is in the cyclic form and only 0.2% is in the open-chain aldehyde form.

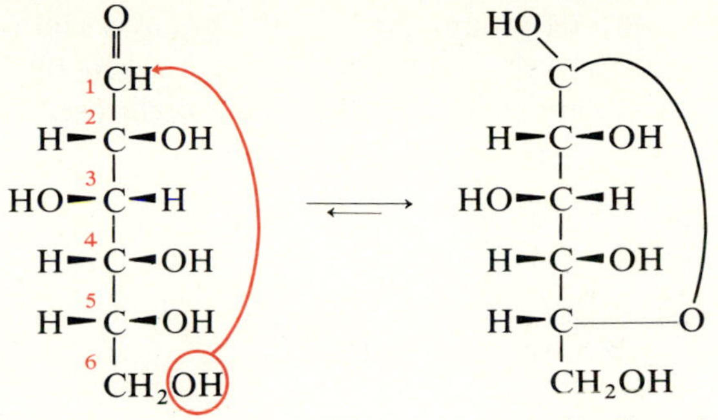

open-chain aldehyde form   a cyclic hemiacetal
of D-glucose   of D-glucose

To emphasize that these hemiacetals are indeed cyclic compounds, we commonly use polygon formulas. The polygon formula is drawn in a condensed form, in which the ring carbons and their hydrogens are implied but are not actually written.

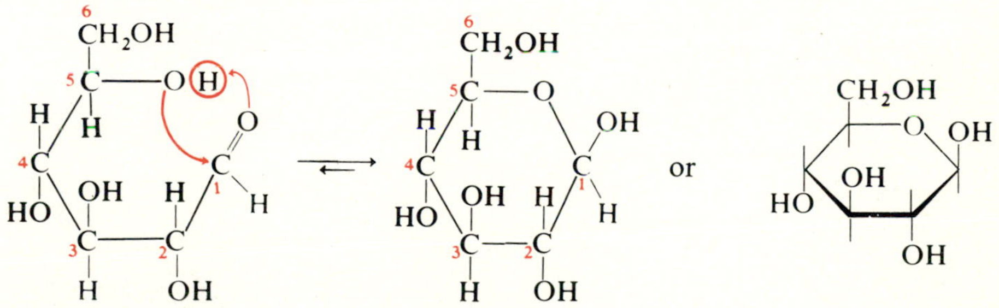

D-glucose as an   D-glucose as a   condensed polygon
open-chain aldehyde   cyclic hemiacetal   formula

α and β Monosaccharides

When a molecule of glucose cyclizes to form a hemiacetal, the hydroxyl group at carbon 5 can attack the carbonyl group from the top or bottom, as we show the structure. The following structures show only the reacting functional groups:

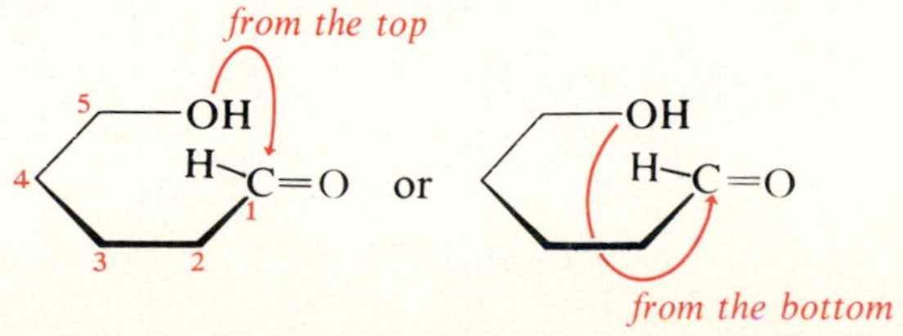

from the top

from the bottom

Because of the two modes of attack, glucose can form two different cyclic hemiacetals. The new —OH group formed at carbon 1 can be either up or down, as we show the formula. If the —OH group is down (*trans* to the —CH$_2$OH group), the hydroxyl group is said to be **alpha** (**α**). If it is above the ring (*cis* to the —CH$_2$OH group), it is **beta** (**β**). These two cyclic hemiacetals of glucose are referred to as α- and β-D-glucose.

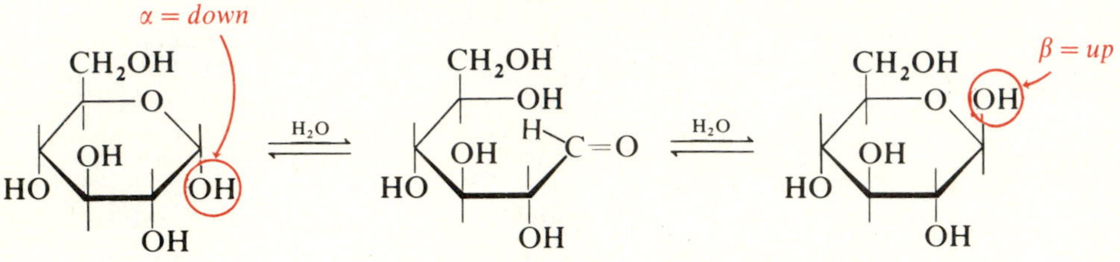

α-D-glucose   D-glucose *open-chain aldehyde form*   β-D-glucose

The α and β forms of D-glucose can be separated from each other and are stable in the solid state. In water solution, however, the two forms are in equilibrium with the open-chain aldehyde form and thus with each other, as shown in the preceding equation.

Figure 18.6. *Cyclic forms of five important monosaccharides.*

cyclic forms of the aldoses:

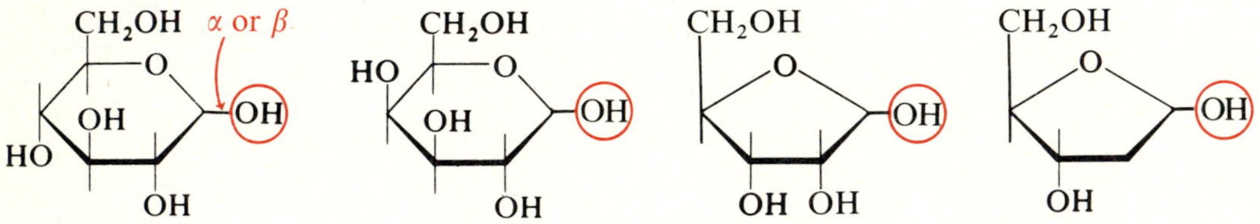

D-glucose   D-galactose   D-ribose   2-deoxy-D-ribose

In each case, the circled OH group can be down (α) or up (β), as shown in the text.

cyclic forms of the ketose fructose:

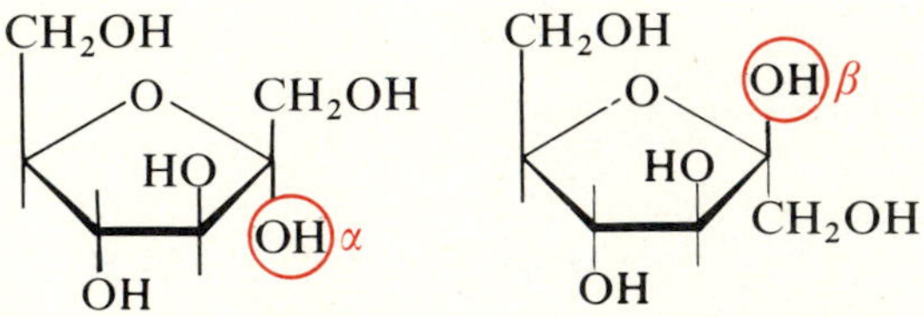

α-D-fructose   β-D-fructose

The equilibrium between the cyclic forms of glucose and its aldehyde form does not affect the ability of glucose to undergo aldehyde reactions, such as the reaction with Tollens reagent. As the aldehyde form is consumed, the equilibrium shifts so that more aldehyde is formed (Le Châtelier's principle).

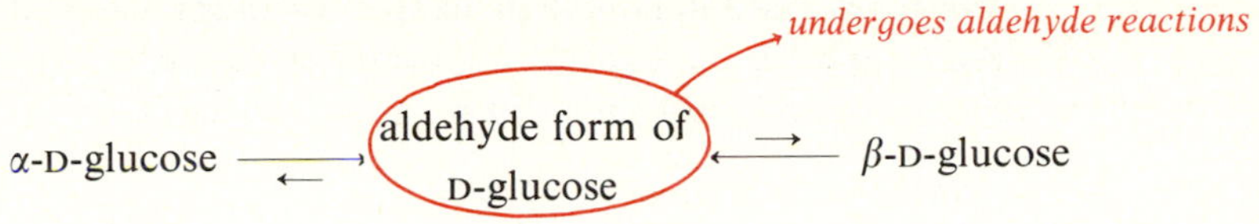

undergoes aldehyde reactions

α-D-glucose ⟶← aldehyde form of D-glucose →← β-D-glucose

The other monosaccharides also undergo cyclization. Figure 18.6 shows the most commonly encountered cyclic forms of the five monosaccharides discussed in this text.

Problem 18.5. Referring to Figure 18.6, draw the polygon formulas for α- and β-D-galactose.

Problem 18.6. Referring to Figure 18.6, write an equation that shows the equilibrium of open-chain D-ribose and its α and β forms. Use the type of equation that we have used for glucose.

18.6 GLYCOSIDE FORMATION

A hemiacetal can react with an alcohol in acidic solution to form an acetal. The hemiacetal-hydroxyl group of a monosaccharide (the —OH at carbon 1 of glucose) can similarly react with an alcohol to form an acetal, which, in carbohydrate chemistry, is called a **glycoside**, *gly-* referring to sugar and *-oside* referring to the acetal form.

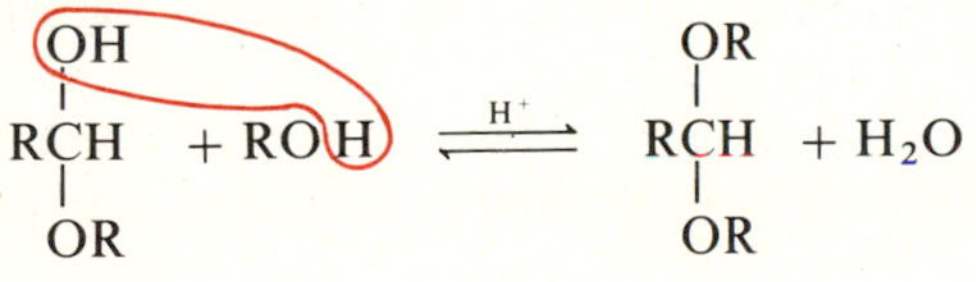

a hemiacetal   *an acetal*

formation of glycosides:

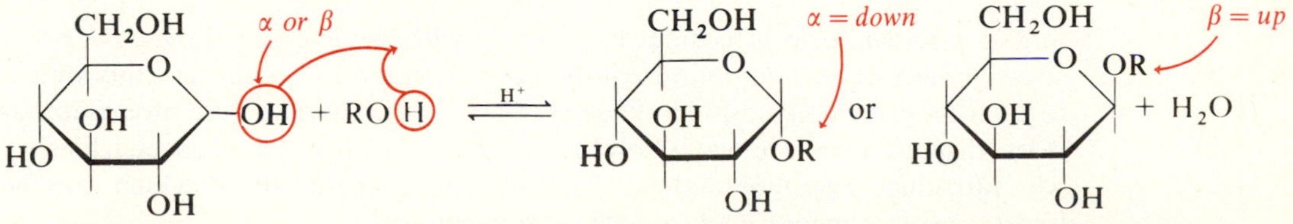

D-glucose   an alkyl α-D-glucoside   an alkyl β-D-glucoside

Like other acetals, glycosides are stable in neutral or alkaline solution. Because glycosides are not in equilibrium with an aldehyde in alkaline solution, they do not react with Tollens, Benedict, or Fehling reagents. Glycosides are, therefore, **nonreducing sugars**.

Glycosides are common in nature. In many cases, a sugar molecule is bonded to another biologically important molecule. Glucovanillin, from which vanillin is obtained, is a simple example of a naturally occurring glycoside.

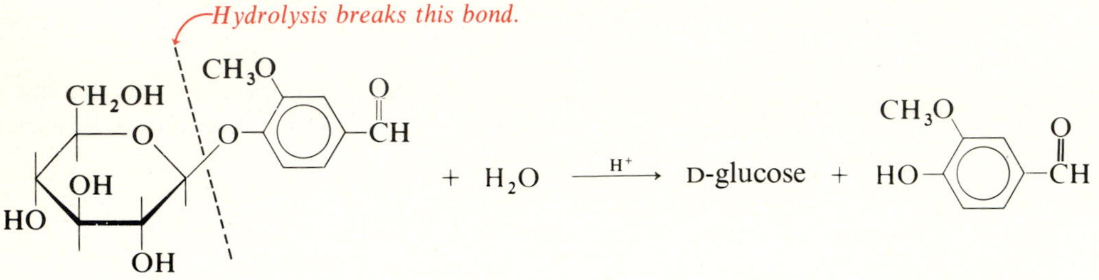

Hydrolysis breaks this bond.

$$\text{glucovanillin} + H_2O \xrightarrow{H^+} \text{D-glucose} + \text{vanillin}$$

glucovanillin
a glycoside in vanilla beans

vanillin
used in vanilla flavoring

Problem 18.7. Predict the hydrolysis products of the following glycoside in aqueous acid. Use the type of equation that we have used for the hydrolysis of glucovanillin.

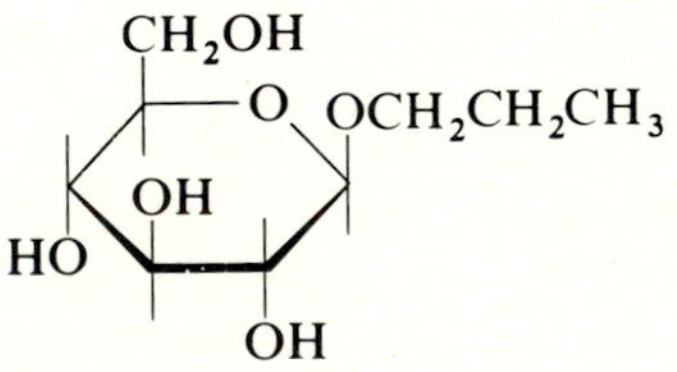

Problem 18.8. (a) Write an equation for the reaction of α-D-glucose with methanol in aqueous acid.
(b) Predict whether the α or β glycoside or a mixture would be formed in this reaction, and explain why.

18.7   DISACCHARIDES

Sugars also can form glycosides by reaction with the hydroxyl groups of other sugar molecules. A compound composed of two monosaccharide units joined to each other by a glycoside linkage is called a **disaccharide**. Like monosaccharides, disaccharides are water soluble and sweet tasting. The disaccharides we will introduce here are maltose, lactose, and sucrose, all of which can be hydrolyzed in aqueous acid to yield monosaccharides.

Hydrolysis of disaccharides yields monosaccharides:

$$\text{maltose} + H_2O \xrightarrow{H^+} \text{glucose} + \text{glucose}$$

$$\text{lactose} + H_2O \xrightarrow{H^+} \text{glucose} + \text{galactose}$$

$$\text{sucrose} + H_2O \xrightarrow{H^+} \text{glucose} + \text{fructose}$$

Maltose (Malt Sugar)

Maltose, a sugar used in baby food and malted milk, is a disaccharide formed from two units of D-glucose. The acetal link joins carbon 1 of one glucose unit to carbon 4 of the other glucose unit.

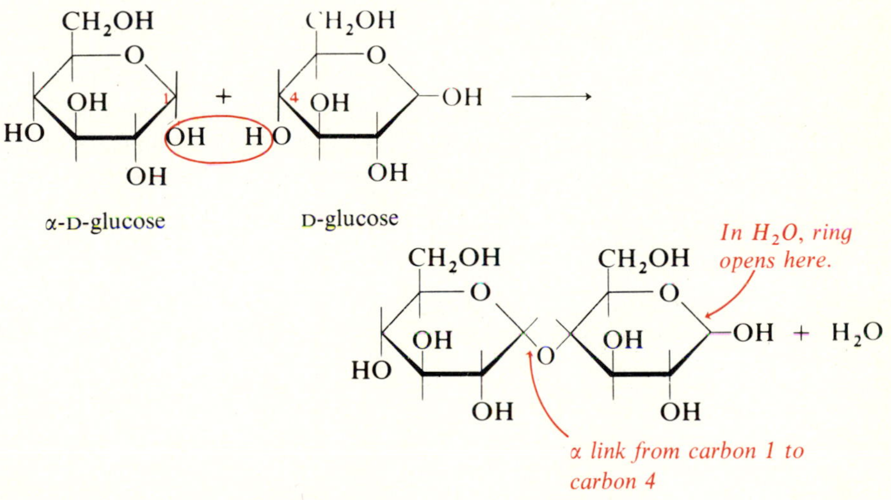

α-D-glucose   D-glucose

In H_2O, ring opens here.

α link from carbon 1 to carbon 4

maltose
a disaccharide

In maltose, the oxygen joining the two rings is in the α position. This α link between the two rings is stable in water solution. However, because the right-hand ring of maltose, as we have shown the structure, is in equilibrium with an aldehyde, maltose is a reducing sugar, just as glucose is.

Lactose (Milk Sugar)

An example of a disaccharide with a *β* link between the two rings is lactose, or milk sugar, found in milk and milk products. Lactose contains two different monosaccharides, D-galactose and D-glucose, that are joined together by a *β* link from carbon 1 of the galactose unit to carbon 4 of the glucose unit. Like maltose, lactose is a reducing sugar because in water the glucose unit is in equilibrium with the aldehyde form.

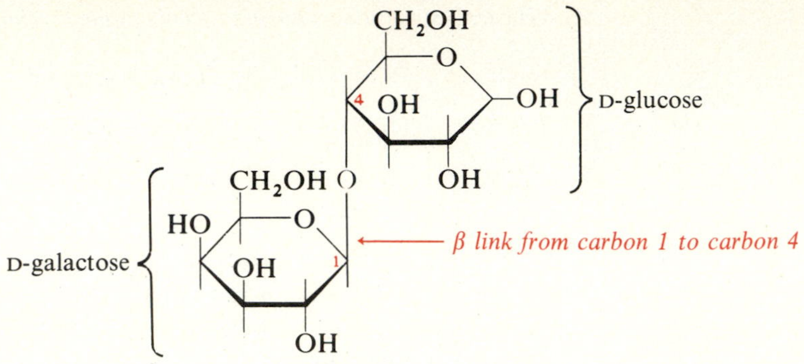

β link from carbon 1 to carbon 4

lactose

Sucrose (Table Sugar)

The disaccharide sucrose is found in all plants and in large amounts in sugar cane and sugar beets. Our commercial supply comes from these two plants. (The processed sugars from both sources are chemically identical.) Sucrose is commercially processed in greater quantity than any other pure organic chemical. The average person in the United States consumes over one hundred pounds of sucrose per year.

Sucrose is a disaccharide of glucose and fructose. The bonding in sucrose is different from that in maltose or lactose, in that the link between the two monosaccharide units is from carbon 1 of glucose to carbon 2 (the "hemiacetal" carbon) of fructose. Therefore, *both* units are glycosidic, and neither ring is in equilibrium with an aldehyde form in neutral or alkaline solution. Sucrose is a nonreducing sugar.

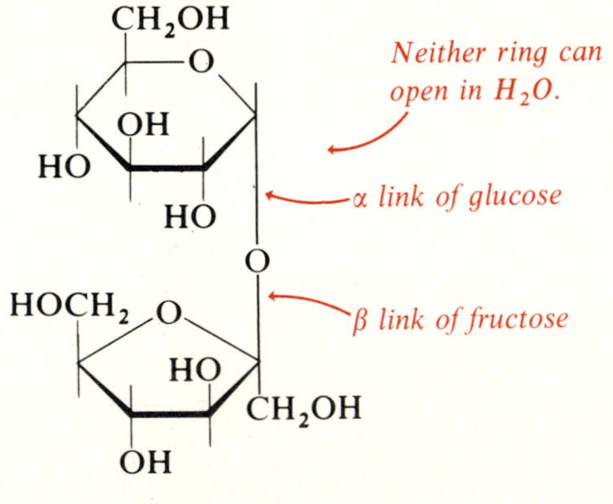

Neither ring can open in H_2O.

α link of glucose

β link of fructose

sucrose

Invert sugar is a mixture of glucose and fructose formed by the acid- or enzyme-catalyzed hydrolysis of sucrose. Because fructose is the sweetest sugar, invert sugar is sweeter than sucrose. Because of its sweetness, invert sugar is used in ice cream and other sweet processed foods; it also occurs naturally in honey.

Problem 18.9.   Referring to the formulas in this section, predict the organic products of the following reactions. If no reaction occurs, write "no reaction." (*Hint:* Do not be intimidated by the complexity of the molecules. Look only at the portion that might undergo reaction.)

(a) maltose + Tollens reagent $\longrightarrow$

(b) sucrose + Tollens reagent $\longrightarrow$

(c) lactose + H$_2$ $\xrightarrow[\text{heat, pressure}]{\text{catalyst}}$

18.8   POLYSACCHARIDES

Starch and cellulose are **polysaccharides** composed of repeating units of glucose. Because of the very size of starch and cellulose molecules, these substances are not soluble in water as are the mono- and disaccharides, and they are not sweet tasting. Polysaccharides are nonreducing carbohydrates.

$$\left.\begin{array}{l} \text{starch} + H_2O \\ \text{cellulose} + H_2O \end{array}\right\} \xrightarrow{H^+} \text{many units of D-glucose}$$

Starch

Starch such as is found in cornstarch, bread, or potatoes is a mixture of two types of polysaccharides: **amylose** and **amylopectin**. The structures of these compounds are shown in Figure 18.7. Amylose is a polymer composed of D-glucose units that are joined by an α link from carbon 1 of one glucose to carbon 4 of the next glucose. Amylopectin also contains units of 1,4-α-linked D-glucose. In addition, amylopectin has cross-linking at the 6 position of some of the glucose units. The cross-linking joins chains together into branched structures.

Amylose forms a deep blue complex with iodine (I$_2$). If you drop tincture of iodine (an alcohol solution of I$_2$) on the cut surface of a raw potato, you can observe a deep blue coloration on the potato. Starch is commonly used as a test for the presence of iodine; conversely, iodine is used to detect the presence of starch.

Starch digestion begins in the mouth. When a starchy food such as corn or bread is chewed, the enzyme *amylase*, which is found in saliva, begins to break down the large starch molecules into maltose, which has a sweet taste. Maltose molecules are too large to pass through membranes. Further breakdown to the monosaccharide glucose occurs in the intestines. The glucose is then absorbed into the bloodstream and carried to cells throughout the body for use as food or to be stored.

Figure 18.7.   *The two types of polysaccharides in starch.*

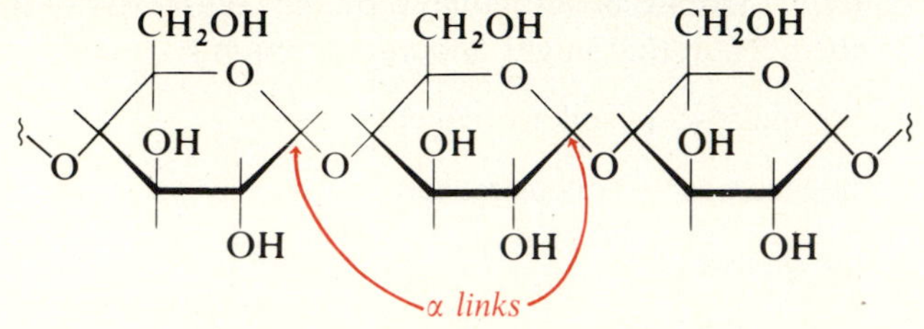

amylose
a linear polymer of α-linked D-glucose

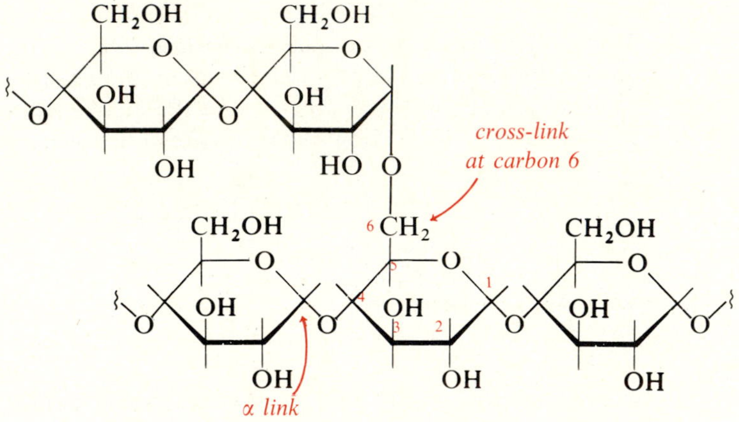

amylopectin
a polymer of α-linked D-glucose with branching

$$\text{starch} \xrightarrow[\text{amylase}]{\text{H}_2\text{O}} \text{maltose} \xrightarrow[\text{other enzymes}]{\text{H}_2\text{O}} \text{glucose}$$

Problem 18.10.   Another disaccharide besides maltose, called *isomaltose*, can be isolated from the partial hydrolysis of amylopectin. What is the structure of isomaltose? (Refer to Figure 18.7.)

Cellulose

Cellulose is one of the main structural components of wood and other plant materials. Cotton is about 98% pure cellulose. Cellulose consists of chains of glucose units, with up to 14,000 units in each chain. These chains are twisted into ropelike bundles, which lend strength to the plant tissues.

Figure 18.8.   *The structure of cellulose, a linear β-linked polymer of* D-*glucose*

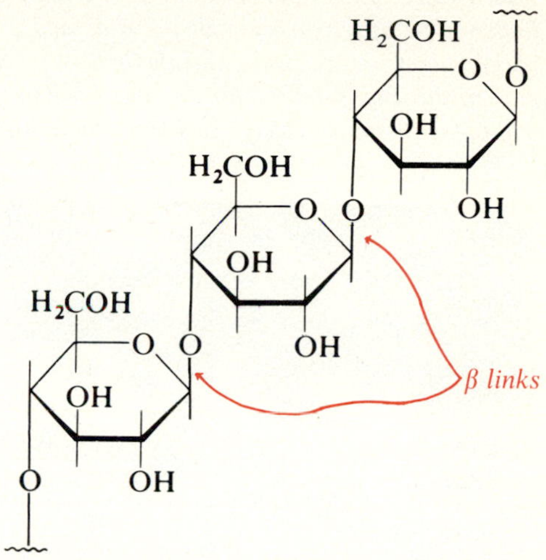

β links

Instead of the α-linked glucose found in starch, the glucose in cellulose is bonded by β links (see Figure 18.8). Enzymes produced by certain bacteria and protozoa are capable of hydrolyzing the β links in cellulose. Higher animals do not have the enzymes that can break β-linked polymeric glucose into monomeric glucose; therefore, humans cannot utilize cellulose as food. Deer, cows, termites, and other animals that use cellulose for food must have colonies of the proper bacteria and protozoa in their digestive tracts to break the cellulose down into glucose.

Glycogen

Plants use starch as a storehouse for glucose. When plant cells require glucose, enzymes catalyze the hydrolysis of starch into glucose. Animals also require glucose storage. Instead of starch, *glycogen* is the glucose-storing polysaccharide of animals. Glycogen is synthesized from excess glucose in the body in a process called **glycogenesis** (from *glycogen* and *genesis*, "creation") and is stored principally in the liver and muscles. When the organism requires glucose, glycogen is hydrolyzed enzymatically back to glucose in a process called **glycogenolysis** (from *glycogen* and *hydrolysis*).

$$\text{D-glucose} \underset{\text{glycogenolysis}}{\overset{\text{glycogenesis}}{\rightleftharpoons}} \text{glycogen} + H_2O$$

D-glucose *in the blood* ⟶ glycogen + H₂O *stored in liver and muscles*

Glycogen is composed principally of 1,4-α-linked glucose units with 1,6-branching. Its structure is very similar to that of amylopectin (Figure 18.7), except that glycogen molecules contain more branching.

Photo 18.1.   *Electron micrographs of a cotton fiber. A cotton fiber, shown in the top left-hand photo, is a hollow tube approximately 20 μm wide and 2.5 cm long. The higher magnification in the top right-hand photo shows a cross section of the hollow tube. The highest magnification in the bottom photo shows the microfibrils that form the fiber's tube. Each microfibril is composed of cellulose molecules that contain over 3000 glucose units apiece. (U.S.D.A. Photo by Southern Research Center, New Orleans, Louisiana.)*

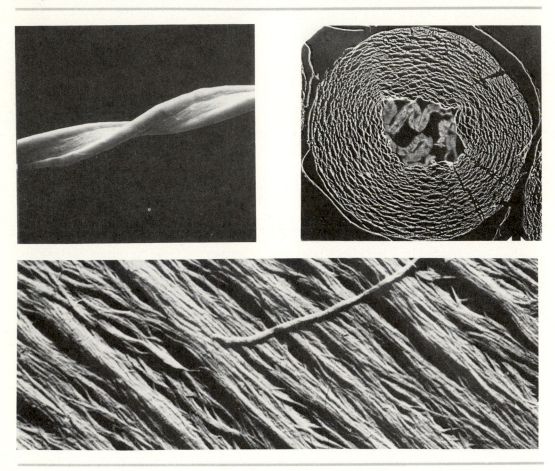

Problem 18.11.   If glycogen were partially hydrolyzed, what disaccharides would be formed? Use words in your answer, not formulas.

SUMMARY

Monosaccharides are carbohydrates composed of one sugar unit. Important examples are glucose, galactose, fructose, ribose, and 2-deoxyribose. Monosaccharides may be classed as *aldoses* or *ketoses* and as *trioses*, *tetroses*, *pentoses*, or *hexoses*.

Stereoisomers are isomers that differ only by the arrangement of their atoms in space. Most naturally occurring monosaccharides belong to the D *series*:

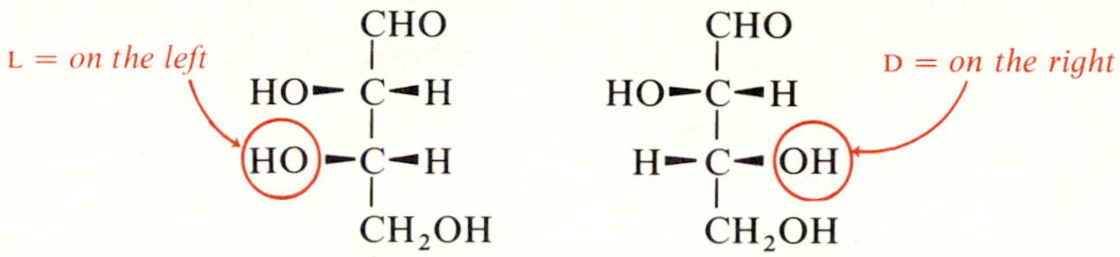

CHO   CHO

L = *on the left*   HO—C—H   HO—C—H   D = *on the right*

HO—C—H   H—C—OH

CH$_2$OH   CH$_2$OH

Optical isomers are nonsuperimposable mirror-image isomers of chiral molecules, generally molecules that contain at least one carbon atom with four different substituents.

Glucose and fructose can undergo enzymatic fermentation to ethanol and CO_2. The aldoses and fructose are *reducing sugars*—they are easily oxidized by Tollens reagent and other mild oxidizing agents. These monosaccharides can also be reduced by reducing agents such as H_2.

In aqueous solution, monosaccharides form cyclic hemiacetals, which we usually represent by polygon formulas.

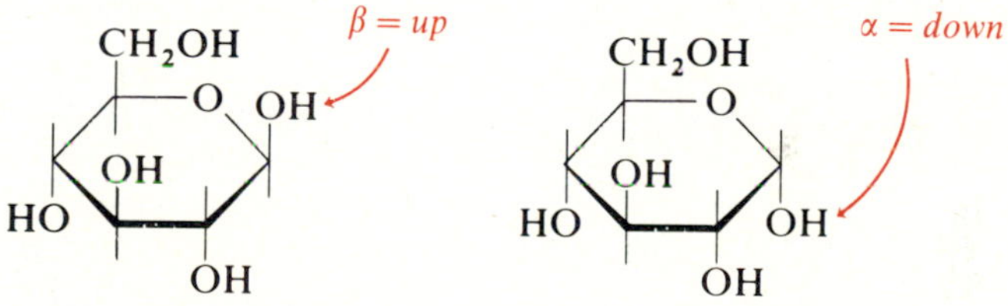

β = *up*   α = *down*

(CH$_2$OH, O, OH, OH, HO, OH structures shown)

The hemiacetal group in cyclic aldoses or ketoses can react with an alcohol to yield acetals called *glycosides*. Glycosides are stable in neutral or alkaline solution and are *nonreducing sugars*.

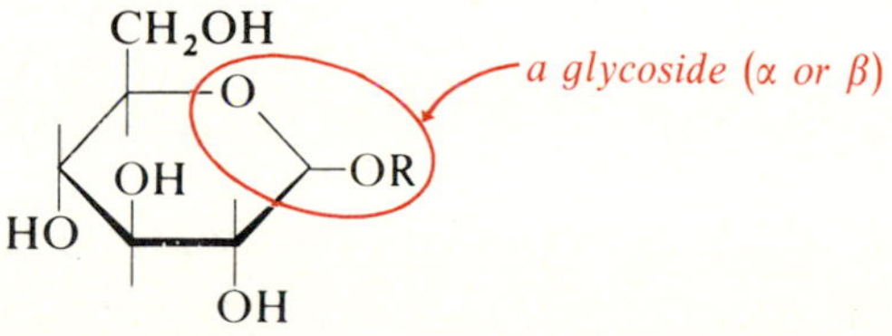

a glycoside (α or β)

(CH$_2$OH, O, OH, OR, HO, OH structure shown)

A *disaccharide* is composed of two monosaccharide units joined by a glycoside link. Common disaccharides include:

maltose: α-D-glucose + D-glucose
lactose: β-D-galactose + D-glucose
sucrose: β-D-fructose + α-D-glucose

Common *polysaccharides*, polymers composed of many monosaccharide units, follow.

starch:
>   amylose: α-linked D-glucose
>   amylopectin: α-linked D-glucose with branching

cellulose: β-linked D-glucose

glycogen: α-linked D-glucose with branching

The formation of glycogen from glucose is called *glycogenesis*; the hydrolysis of glycogen back to glucose is called *glycogenolysis*.

KEY TERMS

carbohydrates	fermentation	starch
monosaccharide	reducing sugar	cellulose
glucose	α- and β-monosaccharides	glycogen
aldohexose	glycoside	glycogenesis
D- and L-stereoisomers	disaccharide	glycogenolysis
*optical isomers	polysaccharide	

STUDY PROBLEMS

18.12. Classify each of the following monosaccharides as an aldose or a ketose:

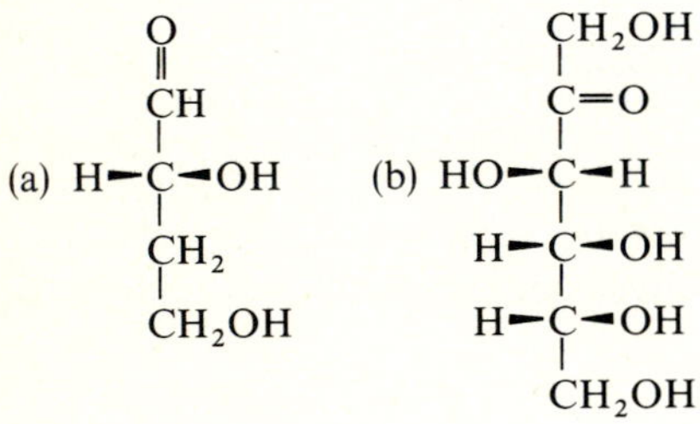

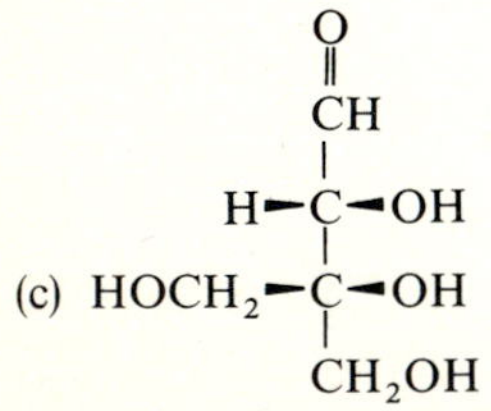

(a)
$$
\begin{array}{c}
O \\
\parallel \\
CH \\
| \\
H-C-OH \\
| \\
CH_2 \\
| \\
CH_2OH
\end{array}
$$

(b)
$$
\begin{array}{c}
CH_2OH \\
| \\
C=O \\
| \\
HO-C-H \\
| \\
H-C-OH \\
| \\
H-C-OH \\
| \\
CH_2OH
\end{array}
$$

(c) $HOCH_2-$
$$
\begin{array}{c}
O \\
\parallel \\
CH \\
| \\
H-C-OH \\
C-OH \\
| \\
CH_2OH
\end{array}
$$

18.13. Classify each of the monosaccharides in the preceding problem as a triose, tetrose, pentose, or hexose.

18.14. Why does the presence of a large number of hydroxyl groups impart water solubility to a compound?

18.15. There are eight stereoisomeric aldohexoses in the D series and eight more in the L series. In this chapter, we considered two D aldohexoses, D-glucose and D-galactose. Write the formula for a D-aldohexose that is a stereoisomer of these two sugars.

18.16. Use "wedge" formulas, such as we use in Figure 18.1, to show the D and L isomers of the following compounds:

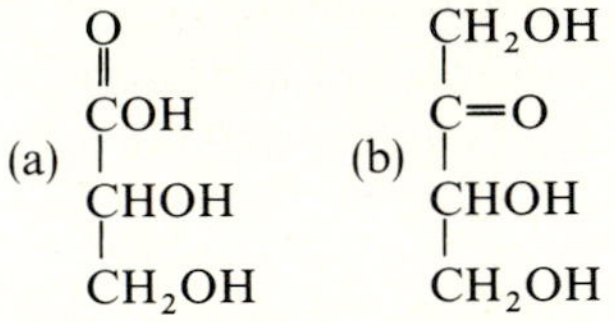

(a)
$$
\begin{array}{c}
O \\
\parallel \\
COH \\
| \\
CHOH \\
| \\
CH_2OH
\end{array}
$$

(b)
$$
\begin{array}{c}
CH_2OH \\
| \\
C=O \\
| \\
CHOH \\
| \\
CH_2OH
\end{array}
$$

*18.17. Carbohydrates are not the only class of compounds containing chiral molecules. Which of the following structures are chiral? (*Hint*: Look for a carbon atom with four different groups bonded to it.)

(a) $(CH_3)_2CHBr$

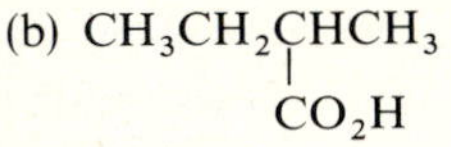

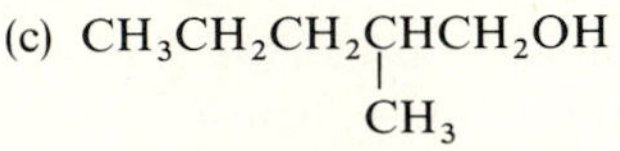

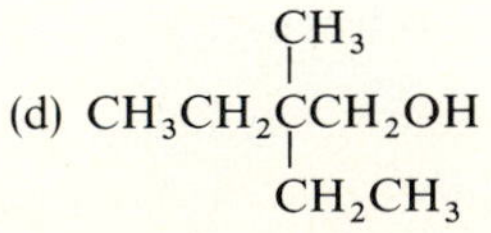

(b) $CH_3CH_2CHCH_3$
$\qquad\qquad\quad |$
$\qquad\qquad\quad CO_2H$

(c) $CH_3CH_2CH_2CHCH_2OH$
$\qquad\qquad\qquad\quad |$
$\qquad\qquad\qquad\quad CH_3$

(d) $CH_3CH_2CCH_2OH$
$\qquad\quad\ \ \overset{\displaystyle CH_3}{\overset{|}{\underset{|}{}}}$
$\qquad\qquad\ \ CH_2CH_3$

18.18. Write the formula for the organic product when each of the following compounds is treated with Tollens reagent:

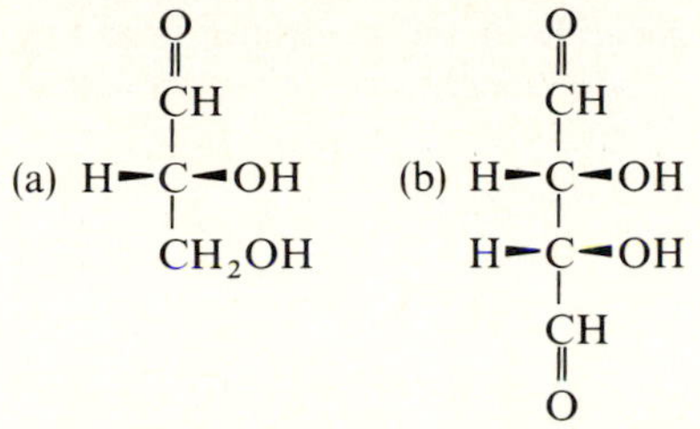

(a)
$$H-\overset{\overset{\displaystyle O}{\|}}{\underset{\overset{\displaystyle |}{CH_2OH}}{\underset{\displaystyle |}{C}}}\!\!\!\!-OH$$

(b)
$$H-\overset{\overset{\displaystyle O}{\|}}{\underset{\overset{\displaystyle |}{\underset{\displaystyle CH}{\underset{\displaystyle \|}{O}}}}{\underset{\displaystyle |}{\underset{\displaystyle H-C}{\underset{\displaystyle |}{C}}}}}\!\!\!\!-OH$$

18.19. Which of the following are true of a reducing sugar?
(a) gives a positive Tollens test
(b) gives a positive Benedict test
(c) is easily oxidized
(d) can be reduced

18.20 Predict the products when the following compounds are treated with an excess of H_2 and a catalyst under heat and pressure:

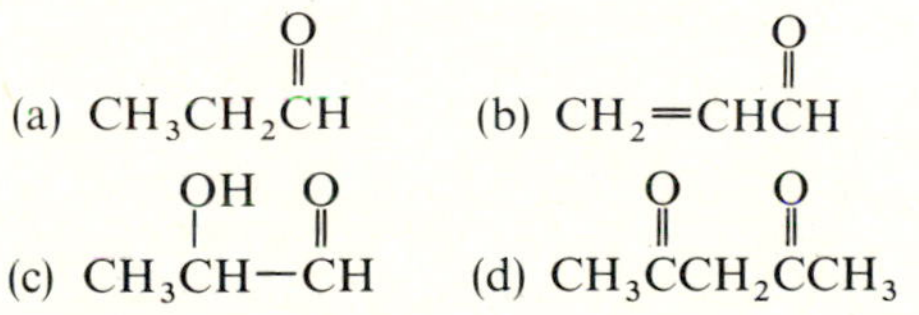

(a) $CH_3CH_2\overset{\overset{\displaystyle O}{\|}}{C}H$

(b) $CH_2{=}CH\overset{\overset{\displaystyle O}{\|}}{C}H$

(c) $CH_3\overset{\overset{\displaystyle OH}{|}}{C}H{-}\overset{\overset{\displaystyle O}{\|}}{C}H$

(d) $CH_3\overset{\overset{\displaystyle O}{\|}}{C}CH_2\overset{\overset{\displaystyle O}{\|}}{C}CH_3$

18.21. (a) Referring to Figure 18.6, write formulas for the α and β forms of 2-deoxy-D-ribose, and label each.
(b) Are the α- and β-2-deoxy-D-ribose structural isomers, stereoisomers, or neither of these? Explain.
(c) If pure α-2-deoxy-D-ribose is dissolved in water, what happens to it? Use a word equation in your answer.

18.22. Identify and circle any hemiacetal or acetal groups in the following formulas:

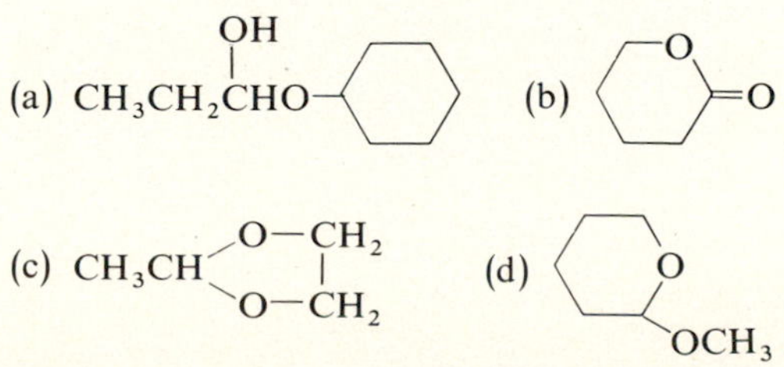

(a) $CH_3CH_2\overset{\overset{\displaystyle OH}{|}}{C}HO-$⬡

(b) ⬡$=O$

(c) $CH_3CH\overset{\displaystyle O-CH_2}{\underset{\displaystyle O-CH_2}{|}}$

(d) ⬡ with O, OCH$_3$

18.23. Complete the following equations, showing the hemiacetal products. Remember that $-CHO$ represents an aldehyde group.

(a) $CH_3CHO + CH_3OH \overset{H^+}{\rightleftharpoons}$

(b) $HOCH_2CH_2CH_2CH_2CHO \overset{H^+}{\rightleftharpoons}$

18.24. Choose one or more correct answers: A glycoside is
(a) a hemiacetal.
(b) an acetal.
(c) stable in a dilute solution of HCl.
(d) stable in a dilute solution of NaOH.

18.25. Which of the following compounds contain at least one glycoside group?
(a) α-D-glucose   (b) maltose
(c) sucrose   (d) lactose
(e) amylose   (f) glycogen

18.26. (a) Which of the compounds in Problem 18.25 are reducing sugars?
(b) Which would undergo reaction with H_2 (plus catalyst, heat, and pressure)?

18.27. Cellobiose is a disaccharide formed in the hydrolysis of cellulose. Draw its structure.

18.28. Write word equations that illustrate each of the following reactions:
(a) Starch is digested.
(b) Glycogenesis
(c) Glycogenolysis
(d) Sucrose is converted to invert sugar.
(e) Lactose is hydrolyzed.
(f) Invert sugar is subjected to fermentation.

18.29. *Hyaluronic acid* is a polysaccharide that is found in the connective tissues, such as tendons, of the body. It is formed from repeating units of the following disaccharide:

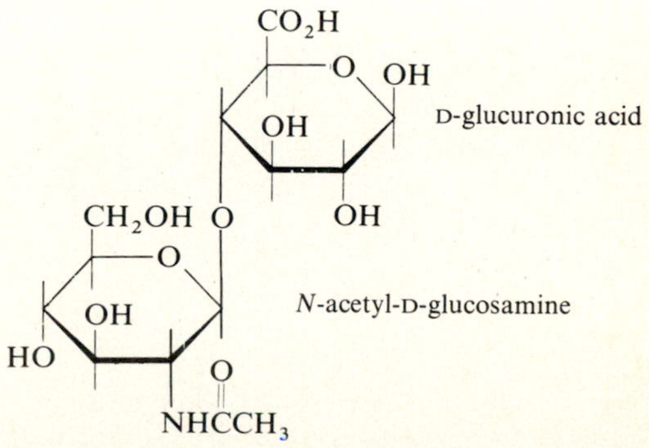

D-glucuronic acid

N-acetyl-D-glucosamine

Identify and circle the following features of this disaccharide.

(a) the hemiacetal group, if any
(b) the acetal group, if any
(c) the carboxyl group, if any
(d) the amine group, if any
(e) the amide group, if any
(f) the glycoside link, if any
(g) any α or β links

18.30.   What would be the hydrolysis products of the disaccharide in the preceding problem? Be careful; amides undergo hydrolysis as well.

19 Lipids

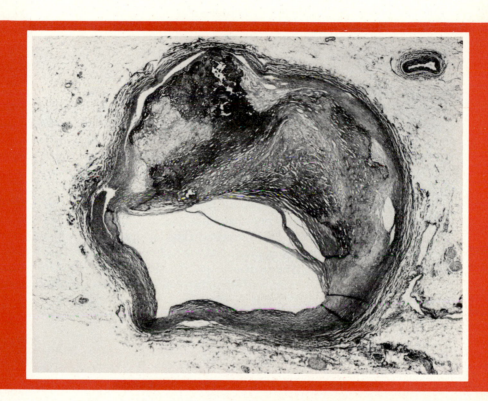

A magnified cross section of a hardened coronary artery. (Martin M. Rotker/Taurus Photos.) Atherosclerosis (a type of hardening of the arteries) is the most common disease of the arteries and is a primary cause of heart attacks and strokes. This magnified cross section of a hardened coronary artery shows extensive fatty deposits on the artery's lining. The deposits are composed of triacylglycerols and cholesterol— compounds called lipids, which are described in this chapter.

Objectives Define and list some classes of lipids. □ Write an equation for triacylglycerol formation. □ Differentiate between an animal fat and a vegetable oil. □ Write a formula for a wax. □ Describe with an equation the hydrogenation of a vegetable oil. □ List the causes and preventions of rancidity. □ Write an equation for the saponification of a triacylglycerol. □ Use general or specific formulas to define phosphoglycerides, lecithins, cephalins, sphingolipids, phosphosphingosides, and cerebrosides, and describe their occurrences (optional). □ Describe the structures, occurrences, and functions of prostaglandins (optional). □ Draw the formula for the steroid ring system, and describe the functions of some common steroids (optional).

Lipids are defined as naturally occurring organic compounds that are *insoluble in water but soluble in relatively nonpolar organic solvents*, such as diethyl ether. (This definition for a group of compounds is unusual because it is based on a physical property instead of on any structural feature.) Carbohydrates, proteins, and many other natural products are insoluble in nonpolar solvents. Consequently, lipids can be extracted from crude cellular mixtures, while carbohydrates and proteins remain behind.

A large variety of very different compounds are classified as lipids. In this chapter, we will discuss only the more important ones. Triacylglycerols (animal fats and vegetable oils) are one class of lipids. We will discuss the similarities and differences between fats and oils and some of their functions and chemical reactions. The structures and functions of phospholipids (such as the lecithins), prostaglandins, and steroids (such as cholesterol), which are also lipids, will be introduced later in the chapter.

19.1   EDIBLE FATS AND OILS: TRIACYLGLYCEROLS

Edible fats and oils (not to be confused with petroleum oils such as mineral oil) are triesters of the triol glycerol (also called glycerine) and long-chain carboxylic acids called fatty acids. Fats and oils are referred to as **triacylglycerols** or **triglycerides**. (The latter is an older term.)

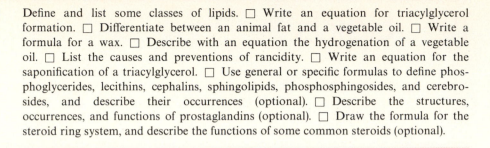

glycerol
a triol

a fatty acid,
which we will represent as RCO_2H

formation of a triacylglycerol:

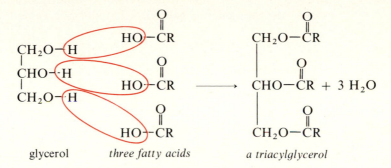

| glycerol | *three fatty acids* | *a triacylglycerol* |

general formula for a triacyglycerol:

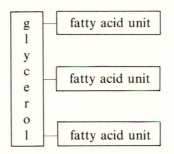

Variations in the structures, and thus in the properties, of fats and oils occur in the fatty acid portion of the ester. These long-chain acids vary in the number of carbon atoms in the hydrocarbon chain and may or may not contain carbon-carbon double bonds. (See Table 19.1 for some representative fatty acids.) In nature, most triacylglycerols are mixed esters; that is, the three fatty acid groups in the ester are not the same.

Table 19.1.   *Some Common Fatty Acids in Fats and Oils*

Structure	Name of acid	Number of carbons	Number of double bonds
$CH_3(CH_2)_{12}CO_2H$	myristic acid	14	0
$CH_3(CH_2)_{14}CO_2H$	palmitic acid	16	0
$CH_3(CH_2)_{16}CO_2H$	stearic acid	18	0
$CH_3(CH_2)_5CH=CH(CH_2)_7CO_2H$	palmitoleic acid	16	1
$CH_3(CH_2)_7CH=CH(CH_2)_7CO_2H$	oleic acid	18	1
$CH_3(CH_2)_4CH=CHCH_2CH=CH(CH_2)_7CO_2H$	linoleic acid[a]	18	2

[a] Linoleic acid is considered to be the only essential fatty acid; the body can convert it and other compounds to other necessary fatty acids.

Example   What are the two possible structures for a triacylglycerol formed from two myristic acid units and one palmitic acid unit? (See Table 19.1.)

Solution

$$CH_2O_2C(CH_2)_{12}CH_3$$
$$|$$
$$CHO_2C(CH_2)_{12}CH_3 \quad and$$
$$|$$
$$CH_2O_2C(CH_2)_{14}CH_3$$

palmityl group
on an end carbon

$$CH_2O_2C(CH_2)_{12}CH_3$$
$$|$$
$$CHO_2C(CH_2)_{14}CH_3$$
$$|$$
$$CH_2O_2C(CH_2)_{12}CH_3$$

palmityl group
on center carbon

Composition and Properties of Fats and Oils

By definition, a **fat** is a triacylglycerol that is solid or semisolid at room temperature; an **oil** is a triacylglycerol that is liquid at room temperature. Since room temperature varies from place to place and from season to season, the distinction between fats and oils is not sharp. Although coconut oil is a solid in temperate climates, it is a liquid in the tropics where it is prepared; thus, it is called an oil.

Saturated fatty acids (those with no carbon-carbon double bonds) form higher-melting triacylglycerols than do the unsaturated fatty acids. The saturated triacylglycerols tend to be solid fats, while the unsaturated triacylglycerols, particularly those that are polyunsaturated (have more than one carbon-carbon double bond), tend to be oils. This is because the molecules of saturated fats are of such a shape that they can be packed closer together. The unsaturated molecules, which contain *cis* double bonds, are more unwieldy in their shapes, remain farther apart, and thus are less attracted by their neighbors.

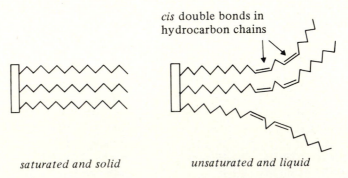

cis double bonds in hydrocarbon chains

saturated and solid   *unsaturated and liquid*

In general, the triacylglycerols found in animals contain few double bonds and are solid; therefore, these are referred to as **animal fats**. On the other hand, triacylglycerols found in plants tend to be polyunsaturated and liquid; hence, we refer to these compounds as **vegetable oils**. The vegetable oils are

Table 19.2.   *Approximate Fatty Acid Composition in Plant and Animal Triacylglycerols*

	Fatty acid composition (%)[a]				
Source	Palmitic	Stearic	Palmitoleic	Oleic	Linoleic
corn	10	5	1	45	38
soybean	10	5	—	25	55
lard	30	15	1	45	5
butter	30	10	—	25	3

[a] The percentages may not add up to 100 because other fatty acids may also be present.

found primarily in fruits and seeds. Such origins are apparent in the names of some of the common oils: olive oil, cottonseed oil, peanut oil, soybean oil.

The composition of the fatty acid portions of fats and oils is exceedingly variable from species to species and within any given species. Dietary lipids are a major factor in the composition of the fat deposits in an animal's body. Thus, the composition of fats in an animal depends somewhat on the season of the year and on the geographic region where the animal is found. Table 19.2 lists the fatty acid composition of some typical plant and animal triacylglycerols.

Waxes are somewhat related to the triacylglycerols in structure. A wax is an ester of a long-chain fatty acid and a simple long-chain alcohol. Beeswax, from which bees make their honeycombs, is composed of a mixture of high-formula-weight esters such as $C_{15}H_{31}CO_2C_{30}H_{61}$ and $C_{25}H_{51}CO_2C_{28}H_{57}$. Waxes are moldable, spreadable, and water insoluble; thus, they are good protective coatings.

19.2   REACTIONS OF TRIACYLGLYCEROLS

Hydrogenation

The carbon-carbon double bonds in an unsaturated triacylglycerol are easily hydrogenated (more easily than are carbonyl groups), and in this way an oil may be converted to a fat. The partial hydrogenation, or hardening, of oils to fats, such as in the preparation of commercial peanut butter or oleomargarine, is an important process that was discussed in Section 15.4.

a general hydrogenation reaction:

$$\underset{\text{unsaturated}}{RCH{=}CHR} \quad + \quad H_2 \quad \xrightarrow{\text{catalyst}} \quad \underset{\text{saturated}}{R\overset{\overset{\displaystyle H}{|}}{C}H{-}\overset{\overset{\displaystyle H}{|}}{C}HR}$$

Problem 19.1. What are the structures of the possible organic products from the reaction of 1.0 mole of triolein (the triester of oleic acid and glycerol) with 2.0 moles of H_2 in the presence of an appropriate catalyst? Refer to Table 19.1 for the structure of oleic acid.

Saponification

In addition to being found in foodstuffs, fats and oils are used in the manufacture of soaps such as $CH_3(CH_2)_{14}CO_2^- Na^+$, which are obtained by the saponification of triacylglycerols. Soaps are discussed in Section 17.7. Because triacylglycerols can be saponified, they are often called *saponifiable lipids*. Steroids, by contrast, contain no ester groups and are called *nonsaponifiable lipids*.

a general equation for a saponification reaction:

$$\underset{\text{RCOR}'}{\overset{O}{\overset{\|}{}}} + NaOH \xrightarrow{H_2O} \underset{\text{RCO}^- Na^+}{\overset{O}{\overset{\|}{}}} + R'OH$$

Example Write the equation for the complete saponification of triolein.

Solution

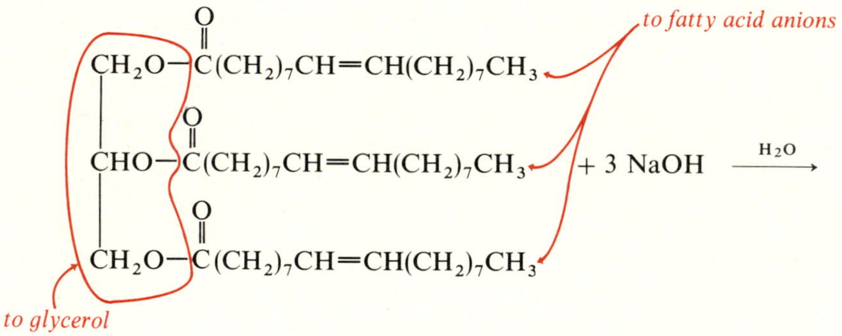

$$
\begin{array}{l}
CH_2O-\overset{O}{\overset{\|}{C}}(CH_2)_7CH{=}CH(CH_2)_7CH_3 \\
\ | \qquad \overset{O}{\overset{\|}{}} \\
CHO-\overset{}{C}(CH_2)_7CH{=}CH(CH_2)_7CH_3 \quad + \ 3\ NaOH \quad \xrightarrow{H_2O} \\
\ | \qquad \overset{O}{\overset{\|}{}} \\
CH_2O-\overset{}{C}(CH_2)_7CH{=}CH(CH_2)_7CH_3
\end{array}
$$

to fatty acid anions

to glycerol

$$
\begin{array}{l}
CH_2OH \\
| \\
CHOH \quad + \ 3\ CH_3(CH_2)_7CH{=}CH(CH_2)_7\overset{O}{\overset{\|}{C}}O^- Na^+ \\
| \\
CH_2OH
\end{array}
$$

Rancidity

Animal fats and vegetable oils for human consumption can be oxidized by atmospheric oxygen, hydrolyzed by moisture, and attacked by microorganisms to yield low-formula-weight, foul-smelling carboxylic acids, such as butyric acid, and other volatile, odorous compounds. We say that the fat or oil has become *rancid* (Latin *rancere*, "to be rank").

Rancidity can be prevented, or the process can be slowed, by keeping a fat such as butter covered and refrigerated or by including an **antioxidant**, or **preservative**. α-Tocopherol (vitamin E) is a naturally occurring preservative found in many vegetable oils, especially wheat germ oil.

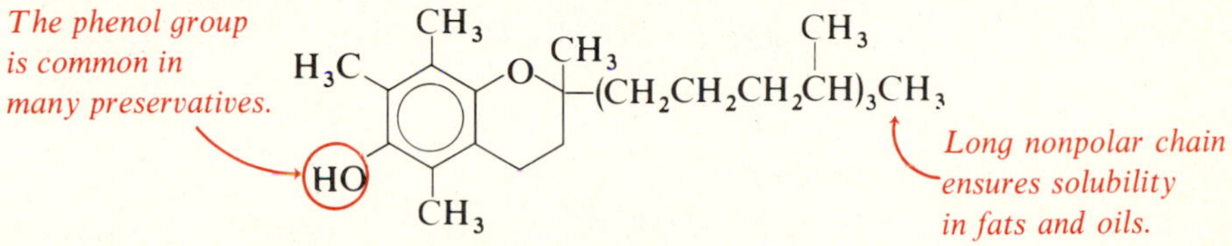

The phenol group is common in many preservatives.

Long nonpolar chain ensures solubility in fats and oils.

α-tocopherol (vitamin E)

■ 19.3 PHOSPHOLIPIDS (optional)

Phosphoglycerides

Phospholipids are phosphate esters of certain lipids. Phospholipids can be isolated from many sources—for example, the brain, nerve cells, egg yolks, yeast, wheat germ, and soybeans. A phospholipid that contains glycerol is called a **phosphoglyceride**. Lecithins and cephalins are examples. Their structures contain glycerol esterified with two long-chain fatty acids plus a phosphate group with an attached alcohol-amine salt. Compare the following general structure of a phosphoglyceride with that of a triacylglycerol (Section 19.1). Note that only the third substituent on glycerol is different.

general formula for a phosphoglyceride:

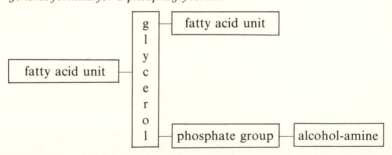

The three alcohol-amines found in the phosphoglycerides are as follows:

$$HO-CH_2CH_2\overset{+}{N}(CH_3)_3 \qquad HO-CH_2CH_2\overset{+}{N}H_3 \qquad HO-CH_2\overset{\overset{\overset{+}{N}H_3}{|}}{C}HCO_2{}^-$$

choline cation   ethanolamine cation   serine

found in lecithins, or phosphatidylcholines   *found in cephalins (phosphatidylethanolamines and phosphatidylserines)*

Examples of a lecithin and a cephalin follow:

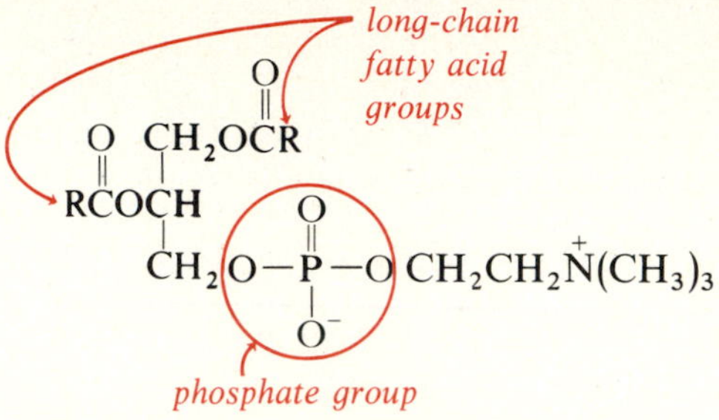

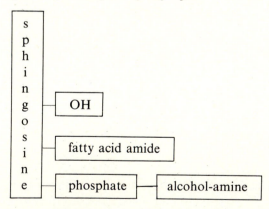

$$\text{a lecithin,}$$
a lecithin,
or phosphatidylcholine

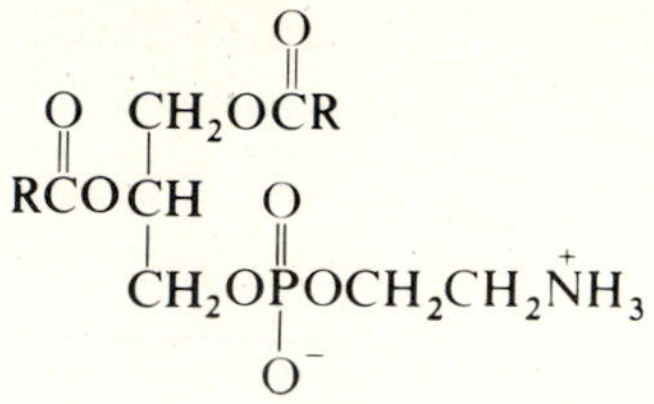

a cephalin,
or phosphatidylethanolamine

A lecithin or cephalin molecule contains two long hydrocarbon chains with an ionic "head." For this reason, these compounds are excellent emulsifying agents and surfactants. The lecithin containing two palmitic acid units esterified to glycerol is the surfactant in the lungs that prevents the alveoli from collapsing. This type of structure also permits the formation of the membrane that encloses an animal cell (Section 22.2).

Sphingolipids

Sphingolipids are fatty acid amides (not esters) of sphingosine, a long-chain alcohol-amine.

general formula for a sphingolipid:

```
s
p
h
i
n
g       ─── [ OH ]
o
s       ─── [ fatty acid amide ]
i
n
e       ─── [ phosphate ]─── [ alcohol-amine ]
```

One type of sphingolipid, called a *phosphosphingoside*, contains a phosphate group bonded to choline at one end. In the following formulas, note that, like the phosphoglycerides, a phosphosphingoside contains a phosphate-amine ionic end and two long hydrocarbon chains—one chain is part of the fatty acid unit, and the other is part of sphingosine itself.

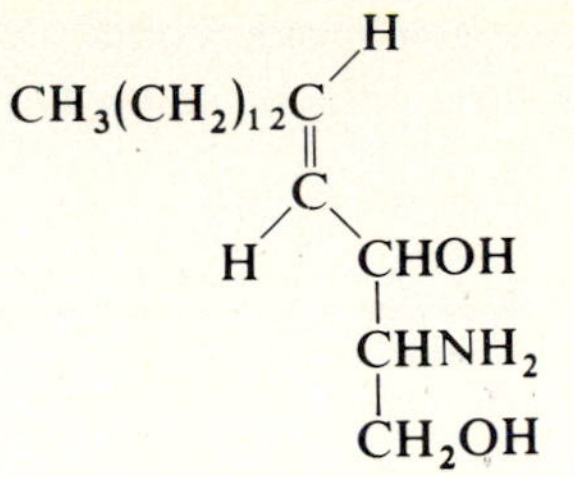

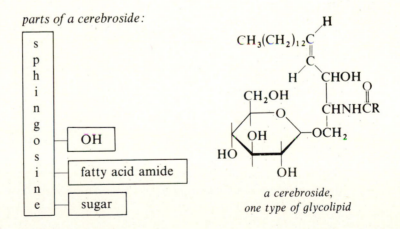

CH$_3$(CH$_2$)$_{12}$C

CHOH

CHNH$_2$

CH$_2$OH

sphingosine

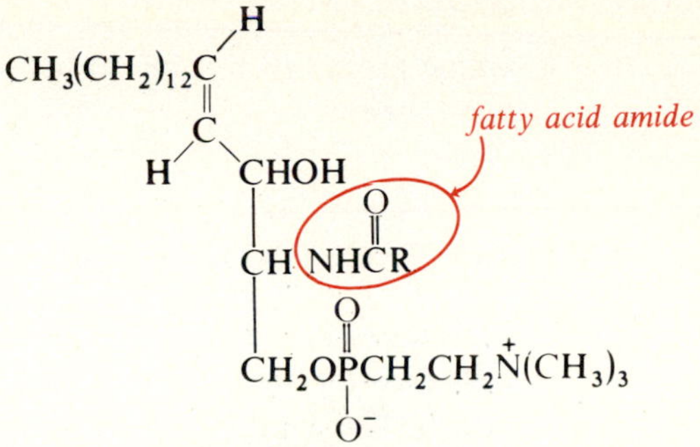

fatty acid amide

CH$_3$(CH$_2$)$_{12}$C

CHOH

CH NHCR

CH$_2$OPCH$_2$CH$_2$N$^+$(CH$_3$)$_3$

*a sphingomyelin, or
phosphosphingoside*

Brain and nerve tissues are rich in sphingomyelins, the type of phosphosphingoside shown in the preceding formula. The intertwining of the sphingomyelins' long hydrocarbon tails strengthens nerve cell membranes, or sheaths. Patients suffering from multiple sclerosis may have cell-membrane phospholipids with hydrocarbon chains that are too short to intertwine effectively.

Cerebrosides are a second type of sphingolipid found in the brain. These compounds have a polar sugar molecule (D-glucose or D-galactose) in place of the phosphate-amine ionic end and thus belong to the general class of compounds called *glycolipids* (*glyco* meaning sugar).

parts of a cerebroside:

s
p
h
i
n
g
o
s
i
n
e

OH

fatty acid amide

sugar

CH$_3$(CH$_2$)$_{12}$C

CHOH

CH$_2$OH

CHNHCR

OCH$_2$

OH

HO

OH

*a cerebroside,
one type of glycolipid*

Tay-Sachs disease is a condition leading to mental retardation, blindness, and death in early childhood. It is caused by the hereditary lack of the enzyme *N*-acetylhexosaminidase, which catalyzes one hydrolysis step in the degradation of a particular cerebroside. The symptoms of Tay-Sachs disease result from the buildup of deposits of unhydrolyzed cerebrosides in the brain cells and eye cells.

Problem 19.2. Would you expect a glycolipid (containing no phosphate-amine group) to be able to act as a surfactant or emulsifying agent? Explain your answer. ■

■ **19.4 PROSTAGLANDINS (optional)**

Hormones are chemical messengers transported by the blood to the body's cells. **Prostaglandins** are hormone-like compounds that were first discovered in semen. The name *prostaglandin* is derived from prostate gland. We now know that prostaglandins are synthesized in virtually all tissues and organs and thus occur throughout the body. Prostaglandins have a variety of functions in mammals. For example, they can raise or lower blood pressure, regulate gastric secretions, induce labor or abortion, open air passages in congested lungs, stimulate insulin release from the pancreas, cause inflammation and pain, and cause diarrhea.

Prostaglandins, which are biosynthesized from long-chain unsaturated fatty acids, are unsaturated 20-carbon acids that contain a cyclopentane ring. In the following flow equation, we show two typical prostaglandins. Other prostaglandins have similar structures.

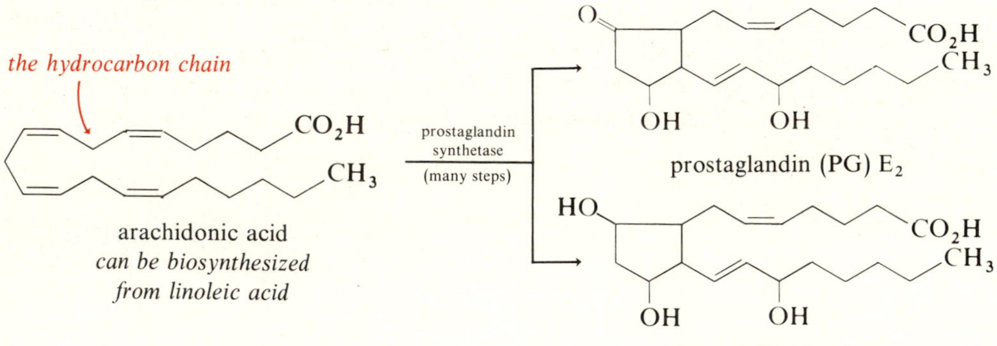

the hydrocarbon chain

arachidonic acid
*can be biosynthesized
from linoleic acid*

prostaglandin
synthetase
(many steps)

prostaglandin (PG) E_2

prostaglandin (PG) $F_{2\alpha}$

Aspirin acts to reduce inflammation, pain, and fever by inhibiting the action of the enzyme *prostaglandin synthetase*, necessary for the biosynthesis of prosta-glandins. ■

■ **19.5 STEROIDS (optional)**

The **steroids** are a class of natural products in both plants and animals that contain a distinctive ring system of three six-membered rings plus one five-membered ring. Branches and functional groups on these rings differentiate the steroids from one another.

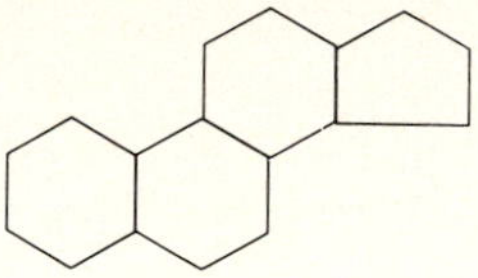

the steroid ring system

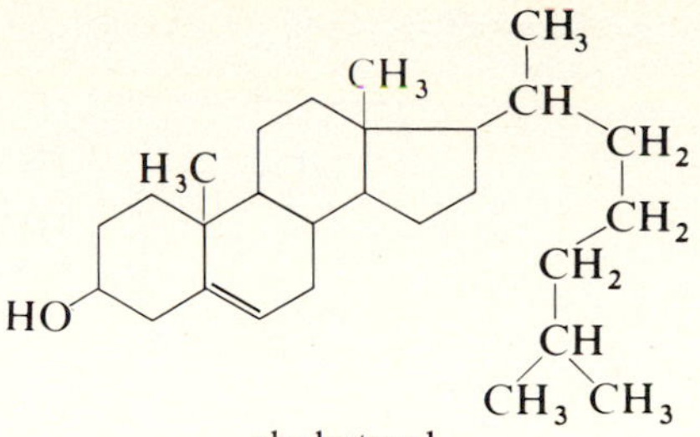

cholesterol
a steroid

Cholesterol is a steroid found in animals but not in plants. It is present in the blood and is concentrated in the spinal cord and brain. It is also the principal constituent of gallstones, and its buildup in the arteries is a factor in hardening of the arteries.

Some nutritionists recommend that foods high in cholesterol (eggs and shellfish, for example) be eaten only in moderation. The theory is that if less cholesterol is ingested, less cholesterol will be deposited in the arteries. However, a low-cholesterol diet does not ensure a low cholesterol level in the blood, because the body can synthesize its own cholesterol.

7-Dehydrocholesterol, a close relative of cholesterol, is found in the skin. When exposed to sunlight, 7-dehydrocholesterol is converted to a D vitamin. Before it was realized that lack of vitamin D is one cause of rickets, many children in temperate climates developed rickets from lack of sunlight in the winter. Today, vitamin D supplements (or milk containing irradiated ergosterol, another form of vitamin D) are used to prevent rickets.

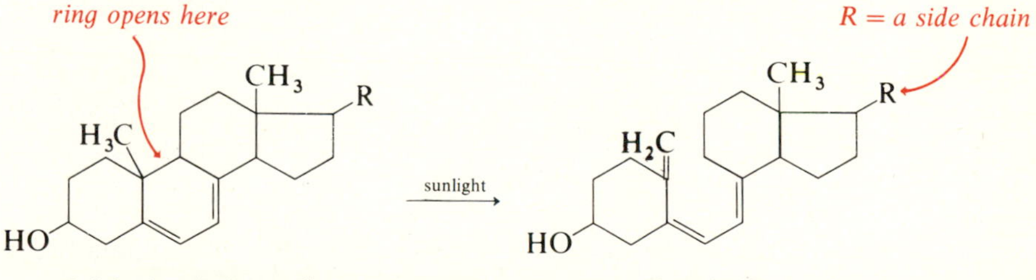

ring opens here

R = a side chain

$\xrightarrow{\text{sunlight}}$

7-dehydrocholesterol

vitamin D_3

Bile salts are emulsifying agents for fats in the small intestine. These steroids are amides of cholic acid (Figure 19.1), which is biosynthesized from cholesterol. The added amide-ionic group and the relatively large steroid ring system are the structural features that allow the bile salts to act as emulsifying agents.

R = steroid ring system

ionic end

$$RC-OH + H-NCH_2CO^- Na^+ \longrightarrow RC-NHCH_2CO^- Na^+ + H_2O$$

cholic acid   sodium glycine

a bile salt,
an emulsifying agent for fats

Figure 19.1.   *Some important steroids.*

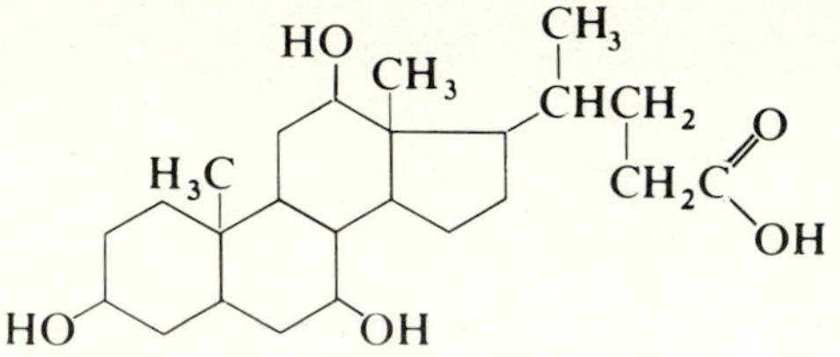

cholic acid
used to form bile salts

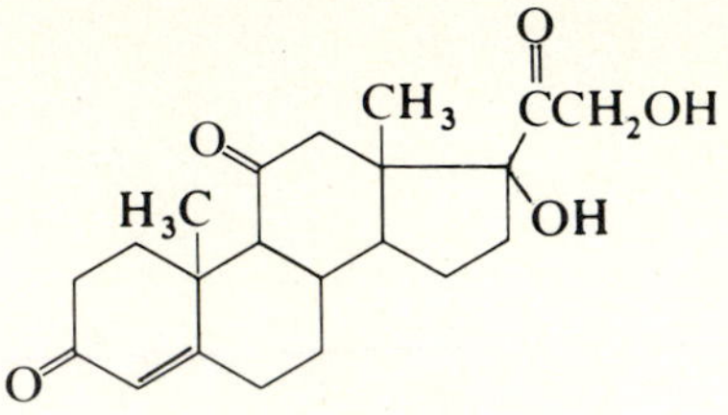

cortisone
*an adrenal steroid with
anti-inflammatory action*

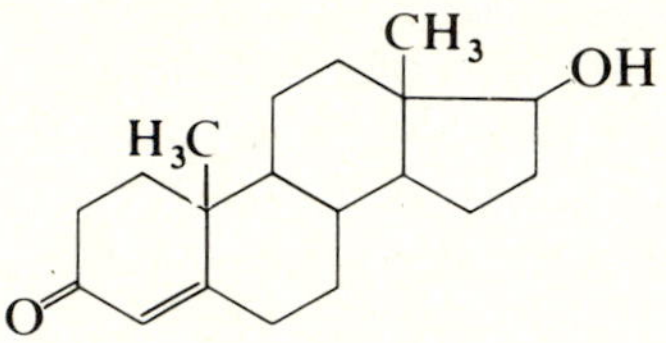

testosterone
an androgen, or male hormone

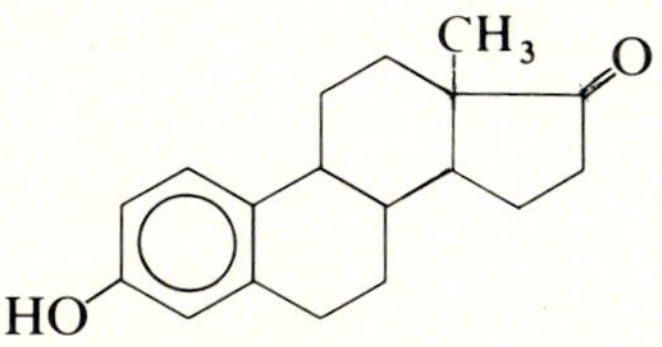

estrone
an estrogen, or female hormone

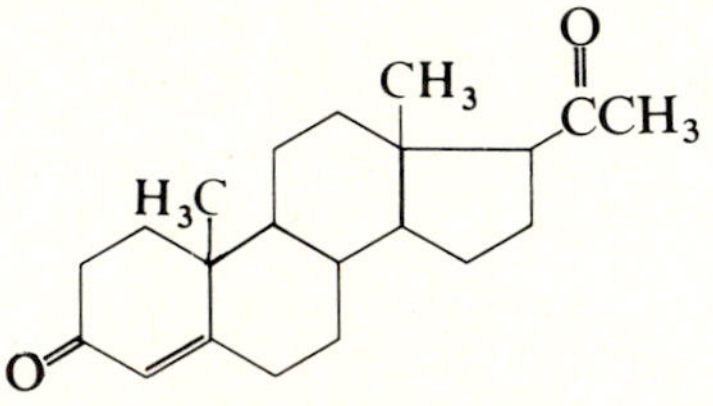

progesterone
a pregnancy hormone

Adrenal hormones are steroids secreted by the adrenal cortex; over 30 of these steroids have been identified. Deficiency in the secretion of the adrenal hormones results in Addison's disease, a rare disorder characterized by a bronzing of the skin, weight loss, and lethargy. One of the better-known adrenal steroids is cortisone (see Figure 19.1). Cortisone alters protein metabolism and is used to treat inflammation and allergies. It is particularly useful for the symptomatic treatment of rheumatoid arthritis.

Sex hormones are produced primarily by the testes of the male or ovaries of the female. Their formation is regulated by pituitary hormones, and the sex hormones themselves regulate the sexual functions and give rise to the secondary sex characteristics. The male hormones are frequently referred to by the general term *androgens* and the female hormones by the term *estrogens*. Figure 19.1 shows the structures of three sex hormones.

The estral cycle is controlled, in part, by estrogens. This cycle is directed by the gonadotropin hormones, which are glycoproteins (sugar-protein complexes) released by the pituitary gland. One of these hormones stimulates the ovaries to produce estrogens, which prepare the system for ovulation and possible pregnancy. Another gonadotropin stimulates the ovaries to produce progesterone, which prevents other ova from ripening and later helps maintain a pregnancy to full term by preventing menstrual periods. The oral contraceptives are steroids related to progesterone; they function by suppressing the gonadotropins that stimulate ovulation.   ■

SUMMARY

A *lipid* is a water-insoluble, naturally occurring compound that is soluble in nonpolar organic solvents. *Fatty acids* are naturally occurring, long-chain carboxylic acids, which may contain carbon-carbon double bonds in their chains. Animal fats and vegetable oils are *triacylglycerols* (triglycerides), esters of glycerol with three fatty acid units. Vegetable oils (liquid) contain more unsaturation than fats (solid). A *wax* is a simple monoester of a fatty acid and a long-chain alcohol.

One or more of the carbon-carbon double bonds in a triacylglycerol can be *hydrogenated*. Triacylglycerols can also be *saponified* with a base or *hydrolyzed* enzymatically:

$$\begin{array}{c} CH_2O_2CR \\ | \\ CHO_2CR \\ | \\ CH_2O_2CR \end{array} \xrightarrow[\text{OH}^- \text{ or enzymes}]{H_2O} \begin{array}{c} CH_2OH \\ | \\ CHOH \\ | \\ CH_2OH \end{array} + 3\ R\overset{\displaystyle O}{\overset{\|}{C}}O^- \ \text{or}\ 3\ R\overset{\displaystyle O}{\overset{\|}{C}}OH$$

The class of *phospholipids* includes phosphate esters of glycerol and sphingosine. A molecule of one of these compounds contains two long hydrocarbon chains and an ionic phosphate-amine group, structural features that allow these compounds to act as surfactants or emulsifying agents. *Cerebrosides* are sphingolipids that contain a sugar instead of a phosphate ester grouping; these compounds are thus called *glycolipids*.

Prostaglandins are 20-carbon unsaturated acids that exhibit a wide range of physiological activity. *Steroids* are compounds that contain a ring system of three six-membered rings plus one five-membered ring. Important steroids are cholesterol, cholic acid, cortisone, and the sex hormones.

KEY TERMS

lipid	*phosopholipid	*cholesterol
triacylglycerol	*sphingolipid	*bile salt
fatty acid	*cerebroside	*adrenal hormone

animal fat   *glycolipid   *sex hormone
vegetable oil   *prostaglandin
wax   *steroid

STUDY PROBLEMS

19.3. Define the term *lipid*.

19.4. Which of the following compounds are lipids?

(a) $C_{17}H_{35}CO_2CH_2$
    $C_{15}H_{31}CO_2CH$
    $C_{17}H_{35}CO_2CH_2$

(b) $C_{17}H_{35}CO_2(CH_2)_{29}CH_3$

(c)

(d)

19.5. Define the following terms and give an example of each:
(a) a saturated fatty acid
(b) an unsaturated fatty acid
(c) a triol
(d) an ester

19.6. What is the difference between a petroleum oil and a vegetable oil?

19.7. The structure of a triacylglycerol follows:

$CH_2OC(CH_2)_7CH{=}CH(CH_2)_7CH_3$

$CHOC(CH_2)_7CH{=}CH(CH_2)_5CH_3$

$CH_2OC(CH_2)_7CH{=}CHCH_2CH{=}CH(CH_2)_4CH_3$

(a) Circle the three ester groups.
(b) Circle the unsaturation, aside from the ester groups.
(c) Circle the glycerol portion of the structure.
(d) Is this triacylglycerol saturated, unsaturated, or polyunsaturated?
(e) Would it be a fat or an oil?
(f) Would it be water soluble or water insoluble?
(g) Would it be a good emulsifying agent?

19.8. (a) Write the equation for the formation of the triacylglycerol in the preceding problem.
(b) Write the equation for its acidic or enzymatic hydrolysis.
(c) Write the equation for its saponification.

19.9. One of the compounds found in carnauba wax (used in automobile waxes, etc.) is $C_{27}H_{55}CO_2C_{32}H_{65}$. Write an equation for the saponification of this ester.

19.10. Write an equation for the conversion of trilinolein (the triacylglycerol of linoleic acid) in corn oil to a semisolid fat that could be used for margarine.

*19.11. Write the general formula (using R for the fatty acid chains) for the lecithin shown in Section 19.3.
(a) Circle the ester groups.
(b) Circle the glycerol part of the structure.
(c) Explain how the lecithins in egg yolks act as an emulsifying agent in mayonnaise.

*19.12. Define the term *glycolipid*.

*19.13. (a) Write a condensed structural formula for arachidonic acid (Section 19.4), in the form shown in Table 19.1.
(b) Write an equation showing the reaction of this compound with dilute aqueous NaOH.

*19.14. (a) Describe the functions of prostaglandins.
(b) Describe the effects of aspirin on prostaglandin activity.

*19.15. Referring to Figure 19.1, match the following. More than one answer may be correct.
(a) cortisone
(b) testosterone
(c) estrone

(1) a female hormone
(2) could cause facial hair growth in females
(3) increases during pregnancy
(4) could cause breast enlargement in males
(5) contains an aromatic ring
(6) contains a ketone group

*19.16. Would you expect each of the following steroids to be an androgen or an estrogen? Explain your answer.

(a)

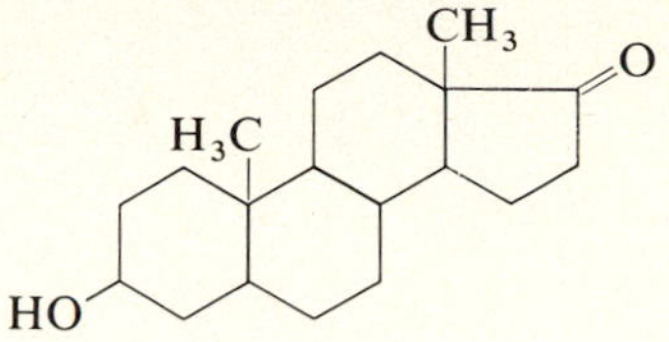

(b)

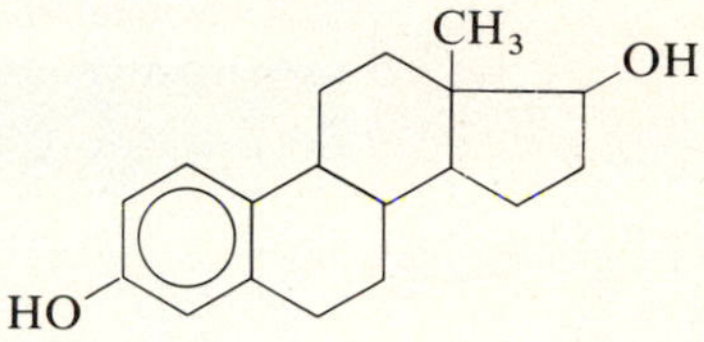

20 Proteins

Collagen in human skin. (Micrograph courtesy of Dr. Jerome Gross.) The connective tissue in the human body is composed primarily of collagen. *This fibrous material is formed from* tropocollagen *subunits—twisted triple strands of protein molecules. The coils visible in the micrograph arise from overlap between tropocollagen subunits in the collagen fibril.*

Objectives   Write general formulas for a protein and an α-amino acid. ☐ Define essential amino acid. ☐ Show how amino acids form dipolar ions and why they are amphoteric. ☐ Define isoelectric point and electrophoresis (optional). ☐ Write the formula for a peptide and show the *N* and *C* terminals. ☐ Describe the functions of a few peptides. ☐ Describe the classes of proteins. ☐ Define the primary and higher structures of proteins, show some ways that the higher structures are maintained, and describe some examples. ☐ Describe the structure and function of hemoglobin (optional). ☐ List some properties of colloidal proteins and some causes of denaturation.

The word protein is derived from the Greek *proteios,* which means "of first importance." Proteins are aptly named, for about 50% of the dry weight of an animal cell is protein material. It is commonly known that muscles are composed chiefly of protein. It is probably less commonly known that hair, skin, fingernails, and the internal organs are principally protein. Enzymes, some hormones, antibodies, serum albumins (blood proteins), and hemoglobin are also proteins.

An amide is a compound containing the $\overset{\overset{\text{O}}{\|}}{\text{RC}}-\text{NR}_2$ functional group. All proteins are *polyamides* with formula weights in the tens or hundreds of thousands. The hydrolysis of a protein yields *amino acids,* compounds that contain an amino group ($-\text{NH}_2$) and a carboxyl group ($-\text{CO}_2\text{H}$) in the same molecule.

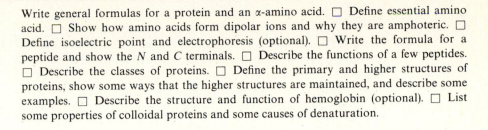

In the process of digestion, proteins are hydrolyzed enzymatically to amino acids, which are absorbed into the bloodstream, carried to the cells, and converted to the animal's own proteins or to other compounds.

Before we discuss how polyamides can exist in such diverse forms as muscle tissue and enzymes, we must consider the structures and properties of the amino acids that are used to build proteins. Then we will consider peptides (small proteins) and finally the major classes of proteins themselves. You will learn how the overall structure and shape of a protein determines its properties and how the shape can be destroyed by denaturation. In subsequent chapters, we will consider the action of enzymes, the metabolism of proteins, and the biosynthesis of proteins by nucleic acids.

20.1   AMINO ACIDS

About 25 amino acids have been identified as products from the hydrolysis of proteins. Of these, 20 are widespread and are found in a variety of proteins (see Table 20.1). Although the amino acids could be named by the IUPAC system, the common names are more frequently encountered. These common names are often abbreviated. For example, gly and ala are abbreviations for glycine and alanine.

Each of the amino acids in Table 20.1 is an **α-amino acid**; that is, the amino group is bonded to the carbon atom adjacent, or alpha, to the carboxyl group. Variations in the structures of the amino acids occur in the side chain (R) bonded to the α carbon.

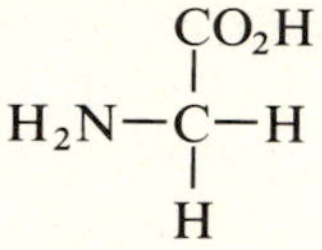

glycine
(*abbreviated* gly)

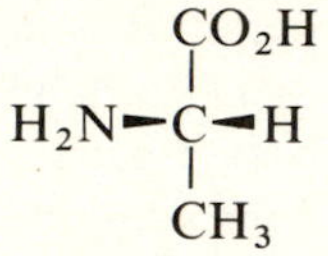

alanine
(*abbreviated* ala)

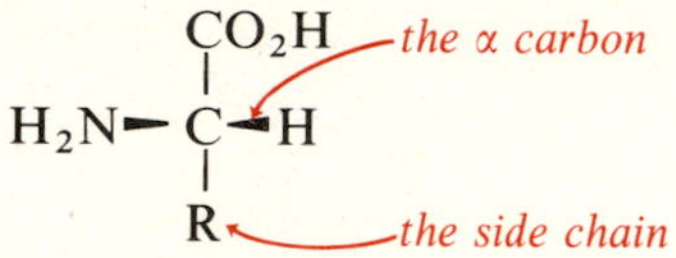

*general formula for
an α-amino acid*

Except in glycine, the alpha carbon of each amino acid is bonded to four different groups (in general: $-CO_2H$, $-NH_2$, $-H$, and $-R$). Glycine does not have any stereoisomers. Alanine and all the other amino acids derived from proteins can exist in one of two stereoisomeric forms: a D form or an L form. Both D- and L-amino acids are found in nature; however, most proteins contain only the L-amino acids. If we write the formula of an amino acid with carbon 1 (the carboxyl group) at the top, similar to the way we wrote the formulas for the monosaccharides, the L-amino acid is the stereoisomer with the amino group on the left.

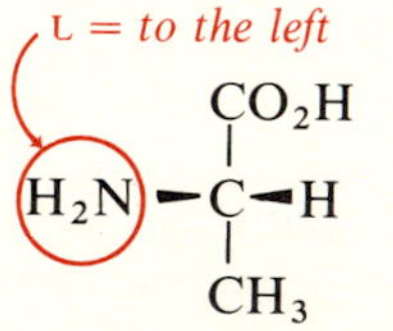

L-alanine
*important in
human metabolism*

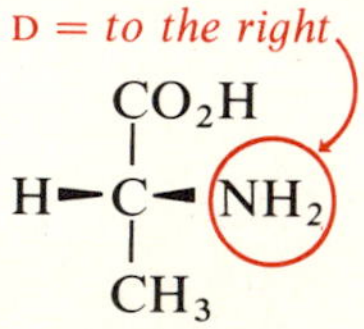

D-alanine
*not important in
human metabolism*

Example   Referring to Table 20.1, draw the formula for phenylalanine as an L-amino acid.

Table 20.1. *The Twenty Common Amino Acids Found in Proteins*

Name	Structure	Name	Structure
alanine (ala)	CH_3CHCO_2H $\underset{\|}{NH_2}$	leucine (leu)[a]	$(CH_3)_2CHCH_2CHCO_2H$ $\underset{\|}{NH_2}$
arginine (arg)[a]	$H_2NCNHCH_2CH_2CH_2CHCO_2H$ $\underset{\|}{NH} \qquad \underset{\|}{NH_2}$	lysine (lys)[a]	$H_2NCH_2CH_2CH_2CH_2CHCO_2H$ $\underset{\|}{NH_2}$
asparagine (asp-NH_2)	$H_2NCCH_2CHCO_2H$ $\underset{\|}{O} \qquad \underset{\|}{NH_2}$	methionine (met)[a]	$CH_3SCH_2CH_2CHCO_2H$ $\underset{\|}{NH_2}$
aspartic acid (asp)	$HO_2CCH_2CHCO_2H$ $\underset{\|}{NH_2}$	phenylalanine (phe)[a]	$\text{(benzene ring)}-CH_2CHCO_2H$ $\underset{\|}{NH_2}$
cysteine (cys)	$HSCH_2CHCO_2H$ $\underset{\|}{NH_2}$	proline (pro)	$\text{(pyrrolidine ring)}-CO_2H$
glutamic acid (glu)	$HO_2CCH_2CH_2CHCO_2H$ $\underset{\|}{NH_2}$	serine (ser)	$HOCH_2CHCO_2H$ $\underset{\|}{NH_2}$
glutamine (glu-NH_2)	$H_2NCCH_2CH_2CHCO_2H$ $\underset{\|}{O} \qquad \underset{\|}{NH_2}$	threonine (thr)[a]	$\underset{\|}{OH}$ $CH_3CHCHCO_2H$ $\underset{\|}{NH_2}$
glycine (gly)	CH_2CO_2H $\underset{\|}{NH_2}$	tryptophan (try)[a]	$\text{(indole ring)}-CH_2CHCO_2H$ $\underset{\|}{NH_2}$
histidine (his)[a]	$\text{(imidazole ring)}-CH_2CHCO_2H$ $\underset{\|}{NH_2}$	tyrosine (tyr)	$HO-\text{(benzene ring)}-CH_2CHCO_2H$ $\underset{\|}{NH_2}$
isoleucine (ile)[a]	$\underset{\|}{CH_3}$ $CH_3CH_2CHCHCO_2H$ $\underset{\|}{NH_2}$	valine (val)[a]	$(CH_3)_2CHCHCO_2H$ $\underset{\|}{NH_2}$

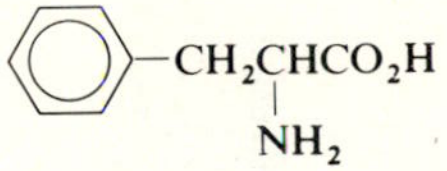

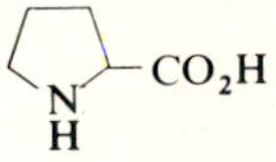

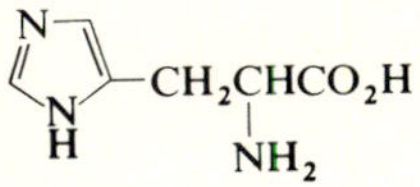

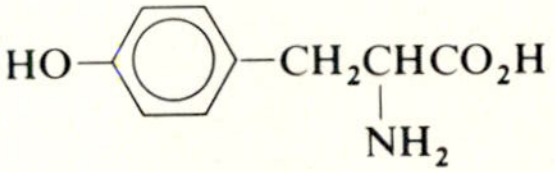

[a] Commonly referred to as an essential amino acid.

371

Solution

$$H_2N-\overset{\displaystyle CO_2H}{\underset{\displaystyle CH_2-\bigcirc}{\overset{\displaystyle |}{\underset{\displaystyle |}{C}}-H}}$$

L ⌐

Essential Amino Acids

The body can synthesize many of the amino acids it needs from other compounds in the body. The amino acids that the body cannot synthesize must be supplied in the diet and are called the **essential amino acids**. The need for these amino acids varies not only from species to species, but also from individual to individual. What may be essential for a rat may not be essential for a human; what may be essential for one person may not be essential for another. Ten amino acids commonly referred to as essential are marked in the list of amino acids in Table 20.1.

Of the essential amino acids, the body can convert a few (tryptophan, phenylalanine, methionine, and histidine) from the D form to the required L form. Therefore, a mixture of the D and L stereoisomers of these amino acids is as nutritious as the pure L form. Only the L forms of the other essential amino acids can be utilized; if the D form is supplied, it is not used by the body to synthesize new protein molecules.

Amino Acids as Dipolar Ions

Amino acids are crystalline, high-melting solids that are more soluble in water than in organic solvents. They are more like inorganic salts than typical organic compounds. The reason for this anomalous behaviour is that an amino acid contains in the same molecule a carboxylic-acid group, which can lose a proton, and a basic amino group, which can gain a proton.

An amino acid undergoes an internal acid-base reaction to yield a **dipolar ion**, an ion that has both a positive and a negative ionic charge. Because the atoms in a dipolar ion are held together by covalent bonds, the ionic charges cannot be separated in water solution as can the ions of sodium chloride; these ionic charges must remain together in the same molecule.

$$\underset{\displaystyle \underset{\displaystyle NH_2}{\overset{\displaystyle |}{}}}{\overset{\displaystyle O}{\overset{\displaystyle \|}{RCHC\ddot{O}H}}} \quad \rightleftharpoons \quad \underset{\displaystyle \underset{\displaystyle {}^+NH_3}{\overset{\displaystyle |}{}}}{\overset{\displaystyle O}{\overset{\displaystyle \|}{RCHC\ddot{O}:^-}}} \quad \textit{behaves like a salt}$$

    an amino acid   *a dipolar ion*

Problem 20.1.   Write the equations for the formation of the dipolar ions of (a) phenylalanine and (b) proline (Table 20.1).

The dipolar ion of an amino acid is *amphoteric*: it can react with either an acid or a base to yield a positive ion or a negative ion, respectively.

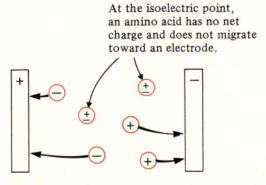

$$\underset{\substack{\text{a neutral dipolar ion}\\ \text{(no net charge)}}}{\overset{\displaystyle\text{O}}{\underset{^+\text{NH}_3}{\text{RCHCO}^-}}}$$

with H^+ →

a positively charged ion
$$\underset{^+\text{NH}_3}{\overset{\displaystyle\text{O}}{\text{RCHCOH}}}$$

with OH^- →

a negatively charged ion
$$\underset{\text{NH}_2}{\overset{\displaystyle\text{O}}{\text{RCHCO}^-}}$$

■ Isoelectric Point and Electrophoresis (optional)

If a pair of electrodes, one + and one −, are placed in a solution of amino acids, positively charged amino-acid ions migrate toward the negative electrode, while negatively charged amino-acid ions migrate toward the positive electrode. At a particular pH, an amino acid carries no net charge (the neutral dipolar ion in the preceding equation) and does not migrate toward either electrode. The pH at which an amino acid carries no net charge is called the **isoelectric point** for that amino acid.

At the isoelectric point, an amino acid has no net charge and does not migrate toward an electrode.

Some amino acids contain side chains with additional carboxyl or amino groups. These side-chain groups can *also* react with an acid or a base. For example:

acidic H on side chain

$$\underset{\substack{\;\\\text{+NH}_3}}{\overset{\displaystyle\text{O}\qquad\qquad\text{O}}{\text{(H)OCCH}_2\text{CH}_2\text{CHCO}^-}} + \text{NaOH} \xrightarrow{\text{H}_2\text{O}} \text{Na}^+\; \underset{\substack{\;\\\text{+NH}_3}}{\overset{\displaystyle\text{O}\qquad\qquad\text{O}}{^-\text{OCCH}_2\text{CH}_2\text{CHCO}^-}} + \text{H}_2\text{O}$$

ionic charge on side chain

glutamic acid

monosodium glutamate
(MSG)

Because carboxyl groups or amino groups in side chains can carry ionic charges, as shown in the preceding equation, the value for the isoelectric point of a particular amino acid depends on what functional groups, if any, are present in the side chain. At a specific pH, one amino acid may be neutral (no net charge), while a different amino acid may be positively or negatively charged.

Electrophoresis is a technique by which amino acids can be separated or identified, based on differences in isoelectric points. The pH is adjusted to the desired value. Electrodes are inserted into the solution. Then, positively or negatively charged amino acids migrate to one electrode or the other, while uncharged amino acids show no net migration.

In proteins, the principal carboxyl groups and α-amino groups of the amino acids are tied up as amide groups. However, the side chains of the component amino acids may contain $-NH_2$ or $-CO_2H$ groups, which can react with H^+ or OH^-. Thus, protein molecules can be positively or negatively charged, or neutral. Proteins, too, can be analyzed by electrophoresis.

Problem 20.2.   Referring to Table 20.1, write equations for the following reactions:
(a) Alanine is dissolved in aqueous HCl.
(b) Alanine is dissolved in dilute aqueous NaOH.
(c) Glycylalanine (Section 20.2) is dissolved in dilute aqueous NaOH.   ■

20.2   PEPTIDES

Peptides and proteins are formed from amino acids joined together by amide bonds. A peptide is differentiated from a protein by the number of amino-acid units, or residues, that would be formed by hydrolysis. Generally, a polyamide with fewer than fifty amino-acid residues is classified as a peptide, and one larger is classified as a protein. Thus, a peptide is really just a small protein.

Peptides are further classified by the number of amino-acid units they contain. A peptide composed of two amino-acid units is a *dipeptide*; three amino-acid units, a *tripeptide*; and so on. A large peptide is referred to as a *polypeptide*.

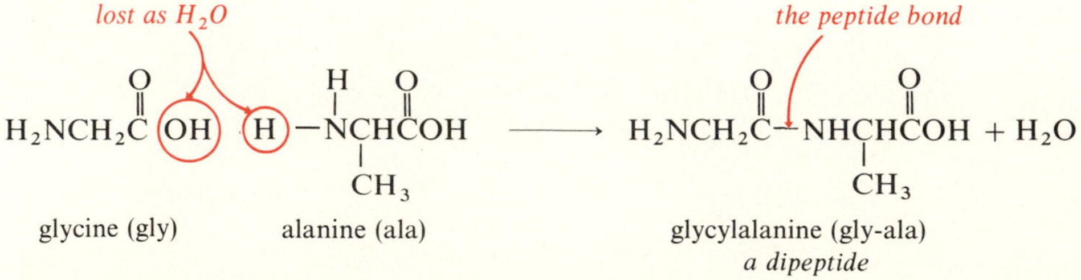

glycine (gly)   alanine (ala)   glycylalanine (gly-ala)
                                                  a dipeptide

The linkage between the α-amino group of one amino acid and the carbonyl group of the other is frequently called a **peptide bond**. This term refers specifically to the amide linkage between two α-amino acids. A peptide bond is relatively nonreactive, but it can be hydrolyzed enzymatically or by heating with aqueous acid or base.

A typical peptide has a free amino group at one end of the chain. The amino acid at that end is called the **N-terminal amino acid** and is usually shown on the left side of the structure. The other end of the chain has a free carboxylic acid, and the amino acid at that end is called the **C-terminal amino acid**.

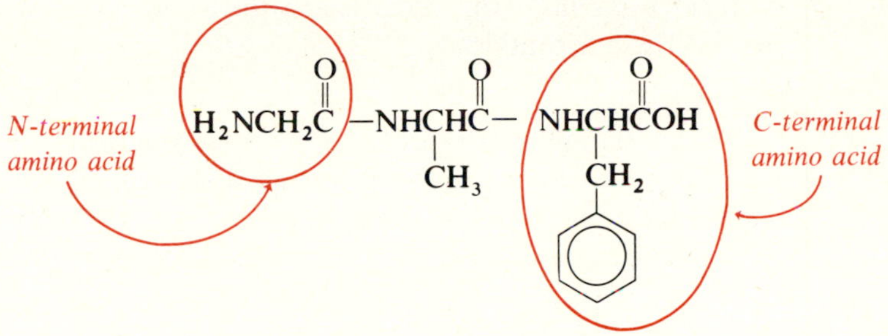

$$\underset{\substack{N\text{-terminal} \\ amino\ acid}}{}\ H_2NCH_2\overset{O}{\underset{}{C}}-NH\underset{CH_3}{CH}\overset{O}{\underset{}{C}}-NH\underset{CH_2}{CH}\overset{O}{\underset{}{C}}OH\ \underset{\substack{C\text{-terminal} \\ amino\ acid}}{}$$

glycylalanylphenylalanine (gly-ala-phe)
a tripeptide

There are two ways of joining two different amino acids to form a dipeptide. Glycine and alanine may be joined into the dipeptide gly-ala, shown earlier in this section, or into a different dipeptide, ala-gly. In the first dipeptide, glycine is the *N* terminal and alanine is the *C* terminal. In the second dipeptide, alanine is the *N* terminal and glycine is the *C* terminal.

Example   Write the structural formula of ala-gly, and show the *N*-terminal and *C*-terminal amino acids.

Solution

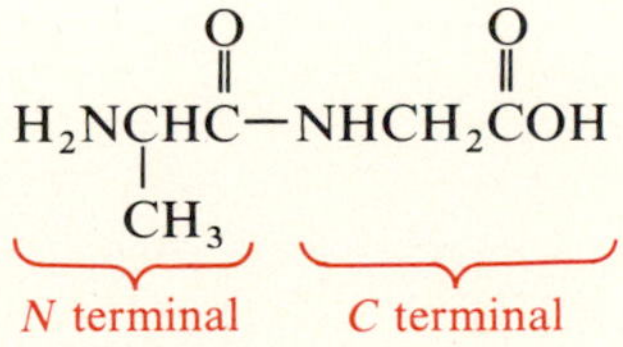

$$H_2N\underset{CH_3}{CH}\overset{O}{\underset{}{C}}-NHCH_2\overset{O}{\underset{}{C}}OH$$

N terminal   *C* terminal

Problem 20.3. Peptides, like amino acids, exist as dipolar ions. Write the formula for the dipolar ion of ala-gly.

As the size of a peptide increases, the number of possibilities for variation in structure increases dramatically. A decapeptide contains ten amino-acid units and has a formula weight of something over 1000. It is far smaller than a protein. Even so, the number of different decapeptides that can be constructed, using each amino acid no more than one time in each structure, is over 4,000,000,000,000. Most proteins have formula weights between 12,000 and 1,000,000, and proteins may contain the same amino acid more than once in a molecule. The number of possible variations is beyond comprehension.

Problem 20.4. The amino acids gly, ala, and phe can be combined in six different ways in a tripeptide. Using the abbreviated names, show the structures of these tripeptides.

Even though most proteins are polypeptides of high formula weight, many smaller peptides are of profound biological significance. Table 20.2 lists a few interesting peptides. For example, the *enkephalins*, peptides containing as few as five amino-acid residues, act as the body's own painkillers. It is thought

Table 20.2. Some Important Peptides

Name	Formula weight	Structure or example	Function
aspartame	280	asp-phe	an artificial low-calorie sweetener
enkephalins	560	tyr-gly-gly-phe-met	a brain peptide with opiate activity
oxytocin	1000	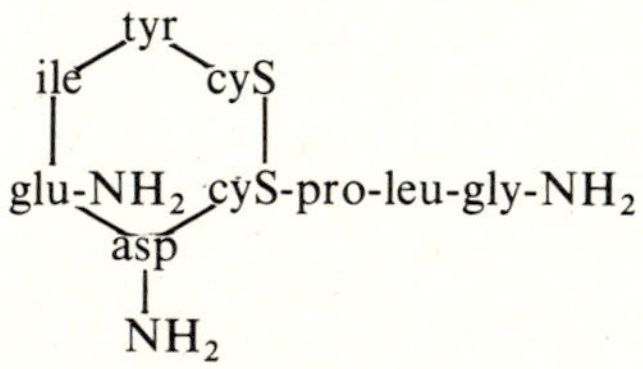 (see structure below)	a hormone that stimulates uterine contractions during parturition
insulin	5700	—	a pancreatic hormone that regulates glucose metabolism
interferon	26,000	—	a peptide produced by virus-infected cells that protects noninfected cells from viral multiplication

Oxytocin structure:

$$
\begin{array}{c}
\text{tyr} \\
\text{ile} \qquad \text{cyS} \\
\text{glu-NH}_2 \quad \text{cyS-pro-leu-gly-NH}_2 \\
\text{asp} \\
\text{NH}_2
\end{array}
$$

that the opiates, such as morphine and codeine, act as analgesics because their shapes and structures mimic those of the enkephalins, and thus these compounds can bind at the enkephalin receptor sites in the brain.

20.3   CLASSIFICATION OF PROTEINS

Proteins are found in different sizes and shapes and with different physical and chemical properties, according to their function. The *fibrous proteins* are long and threadlike, tough and water insoluble; they form muscles, tendons, skin, hair, fur, fingernails, and feathers. The *globular proteins* are named because they are somewhat spherical in shape. Globular proteins are more soluble than fibrous proteins and carry on much of the work of a living system. Insulin, hemoglobin, albumins, enzymes, antibodies, and many hormones are globular proteins. Several types of proteins and their classifications are listed in Table 20.3.

20.4   SHAPES OF PROTEINS

Primary and Higher Structures

The ability of a protein to perform its prescribed function depends partly on the number and sequence of amino-acid residues in the protein's molecules. The amino-acid sequence is called the **primary structure** of the protein.

The sequence of amino acids is not the only important feature of protein structures. Proteins obtain their individual properties partly by the ordered shapes into which the molecules arrange themselves, such as those shown in Figure 20.1. These shapes are referred to as the **higher structures** of a protein and are classified as *secondary*, *tertiary*, and *quaternary*.

When a protein molecule with a particular primary structure is synthesized in its proper environment (correct pH, electrolyte concentration, tem-

Figure 20.1.   *Higher structures of fibrous and globular proteins.*

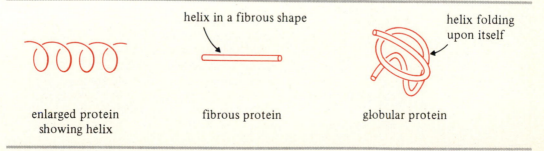

helix in a fibrous shape

helix folding upon itself

enlarged protein showing helix

fibrous protein

globular protein

Table 20.3.   *Classification of Proteins*

Type or example of protein	Occurrence or purpose
fibrous (long-chain molecules; tough, water insoluble; used as structural components):	
collagens	form connective tissues; also found in bone, teeth, and tendons
elastins	in tendons and walls of blood vessels
myosins	contractile proteins found in muscles
fibrin	formed from the soluble protein *fibrinogen* as part of the blood-clotting mechanism
globular (folded globular molecules; water-soluble):	
enzymes	biological catalysts in body cells and fluids
albumins	in egg whites (*egg albumins*) and blood (called *serum albumins* or *plasma proteins*)
globulins	in blood and lymph as antibodies and as carriers of some metal ions
hormones	biological messengers in body cells and fluids
hemoglobin	a transport protein that carries O_2 and CO_2 in the blood
fibrinogen	forms fibrin in blood clots
histones	in some glands, such as the thymus; associated with nucleic acids
protamines	associated with nucleic acids
conjugated (associated with nonprotein organic molecules; may be fibrous or globular):	
nucleoproteins	proteins such as histones and protamines associated with *nucleic acids*; found in all cells
glycoproteins and mucoproteins	proteins associated with *carbohydrates*; interferon and proteins in mucus are examples
lipoproteins	proteins associated with *lipids* such as cholesterol or phosphoglycerides; involved in lipid transport in blood and lymph
chromoproteins	proteins such as hemoglobin associated with *chromophores* (color-producing groups)
phosphoproteins	proteins associated with *phosphates*; casein (a protein in milk) is an example
metalloproteins	proteins associated with *metal ions* such as Fe^{2+} or Zn^{2+}

Figure 20.2.   *Three factors that help hold protein molecules in their proper shapes.*

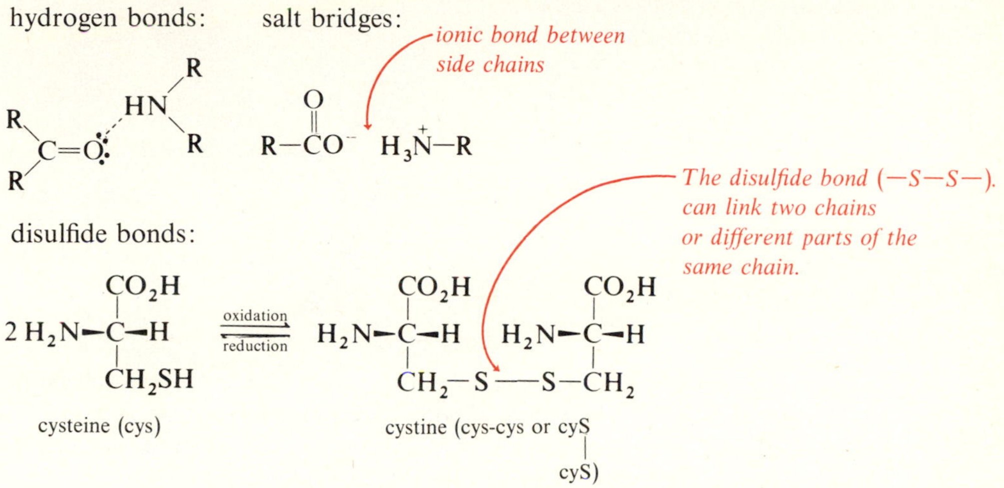

hydrogen bonds: salt bridges:

ionic bond between side chains

disulfide bonds:

The disulfide bond (—S—S—). can link two chains or different parts of the same chain.

$$2\ H_2N{-}\overset{\displaystyle CO_2H}{\underset{\displaystyle CH_2SH}{C}}{-}H \underset{\text{reduction}}{\overset{\text{oxidation}}{\rightleftarrows}} H_2N{-}\overset{\displaystyle CO_2H}{\underset{\displaystyle CH_2{-}S}{C}}{-}H \quad H_2N{-}\overset{\displaystyle CO_2H}{\underset{\displaystyle S{-}CH_2}{C}}{-}H$$

cysteine (cys)   cystine (cys-cys or cyS
                                                    |
                                                 cyS)

perature, and so forth), it automatically assumes its proper shape because of intramolecular (within-the-molecule) hydrogen bonding; other attractions (or repulsions), such as salt bridges; and disulfide bonds. Figure 20.2 shows some examples of these intramolecular interactions.

Secondary Structures

The *secondary structure* of a protein molecule is the ordered shape in which a protein molecule arranges its long chain, or "backbone," because of inter- actions among different parts of the molecule or molecules. Let us consider some typical secondary structures.

The α-Helix.   It has been discovered by x-ray analysis (which gives a three- dimensional map of molecular shape) that most fibrous protein molecules are helical, or spiral, in shape. This shape, called an *α-helix*, is maintained by hydrogen bonding between the carbonyl group from one amino-acid unit and the hydrogen attached to the amide nitrogen of another unit in the helix. Although one hydrogen bond is weak, the collective effect of many hydrogen bonds holds the helix in its shape quite strongly. Because proteins contain only ʟ-amino acids, the helix is "right handed," like a right-handed screw. Figures 20.1 and 20.3 depict the helical structure.

The helical secondary structure gives rise to a rather flexible and elastic structural material. *Keratin* (a protein in hair, skin, fur, and feathers) is an example of a fibrous protein with this type of helical structure.

The fundamental structure of collagen (which forms cartilage, bone, tendons, ligaments, and skin) appears to be that of a *tropocollagen molecule,*

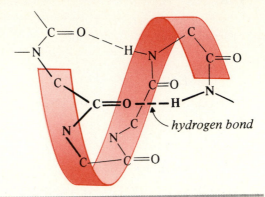

Figure 20.3. *Hydrogen bonds hold a protein molecule in a helix. (The nonparticipating groups attached to carbon atoms have been omitted.)*

which is composed of superhelices from the intertwining of three helical polypeptides in a single chain (see Figure 20.4). The three-strand helix is also stabilized by hydrogen bonding. The intertwining of the three strands gives extra strength where it is needed in the structure of an organism.

The β-Pleated Sheet. Another type of secondary structure is the *pleated sheet*, or *β-pleated sheet*. The term is derived from the general appearance of this type of protein, which resembles corrugated metal roofing. Silk fibroin, a fibrous protein found in silk, is an example of a pleated-sheet protein. The protein chains in silk fibroin are lined up side by side and held together by hydrogen bonds, as shown in Figure 20.5. This arrangement lends strength and flexibility, but little elasticity.

Tertiary Structures

Different polar and nonpolar parts of a coiled protein molecule can interact with one another so that the coil folds and bends upon itself. The folded shape of a particular protein molecule is referred to as its *tertiary structure*.

Globular proteins, such as enzymes, depend completely on their tertiary structures in order to carry out their roles in an organism. The folded molecule presents a unique surface, usually with polar side chains on the outside and nonpolar groups on the inner surfaces. The nonpolar inner surface allows an enzyme to catalyze organic reactions in a nonaqueous medium, while the polar outer surface ensures solubility.

The positioning of functional groups is crucial for the proper functioning

Figure 20.4. *A tropocollagen molecule consists of three helical protein chains.*

Figure 20.5.   *In silk fibroin, the protein molecules are lined up and held together by hydrogen bonding, giving the general impression of corrugated roofing.*

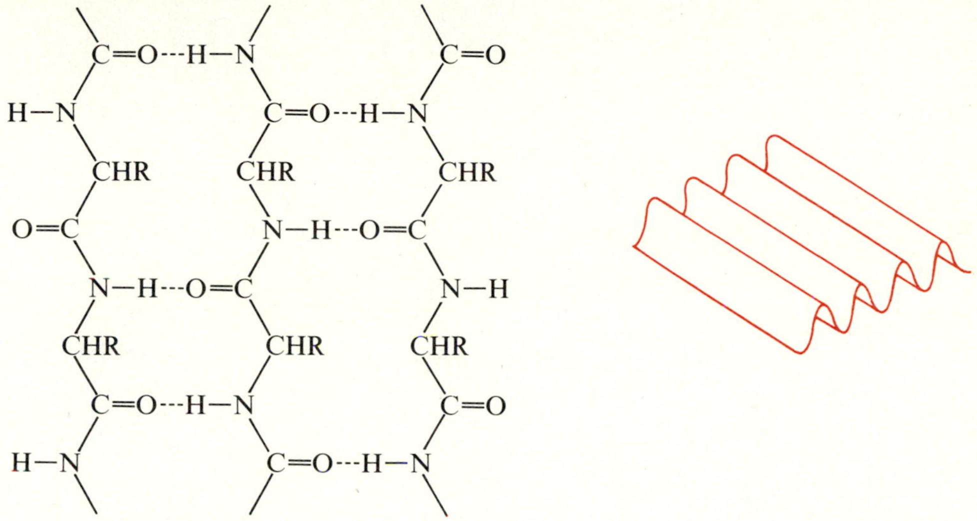

of a globular protein. A properly folded molecule with its unique surface is capable of "recognizing" other specific organic molecules by virtue of their sizes, shapes, polar parts, and nonpolar parts. This recognition means that an enzyme can be used to catalyze specific reactions of specific compounds.

Quaternary Structures

Certain enzymes and other proteins cannot perform their functions unless two or more protein molecules (each with its own primary, secondary, and tertiary structures) are associated with each other by hydrogen bonds and other inter-actions. Thus, the *quaternary structure* of a protein is the unique shape of the aggregate of two or more protein molecules.

■   **Hemoglobin (optional)**

To illustrate the structure and function of a globular protein, let us consider hemoglobin, the protein in red blood cells responsible for transporting oxygen from the lungs to the cells of the body. A model of hemoglobin is shown in Figure 20.6. A unit of hemoglobin consists of four protein chains called **globins**. Each of the four globins is folded around a pocket that holds a molecule of **heme**, which is a porphyrin ring system holding an encaged, or chelated, iron(II) ion (Section 16.7).

The surface of each globin has functional groups that hold the heme in position and hold the four units together. The outer surfaces of the globins contain amino acids with polar side chains to ensure the solubility of hemoglo-

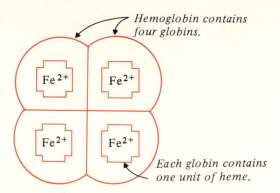

Hemoglobin contains four globins.

Each globin contains one unit of heme.

bin. The inner surfaces of the globins are nonpolar; this nonpolarity protects the Fe^{2+} ion so that it is not oxidized to Fe^{3+}.

Each iron(II) ion in a heme unit can carry an oxygen molecule. Therefore, each hemoglobin structure can carry four oxygen molecules from the lungs by way of the bloodstream to the cells of the body. It is interesting that hemoglobin picks up the fourth oxygen with greater ease than it does the first three, so hemoglobin does not leave the lungs without a full load.

In **carbon monoxide poisoning**, carbon monoxide molecules replace

Figure 20.6.   *A model of hemoglobin: four globins, each with a unit of heme, folded together.*

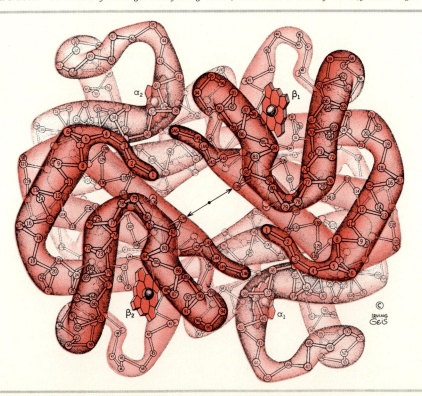

oxygen molecules in the hemoglobin. The affinity of hemoglobin for carbon monoxide is more than 200 times greater than its affinity for oxygen. The hemoglobin with carbon monoxide cannot carry as much oxygen as can hemoglobin without carbon monoxide, and hemoglobin carrying both does not release its oxygen to the cells. The result is that the hemoglobin cannot function as an oxygen transporter if it is carrying carbon monoxide.

Sickle-cell anemia is a hereditary disease caused by a defect in hemoglobin. Severe cases, in which the sufferer has inherited the disease from both parents, are usually fatal. The difference between normal hemoglobin and the hemoglobin of a person affected with sickle-cell anemia is that, in the latter, one glutamic acid unit in a chain of 146 amino-acid residues has been replaced by a valine unit.

$$
\underset{\text{glutamic acid}}{
\begin{array}{c}
CO_2H \\
| \\
H_2N-C-H \\
| \\
CH_2CH_2CO_2H
\end{array}}
\qquad\qquad
\underset{\text{valine}}{
\begin{array}{c}
CO_2H \\
| \\
H_2N-C-H \\
| \\
CH(CH_3)_2
\end{array}}
$$

acidic   *nonpolar*

Where normal hemoglobin has an acidic (polar) side chain, the hemoglobin of

Figure 20.7.   *On the left are normal red blood cells; on the right are sickled red blood cells from a patient with sickle cell anemia. (Courtesy of the Sickle Cell Association of Massachusetts, Inc.)*

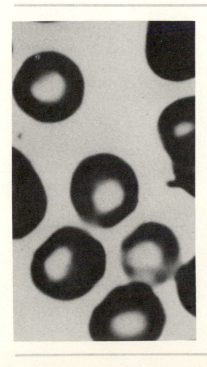

a victim of sickle cell anemia has a nonpolar side chain. Because the abnormal hemoglobin is slightly less soluble than normal hemoglobin, it may precipitate and alter the shape of the red blood cell from round to sickle shaped. (See Figure 20.7.) The lifetimes of these abnormal cells are shorter than those of normal erythrocytes. The abnormal cells may precipitate and clog capillaries. Thus, the oxygen-carrying capacity of the blood is decreased, and the work load on the heart is increased.   ∎

20.5   PROPERTIES OF PROTEINS

Colloidal Proteins

Serum albumins and other proteins in the blood and body fluids are said to be soluble proteins. These proteins are not truly soluble because of the large sizes of protein molecules; they are actually *colloidal* in nature. (You may want to review the properties of colloids in Chapter 8.)

Solubility and pH.   Colloidal proteins are more soluble when they carry a net ionic charge because substances with ionic charges are more attracted to water and because ionic particles with the same charge repel each other. Colloidal proteins are less soluble at their isoelectric points because their molecules carry no charge. Serum albumins are found in the blood (pH 7.3, slightly alkaline) as negatively charged anions; for this reason, they are often referred to as *proteinates.*

When fresh milk (pH 6.3–6.6) becomes sour, the action of microorganisms results in the formation of lactic acid and the lowering of the pH to more acidic values. The curdling that accompanies the souring of milk is the precipitation of the milk protein *casein* as the pH approaches the isoelectric point.

Lack of Dialysis.   Another property of colloidal proteins is that, although they can pass through filter paper, they cannot pass through semipermeable membranes. This is why normal urine does not contain albumins from the blood. The presence of albumins in the urine may signify a kidney infection or some other renal problem.

Denaturation and Renaturation

One of the great difficulties in studying the structure of proteins is that, if the normal environment of a living protein molecule is changed even slightly (such as by a change in pH, temperature, or ionic concentration), the hydrogen bonds are disturbed and broken. When attractions between and within protein molecules are destroyed, the chains separate from each other, globules unfold, and helices uncoil. We say that the protein has been *denatured.* Denaturation

Figure 20.8.   *Denaturation causes loss of the higher structure of the protein.*

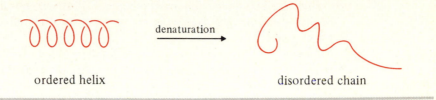

ordered helix   disordered chain

might be compared to stretching a spring out of shape, causing the neat coils to become a tangled mass of metal. In the case of a denatured protein, the chemist is left with a long-chain polymer that has lost its important characteristics. (See Figure 20.8.)

Denaturation is seen in our daily life in many forms. The curdling of milk is caused by a change in pH. Similarly, cooking an egg causes precipitation of the albumin proteins in the egg white. A denatured protein is easier to digest than a nondenatured protein; cooking food increases its digestibility by denaturation. Some proteins (such as those in skin, fingernails, and the stomach lining) are extremely resistant to denaturation.

The examples of denaturation cited are instances of *irreversible denaturation.* If the protein were placed back in its natural environment, it would not return to its normal state. Some cases of denaturation are *reversible.* Depending on the protein and the denaturation conditions, some proteins can spon-

Table 20.4.   *Factors That Can Cause Denaturation*

Factor	Result
changes in pH	disrupt hydrogen bonds and salt bridges, causing change in shape or precipitation
detergents	cause unfolding, followed by interactions between hydrocarbon portion of detergent with nonpolar portion of protein
heat	causes precipitation of globular proteins, such as egg albumins or abnormal albumins in the urine
polar organic solvents such as rubbing alcohol	disrupt attractive forces; can act as an antiseptic by denaturing bacterial protein
oxidizing and reducing agents	disrupt —SH groups and —S—S— cross links
salts of heavy metals such as lead, mercury, or silver	precipitate proteins containing —SH groups
radiation such as ultraviolet or x rays	disrupts hydrogen bonds and initiates free-radical reactions (see Section 12.7)

taneously refold themselves back into their ordered state when placed into a solution identical to their natural environment. The process, called *renaturation*, is usually very slow compared to denaturation, which is fast.

A list of some of the causes of denaturation and their effects on proteins is shown in Table 20.4.

Poisoning by salts of heavy metals deserves special mention. The ions of these metals precipitate proteins, such as enzymes, that contain $-SH$ groups.

R = *rest of protein*

$$Hg^{2+} \quad + \quad 2\,H-SR \quad \longrightarrow \quad Hg(SR)_2 \quad + \quad 2\,H^+$$
$$\text{insoluble}$$

Antidotes for heavy metal poisoning are chelating agents such as EDTA (Section 16.7) or egg whites. Egg albumins can precipitate the metal ions in the stomach before they react with the body's proteins; however, an emetic must be given to remove the metal-albumin precipitate from the stomach before it is digested and the metal ions are freed.

SUMMARY

Proteins are polyamides; hydrolysis of proteins yields *α-amino acids*, which generally have an L configuration at the α carbon. *Essential amino acids* are those required in the diet because the body cannot synthesize them.

Amino acids are *dipolar ions*, which are *amphoteric*.

$$\underset{\textit{a dipolar ion}}{\overset{\overset{\displaystyle CO_2^-}{|}}{\underset{\underset{\displaystyle R}{|}}{H_3\overset{+}{N}-C-H}}} \quad \xrightarrow{\text{H}^+ \text{ or OH}^-} \quad \underset{\textit{in acid}}{\overset{\overset{\displaystyle CO_2H}{|}}{\underset{\underset{\displaystyle R}{|}}{H_3\overset{+}{N}-C-H}}} \quad \text{or} \quad \underset{\textit{in base}}{\overset{\overset{\displaystyle CO_2^-}{|}}{\underset{\underset{\displaystyle R}{|}}{H_2N-C-H}}}$$

In *electrophoresis*, amino acids carrying a net ionic charge migrate to the electrode of opposite charge. At the proper pH, the *isoelectric point*, an amino acid is neutral and does not migrate.

A *peptide* is a polyamide containing 2–50 amino acid residues. Peptides are classified as dipeptides, tripeptides, etc. The *N*-terminal amino acid of a peptide is at the end of the chain that contains an alpha $-NH_2$ group, while the *C*-terminal amino acid, at the other end of the chain, contains a $-CO_2H$ group.

The general classifications of proteins are shown in Table 20.3. The sequence of amino acids in a protein is called the *primary structure*. Group interactions, such as hydrogen bonding, hold protein molecules in their unique shapes; these shapes are referred to as the *secondary*, *tertiary*, and *quaternary* structures.

Fibrous proteins are found as *α-helices*. Collagen is formed by the intertwining of three helices. Globular proteins are formed by the folding of helices. Silk fibroin is found as *β-pleated sheets*, instead of as helices.

Albumins are *colloidal* proteins, least soluble at their isoelectric points and generally unable to pass through a semipermeable membrane.

Denaturation is the disruption of hydrogen bonding and other group interactions and the loss of the higher structure of the protein. Some causes of denaturation are shown in Table 20.4.

KEY TERMS

protein	*isoelectric point	higher structures
amino acid	peptide	*hemoglobin
essential amino acid	*C* and *N* terminals	denaturation
dipolar ion	primary structure	

STUDY PROBLEMS

20.5. Which of the following structures are α-amino acids?

(a)

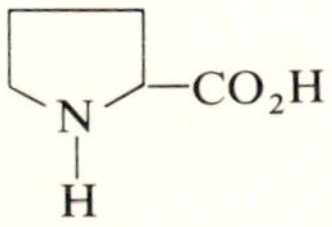

(b) $NH_2CH_2CH_2CO_2H$

(c)

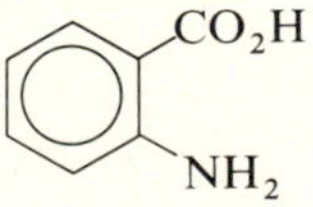

(d)

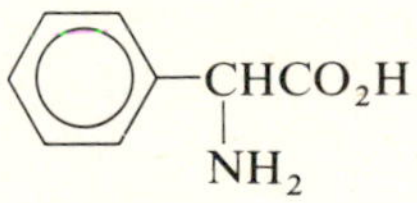

20.6. Write formulas for D- and L-serine:

$$HOCH_2 \underset{\underset{NH_2}{|}}{CH}CO_2H$$

20.7. (a) Define essential amino acid.
(b) Suggest a reason that a mixture of dried beans and corn is considered a complete protein, while dried beans alone are considered incomplete.

20.8. Show by an equation how each of the following amino acids becomes a dipolar ion:
(a) alanine
(b) methionine (see Table 20.1)

20.9. Complete the following equations:

(a) $CH_3\underset{\underset{NH_3^+}{|}}{CH}CO_2^- + H^+ \rightleftharpoons$

(b) $CH_3\underset{\underset{NH_3^+}{|}}{CH}CO_2^- + OH^- \rightleftharpoons$

20.10. Explain why amino acids are water soluble.

20.11. The formula for a peptide follows:

$$H_2N\overset{\overset{O}{\|}}{C}H\overset{\overset{O}{\|}}{C}NHCH_2\overset{\overset{O}{\|}}{C}NHCHCH\overset{\overset{O}{\|}}{C}NHCH_2CO_2H$$

(a) Rewrite the formula to indicate the peptide bonds.
(b) Circle the *N*-terminal and *C*-terminal amino-acid portions of the structure.

20.12. (a) Write the equation for the hydrolysis of acetamide ($CH_3\overset{\overset{O}{\|}}{C}NH_2$) in aqueous acid.
(b) Write the equation for the digestion of the peptide in Problem 20.11.

20.13. (a) Write the equation for the reaction

that leads to phe-leu from its amino acids; circle the OH and H that are lost as H_2O.

(b) One of the problems with synthesizing peptides in the laboratory is that mixtures of products result if the component amino acids are just mixed and polymerized. (Today, genetic engineering is used to force bacteria to synthesize a desired peptide.) What peptides would be obtained if phenylalanine and leucine were mixed and polymerized? Use the abbreviated names, not the structures, in your answer.

20.14. Based on the following partial structures alone (see Table 20.1), decide which peptide would be more water soluble. Explain your answer.

(a) -gly-ala-ser-phe-
(b) -gly-pro-ser-asp-

20.15. Define the following terms as related to proteins:

(a) primary structure
(b) secondary structure
(c) tertiary structure
(d) quaternary structure

20.16. (a) List three features that hold a protein in its higher structures.
(b) Of these, which is the most important?

20.17. Show how a salt bridge could form between the following two protein molecules:

$$-NHCH_2\overset{O}{\overset{\|}{C}}-NHCHC-NHCH_2\overset{O}{\overset{\|}{C}}-$$
$$CH_2CH_2CO_2H$$

$$(CH_2)_4NH_2$$
$$-NHCHC-NHCH_2C-$$

20.18. Describe briefly in words the following higher structures of proteins:

(a) α-helix
(b) β-pleated sheet
(c) collagen triple helix
(d) the shape of a globular protein

***20.19.** Describe briefly the structure and function of hemoglobin. Include in your answer the structural features that allow hemoglobin to be relatively soluble and that protect the iron(II) ion from oxidation.

20.20. Explain why adding lemon juice to milk causes the milk to curdle, or precipitate.

20.21. (a) Define denaturation. (b) List the factors that can cause denaturation.

20.22. Gelatin is obtained by boiling collagen-containing animal skins and other scraps. It has been determined that the formula weight of gelatin is one-third that of collagen. Explain.

20.23. The blood contains colloidal proteins called *serum albumins*. Explain why these proteins do not normally pass through the capillary walls and thus are not found in urine or other body fluids.

21 Enzymes

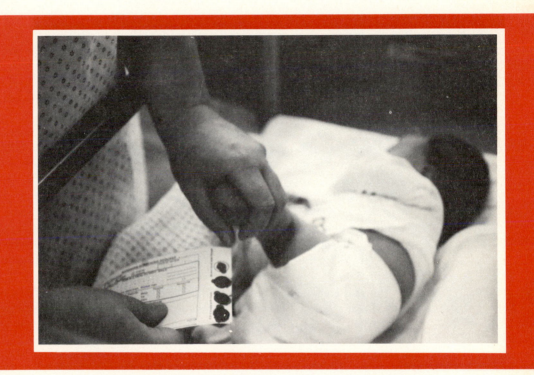

Testing an infant for phenylketonuria (PKU). (c Grace Moore/Taurus Photos.) Phenylketonuria, meaning "a phenylketone in the urine," is an hereditary disorder in which a baby lacks the enzyme phenylalanine hydroxylase. This enzyme normally catalyzes the hydroxylation of the amino acid phenylalanine as the first step in its catabolism (see Section 15.5). When the enzyme is in short supply, the blood level of phenylalanine rises, and transaminase enzymes catalyze its conversion to phenylpyruvic acid (a phenyl ketone). Phenylpyruvic acid is toxic and causes rapid and irreversible brain damage in an infant; the result is severe mental retardation or death.

If detected early, PKU can be controlled by diet and its tragic effects prevented. Today, the blood or urine of newborns is routinely tested for the presence of phenylalanine or phenylpyruvic acid. The infant in this photograph has had its heel pricked to provide a small sample of blood for testing.

Objectives   Describe the general structure, function, and properties of an enzyme. ☐ List the general classes of enzymes, and determine the function from the name. ☐ Define and describe apoenzyme, cofactor, coenzyme, and holoenzyme. ☐ Define vitamins, and describe their importance. ☐ Write the general flow equation that describes enzyme activity. ☐ Describe the induced fit theory. ☐ List and describe some types of enzyme inhibitors and enzyme regulatory mechanisms (optional). ☐ Describe how blood serum enzyme levels are used in medical diagnoses (optional). ☐ Define and give examples of chemotherapy, antimetabolites, and antibiotics (optional).

One of the very important classes of compounds in living systems is the class called **enzymes**. Every known enzyme is either a globular protein alone or a globular protein associated with an ion, a nonprotein organic molecule, or both.

Enzymes are catalysts. Virtually every chemical reaction that occurs in a living system is catalyzed by an enzyme. Many hereditary diseases, such as Tay-Sachs disease (Section 19.3) or phenylketonuria (Section 15.5), are a direct consequence of the lack of a particular enzyme and the resultant failure of a biochemical reaction to proceed.

Just like chemical catalysts, enzymes are needed in only small quantities and act by lowering the energy barrier of a chemical reaction. For example, *peptidase* enzymes catalyze the hydrolysis of peptide bonds. With the proper peptidase, hydrolysis can occur at nearly neutral pH and at moderate temperature. Without the enzyme, the reaction requires boiling in aqueous acid. (An energy diagram for the hydrolysis of a protein is shown in Figure 5.3, in Section 5.3.)

In this chapter, we will discuss the general structures, properties, naming, and classification of enzymes; enzyme activity and how it is regulated; and then briefly the role of enzymes in medical diagnosis and chemotherapy.

21.1 PROPERTIES OF ENZYMES

Because enzymes are proteins, they are subject to the same chemical laws as are other proteins. For example, enzymes are susceptible to denaturation by heating, changes in pH, and so forth.

Because enzymes act in various parts of the body, they have evolved to be most efficient under specific sets of conditions appropriate for their locations in the body. For example, the *optimum pH range* for enzyme activity is not the same for every enzyme. The optimum pH range for the enzyme pepsin, which catalyzes protein hydrolysis in the stomach, is 1.5–2.0, the usual acidic pH range found in the stomach. Trypsin, a protein-hydrolysis enzyme in the intestines (nonacidic), has an optimum pH range of 7–8. Each enzyme also has an *optimum temperature range*, which, in humans, is usually 37–40°C (body temperature).

The compound on which an enzyme acts is called the **substrate**. One enzyme molecule can catalyze the reaction of from 1 to about 10,000 substrate molecules per second. The actual rate of reaction depends on the enzyme-substrate system, the temperature, and the pH. The rate also depends on the concentrations of the enzyme and the substrate—a greater concentration of either generally increases the rate. Regulation of an enzyme's concentration is one way the body controls enzyme activity; this topic will be discussed in Section 21.6.

21.2   NAMING AND CLASSIFICATION OF ENZYMES

Enzymes are named both by common, or trivial, names and by more systematic names. In general, the enzymes that were discovered early in the history of biochemistry have trivial names ending in -in. For example,

pepsin and *trypsin*: catalyze protein hydrolysis
ptyalin: catalyzes carbohydrate hydrolysis
rennin: catalyzes the hydrolysis of casein (milk protein)

Today, enzymes are named after the substrate they work upon or after the reaction they catalyze, with the ending -ase. For example,

maltase: catalyzes the hydrolysis of maltose
amylase: catalyzes the hydrolysis of amylose

The ending -ase is also used in the names for classifying *groups* of enzymes, according to the type of reaction they catalyze. A *hydrolase* is any enzyme, such as pepsin or maltase, that catalyzes a hydrolysis reaction. Table 21.1 summarizes the general classifications of enzymes.

Example   What would be the function of each of the following enzymes?
(a) phenylalanine hydroxylase   (b) acetyltransferase

Solution   (a) Oxidation of phenylalanine by insertion of a hydroxyl group into the molecule.
(b) Transfer of an acetyl group between molecules.

Problem 21.1.   Referring to Table 21.1, suggest a classification name for the enzyme that catalyzes each of the following biochemical reactions:

$$
\begin{array}{ccc}
\text{CH}_2\text{CO}_2\text{H} & & \text{CH}_2\text{CO}_2\text{H} \\
| & & | \\
\text{(a)} \quad \text{CHCO}_2\text{H} & \longrightarrow & \text{CHCO}_2\text{H} \\
| & & | \\
\text{HOCHCO}_2\text{H} & & \text{O}{=}\text{CCO}_2\text{H} \\
\text{isocitric acid} & & \text{oxalosuccinic acid}
\end{array}
$$

Table 21.1. *Classification of Enzymes*

Name of class	Type of reaction catalyzed	Comments
hydrolases:	hydrolysis	found in digestive juices and lysosomes (cellular sites of hydrolysis reactions)
carbohydrases	hydrolysis of carbohydrates	examples are *ptyalin, sucrase, maltase, pancreatic amylase*
lipases	hydrolysis of lipids	act on triacylglycerols
esterases	hydrolysis of esters	*simple esterases* act on small esters and triacylglycerols; *phosphatases* act on phosphate esters
proteases	hydrolysis of proteins	*proteinases* (pepsin, trypsin, chymotrypsin) act on proteins; *peptidases* act on small peptides
nucleases	hydrolysis of nucleic acids	—
oxidoreductases:	oxidation-reduction reactions	in mitochondria (cellular sites of nutrient oxidation)
oxidases	add O atoms by oxidation	—
dehydrogenases	remove H atoms by oxidation	—
transferases:	transfer of functional groups	—
transaminases	transfer of amino groups	—
kinases	transfer of phosphate groups	—
lyases	remove functional groups (not by hydrolysis) to form a double bond; or add functional groups to a double bond	example: $$\underset{\underset{R}{\displaystyle\mid}}{\overset{\overset{OH}{\displaystyle\mid}}{R}}CHCH_2R \rightleftharpoons RCH{=}CHR$$
isomerases	isomerization, or rearrangements of atoms within a molecule	examples: 11-*trans*-retinal $\longrightarrow$ 11-*cis*-retinal glucose $\longrightarrow$ fructose
ligases (synthetases)	coupling, or combination, reactions	example: $$CH_3\overset{O}{\overset{\|}{C}}CO_2H + CO_2 \longrightarrow$$ $$HO\overset{O}{\overset{\|}{C}}CH_2\overset{O}{\overset{\|}{C}}CO_2H$$

$$\text{(b)} \quad \underset{\displaystyle \overset{O}{\parallel}}{R}CCO_2H \; + \; \underset{\displaystyle \overset{NH_2}{|}}{R'}CHCO_2H \; \longrightarrow \; \underset{\displaystyle \overset{NH_2}{|}}{R}CHCO_2H \; + \; \underset{\displaystyle \overset{O}{\parallel}}{R'}CCO_2H$$

21.3 STRUCTURES OF ENZYMES

Some enzymes are globular protein molecules alone. However, many enzymes are *conjugated* proteins; they contain both a globular protein and a non-protein component, which can be either an organic molecule or a metal cation, or both. In a conjugated system of this sort, the protein part of the enzyme is called an *apoenzyme*, while the nonprotein part is called a *cofactor* or *prosthetic group*. If the cofactor is an organic molecule, it is called a *coenzyme*. A complete active enzyme system containing one or more cofactors is called a *holoenzyme* (from *holo-*, meaning whole).

globular protein   *metal ion (inorganic)*
                                              or coenzyme (organic)

apoenzyme + cofactor ———→ holoenzyme

Cofactors

Metal Ions.   Metal ions are essential cofactors in many enzyme systems. Sometimes these cations are referred to as **enzyme activators**. Table 21.2 lists some metal ions that are essential cofactors.

The essential metal ions are obtained from dietary sources. Some metal ions (such as calcium, magnesium, and iron) are needed in relatively large

Table 21.2.   *Metal Ions Associated with Enzyme Action*[a]

Ion	Occurrence or function	Dietary sources
chromium (Cr^{3+})	activates insulin	whole-grains; meat
cobalt (Co^{2+})	in vitamin B_{12} (part of a coenzyme)	green, leafy vegetables
copper (Cu^{2+})	in oxidases; required for synthesis of red blood cells and collagen	liver; seafood; whole grains; nuts; raisins; legumes
iron (Fe^{2+})	in oxidation enzymes and other enzymes	red meat; liver; legumes
magnesium (Mg^{2+})	in many enzymes	green, leafy vegetables; meats; grains; fruits
manganese (Mn^{2+})	in phosphatases, peptidases, decarboxylases, transferases	green, leafy vegetables; whole grains; fruits; nuts
molybdenum (Mo^{6+})	in nucleases, oxidases	liver; yeast; milk
zinc (Zn^{2+})	in dehydrogenases and other enzymes	meat; seafood; grains; fruits; nuts; eggs

[a] It is not known if some other trace elements, such as silicon and selenium, are essential.

Table 21.3. *The Important Vitamins*

Vitamin	Occurrence or function	Deficiency	Dietary sources
water soluble:			
B$_1$ (thiamine)[a]	coenzyme for conversion of pyruvic acid to acetyl CoA	beriberi; nervous-system disorders	whole grains; liver; vegetables
B$_2$ (riboflavin)[a]	in FAD	skin and visual disorders	milk; eggs; liver; vegetables
B$_6$ (pyridoxine)[a]	in transaminases	convulsions in infants; skin disorders	whole grains; liver; pork; legumes
B$_{12}$ (cyanocobalamin)[a]	needed for synthesis of nucleic acids and red blood cells; for nerve growth	pernicious anemia	milk; eggs; meat
niacin (nicotinic acid)[a]	in NAD$^+$	pellagra; skin and intestinal disorders	meat; whole grains; yeast
C (ascorbic acid)[a]	coenzyme for collagen synthesis and hydroxylation reactions	scurvy; bleeding gums	citrus fruits; potatoes; tomatoes
folic acids[a]	needed for cell replacement and nucleoprotein synthesis	anemia; inhibition of cell division	green, leafy vegetables; liver; ye .t; wheat germ

pantothenic acid[a]	in coenzyme A	disorders of the nervous system, muscles, heart, digestive tract	liver; eggs; meat; milk; whole grains
biotin[a]	in ligase enzymes	skin disorders	liver; meat; yeast; milk; vegetables; synthesized by bacteria in intestines
fat soluble:			
A	necessary for synthesis of rhodopsin and for skin formation and maintenance	night blindness; skin disorders; sterility; retarded growth	fish liver oils; green and yellow vegetables
D	instrumental in calcium metabolism and bone formation	rickets	fish liver oils; enriched milk; synthesized by the action of sunlight on steroids in skin
E	maintenance of cell-membrane lipids	scaly skin; liver degeneration; sterility in rats	green, leafy vegetables; nuts; liver; whole grains; vegetable oils
K[a]	necessary for blood clotting	low blood-coagulation rate	green, leafy vegetables; fish; synthesized by intestinal bacteria

[a] The vitamins known to act as coenzymes or as part of coenzymes. It is not known if the other vitamins also function as coenzymes.

Figure 21.1. *Structures of two coenzymes important in oxidation-reduction reactions.*

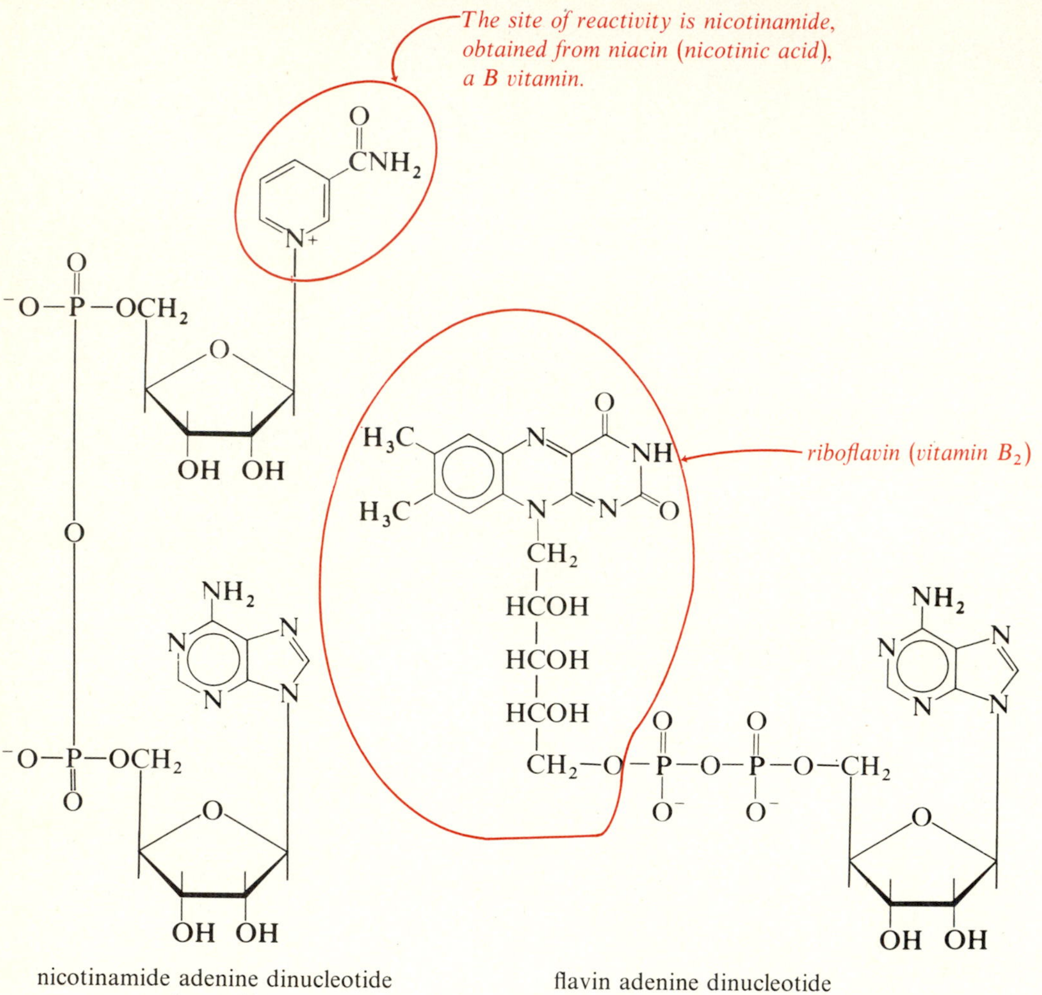

The site of reactivity is nicotinamide, obtained from niacin (nicotinic acid), a B vitamin.

riboflavin (vitamin B$_2$)

nicotinamide adenine dinucleotide
(NAD$^+$)

flavin adenine dinucleotide
(FAD)

quantities. However, most essential metal ions are necessary in the diet in only trace amounts; in many cases, such as cobalt or chromium salts, an overdose can be lethal.

We do not entirely understand how metal ions activate enzymes. In some instances, such as the cobalt ion in vitamin B$_{12}$, the metal ion is an integral part of a coenzyme structure. In other cases, the metal ion is only loosely associated. In either type of enzyme system, the metal ion may be instrumental in attracting the substrate to the enzyme or in holding the substrate to the enzyme. The cation may also participate in catalysis of a biochemical reaction by interaction with a negatively charged portion of the substrate.

Coenzymes and Vitamins.   Coenzymes are nonprotein organic molecules that work as part of an enzyme system in catalyzing biochemical reactions. The structure of coenzyme A is shown in Figure 17.1 (Section 17.8). Figure 21.1 shows the structures of two other very important coenzymes, NAD^+ and FAD.

A **vitamin** is an organic compound usually found in food and necessary in the diet for an organism to function normally. A vitamin is not a major nutrient as are carbohydrates, proteins, and lipids; it is needed in only small amounts. Many coenzymes (including NAD^+ and FAD) contain vitamins, especially the B vitamins, as part of their structures.

Vitamins are classified as fat soluble or as water soluble. The water-soluble vitamins, such as vitamin C and the B vitamins, are not stored to any appreciable extent in the body. These vitamins should be supplied daily in the diet. The fat-soluble vitamins, such as vitamins A and D, can be stored in the tissues of the body to be drawn upon when needed. Table 21.3 lists some vitamins, along with their functions and sources.

21.4   ENZYME ACTIVITY

Enzymes are selective catalysts. An enzyme may be specific for one chemical change in one particular compound, or for one change in only one isomer (*cis*, *trans* or D, L) of a compound. For example, the enzyme *aspartase* catalyzes the reversible addition of ammonia to the *trans* diacid fumaric acid but is inactive toward the *cis* isomer (maleic acid) and other compounds.

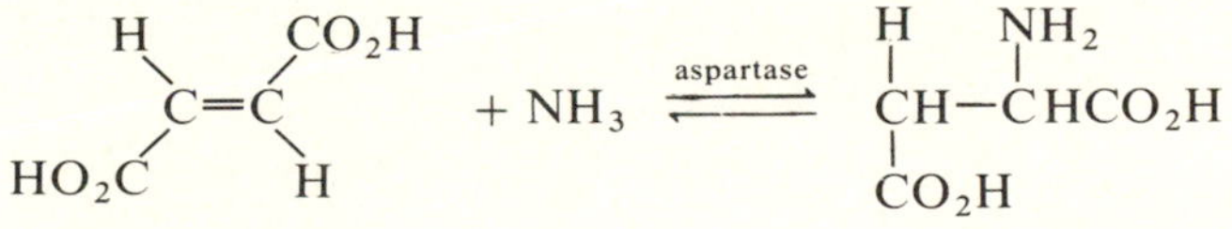

fumaric acid
(*trans*)

aspartic acid

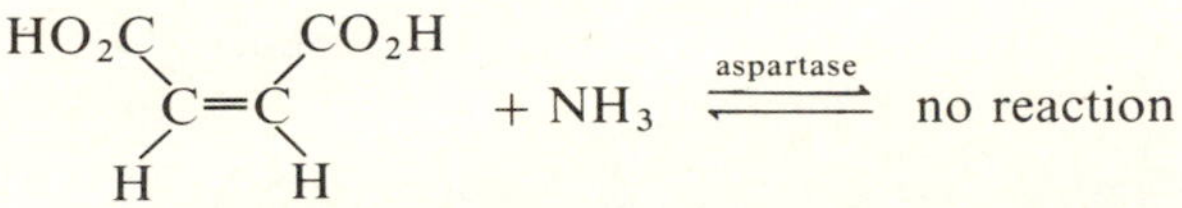

maleic acid
(*cis*)

Other enzymes are more general and catalyze reactions for whole classes of compounds containing the same functional group. Typical of this group are the digestive enzymes trypsin, chymotrypsin, and pepsin, all of which catalyze the hydrolysis of peptide bonds.

Enzymes are much larger than most substrates. In many enzyme-catalyzed reactions, the enzyme along with its cofactor, if any, fits itself around the substrate molecule to form an enzyme-substrate complex, much as a lock fits around a key. In some cases, the enzyme undergoes a change in its shape to entrap the substrate. In other words, the enzyme is induced to fit the substrate. This view of enzyme activity is, therefore, referred to as the **induced fit theory**.

$$E \; + \; S \; \rightleftharpoons \; E\!-\!S$$

enzyme   substrate   enzyme-substrate
                        complex

The attractions that hold the enzyme and substrate together may be due to unique and specific electronic attractions or may be actual covalent bonds. Regardless of how the complex is formed, the shape and polarity of the substrate determine whether or not a complex can be formed with a particular enzyme.

The portion of the enzyme that actually catalyzes reaction is called the **active site**. Often this active site is the cofactor. Other parts of the enzyme molecule serve as binding sites for the rest of the substrate molecule. The enzyme can lower the energy of activation and thus catalyze reaction of the substrate in any of a number of ways: by helping to bring the reactants together or orienting them correctly; by causing strain in the bonds of the reactant (by polar attractions, for example); or by actually reacting with the reactant temporarily.

As the reaction takes place, the substrate is converted to the product. The product's shape is different from that of the substrate, and the product molecule no longer fits the enzyme. Therefore, the enzyme releases the product and becomes ready to work again. Figure 21.2 depicts how these steps may occur.

$$E\!-\!S \; \rightleftharpoons \; E\!-\!P \; \rightleftharpoons \; E \; + \; P$$

enzyme-substrate   enzyme-product   enzyme   product
complex   complex

Figure 21.2.   *The induced fit theory of enzyme activity.*

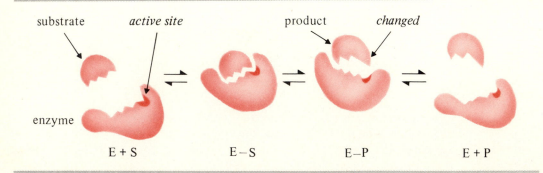

substrate   *active site*   product   *changed*

enzyme

E + S   E−S   E−P   E + P

Figure 21.3. *One isomer fits onto an enzyme surface; its mirror image may not fit.*

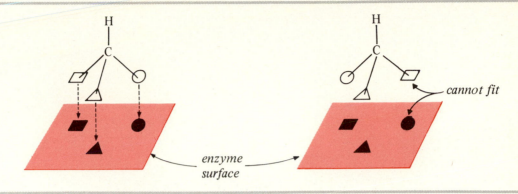

The induced fit theory of enzyme activity explains why enzymes are so selective toward substrates. If an enzyme cannot be induced to fit a substrate and form a complex, the reaction occurs slowly, if at all. This theory explains why one mirror-image isomer may be biologically active and the other isomer inactive; see Figure 21.3.

■ 21.5   ENZYME INHIBITION (optional)

Enzymes can be inhibited from performing their role as biological catalysts by a variety of agents called **enzyme inhibitors**. An inhibitor that is general for all enzymes is called a *nonspecific inhibitor*. Denaturing agents, such as acids and heat, are nonspecific inhibitors. Figure 21.4 shows how nonspecific inhibition can occur.

An inhibitor that is specific for only one enzyme or group of enzymes is called a *specific inhibitor*. Sulfanilamide is an example of a specific inhibitor. In many infectious microorganisms, *p*-aminobenzoic acid (PABA) is required for the synthesis of the coenzyme folic acid. The sulfa drugs can be biologically hydrolyzed to sulfanilamide, which is similar in shape and structure to PABA.

Figure 21.4. *Denaturation inhibits enzyme activity by changing the shapes of the enzymes.*

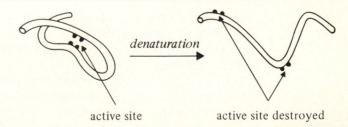

Figure 21.5. *A competitive enzyme inhibitor mimics the shape and polarity of the substrate; thus, it can block the active site of the enzyme so that the substrate cannot be acted upon.*

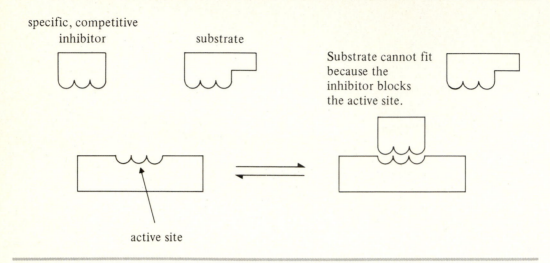

In the microorganism, the sulfanilamide mimics PABA and competes for one of the enzymes necessary to the synthesis of folic acid. (See Figure 21.5.) The synthesis is thus blocked, because PABA cannot be incorporated, and the microorganism cannot grow or reproduce.

When two different molecules, such as sulfanilamide and PABA, compete to occupy the active site of an enzyme, the inhibition of the enzyme is said to be **competitive inhibition**. Competitive inhibition is a common mode of action of many poisons and drugs. Competitive inhibition is a *reversible process*; therefore, the extent of inhibition depends on the relative concentrations of inhibitor and substrate. Because sulfanilamide is a competitive inhibitor, a high and constant concentration is required in an animal's blood for effective antibacterial action.

Competitive inhibition requires that the inhibitor mimic the normal sub-strate. However, one substance can block the action of an enzyme by other means. A **noncompetitive inhibitor** is one that does not mimic the substrate but reacts with the enzyme to change the active site, to change the shape of other parts of the enzyme molecule, or perhaps to precipitate the enzyme. Noncompetitive inhibition is often irreversible. Many heavy metal ions, such as Hg^{2+} or Pb^{2+}, are poisonous because they act as irreversible, noncompetitive inhibitors for numerous enzymes. One way these ions deactivate an enzyme is by forming insoluble salts with the enzyme's —SH groups, as shown in Section 20.6.

Problem 21.2. The normal action of succinic acid dehydrogenase is to catalyze the loss of two hydrogen atoms from succinic acid. Malonic acid inhibits this reaction by competitive inhibition.

removed by oxidizing agent

$$\begin{array}{c} CO_2H \\ | \\ CH_2 \\ | \\ CH_2 \\ | \\ CO_2H \end{array} \xrightarrow[\text{dehydrogenase}]{\text{succinic acid}} \text{enzyme-succinic acid complex} \rightleftharpoons \begin{array}{c} H \quad CO_2H \\ \diagdown \diagup \\ C \\ || \\ C \\ \diagup \diagdown \\ HO_2C \quad H \end{array} + \boxed{2\,H^+ + 2\,e^-}$$

succinic acid   fumaric acid

$$\begin{array}{c} CO_2H \\ | \\ CH_2 \\ | \\ CO_2H \end{array} \xrightleftharpoons[\text{dehydrogenase}]{\text{succinic acid}} \text{enzyme-malonic acid complex} \longrightarrow \text{no reaction}$$

malonic acid

Which of the following compounds might also be a competitive inhibitor of succinic acid dehydrogenase? Explain.
(a) HO_2CCO_2H
(b) $HO_2CCH_2CH_2CH_2CO_2H$
(c) $HOCH_2CH_2CH_2CH_2OH$

■

___

■ 21.6 ENZYME REGULATION (optional)

The various chemical reactions in an organism must be regulated; otherwise, the system would be completely out of control. Biological reactions are generally controlled by *regulation of the enzymes* that catalyze the reactions. Enzymes can be controlled in a number of ways. Because enzymes are continuously broken down and resynthesized in an organism, one way to regulate biochemical reactions is by altering the rate of synthesis and thus the *concentration* of the enzyme. Another technique for enzyme regulation is control of its *activity*. The following factors affecting enzymes all depend on control of concentration or activity.

diet: A diet rich or poor in a particular essential amino acid, vitamin, or metal ion can affect the amount of active enzyme synthesized and thus its concentration in the cells.

genetic control: The synthesis of enzymes is directed by the genes. The genes have their own regulatory mechanisms that signal whether or not a particular enzyme is to be synthesized. (See Chapter 27.)

hormones: These are chemical messengers secreted by the endocrine glands, such as the pituitary gland or adrenal glands. They travel through the body fluids to target cells where they can activate a gene or an enzyme. They can also change the permeability of the cell membrane, allowing other activators to enter the cell.

prostaglandins: These are hormone-like substances that are synthesized in most tissues and that regulate many biochemical reactions. (See Section 19.4.)

regulatory enzymes: A discussion of this mode of enzyme regulation follows.

Regulatory enzymes are enzymes that are needed to catalyze part of a reaction sequence but that are deactivated by a product of the sequence. Consider the following hypothetical sequence of enzyme-catalyzed reactions:

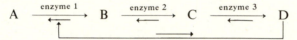

The product of the sequence, D, undergoes reversible reaction with enzyme 1. Thus, as compound D increases in concentration, enzyme 1 *decreases* in concentration, and the entire reaction sequence slows down.

$$D \ + \ \text{enzyme 1} \ \underset{\textit{active}}{\rightleftharpoons} \ \underbrace{\text{D-enzyme 1 complex}}_{\textit{inactive}}$$

When the reaction sequence becomes sufficiently slow, the concentration of D diminishes, and enzyme 1 is again activated in the reverse of the preceding reaction. Thus, the reaction sequence (A → B → C → D) begins anew. In this way, a cell can hold the concentrations of enzyme 1 and compound D within certain limits. A regulatory mechanism such as this one, in which a product can inhibit its own formation, is called **feedback inhibition**. ∎

■ **21.7 ENZYMES IN MEDICINE** (optional)

Enzymes as Diagnostic Tools

The cells in different tissues of the body contain differing amounts of the various enzymes. When a cell is injured, it can release its enzymes into the bloodstream. Therefore, in many diseases, unusually high levels of particular enzymes can be found in the blood serum. The analysis of a patient's blood serum for abnormal levels of these specific enzymes is an extremely useful medical tool.

One enzyme diagnostic test is that for *myocardial infarction*, a heart attack in which part of the heart is damaged because of a blockage in its blood supply. After even a mild heart attack, several enzymes normally confined in the heart-muscle cells are released into the bloodstream. The principal enzymes released are creatine phosphokinase, glutamate-oxaloacetate transaminase, and several forms of lactate dehydrogenase called **isoenzymes**, or **isozymes**, structurally similar proteins that exhibit the same or similar enzyme activity. Analysis of the serum for the presence or absence of these enzymes can aid the physician in making a proper diagnosis. Table 21.4 lists a few diseases that cause abnormal blood levels of certain enzymes.

Table 21.4.   *A Few Diseases Characterized by Altered Blood Enzyme Levels*

Disease	Enzymes released into blood
infectious hepatitis and other liver diseases	glutamate-pyruvate transaminase (GPT)[a] lactate dehydrogenase (LDH-5)
pancreatic disease	trypsin amylase
myocardial infarction	creatine phosphokinase (CPK) glutamate-oxaloacetate transaminase (GOT)[a] lactate dehydrogenase (principally LDH-1 and LDH-2)
carcinoma of the prostate	acid phosphatase

[a] These abbreviations are sometimes preceded by S (SGPT, SGOT) to denote serum.

A number of techniques have been developed for enzyme assay. One technique involves treating an aliquot (a small sample) of the serum with a sample of the enzyme's own substrate and checking the rate or extent of enzyme-catalyzed reaction. Another technique to measure an enzyme's concentration in the blood is to subject it to a reaction that yields a colored product, then measure the depth of color of the product (a process called *colorimetry*) or the amount and wavelength of light absorbed by the product (*spectrophotometry*). In the case of the enzyme amylase, a variation of the starch-iodide test (Section 18.8) can be used.

Chemotherapy

Chemotherapy involves treating a patient with drugs to destroy an infectious organism or cancer cells without harming the patient, or at least without harming the patient to the extent to which the infectious organisms are harmed. Many chemotherapeutic agents are *antimetabolites* (compounds that work against substances involved in metabolism), or specific enzyme inhibitors. Ideally, an antimetabolite will be specific for an enzyme of the infectious organism and will not affect any of the patient's enzymes. The sulfa drugs (Section 21.5) are examples of antimetabolites specific for bacteria that require PABA to synthesize folic acids. (Other bacteria and animals use preformed folic acids and do not synthesize them from PABA.)

The *antibiotics* (meaning "anti-life"), compounds produced by one organism that are toxic to another organism, are a special class of antimetabolites. Some examples of antibiotics used today are the penicillins, terramycin, aureomycin, and tetracycline, which are produced by certain fungi.

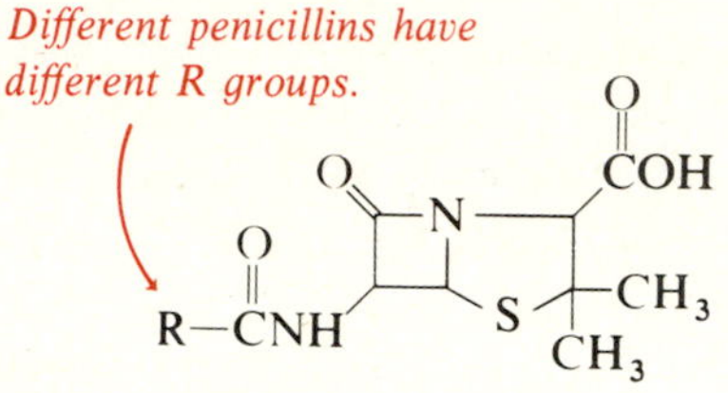

Different penicillins have different R groups.

a penicillin

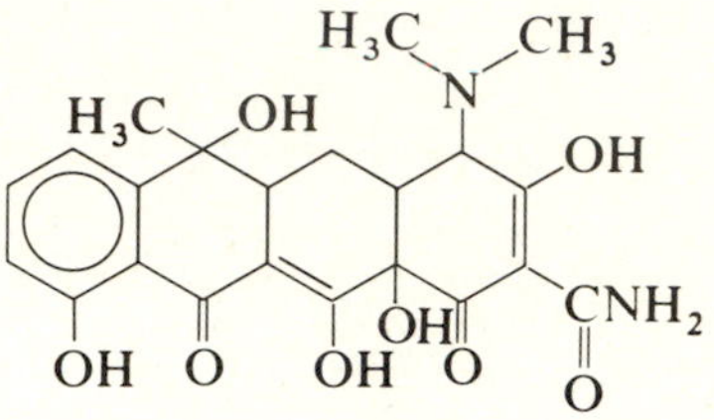

a tetracycline

SUMMARY

Enzymes are globular proteins or conjugated globular proteins that are biological catalysts. Enzyme function has an *optimum temperature range* and an *optimum pH range* that depend on the enzyme. The rate at which an enzyme catalyzes reaction of the substrate also depends on the enzyme-substrate system and the concentrations of the enzyme and substrate.

Enzymes and enzyme classes are named after the substrate or reaction type with the ending *-ase*. Older enzyme names end in *-in*. Table 21.1 summarizes the classification of enzymes.

An enzyme may be a simple protein or it may be conjugated.

$$\text{apoenzyme} \quad + \quad \text{cofactor} \quad \longrightarrow \quad \text{holoenzyme}$$

(*protein*)   (*metal ion or coenzyme*)   (*a conjugated enzyme*)

Many *vitamins* are coenzymes or part of coenzyme structures.

An enzyme contains an *active site* and a *binding site* that interact with the substrate. The unique shape and polarity of these sites determine which substrate is acted upon.

$$\underbrace{E + S \rightleftharpoons E-S} \rightleftharpoons \underbrace{E-P \rightleftharpoons E + P}$$

E is induced to fit around S.   *E cannot be induced to fit P, which leaves.*

The action of enzymes can be inhibited

by a *nonspecific inhibitor* (inhibits all enzymes);

by a *specific inhibitor* (inhibits one enzyme);

in *competitive inhibition* (usually reversible, with the inhibitor mimicking the substrate at the active site);

in *noncompetitive inhibition* (usually irreversible, with the inhibitor reacting with some portion of the enzyme).

Factors affecting an enzyme's concentration and activity, and thus the rate of enzyme action, are diet, regulatory genes, hormones, prostaglandins, and regulatory enzymes.

The presence of abnormal amounts of enzymes in blood serum can be used as an aid in diagnosing certain diseases. In some cases, the relative quantities of *isoenzymes* are measured. Some diseases can be treated with *antimetabolites* (specific enzyme inhibitors) or *antibiotics* (antimetabolites synthesized by another organism).

KEY TERMS

enzyme	holoenzyme	*isoenzymes
substrate	vitamin	*chemotherapy
apoenzyme	induced fit theory	*antimetabolites
cofactor	*enzyme inhibitor	*antibiotics
coenzyme	*feedback inhibition	

STUDY PROBLEMS

21.3. Which of the following phrases apply to enzymes?
(a) are chemical messengers in the body
(b) are biochemical catalysts
(c) are globular proteins, sometimes associated with other ions or molecules
(d) can precipitate if the pH is changed

21.4. Insulin is administered to a diabetic intravenously. Explain why insulin cannot simply be administered orally.

21.5. Match the following enzyme names with the functions:

(a) DNA polymerase   (1) catalyzes the hydrolysis of proteins
(b) transaminase   (2) catalyzes the hydrolysis of esters
(c) proteinase   (3) catalyzes the polymerization of DNA units
(d) esterase   (4) catalyzes the transfer of an amino group
(e) hydrolase   (5) catalyzes the hydrolysis of lactose
(f) lactase   (6) catalyzes the hydrolysis of reactions in general

21.6. Which of the following factors could slow the rate of enzyme catalysis?
(a) dilution of the medium
(b) change in pH
(c) increase in temperature
(d) decrease in temperature

21.7. An enzyme is isolated and found to contain

(a) a protein molecule   (b) a B vitamin

(c) magnesium ions

Label each of these components as an apo-enzyme, a cofactor, a holoenzyme, or a coenzyme.

21.8. *Hypervitaminosis* is the toxic over-accumulation of a vitamin in the body. For example, too much vitamin D can lead to disso-lution of the bones. Suggest a reason why hyper-vitaminosis can be a problem with vitamins A and D, but not with vitamin C.

21.9. Using the letters E and S, write and label the equation that describes enzyme activity.

21.10. Define and describe the terms (a) active site and (b) induced fit theory of enzyme activity.

21.11. Suggest a reason, based on enzyme structure, that L-amino acids can be incorpo-rated into proteins, while D-amino acids are oxi-dized to other compounds in the body.

21.12. The digestive enzyme chymotrypsin catalyzes the hydrolysis of phe-ala, phe-gly, phe-val, and the following compounds:

$$\text{C}_6\text{H}_5\text{—CH}_2\text{CH}_2\overset{\displaystyle\text{O}}{\overset{\|}{\text{C}}}\text{NH}_2$$

$$\text{C}_6\text{H}_5\text{—CH}_2\text{CH}_2\overset{\displaystyle\text{O}}{\overset{\|}{\text{C}}}\text{OCH}_3$$

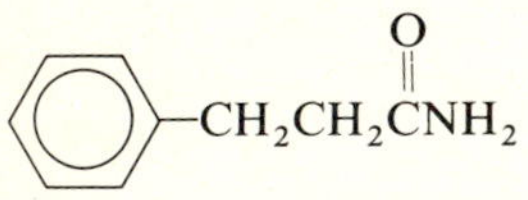

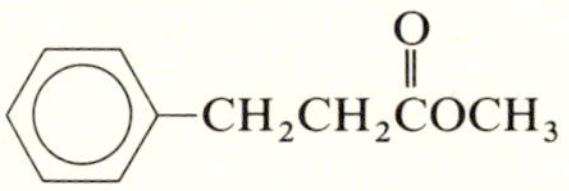

Based on this information, decide which of the following compounds might be hydrolyzed with chymotrypsin as a catalyst. Write formulas for the products of their hydrolysis.

$$\text{(a)} \quad \text{C}_6\text{H}_5\text{—CH}_2\underset{\underset{\text{NH}_2}{|}}{\text{CH}}\overset{\displaystyle\text{O}}{\overset{\|}{\text{C}}}\text{NH}_2$$

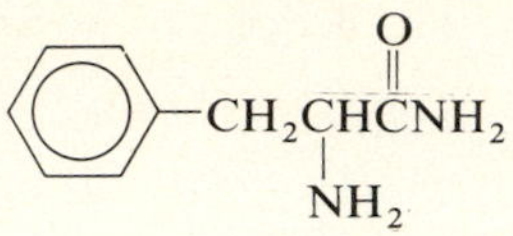

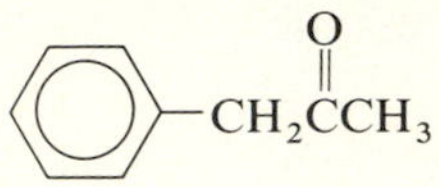

$$\text{(b)} \quad \text{C}_6\text{H}_5\text{—CH}_2\overset{\displaystyle\text{O}}{\overset{\|}{\text{C}}}\text{CH}_3$$

$$\text{(c)} \quad \text{C}_6\text{H}_5\text{—CH}_2\underset{\underset{\text{NH}_2}{|}}{\text{CH}}\overset{\displaystyle\text{O}}{\overset{\|}{\text{C}}}\text{OCH}_3$$

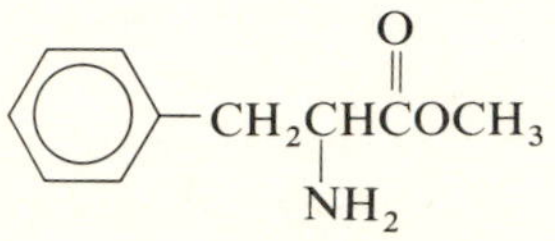

*21.13. Label each of the following reactions as an example of competitive or noncompetitive enzyme inhibition.

(a) enzyme + Ag^+ $\longrightarrow$ enzyme-Ag^+ precipitate

(b) enzyme + substrate-mimic $\rightleftharpoons$ enzyme-mimic complex

*21.14. Label the reactions in the preceding problem as being reversible or irreversible enzyme inhibition.

*21.15. (a) List ways in which the concentra-tion or activity of an enzyme can be *decreased* in the body.

(b) In general terms, describe the process of feedback inhibition of enzyme activity.

*21.16. The blood of a healthy person does not contain the enzyme GPT. Suggest a reason why liver disease is suspected when a patient's blood contains this enzyme.

*21.17. (a) Define antimetabolite and anti-biotic.

(b) Give the name of an example of each.

(c) Describe the general action of these com-pounds.

PART 4

Metabolism of Biological Compounds

22 Introduction to Metabolism

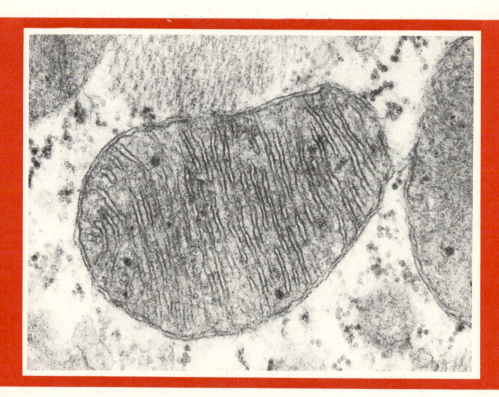

Electron micrograph of a mitochondrion. Sometimes referred to as the powerhouses or power packs of plant and animal cells, the mitochondria are the cellular sites of energy production. These organelles (small organs) are responsible for the metabolism reactions in which nutrients are oxidized to provide the energy needed to sustain life and growth of the cell.

The number of mitochondria in a cell depends on the function of the cell. A muscle cell contains relatively few mitochondria, while a liver cell may contain up to 1,000.

Objectives   Define digestion, metabolism, catabolism, and anabolism. □ Diagram a typical animal cell. □ Describe the functions of the principal organelles. □ Describe the structure and functions of the cell membrane, including the active transport of substances. □ Write an equation to show how ATP can store and provide energy for an organism. □ Using abbreviated formulas, write equations that show the oxidizing action of NAD^+ and of FAD. □ Write an abbreviated biochemical scheme for the cellular respiratory chain, showing where and in what quantity energy is gained by the conversion of ADP to ATP.

A human being ingests food, and this food is converted to compounds used for growth and maintenance. Food also provides the energy for movement, warmth, thought, and life itself.

In this chapter, we will formally define the digestion and metabolism of food. We will consider the structure of an animal cell, where most biochemical reactions occur, and the cell membrane, through which nutrients, ions, and waste products must pass. How a cell gains energy and stores it for later use are the final topics in this chapter.

22.1   DIGESTION AND METABOLISM

The food we eat can be classified in three main groups: *carbohydrates*, *fats*, and *proteins*. When oxidized in the body, all three types of food can yield energy. Carbohydrates yield approximately 4.1 kcal of metabolic energy per gram; fats, 9.3 kcal per gram; and proteins, 4.4 kcal per gram.

Carbohydrates, fats, and proteins undergo many series of chemical reactions in the body. The first set of reactions is *digestion*.

Digestion occurs in the gastrointestinal tract and is the process by which large food molecules are broken down by hydrolysis (cleavage with water) to smaller molecules that can be absorbed into the bloodstream through the intestinal walls.

In the process of digestion, starch molecules are enzymatically hydrolyzed to soluble monosaccharides; protein molecules are hydrolyzed to amino acids; and fats are hydrolyzed to fatty acids and glycerol.

Once the products of digestion are in the bloodstream, they can be transported to and taken in by cells, where they undergo further reactions. These cellular reactions are collectively called *metabolism*.

Metabolism refers to all the enzyme-catalyzed reactions that occur within the cells of living organisms.

The topic of metabolism can be subdivided into *catabolism* and *anabolism*.

Catabolism is the breaking down of relatively large nutrient molecules into smaller molecules, usually with the production of energy. The metabolism of glucose to CO_2, H_2O, and energy is an example of catabolism.

$$\text{glucose} + 6\,O_2 \xrightarrow{\text{many steps}} 6\,CO_2 + 6\,H_2O + 686 \text{ kcal/mole}$$

Anabolism is the building of large molecules, such as proteins, from smaller units, usually with the consumption of energy. Note that anabolism is similar to digestion in reverse.

$$\text{amino acids} + \text{energy} \xrightarrow{\text{many steps}} \text{protein} + H_2O$$

22.2   THE ANIMAL CELL

Metabolic reactions take place in the cells of an organism. These cells are very small; for example, a human liver cell is only about 0.002 cm long, and the cell of the bacterium *Escherichia coli* is about 0.0002 cm long. Despite their small size, typical animal cells have highly organized architectures, which permit biochemical reactions either to be isolated from one another or to be coupled together. This isolation or coupling would not be possible if all the reactants and enzymes were scattered randomly throughout the cell.

Figure 22.1 is a diagram of a typical animal cell. Not all animal cells look like this one—cells vary in size, shape, and general or specific functions. Some cells, such as red blood cells, do not even contain nuclei.

Figure 22.1.   *Diagram of a typical animal cell.*

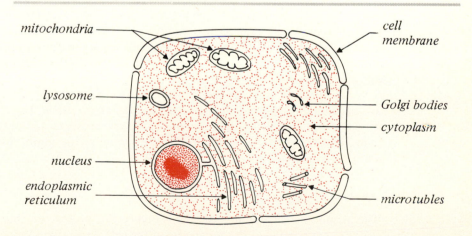

A typical cell is surrounded by and contained in a *cell membrane*, which we will discuss later in this section. Everything inside this membrane is called *protoplasm*. The protoplasm itself can be subdivided into two main parts: the *cell nucleus* and the *cytoplasm*, which is all the material within the cell except the nucleus.

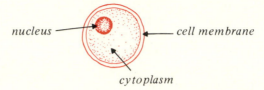

Protoplasm is heterogeneous and contains numerous small bodies called *organelles* (little organs) that carry out various functions within the cell.

The Organelles

The Cell Nucleus and Endoplasmic Reticulum.   The largest organelle, the cell nucleus, is composed of dense aggregates of DNA-protein complexes called *chromatin granules*. In cell reproduction, the chromatin granules become *chromosomes*, the carriers of the genetic code.

The nucleus is surrounded by a double membrane. In many cells, the space between these two membranes is connected to the *endoplasmic reticulum*, a network of membrane-bound tubules extending from the nuclear membrane throughout the cytoplasm and occasionally to the exterior of the cell. Water and various chemicals can be moved through the endoplasmic reticulum. The endoplasmic reticulum also provides the sites for the syntheses of proteins and steroids. The functions of chromatin and the endoplasmic reticulum will be discussed in more detail in Chapter 27.

The Other Organelles.   Like the nucleus, the other organelles within a cell have distinct membranous boundaries, along with filamentous structures (*microfilaments*) and channels (*microtubules*). Both the microfilaments and the microtubules help maintain the cell shape and play a role in cell movement. The microtubules also provide pathways for intracellular transport of water and chemicals.

The various organelles serve a variety of purposes. Many of the organelles are packages of enzymes that catalyze particular reactions. For example, the *lysosomes* are organelles that contain hydrolysis enzymes and are the sites of many hydrolysis reactions. Table 22.1 lists some of the important organelles and their functions.

The Cell Membrane

The membrane surrounding the cell is highly convoluted and thus has a large surface area relative to the actual size of the cell. Typically, a cell membrane is

Table 22.1.   *Some Important Cellular Organelles*

Name	Principal function
nucleus	contains DNA, which carries the genetic code
endoplasmic reticulum	site of protein biosynthesis
lysosomes	contain hydrolysis enzymes and are the sites of hydrolysis reactions
glycogen granules	contain stored glucose as the polymer glycogen (primarily in the liver)
Golgi bodies	collect cell products from the endoplasmic reticulum and discharge these products into the interstitial fluid (fluid between cells)
mitochondria	"power packs" of the cell; contain enzymes that catalyze oxidation and cell-respiration reactions; storehouses of the high-energy compound ATP

composed of two opposing layers, called a *bilayer*, of phospholipid molecules. This lipid bilayer is studded with proteins. The lipid bilayer and the embedded proteins work in concert to carry out the functions of the cell membrane. Let us first consider the structure of the lipid bilayer, then discuss the function of the bilayer and the proteins.

The Lipid Bilayer.   The phospholipid molecules in a cell membrane have a structure resembling that of soap molecules. Each phospholipid molecule contains an ionic "head" that is *hydrophilic* (water loving) and two long hydrocarbon "tails" that are *hydrophobic* (water hating).

$$\begin{array}{c} O^- \\ | \\ CH_2OPOCH_2CH_2\overset{+}{N}R_3 \\ \| \\ O \end{array}$$

$$\underset{\displaystyle RCO-CH}{\overset{\displaystyle O}{\overset{\|}{}}}$$

$$\underset{CH_2OCR}{\overset{O}{\overset{\|}{}}}$$

or

ionic head
(hydrophilic, or water-loving)

hydrocarbon tails
(hydrophobic, or water-hating)

In the cell membrane, the hydrophobic tails of the phospholipid molecules attract one another and form a fatlike layer in the center of the membrane. The ionic, hydrophilic heads of the molecules point either toward the aqueous interior of the cell or toward the aqueous exterior of the cell. Figure 22.2 shows a diagram of the bilayer along with its embedded protein molecules.

Functions of the Cell Membrane.   A cell membrane serves a variety of functions. One is molecular recognition. Cells can recognize other cells or mole-

Figure 22.2.   *The cell membrane is composed of a phospholipid bilayer embedded with protein molecules.*

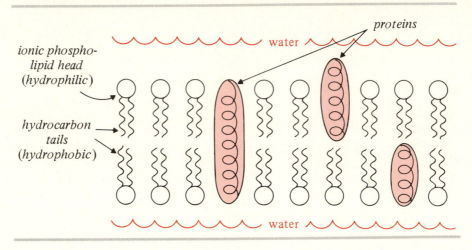

cules because of individualized "antennae," called *recognition sites* and *hormone-reception sites.* These sites ensure that cells can adhere to other similar cells and can receive chemical messages from other parts of the body.

Another function of the cell membrane is controlling the permeability of the cell—determining what ions and molecules can pass in or out of the cell and regulating the rate of their passage. The fatlike layer formed by the hydrocarbon tails of the phospholipid molecules allows some small nonpolar molecules to pass through the cell membrane. However, this fatlike layer is impermeable to ions and polar molecules.

The passage of ions and polar molecules through the cell membrane is controlled by the proteins, which act as pumps, moving nutrients, wastes, and other chemicals in and out of the cell. For example, one type of protein moves sodium ions (Na^+) out of a cell and potassium ions (K^+) into the cell. Recall that the interior of a cell is richer in K^+ than in Na^+, while the blood and interstitial fluids (fluids between cells) are richer in Na^+. Therefore, these proteins can move Na^+ or K^+ from a solution of lower concentration to a solution of higher concentration. We say that these ions are moved *against a concentration gradient*, in the direction opposite to that of the natural, passive flow of substances.

Energy is required to move ions against a concentration gradient. For this reason, the transport of ions through a cell membrane is called *active transport*, in order to differentiate it from passive processes such as diffusion, osmosis, or dialysis.

Diffusion, osmosis, and **dialysis:** Substances flow naturally in the direction that will tend to equalize concentrations.

Active transport: Substances are forced to flow from a lower concentration to a higher concentration.

K⁺ can be forced into a cell, even though its concentration is higher in the cell than in the interstitial fluids.

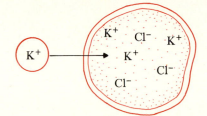

Diversity of Cell Membranes.   All cells are not the same, and the diversity of cellular types is reflected in the structures and functions of their membranes. For example, myelin sheaths (nerve-cell membranes) contain less protein than do the cell membranes surrounding red blood cells. The myelin sheaths also are richer in the steroid cholesterol.

The cell membranes of different organisms also vary. The cell membrane of a bacterium contains a cell wall (*murein*), which is composed of long carbohydrate molecules cemented together by proteins. Penicillin blocks the synthesis of the cell wall in growing bacteria by preventing the protein cross-linking that occurs in the final step of the cell wall synthesis. Penicillin does not have this effect on the cells of higher animals because animal cells do not have cell walls.

22.3   HOW THE CELLS STORE ENERGY

Because we do not eat continuously, the body must have a chemical method of storing energy. Let us consider how the cells store energy to be used when we need it.

The oxidation of glucose and other nutrients, called **cellular respiration**, occurs within the mitochondria of the cells. Each cell contains hundreds of these organelles. It is thought that a mitochondrion is rather like a factory, with the proper enzymes lined up on the inner membrane. (See Figure 22.3.) Nutrients pass through the enzyme lineup and are broken down in assembly-line fashion.

When an organic compound is oxidized in an organism, the released energy is used to synthesize high-energy compounds, which are the storehouses of biological energy. The principal high-energy compound in biological systems is an anhydride (a compound formed by loss of H_2O) called **adenosine triphosphate** and abbreviated **ATP**. Figure 22.4 shows the structure of ATP.

When the body requires energy for movement, warmth, or chemical reactions, ATP is hydrolyzed. The hydrolysis reaction gives off energy which can then be used by the body. In the hydrolysis of ATP, usually only one of the phosphate groups is cleaved. This inorganic cleavage product, a hydrogen

Figure 22.3. A mitochondrion (see also the photograph on page 409).

folded inner membrane where oxidation of nutrients occurs

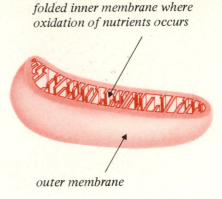

outer membrane

phosphate ion, is referred to as *inorganic phosphate* or P_i. The other cleavage product is adenosine diphosphate, ADP. For our discussion, we will consider only the working end of ATP and will use R to represent the rest of the molecule.

when the organism needs energy:

$$\underset{\text{ATP}}{\text{ROPO}-\text{PO}-\text{PO}^-} + \text{H}_2\text{O} \longrightarrow$$

$$\underset{\text{ADP}}{\text{ROPO}-\text{PO}^-} + \underset{\substack{\text{inorganic} \\ \text{phosphate (}P_i\text{)}}}{\text{H}_2\text{PO}_4^-} + \underset{\substack{\text{energy for activity} \\ \text{and body processes}}}{7.3 \text{ kcal/mole}}$$

or simply

$$\text{ATP} + \text{H}_2\text{O} \longrightarrow \text{ADP} + P_i + 7.3 \text{ kcal/mole}$$

When the concentration of ATP in the cells falls below a certain level, nutrients in the mitochondria are oxidized to supply the energy to reconvert ADP back to ATP, a process called *phosphorylation*, so that the storehouse of energy is replenished.

storing energy for later use:

$$\text{ADP} + P_i + 7.3 \text{ kcal/mole} \longrightarrow \text{ATP} + \text{H}_2\text{O}$$

from the oxidation of nutrients

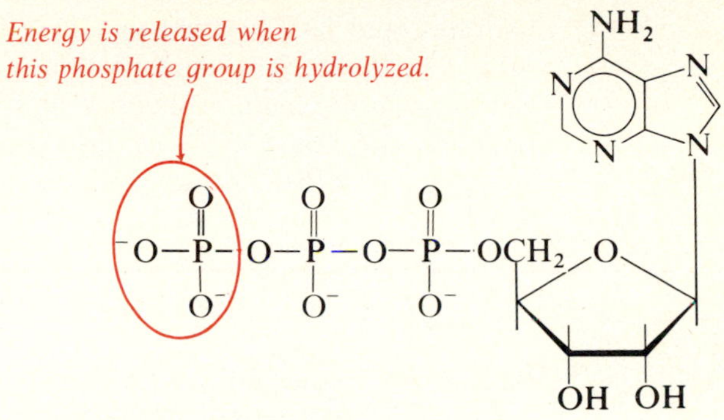

Figure 22.4. *The structure of adenosine triphosphate (ATP), a high-energy compound. In the body, the ATP anion is associated with magnesium ions, Mg^{2+}.*

Energy is released when this phosphate group is hydrolyzed.

Problem 22.1. The following equation shows ATP being hydrolyzed to adenosine monophosphate, AMP. Would this reaction be energy releasing or would it require energy?

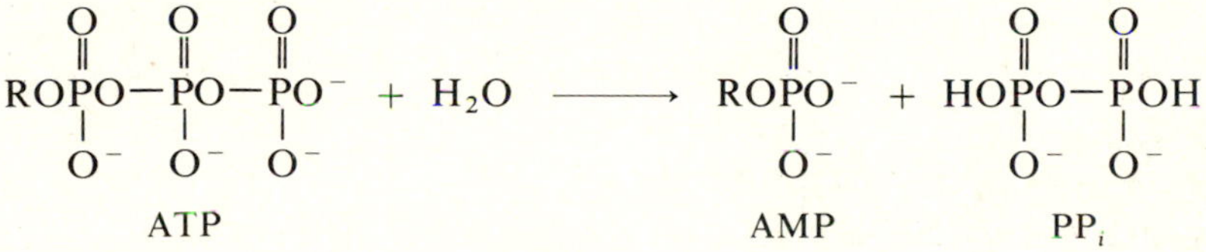

$$ROPO-PO-PO^- + H_2O \longrightarrow ROPO^- + HOPO-POH$$

ATP   AMP PP_i

Figure 22.5. *ADP is converted to ATP in many steps of the abbreviated metabolism schemes shown.*

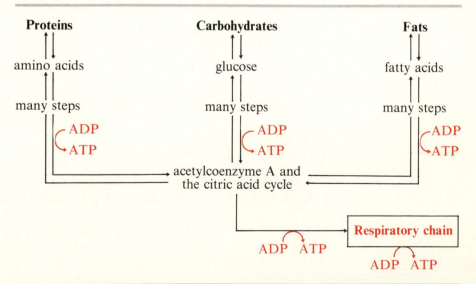

Figure 22.5 outlines the digestion and catabolism of proteins, carbohydrates, and fats, and shows where in the sequences ADP is converted to ATP. The lower right-hand portion of the figure shows the *respiratory chain*. The respiratory chain is involved with virtually all metabolism schemes, not only the citric acid cycle as shown in the figure. For this reason, we will consider it here, before discussing the metabolism of biological compounds.

22.4   THE CELLULAR RESPIRATORY CHAIN

Nutrient molecules do not react directly with oxygen in the mitochondria to yield energy that converts ADP to ATP. Instead, they are catabolized and oxidized by other cellular oxidizing agents. The hydrogen ions and electrons removed by these oxidizing agents are then transferred to a reaction sequence called the **cellular respiratory chain** or **electron-transport chain**, where oxygen accepts electrons and hydrogen ions and is reduced to water.

To see how this sequence of reactions occurs and to introduce the manner in which biochemical reactions are written, let us trace the results of the biological oxidation of a secondary alcohol.

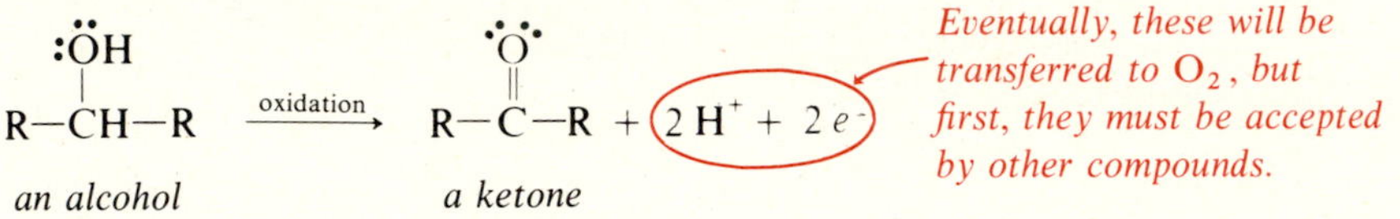

an alcohol   *a ketone*

Eventually, these will be transferred to O_2, but first, they must be accepted by other compounds.

In this oxidation, two hydrogens ions and a pair of electrons are removed from the alcohol. As in all oxidation-reduction reactions, the electrons must be accepted by an oxidizing agent. The principal electron acceptors in the respiratory chain are two coenzymes, **NAD⁺** and **FAD**. The structures of these compounds are shown in Figure 21.1, Section 21.3. Although NAD⁺ and FAD look complex, the reactions they undergo are not. Each of these coenzymes has a working end, which is the only portion of the structure we need to consider. The working end of NAD⁺ is nicotinamide, while that of FAD is riboflavin.

When an alcohol is oxidized, one hydrogen atom and two electrons (H:⁻) are transferred to NAD⁺, while the other hydrogen is lost as H⁺ to the medium. The reaction with actual structures is shown in Figure 22.6.

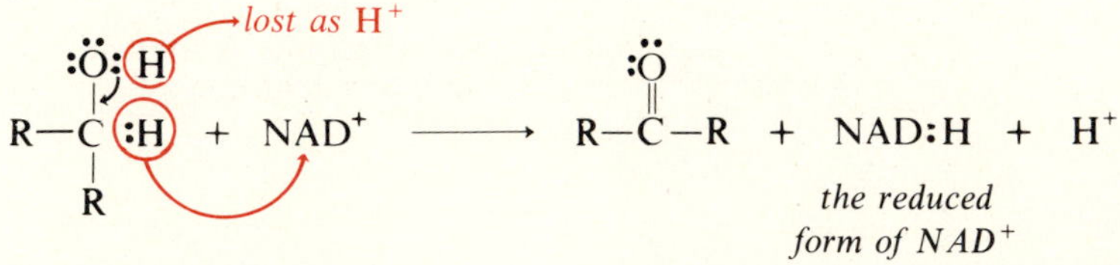

lost as H⁺

the reduced form of NAD⁺

We can describe biochemical processes more simply by depicting these reactions in an abbreviated form. The following symbols are taken to mean

Figure 22.6. *How NAD⁺ acts as an oxidizing agent, then is regenerated by the reaction of NADH with FAD.*

oxidation of an alcohol, in which NAD⁺ is reduced to NADH:

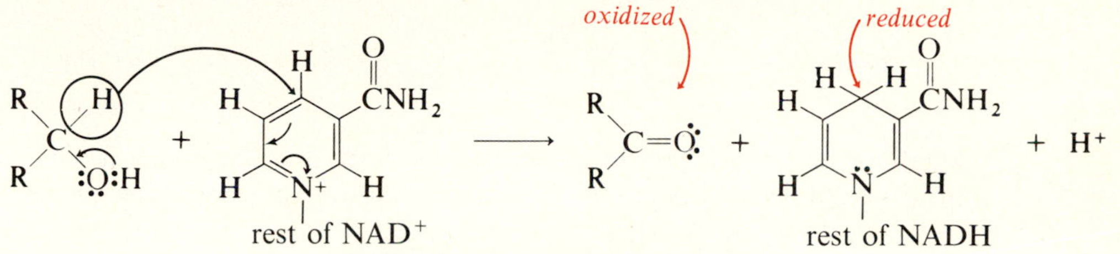

regeneration of NAD⁺ by oxidation of NADH with FAD:

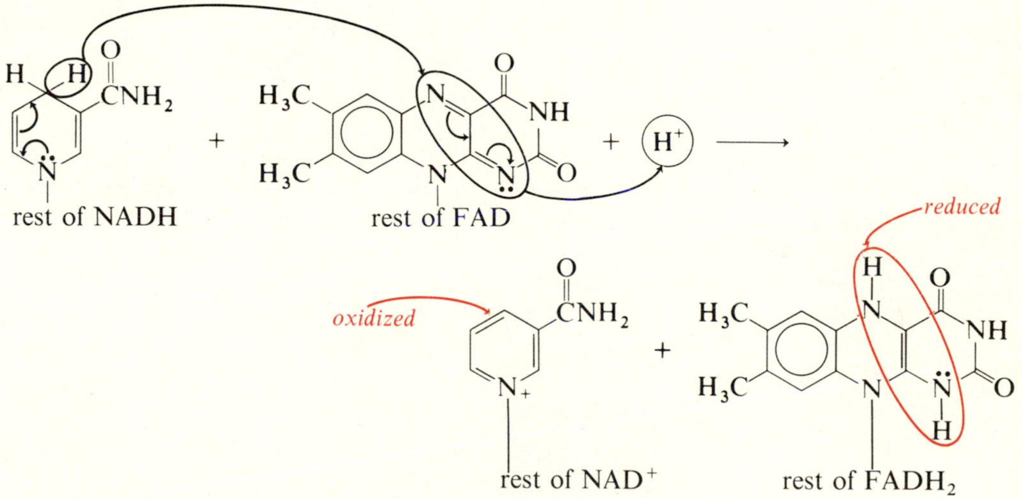

that when (1) the alcohol is oxidized to a ketone, (2) NAD⁺ is reduced to NADH.

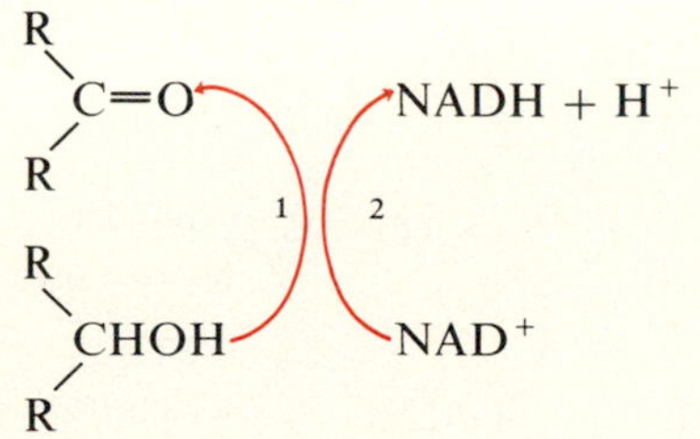

To keep the respiratory chain going, NADH must be oxidized back to NAD⁺. The oxidizing agent in this case is FAD. When FAD reacts with

NADH, the NADH is oxidized to NAD^+, and FAD is reduced to $FADH_2$. The net result is that the two electrons and two hydrogen atoms from the original alcohol are passed from NADH to FAD. Again, the actual structures are shown in Figure 22.6.

$$NAD:H + H^+ + FAD \longrightarrow \underset{oxidized}{NAD^+} + \underset{reduced}{FADH_2}$$

Symbolizing this reaction in an abbreviated form and combining it with our previous abbreviated equation, we have the following biochemical sequence:

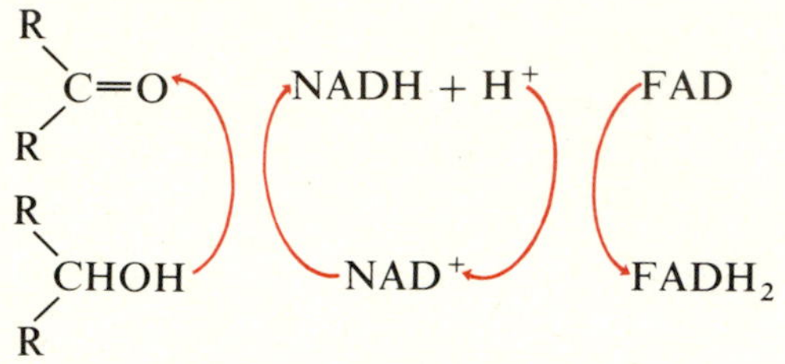

From $FADH_2$, the electrons and hydrogens pass to a compound called ubiquinone (also called coenzyme Q), and FAD is regenerated. From there, only the electrons are passed to compounds called *cytochromes*, which are closely related to hemoglobin in structure. One of the cytochrome enzymes, *cytochrome oxidase*, which contains both copper and iron ions, binds with O_2 so that it can be reduced. (In cyanide poisoning, the cyanide ion forms very stable complex ions with the metal ions; thus, the cytochrome becomes inactive.) When oxygen is reduced, the hydrogen ions produced earlier and the electrons are used in this final step to produce water.

$$\tfrac{1}{2}O_2 + 2\,H^+ + 2\,e^- \longrightarrow H_2O$$

Figure 22.7. *One branch of the respiratory chain, also called the electron-transport chain, showing where energy is released and ATP is synthesized from ADP.*

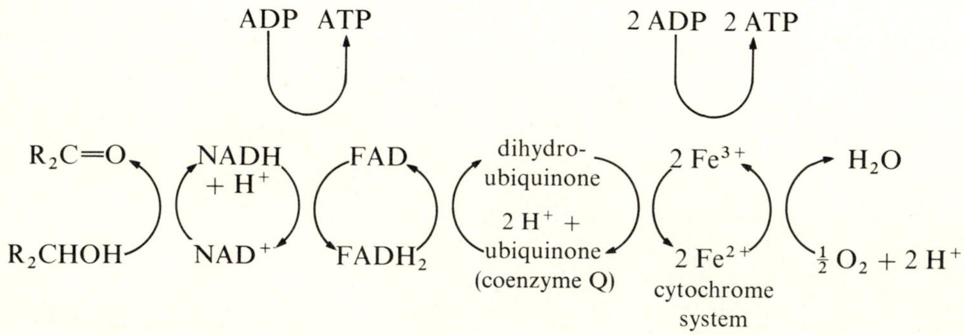

The entire sequence of reactions in the respiratory chain is shown in Figure 22.7. In the sequence, all the coenzymes are regenerated to begin the process anew.

Despite the complexity of the cellular respiratory chain, the net result is that oxygen absorbed by the cell has been used to oxidize an alcohol.

net equation:

$$\underset{\displaystyle R-\overset{\displaystyle OH}{\overset{|}{C}}H-R}{} + \tfrac{1}{2} O_2 \xrightarrow{\text{many steps}} R-\overset{\displaystyle O}{\overset{\|}{C}}-R + H_2O + energy$$

Problem 22.2. How many molecules of R_2CHOH must be oxidized to reduce one entire O_2 molecule?

ATP Synthesis in the Respiratory Chain

The oxidation of an alcohol to a ketone produces energy. This energy is trapped at two steps in the respiratory chain—in the conversion of NADH to NAD^+ and in the cytochrome system, as shown in Figure 22.7. For every H^+ and pair of electrons transported in one unit of NADH, three ATP units are generated from inorganic phosphate and ADP. We can write a net equation that shows three ATP units being produced for each NADH being oxidized. In later chapters, where NAD^+ will be encountered again, we will refer back to this ATP production.

storing energy gained in the respiratory chain:

$$NADH + \tfrac{1}{2} O_2 + H^+ + 3\ ADP + 3\ P_i \xrightarrow{\text{many steps}} NAD^+ + H_2O + 3\ ATP$$

Problem 22.3. Identify each of the following equations as an oxidation reaction, a reduction reaction, or neither.

(a) $Fe^{3+} \xrightarrow{\text{dihydroubiquinone}} Fe^{2+}$ (b) $ATP \rightarrow ADP$

(c) $FADH_2 \rightarrow FAD$ (d) $NAD^+ \rightarrow NADH$

SUMMARY

When food is ingested, it is digested, absorbed, carried to the cells, and metabolized:

metabolism: enzyme-catalyzed cellular reactions
catabolism: the breaking down of nutrient molecules, with the production of energy
anabolism: the building up of larger molecules, with the consumption of energy

A typical animal cell contains a *nucleus*, which contains chromatin granules (DNA-protein), and other *organelles* (endoplasmic reticulum, lysosomes, glycogen granules, mitochondria, etc.). The entire interior of the cell is called *protoplasm*. The interior of the cell excluding the nucleus is called *cytoplasm*. A cell is surrounded by a *cell membrane*, a phospholipid bilayer plus protein, which is instrumental in the active transport of ions and molecules to and from the cytoplasm.

Energy from the oxidation of nutrients is stored in cells by phosphorylation of the lower-energy ADP to the higher-energy *ATP*. When the cells need energy, ATP is hydrolyzed back to ADP, in the reverse of the following reaction.

$$ADP + P_i + energy \rightleftharpoons ATP + H_2O$$

When nutrients are oxidized, the H^+ and electrons produced are transferred to O_2 in the *cellular respiratory chain* (*electron-transport chain*). Two important coenzymes in this process are NAD^+ and FAD. For every NADH that enters the respiratory chain, three ATP units are produced.

KEY TERMS

digestion
metabolism
catabolism
anabolism
cell nucleus
protoplasm

cytoplasm
organelle
cell membrane
lipid bilayer
active transport
cellular respiration

ATP
cellular respiratory chain
NAD^+
FAD

STUDY PROBLEMS

22.4. Label each of the following conversions as digestion or metabolism:

(a) phe-tyr-phe-ala + H_2O ———→

phe-tyr-phe + ala

(b) phe-tyr-phe + H_2O ———→ 2 phe + tyr

(c) sucrose + H_2O ———→ glucose + fructose

(d) glucose ———→ $2\ CH_3\overset{O}{\overset{\|}{C}}-\overset{O}{\overset{\|}{C}}OH$

22.5. Label each of the following conversions as catabolism or anabolism:

(a) glucose ———→ glycogen

(b) glucose → $2\ CH_3\overset{O}{\overset{\|}{C}}-\overset{O}{\overset{\|}{C}}OH$

(c) 2 phe + tyr ———→ phe-tyr-phe

22.6. *Escherichia coli* (*E. coli*) is
(a) an organelle in an animal cell.
(b) a species of bacteria.
(c) part of a cell membrane.

22.7. Define the following parts of an animal cell, and describe the functions of (c)–(f):
(a) protoplasm   (b) cytoplasm
(c) mitochondrion   (d) lysosome
(e) endoplasmic reticulum (f) nucleus

22.8. Describe the structure of a cell membrane. Include in your answer the definition of the lipid bilayer.

22.9. Can aqueous solutions pass through the lipid bilayer? Why or why not?

22.10. Differentiate between dialysis and active transport through a membrane.

22.11. Why can penicillin inhibit bacterial growth without harming human cells?

22.12. In which of the following reactions is energy liberated? In which is energy consumed?

(a)
$$ROP\overset{O}{\overset{\|}{-}}O\overset{O}{\overset{\|}{-}}P\overset{O}{\overset{\|}{-}}PO^- + H_2O \longrightarrow$$
with O^-, O^-, O^- substituents

$$ROP\overset{O}{\overset{\|}{-}}O\overset{O}{\overset{\|}{-}}PO^- + H_2PO_4^-$$
with O^-, O^- substituents

(b)
$$ROP\overset{O}{\overset{\|}{-}}O\overset{O}{\overset{\|}{-}}PO^- + H_2PO_4^- \longrightarrow$$
with O^-, O^- substituents

$$ROP\overset{O}{\overset{\|}{-}}O\overset{O}{\overset{\|}{-}}P\overset{O}{\overset{\|}{-}}O\overset{O}{\overset{\|}{-}}PO^- + H_2O$$
with O^-, O^-, O^- substituents

22.13. (a) Rewrite the equations in the preceding problem, using the terms ATP, ADP, and P_i.

(b) Put the word *energy* on the appropriate side of each equation.

(c) Tell which reaction occurs when nutrients are oxidized in the mitochondria.

(d) Tell which reaction occurs when the cell needs energy.

22.14. Classify each of the following conversions as oxidation or reduction.

(a) $NADH \rightarrow NAD^+$

(b) $FAD \rightarrow FADH_2$

(c)
$$\begin{array}{c} CH_2CO_2H \\ | \\ CHCO_2H \\ | \\ HOCHCO_2H \end{array} \longrightarrow \begin{array}{c} CH_2CO_2H \\ | \\ CHCO_2H \\ | \\ O{=}CCO_2H \end{array}$$

citric acid   oxalosuccinic acid

22.15. Complete the following equations:

(a)
$$HOCCH_2C\overset{OH}{\underset{|}{\overset{|}{H}}}{-}COH + NAD^+ \longrightarrow$$

$$HOCCH_2C{-}COH + \underline{\quad} + \underline{\quad}$$

(b) $NADH + H^+ + FAD \longrightarrow$

22.16. Rewrite the following diagram as an ordinary chemical equation:

$$\begin{array}{cc} H & CO_2H \\ \diagdown & \diagup \\ & C{=}C \\ \diagup & \diagdown \\ HO_2C & H \end{array}$$

$$HO_2CCH_2CH_2CO_2H$$

with FADH$_2$ / FAD

22.17. In a typical biological oxidation, the oxidizing agent NAD^+ is reduced to NADH. NADH enters the respiratory chain and is reoxidized to NAD^+. The energy released in this reaction sequence converts ____ units of ADP to ATP.

23 Metabolism of Carbohydrates

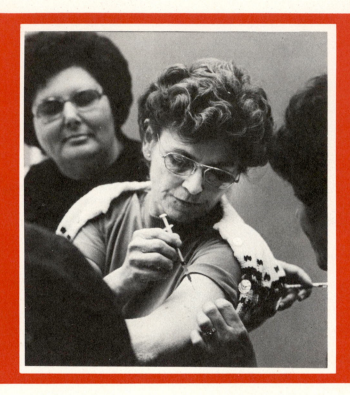

A diabetic learning to inject herself with an insulin solution. (© Jeffrey Grosscup.) In adults, diabetes mellitus is the third leading cause of death; only cancer and cardiovascular disease claim more victims. Diabetes is also the leading cause of blindness in adults. A diabetic either lacks insulin, a hormone necessary for proper glucose metabolism, or has insulin-resistant cell receptors. Some diabetics can live nearly normal lives by carefully balancing their dietary carbohydrates and insulin injections. Because insulin must be supplied fairly regularly, most diabetics learn to give their own injections.

Objectives   Describe the digestion of carbohydrates. ☐ Define glycogenesis, glycogenolysis, glycolysis, and gluconeogenesis. ☐ Describe the conversion of glucose to pyruvic acid. ☐ Write the equations for the conversion of pyruvic acid to lactic acid under anaerobic conditions and to acetyl CoA under aerobic conditions; compare the energy yields of the two conversions. ☐ Define the citric acid cycle and describe its energy yield. ☐ Define the pentose phosphate path and its importance (optional). ☐ Define blood glucose level and glucose tolerance (optional). ☐ List the hormones that control the blood glucose level, and describe their action (optional). ☐ Define and describe the symptoms of hyperglycemia, hypoglycemia, and galactosemia (optional).

Living cells obtain most of their energy from the oxidation of glucose from dietary carbohydrates. The mammalian brain depends almost entirely on glucose for its energy. Glucose is so important to an animal system that, if the animal is subjected to starvation, muscle protein will be depleted and converted to glucose so that the blood glucose level can be maintained.

In this chapter, we will first discuss the digestion of carbohydrates. We will then turn our attention to a discussion of the catabolism of glucose. We will also discuss how the blood glucose level is controlled and some of the pathological conditions that arise from errors in carbohydrate metabolism.

23.1   DIGESTION AND FATE OF CARBOHYDRATES

The digestion of starches takes place in two stages. In the mouth, starches are hydrolyzed to a mixture of small polymeric carbohydrates, called *limit dextrins*, and to maltose by the salivary enzyme *amylase* (ptyalin). Then, in the small intestine (not the stomach), digestion is completed by pancreatic and intestinal enzymes. The intestines also secrete the enzymes sucrase and lactase, which catalyze the hydrolysis of sucrose and lactose. The end products of the digestion of starch and sugars are monosaccharides (primarily glucose), which can be absorbed into the bloodstream through the intestinal wall.

$$\text{starch} \xrightarrow[\text{(mouth)}]{\text{amylase}} \text{dextrins} + \text{maltose} \xrightarrow[\text{(small intestine)}]{\text{maltase and dextrinase}} \text{glucose}$$

Once glucose is in the bloodstream, it can be carried to the cells to provide energy and organic compounds needed for other metabolic sequences. If the blood and cells contain sufficient glucose, excess glucose is carried to the liver, kidney, and muscles, where it is polymerized to *glycogen* (Section 18.8) in the process called **glycogenesis** and is stored until needed. When the body needs glucose, enzymes break the glycogen back down to glucose (**glycogenolysis**).

$$\text{glycogen} \underset{\text{glycogenesis}}{\overset{\text{glycogenolysis}}{\rightleftharpoons}} \text{glucose} \xrightarrow{\text{catabolism}}$$

smaller molecules, CO_2, H_2O, and energy

At least seven diseases have been identified as arising from abnormalities in the mechanism of glycogen formation, storage, and hydrolysis. In one of these diseases, Von Gierke's glycogen-storage disease, the patient is deficient in the enzyme that catalyzes one step in the hydrolysis of liver glycogen to glucose. A similar condition arises when a glycogen-debranching enzyme is missing. In these cases, the blood glucose level drops, while the liver becomes enlarged from excess glycogen.

23.2 CATABOLISM OF GLUCOSE TO PYRUVIC ACID

Two principal paths of glucose catabolism in muscle tissue are the **anaerobic** (without oxygen) and the **aerobic** (with oxygen) sequences. These sequences are summarized in Figure 23.1. Note from the figure that glucose is first converted to pyruvic acid in both the anaerobic and aerobic sequences. (At the pH of the body fluids, pyruvic acid actually exists primarily as its anion, the pyruvate ion.)

Under anaerobic conditions, pyruvic acid is converted to lactic acid (or the lactate ion). The reaction sequences that convert glucose to pyruvic acid or lactic acid are collectively called *glycolysis*.

Figure 23.1.   *The catabolism of glucose in muscle tissue.*

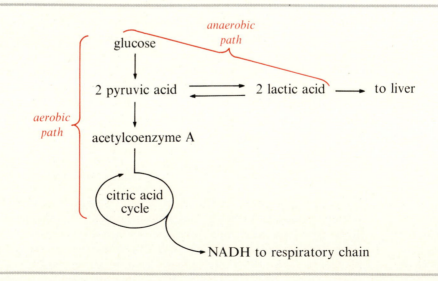

Glycolysis is the catabolism of glucose to pyruvic acid or lactic acid:

$$\text{glucose} \xrightarrow[\text{many steps}]{\text{glycolysis}} 2 \underset{\text{pyruvic acid}}{CH_3 \overset{O}{\underset{||}{C}} - \overset{O}{\underset{||}{C}}OH} \quad \text{or} \quad 2 \underset{\text{lactic acid}}{CH_3 \overset{OH}{\underset{|}{C}}H - \overset{O}{\underset{||}{C}}OH}$$

In this section, we will discuss only the initial conversion of glucose to pyruvic acid. In Section 23.3, we will discuss the anaerobic conversion of pyruvic acid to lactic acid. Then, in Section 23.4, we will discuss the catabolism of pyruvic acid under aerobic conditions to yield CO_2 and H_2O.

Figure 23.2 shows the actual reaction sequence that converts glucose to pyruvic acid under either aerobic or anaerobic conditions. Recall that energy is stored in the high-energy compound ATP. You can see in Figure 23.2 that energy is required to initiate the glycolysis sequence. In the very first reaction, one ATP unit loses a phosphate group to glucose.

step 1:

$$\text{glucose} + \text{ATP} \rightleftharpoons \text{glucose 6-phosphate} + \text{ADP}$$

Then, glucose 6-phosphate is isomerized to fructose 6-phosphate. This compound then removes a phosphate group from a second unit of ATP.

step 2:

$$\text{glucose 6-phosphate} \rightleftharpoons \text{fructose 6-phosphate}$$

step 3:

$$\text{fructose 6-phosphate} + \text{ATP} \rightleftharpoons \text{fructose 1,6-diphosphate} + \text{ADP}$$

Thus, two ATP units are converted to ADP at the start of the sequence.

The fructose diphosphate is cleaved into two three-carbon molecules (step 4), which are both converted to a diphosphate of glyceric acid (step 5). The second phosphate group in this case does not come from ATP but from inorganic phosphate ions. In steps 6 and 7, the conversion of two molecules of the glyceric acid diphosphate to pyruvic acid, the two phosphate groups bonded to glyceric acid are used to convert four units of ADP back to the higher-energy ATP. Thus, a net gain of two ATP units is realized in the entire sequence.

steps 6 and 7:

$$2 \text{ glyceric acid diphosphate} + 4 \text{ ADP} \rightleftharpoons 2 \text{ pyruvic acid} + 4 \text{ ATP}$$

Note that NAD^+ is reduced to NADH in the sequence (step 5). Under aerobic conditions, these NADH units can enter the respiratory chain to yield additional energy.

Figure 23.2. *Conversion of glucose to two molecules of pyruvic acid.* (P) *in these formulas means a phosphate group,* $-PO_3^{2-}$.

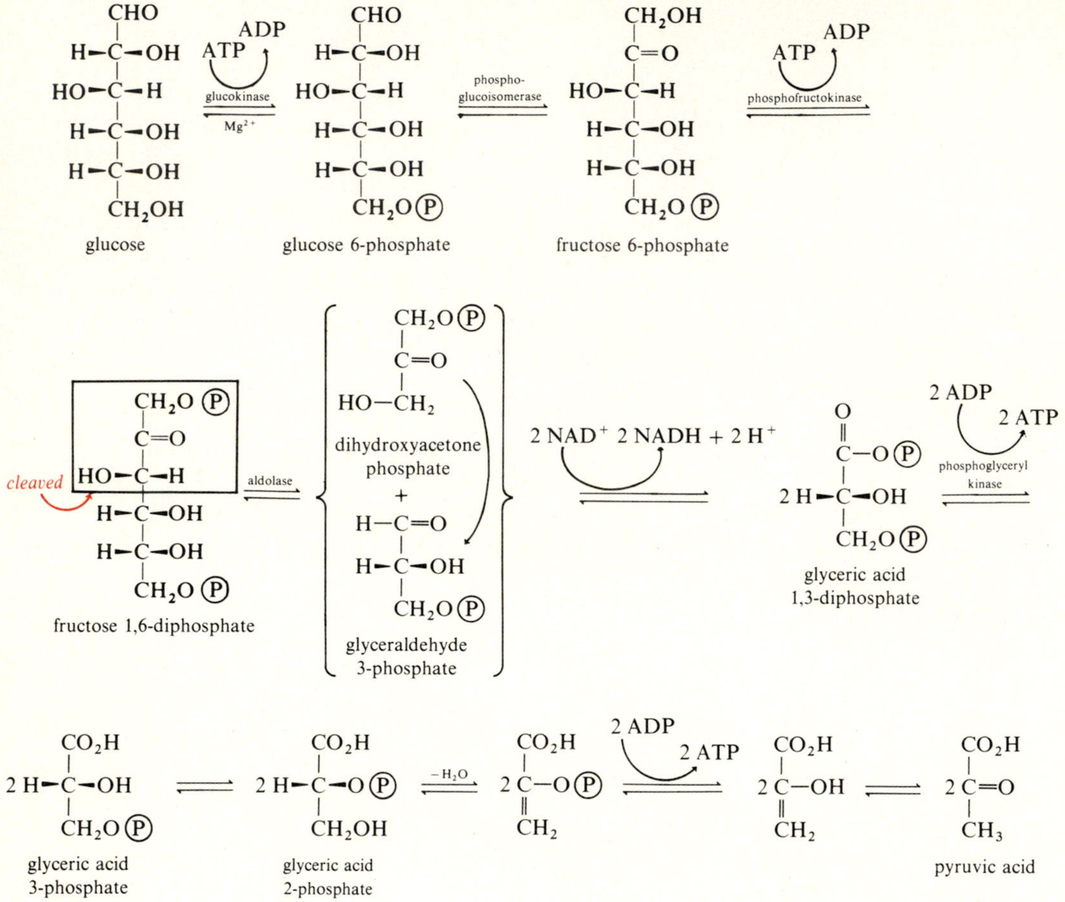

or, in words,

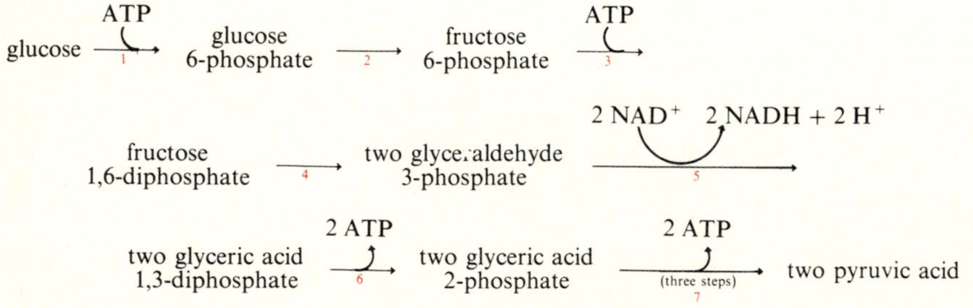

Problem 23.1.   The following equations represent standard organic reactions. Referring to Figure 23.2, identify where these types of reactions (enzyme catalyzed) occur in the steps of the glycolysis of glucose to pyruvic acid.

(a) $\underset{\displaystyle |}{\overset{\displaystyle OH}{}}$
$RCHCH_2R \xrightarrow[\text{heat}]{H_2SO_4} RCH{=}CHR + H_2O$

(b) $\underset{\displaystyle \|}{\overset{\displaystyle O}{}}$
$RCH \xrightarrow[\text{agent}]{\text{oxidizing}} \underset{\displaystyle \|}{\overset{\displaystyle O}{}}RCOH \quad \text{or} \quad \underset{\displaystyle \|}{\overset{\displaystyle O}{}}RCO^-$

(c) $\underset{\displaystyle \|}{\overset{\displaystyle O}{}}$
$RCCH_2OH \xrightarrow[\text{H}_2\text{O}]{\text{OH}^-} \underset{\displaystyle |\quad\|}{\overset{\displaystyle OH\ \ O}{}}RCH{-}CH$

23.3   ANAEROBIC CATABOLISM OF PYRUVIC ACID TO LACTIC ACID

The anaerobic path of glucose catabolism, also called the *Embden-Meyerhof path*, is used by the body in times of physical stress (such as when one is running fast), when energy is needed but not enough oxygen is available for aerobic oxidation. In this pathway, the NADH produced in the formation of pyruvic acid cannot enter the respiratory chain, which requires oxygen. Instead, NAD^+ is regenerated by the reduction of pyruvic acid to lactic acid. No ATP is generated in this oxidation-reduction reaction.

$$2\ \underset{\text{pyruvic acid}}{\underset{\displaystyle \|}{\overset{\displaystyle O}{}} CH_3CCO_2H} + 2\ NADH + 2\ H^+ \longrightarrow 2\ \underset{\text{lactic acid}}{\underset{\displaystyle |}{\overset{\displaystyle OH}{}} CH_3CHCO_2H} + \underset{\textit{regenerated}}{2\ NAD^+}$$

The production of lactic acid by anaerobic glycolysis is a dead end as far as muscular energy is concerned. Anaerobic glycolysis has a low energy yield compared to that of the aerobic oxidation pathway, to be discussed shortly.

anaerobic glycolysis:

$$\text{glucose} \xrightarrow{\overset{\displaystyle \text{2 ADP} \quad \text{2 ATP}}{\curvearrowright}} \text{2 lactic acid}$$

In the anaerobic sequence, lactic acid builds up in the muscle cells. This buildup is part of an *oxygen debt*, or insufficient oxygen. Heavy breathing and cessation of exercise restore the system to normal. The lactic acid is then carried by the blood to the liver, where part of the lactic acid is oxidized to CO_2 and H_2O. Enough energy is produced by this oxidation to reconvert the remaining lactic acid to glucose and then to glycogen. The conversion of small noncarbohydrate molecules, such as lactic acid, to glucose is called *gluconeogenesis*, which literally means the creation of new glucose.

Gluconeogenesis is the synthesis of glucose from small noncarbohydrate molecules. For example,

$$\text{lactic acid} \xrightarrow[\text{gluconeogenesis}]{\text{liver}} \text{glucose} + CO_2 + H_2O$$

23.4   AEROBIC CATABOLISM OF PYRUVIC ACID TO CARBON DIOXIDE

As in the anaerobic sequence, the first step in the aerobic oxidation of glucose is its conversion to pyruvic acid. In the aerobic sequence, the two NADH units produced are sent to the respiratory chain, a transfer that consumes two ATP units but produces six. The result is the net conversion of four ADP units to ATP. Thus, in the overall aerobic glycolysis to pyruvic acid, a net gain of six ATP units is realized: two from the glycolysis, and four from two NADH entering the respiratory chain.

aerobic glycolysis:

$$\text{glucose} \xrightarrow[]{6\ ADP \quad 6\ ATP} \text{2 pyruvic acid}$$

In the aerobic path, the pyruvic acid is converted to *acetylcoenzyme A*, which then enters the *citric acid cycle*. In these reactions, the pyruvic acid is eventually converted to CO_2, H_2O, and additional energy.

Conversion of Pyruvic Acid to Acetylcoenzyme A

In the aerobic metabolism path, pyruvic acid undergoes a complex series of reactions with the thiol coenzyme A (Figure 17.1, Section 17.8) to yield the thioester acetylcoenzyme A. This reaction sequence involves an oxidation of the $-CO_2H$ group of pyruvic acid by NAD^+ to yield CO_2 and NADH.

lost as CO_2 and H^+

$$\underset{\substack{\uparrow \\ \textit{the acetyl group}}}{CH_3}\overset{\overset{O}{\|}}{C}-CO_2H \quad + \quad \underset{\substack{\text{coenzyme A} \\ \textit{a thiol}}}{HSCoA} \quad + \quad NAD^+ \xrightarrow{\overset{Mg^{2+}}{\text{several enzymes}}}$$

$$\underset{\substack{\text{acetylcoenzyme A} \\ \textit{a thioester}}}{CH_3\overset{\overset{O}{\|}}{C}-SCoA} \quad + \quad CO_2 \quad + \quad \boxed{NADH} \quad + \quad H^+$$

to respiratory chain
(3 ADP $\longrightarrow$ 3 ATP)

Each NADH molecule formed in this reaction enters the respiratory chain

Figure 23.3.   *The citric acid cycle.*

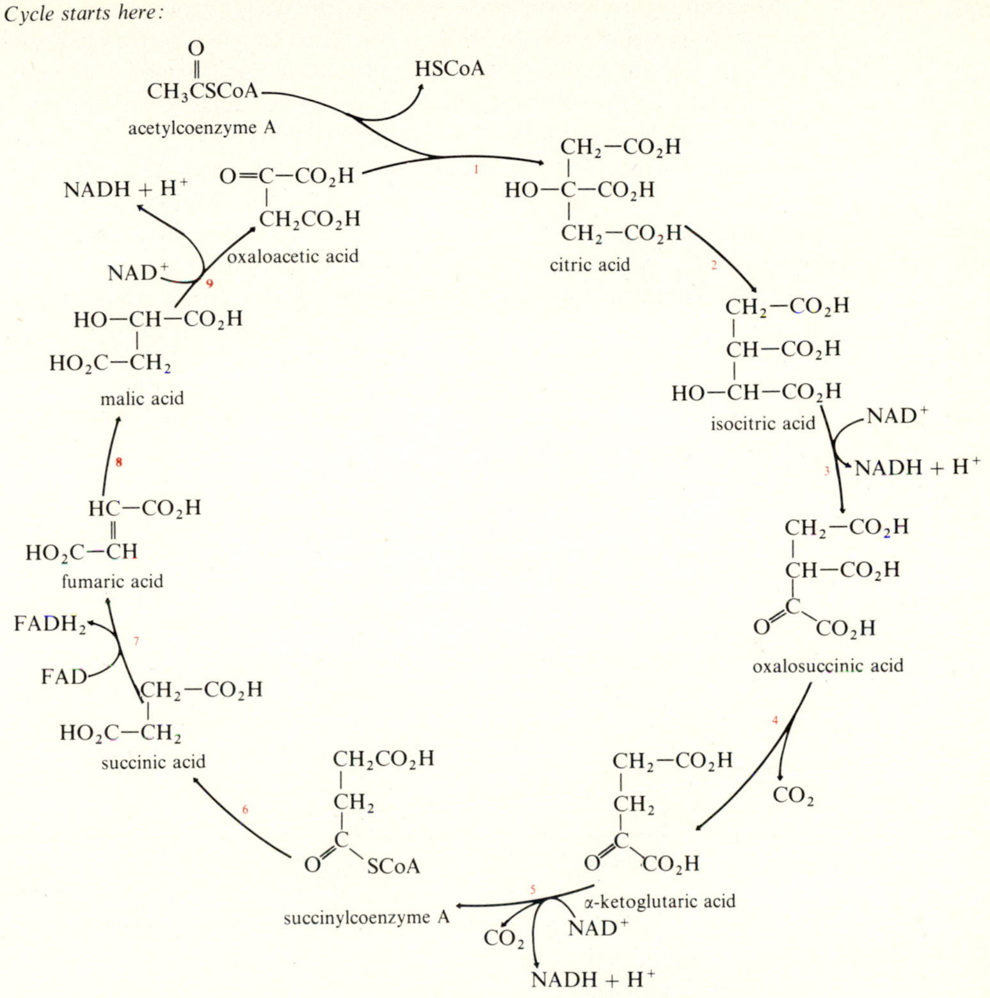

Cycle starts here:

and results in the conversion of three ADP units to ATP. Since one glucose molecule yields two pyruvic acid molecules, a total of six ATP units is produced at this step.

Citric Acid Cycle

Aside from the respiratory chain, the final phase in the aerobic catabolism of glucose is a series of enzyme-catalyzed reactions known as the **citric acid cycle**, the **Krebs cycle**, or the **tricarboxylic acid cycle**. This cycle is outlined in Figure

23.3. In this cycle, the remaining four carbons from the glucose, in the form of two acetyl groups, are oxidized to carbon dioxide.

Proceeding through the cycle stepwise, the acetyl group from the acetyl-coenzyme A is transferred to oxaloacetic acid, to yield citric acid (step 1), which is isomerized and oxidized to oxalosuccinic acid (steps 2 and 3). This acid undergoes loss of CO_2 (step 4), and the product, α-ketoglutaric acid, undergoes oxidation and loss of another molecule of CO_2 (step 5). The rest of the cycle is involved in converting the succinyl group back to oxaloacetic acid so that another acetylcoenzyme A can be accepted and oxidized.

A total of 12 ATP units is produced in the citric acid cycle for each acetylcoenzyme A unit, or a total of 24 ATP units for each unit of glucose. Table 23.1 sums up the ATP production of the entire aerobic glucose-catabolism path.

Table 23.1.   *Total Energy Yield from the Aerobic Catabolism of Glucose*

Reaction	ADP $\longrightarrow$ ATP
glucose $\longrightarrow$ 2 pyruvic acid	6
2 pyruvic acid $\longrightarrow$ 2 acetylcoenzyme A + 2 CO_2	6
2 acetylcoenzyme A $\longrightarrow$ 2 coenzyme A + 4 CO_2	24
Total ATP	36

Problem 23.2.   Which step(s) in the citric acid cycle is
(a) an oxidation of an organic compound?
(b) a reduction of an organic compound?
(c) hydration of an alkene?

23.5   PENTOSE PHOSPHATE PATH (optional)

In addition to the anaerobic and aerobic pathways just described, glucose can follow other minor metabolic paths. One of these is the **pentose phosphate path**, or **phosphogluconate shunt**:

$$\text{glucose 6-phosphate} + 2\ NADP^+ + H_2O \xrightarrow{\text{many steps}}$$

$$\text{ribose 5-phosphate} + CO_2 + 2\ NADPH + 2\ H^+ + \text{energy}$$

The importance of this sequence lies in the conversion of $NADP^+$ (a phosphate of NAD^+) to NADPH (a phosphate of NADH). NADPH is necessary in the biosynthesis of fatty acids. It also helps prevent red blood cells from undergoing hemolysis by protecting the fatty acid groups in the cell membrane. Besides producing NADPH, the pentose phosphate path provides a route to pentoses, such as ribose, which are used in the biosynthesis of nucleic acids and ATP.

■   **23.6   BLOOD GLUCOSE LEVEL (optional)**

Many important tissues of the body, such as the brain, blood cells, and kidney, obtain virtually all of their energy from the oxidation of glucose; therefore, it is imperative that blood glucose levels be maintained. Clinically, the blood glucose level is defined as the milligrams of glucose per 100 mL of whole blood. The normal fasting blood glucose level (after 12 hours with no food) is about 60–100 mg/100 mL.

$$\text{blood glucose level} = \frac{\text{mg glucose}}{\text{100 mL blood}}, \text{ also expressed as mg\% or mg/dL}$$

The rates of glucose metabolism reactions, and thus the blood glucose level, are controlled by a variety of hormones secreted by endocrine glands.

Insulin, a pancreatic hormone, accelerates the oxidation of glucose, the formation of glycogen, and the transport of glucose into the cells. Thus, insulin acts to *lower the blood glucose level*. In a normal person, insulin is secreted as needed to remove excess glucose from the blood.

Epinephrine (also called *adrenaline*) is a hormone secreted by the adrenal glands in times of stress, fear, or danger. Epinephrine catalyzes hydrolysis of glycogen to glucose in the muscles and the liver, and *increases the blood glucose level*. This is just one of the ways in which epinephrine energizes the body to deal with a stressful situation.

Glucagon is a pancreatic hormone that stimulates liver glycogenolysis, but not muscle glycogenolysis, and thus also serves to increase blood glucose levels.

A few other hormones, such as thyroxine (a thyroid hormone) and some steroids, can also increase blood glucose levels. In the normal person, all these hormones work in concert to maintain the blood glucose within its proper range.

Glucose Tolerance

A **glucose tolerance test** can show if the body utilizes glucose normally. After a normal person ingests 100 or more grams of carbohydrates, the blood glucose level rises. Then, as the pancreas supplies insulin to the bloodstream, the blood glucose level drops sharply. Finally, as the insulin is consumed by body processes, the blood glucose level returns to a normal fasting level, as shown in the curve labeled normal in Figure 23.4.

Diabetes mellitus is a metabolic defect that is partially inherited. Approximately 3% of the population, principally the elderly, have some form of diabetes. A juvenile-onset diabetic has a defective pancreas that does not

Figure 23.4.   Blood glucose levels in glucose tolerance tests. (Shaded area indicates the normal fasting blood glucose range.)

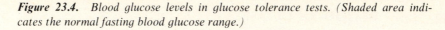

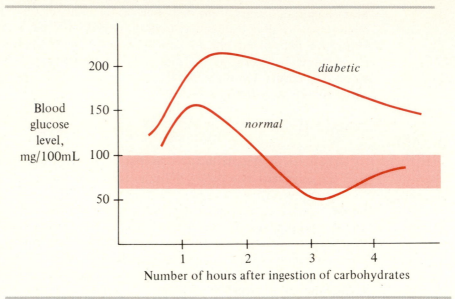

produce enough insulin to stimulate glucose metabolism; insulin must be given by injection. An adult-onset diabetic may also have pancreatic damage or may have developed insulin-resistant cell receptors; sometimes diet alone can control this type of diabetes. Figure 23.4 shows how the blood glucose level of an untreated diabetic remains very high after ingestion of carbohydrates.

Hyperglycemia (*hyper*, high; *gly*, sugar; and *-emia*, of the blood), a high blood glucose level, arises normally after ingestion of a high-carbohydrate meal. A person with diabetes mellitus that is not controlled by diet, insulin, or other drugs has a chronically high blood glucose level. The term *renal threshold* refers to the upper limit of a substance's concentration in the blood before the substance spills over into the urine. The renal threshold of glucose is approximately 150 mg/100 mL. A person with a blood glucose level higher than this exhibits *glucosuria*, or glucose in the urine.

Hypoglycemia (*hypo*, low), an abnormally low level of blood glucose, is characterized by dizziness and light-headedness. The symptoms arise because the brain cells are starved for glucose and thus do not have enough energy to function properly. An extreme case of hypoglycemia occurs when a diabetic has administered insulin but has not eaten sufficient carbohydrates for the insulin to work on. In this case, blood glucose levels become extremely low. The result is *insulin shock*, characterized by fainting, convulsions, shock symptoms, and even death.

Example   A *diabetic coma* can arise from extreme hyperglycemia. How would emergency treatment for a diabetic coma differ from that for insulin shock?

Solution   A person in a diabetic coma needs *insulin* to rid the blood of excess glucose. A person with insulin shock needs additional *glucose* in the blood. ■

■ 23.7 METABOLISM OF GALACTOSE (optional)

Lactose (milk sugar) is a disaccharide of glucose and galactose (see Section 18.7). When lactose is digested, it is hydrolyzed by the intestinal enzyme *lactase* to glucose and galactose. In a normal metabolism, galactose is then phosphorylated, converted to glucose, and metabolized in the usual ways. In the following equations, UDP means *uridine diphosphate*, which is a compound similar to ADP.

galactose 1-phosphate + UDP-glucose $\xrightarrow{\text{UDP-galactose transferase}}$

glucose 1-phosphate + UDP-galactose

$\xrightarrow{\text{many steps}}$ glucose 1-phosphate + UDP

Galactosemia is a hereditary disease observed in about 0.005% of infants and characterized by high levels of galactose in the blood and urine. The symptoms include vomiting when the infant is fed milk, failure to gain weight, and eventually dwarfism, mental retardation, and sometimes death. The cause of galactosemia is a deficiency of the enzyme *UDP-galactose transferase*, which catalyzes the first step in the conversion of galactose 1-phosphate to glucose. The treatment for this condition is the restriction of milk from the diet of affected infants. As these children grow, they may develop alternative metabolic paths so that they can metabolize galactose. ■

SUMMARY

Starches are converted to glucose when they are digested. Glucose is converted in the liver and muscles to glycogen (*glycogenesis*), which can be hydrolyzed to glucose as needed (*glycogenolysis*).

In the cells, glucose is catabolized to smaller molecules and energy (*glycolysis*). Some of these small molecules can be reconverted to glucose in the liver (*gluconeogenesis*).

In the two principal paths of glycolysis, glucose is converted to pyruvic acid. In the *anaerobic path*, which produces only a small amount of energy

(2 ADP → 2 ATP), pyruvic acid is reduced to lactic acid. In the *aerobic path*, which produces a much larger amount of energy (36 ADP → 36 ATP), the pyruvic acid is converted to CO_2 and acetyl groups in acetylcoenzyme A. The acetyl groups are further oxidized to CO_2 and H_2O in the *citric acid cycle*. The *pentose phosphate path* is a minor glucose metabolism path that leads to pentoses and NADPH.

The *blood glucose level* (usually reported as mg/100 mL) is controlled by hormones such as *insulin*, *epinephrine*, and *glucagon*. A diabetic has insufficient levels of insulin or has resistant cell receptors, and has a poor *glucose tolerance*; the blood glucose levels are higher than normal, because the glucose is not used. A high blood glucose level is called *hyperglycemia*, while a low blood glucose level is called *hypoglycemia*.

Galactosemia is a condition in which galactose is not converted to glucose and thus cannot be metabolized.

KEY TERMS

anaerobic catabolism	citric acid cycle	*glucagon
aerobic catabolism	*pentose phosphate path	*glucose tolerance
glycolysis	*blood glucose level	*hyperglycemia
gluconeogenesis	*insulin	*hypoglycemia
acetylcoenzyme A	*epinephrine	*galactosemia

STUDY PROBLEMS

23.3. Describe in words how starch is digested and the various general metabolic paths available to the digested starch.

23.4. Define the following terms:
(a) glycogenesis   (b) glycogenolysis
(c) glycolysis   (d) glyconeogenesis

23.5. Differentiate between (a) the aerobic catabolism of glucose and (b) the anaerobic catabolism of glucose.

23.6. Tell how many units (net) of ADP are converted to ATP as a result of each of the following sets of reactions:

(a) glucose $\xrightarrow{\text{aerobic}}$ 2 pyruvic acid

(b) glucose $\xrightarrow{\text{anaerobic}}$ 2 pyruvic acid
$\longrightarrow$ 2 lactic acid

23.7. Tell how NAD^+ is regenerated from NADH (a) in the aerobic catabolism of glucose

to pyruvic acid and (b) in the anaerobic path. Use an equation in your answer to (b).

23.8. Complete the following equation for the formation of acetylcoenzyme A:

$$CH_3\overset{\overset{\displaystyle O}{\|}}{C}CO_2H + HSCoA + NAD^+ \longrightarrow$$

23.9. In the conversion of pyruvic acid to acetylcoenzyme A, does the system lose or gain energy? How much energy in terms of ATP units?

23.10. Referring to Figure 23.3 but not to the textual description, tell whether each of the following conversions is an oxidation, a reduction, an isomerization, an addition of H_2O, or a decarboxylation (loss of CO_2):
(a) citric acid → isocitric acid
(b) isocitric acid → oxalosuccinic acid
(c) succinic acid → fumaric acid

(d) fumaric acid → malic acid

(e) malic acid → oxaloacetic acid

*23.11. List two important results of the pentose phosphate path of glucose catabolism.

*23.12. What is the effect on an animal organism if its system contains no insulin?

*23.13. Under times of stress, epinephrine is secreted into the bloodstream by the adrenal gland. What is the effect of this secretion on the blood glucose level?

*23.14. Why should a person fast for several hours before a blood glucose test is carried out?

*23.15. Define the following terms:

(a) hypoglycemia   (b) hyperglycemia

(c) glucosuria   (d) renal threshold

*23.16. Which of the labeled steps in the following sequence is abnormal in galactosemia?

$$\text{lactose} \xrightarrow[\text{(a)}]{} \text{galactose} \xrightarrow[\text{(b)}]{\overset{\text{ATP}}{\text{galactokinase}}}$$

$$\text{galactose 1-phosphate} \xrightarrow[\text{(c)}]{\text{many steps}}$$

$$\text{glucose 1-phosphate} \xrightarrow[\text{(d)}]{\text{many steps}} \text{pyruvic acid}$$

24 Metabolism of Lipids

Human fat cells. (Martin M. Rotker/Taurus Photos.) Triacylglycerols are stored in fat cells, which form the adipose tissues of the body. Cellular triacylglycerols are in dynamic equilibrium with the blood lipids and thus are in a continuous state of being absorbed, metabolized, and redeposited. Chapter 24 describes some of the metabolism reactions of triacylglycerols.

Objectives   Write equations that describe lipolysis. ☐ Describe the role of bile salts in lipolysis. ☐ Outline the transport of fatty acids from the intestines to the liver. ☐ Describe the catabolism of glycerol, and outline the fatty acid oxidation cycle. ☐ List the metabolic origins of ketone-body production, and outline the formation of these compounds (optional). ☐ Define ketosis, and list its symptoms (optional). ☐ Define and describe the process of lipogenesis (optional).

An animal's fatty deposits, or adipose tissues, provide padding, insulation, and a very efficient energy reserve. Adipose tissues and other lipids, such as phospholipids, are not stationary deposits; half of the total lipid content in the human body is replaced about every 70 days.

Recall from Chapter 19 that animal fats and vegetable oils are *triacylglycerols* (triglycerides), esters formed from the triol glycerol and long-chain fatty acids. Triacylglycerols are the most important class of lipids in our diet and are found in almost everything we eat—nuts, meat, milk, cereals, and so forth. Dietary triacylglycerols provide the raw materials for replacement of the adipose tissue and for the biosynthesis of many other lipids, such as phosphoglycerides and prostaglandins.

In Chapter 19, we discussed the structures of fats. Here, we will consider the digestion and metabolism of fats, including how fat metabolism is related to carbohydrate metabolism. We will also discuss ketosis, one of the more serious symptoms of diabetes.

24.1   DIGESTION AND FATE OF FATS

Lipolysis

The digestion of fats, **lipolysis**, occurs in the small intestine, where the hydrolysis of triacylglycerols is catalyzed by enzymes called *lipases*. The principal lipolysis products are fatty acids and glycerol, along with some monoacylglycerols. Figure 24.1 shows the stepwise lipolysis. Fats themselves cannot be absorbed through the intestinal wall, but fatty acids, glycerol, and monoacylglycerols can be absorbed.

Problem 24.1.   Write one equation for the complete lipolysis of a fat to glycerol and fatty acids.

The digestion of fats is aided by bile, which is formed in the liver, stored in the gallbladder, and released periodically into the small intestine. Bile is composed principally of bile pigments, bile salts, and cholesterol. The bile pigments arise from the partial breakdown of hemoglobin in the liver. In the in-

Figure 24.1.   *The stepwise lipolysis (hydrolysis) of fats. The principal products, which can be absorbed through the intestinal wall, are fatty acids and glycerol, plus some monoacylglycerols.*

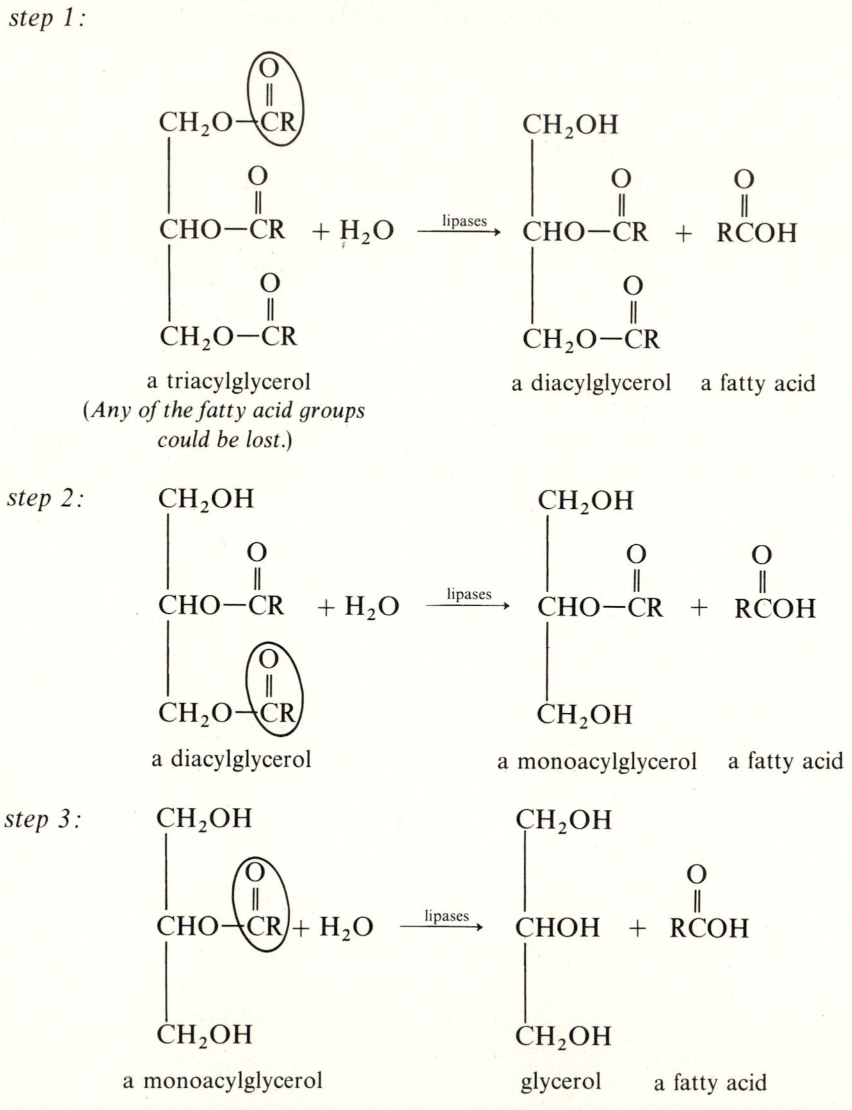

step 1:

a triacylglycerol   a diacylglycerol   a fatty acid
(*Any of the fatty acid groups
could be lost.*)

step 2:

a diacylglycerol   a monoacylglycerol   a fatty acid

step 3:

a monoacylglycerol   glycerol   a fatty acid

testines, bile salts serve to emulsify fats, so that hydrolysis proceeds smoothly. When the products of fat hydrolysis are absorbed, the bile salts are left behind in the intestines.

One of the principal pathological conditions associated with fat digestion is lack of absorption of fatty acids from the intestines. A primary cause of this malabsorption is a deficiency of bile salts needed to emulsify the fats so that they can undergo lipolysis. A lack of bile salts can result from insufficient bile production caused by liver disease or from an obstruction of the bile duct. Deficiency of fat-soluble vitamins, such as A, D, and K, is commonly associated with malabsorption of fatty acids.

Transport of Fatty Acids

As soon as fatty acids are absorbed, they react with glycerol from the system's metabolic pool and re-form new triacylglycerols. These new triacylglycerols become associated with proteins as *lipoproteins*, which in turn become associated with phosphoglycerides and cholesterol. In this form, the triacylglycerols are transported via the lymphatic system into the bloodstream, and then to the organs of the body, primarily the liver. The triacylglycerol level in the blood remains elevated for 4–6 hours after fats are ingested.

$$\text{triacylglycerols} \xrightarrow[\text{(intestines)}]{\overset{\text{H}_2\text{O}}{\underset{}{\text{lipases}}}} \begin{array}{c} \text{fatty acids} \\ \text{monoacylglycerols} \\ \text{glycerol} \end{array} \longrightarrow$$

ingested   *absorbed into bloodstream*

$$\text{reformed triacylglycerols} \xrightarrow{\text{protein}} \text{lipoproteins} \longrightarrow \text{to liver}$$

after absorption   *in lymph and blood*

Fate of the Fatty Acids and Glycerol

In the liver and, to some extent, in the adipose tissue, the triacylglycerols are hydrolyzed again to fatty acids and glycerol. These cleavage products can then be used to synthesize other compounds, such as phospholipids, or catabolized to produce CO_2, H_2O, and energy. Glycerol, for example, can be oxidized to dihydroxyacetone phosphate, a compound that can enter the glycolysis scheme (Figure 23.2, Section 23.2). We will discuss the catabolism of fatty acids in Section 24.2.

catabolism of glycerol:

$$
\begin{array}{ccccc}
\text{CH}_2\text{OH} & & \text{CH}_2\text{OH} & & \text{CH}_2\text{OH} \\
| & \xrightleftharpoons[\text{phosphatase}]{\text{glycerokinase}} & | & \xrightleftharpoons[\text{dehydrogenase}]{\text{glycerophosphate}} & | \\
\text{CHOH} & & \text{CHOH} & & \text{C}=\text{O} \\
| & & | & & | \\
\text{CH}_2\text{OH} & & \text{CH}_2\text{O}\textcircled{P} & & \text{CH}_2\text{O}\textcircled{P} \\
\text{glycerol} & & & & \text{dihydroxyacetone} \\
& & & & \text{phosphate}
\end{array}
$$

with ATP → ADP over glycerokinase/phosphatase, and NAD⁺ → NADH + H⁺ over glycerophosphate dehydrogenase.

Problem 24.2.   Referring to Section 22.4, determine the energy gain (in ATP units) realized by the conversion of glycerol to dihydroxyacetone phosphate.

24.2   THE FATTY ACID CYCLE

Fatty acids are catabolized in the mitochondria of liver cells in a series of enzymatic reactions called the **fatty acid oxidation cycle**. (See Figure 24.2.) In this series of reactions, a fatty acid undergoes cleavage, losing two carbons. These two carbons become an acetyl group of acetylcoenzyme A.

Figure 24.2.   *The fatty acid cycle.*

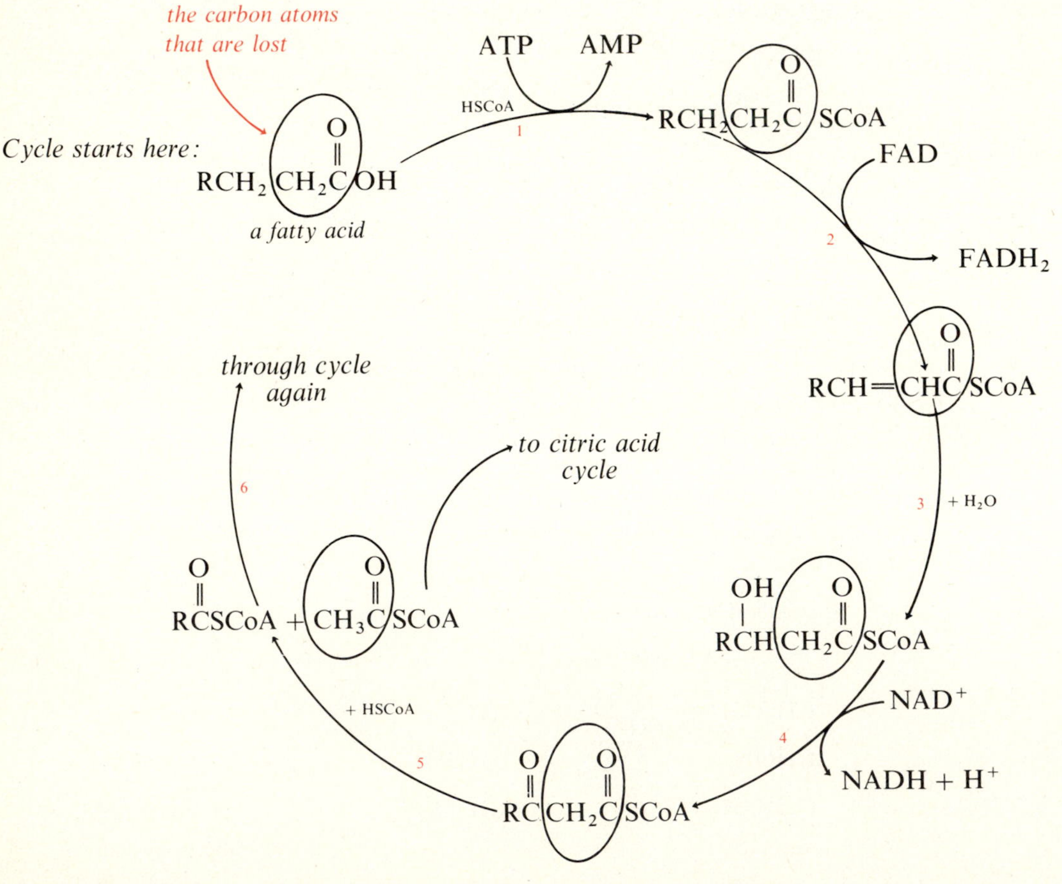

overall reaction:

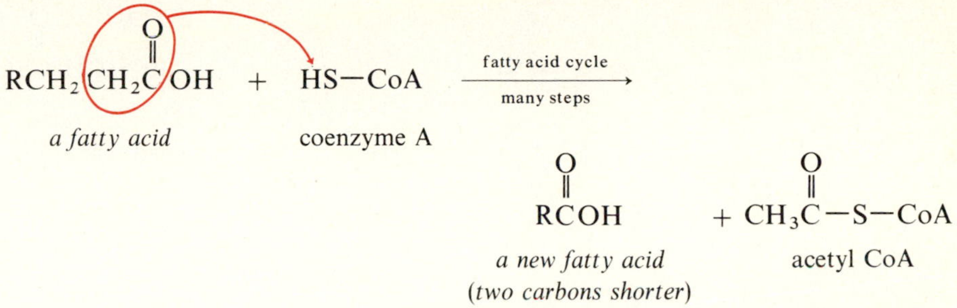

$$RCH_2CH_2COH \ + \ HS-CoA \ \xrightarrow[\text{many steps}]{\text{fatty acid cycle}}$$

a fatty acid   coenzyme A

$$RCOH \ + \ CH_3C-S-CoA$$

a new fatty acid   acetyl CoA
(two carbons shorter)

The shortened fatty acid is recycled through the pathway, losing two more carbons. Most naturally occurring fatty acids contain an *even* number of carbon atoms. Therefore, as the shortened fatty acid goes through the cycle repeatedly, losing two carbons with each pass, all the carbon atoms from the fatty acid are eventually converted to the acetyl groups of acetylcoenzyme A.

Let us consider each step in the fatty acid cycle. The initial reaction (step 1) is the esterification of the fatty acid by coenzyme A. A unit of ATP is needed to provide the energy for the esterification. In this step, the ATP yields AMP (adenosine monophosphate) and pyrophosphate ions (composed of two phosphate units and designated PP$_i$).

step 1:

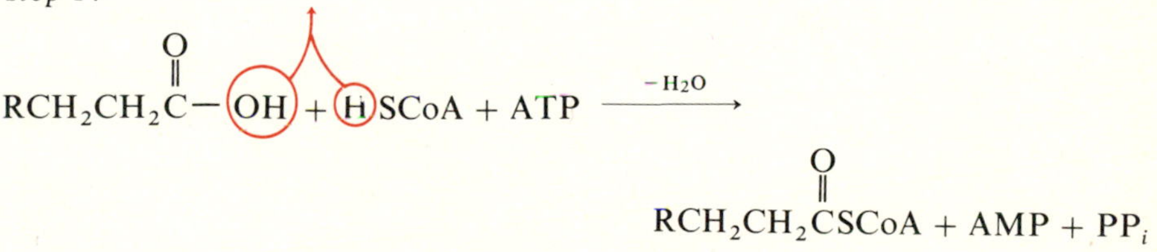

$$RCH_2CH_2C-OH + HSCoA + ATP \xrightarrow{-H_2O}$$

$$RCH_2CH_2CSCoA + AMP + PP_i$$

where

$$PP_i = {}^-OPO-PO^-$$

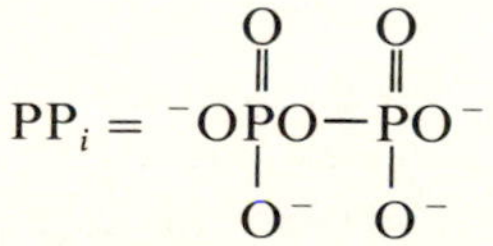

The second step in the fatty acid cycle is a dehydrogenation (loss of 2 H$^+$ and 2 e^-) of the fatty acid chain.

step 2:

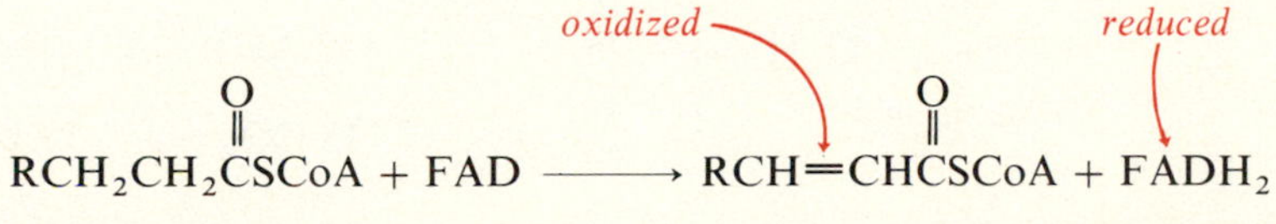

oxidized   reduced

$$RCH_2CH_2CSCoA + FAD \longrightarrow RCH=CHCSCoA + FADH_2$$

Step 3 is the addition of water across the double bond of the dehydrogenated fatty acid. This reaction is similar to the laboratory addition of water to an alkene. (See Chapter 15.)

step 3:

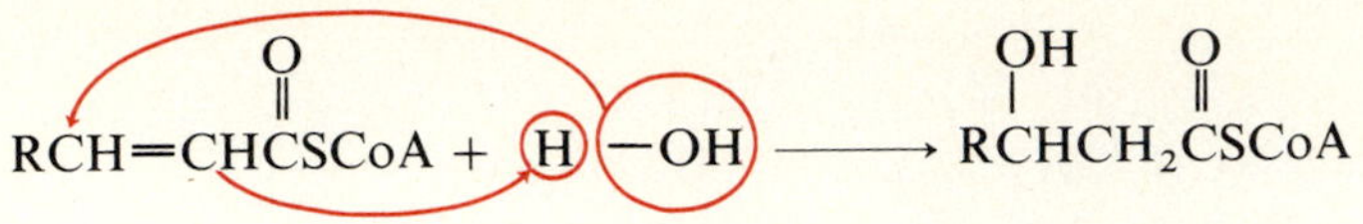

$$RCH{=}CHCSCoA + (H){-}OH \longrightarrow RCHCH_2CSCoA$$

(with O double bonds on the carbonyl carbons; product has OH and O groups)

The newly formed hydroxyl group is then oxidized to a ketone group by NAD^+.

step 4:

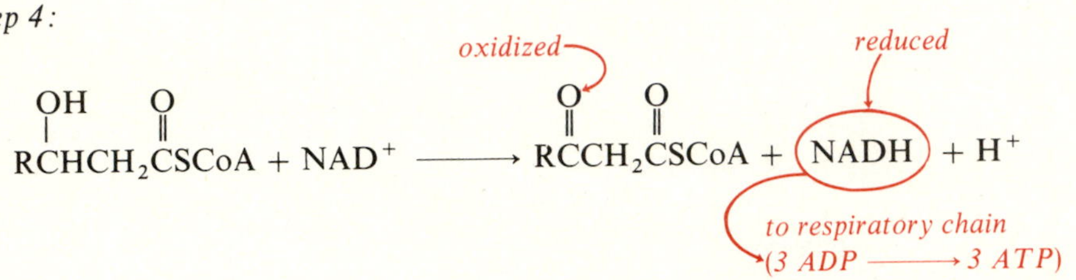

oxidized — *reduced*

$$RCHCH_2CSCoA + NAD^+ \longrightarrow RCCH_2CSCoA + (NADH) + H^+$$

to respiratory chain
(3 ADP $\longrightarrow$ 3 ATP)

Another molecule of coenzyme A now enters the reaction sequence. The oxidized fatty acid molecule is cleaved, and two carbon atoms leave with the original coenzyme A. The new coenzyme A ester is formed from the rest of the fatty acid, which is now two carbons shorter.

step 5:

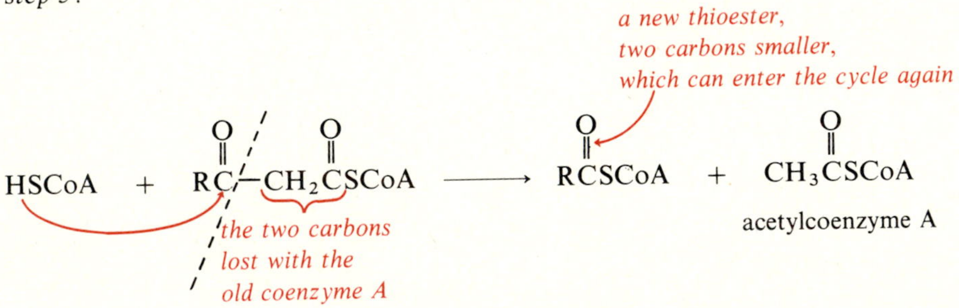

a new thioester,
two carbons smaller,
which can enter the cycle again

$$HSCoA + RC{-}CH_2CSCoA \longrightarrow RCSCoA + CH_3CSCoA$$

the two carbons
lost with the
old coenzyme A

acetylcoenzyme A

The acetylcoenzyme A that is formed by each turn of the cycle can enter the citric acid cycle, where the acetyl groups from the fatty acid are oxidized to carbon dioxide. The oxidation of each acetyl group leads to 12 ATP units. In theory, a fatty acid such as stearic acid, $CH_3(CH_2)_{16}CO_2H$, would yield 147 ATP units. In reality, the fatty acid cycle is only about 40% efficient, because the products of the fatty acid cycle can also be utilized in many other synthetic pathways, and some of the energy produced is lost as heat.

$$CH_3(CH_2)_{16}CO_2H + 26\,O_2 + 147\,ADP + 147\,P_i \longrightarrow$$
stearic acid

$$18\,CO_2 + 18\,H_2O + 147\,ATP$$

Example   In the preceding equations showing the stepwise oxidation of a fatty acid, we used R to indicate a long hydrocarbon chain. (a) Write a condensed structural formula for the thioester of lauric acid, $CH_3(CH_2)_{10}CO_2H$, with coenzyme A (using HSCoA). (b) Write the formula for the thioester remaining after one pass through the fatty acid cycle. (c) Write the formula for the thioester remaining after the second pass through the cycle.

Solution

(a) $CH_3(CH_2)_{10}\overset{\displaystyle O}{\overset{\|}{C}}-SCoA$ (b) $CH_3(CH_2)_8\overset{\displaystyle O}{\overset{\|}{C}}-SCoA$

chain two carbons shorter

(c) $CH_3(CH_2)_6\overset{\displaystyle O}{\overset{\|}{C}}-SCoA$

■ **24.3   KETONE BODIES (optional)**

In cases of starvation, a low-carbohydrate diet, or severe liver damage (resulting in the inability to store glycogen), the body's cells become starved for glucose. Although an untreated diabetic has a high level of blood glucose, this glucose cannot be used by the cells without insulin; therefore, a diabetic's cells are also glucose-starved.

One way in which the liver responds to the body's signals for large amounts of glucose is by *gluconeogenesis*, the synthesis of glucose from non-carbohydrate molecules. In Chapter 23, we stated that lactic acid can be used in gluconeogenesis. Oxaloacetic acid and amino acids can also be used for the synthesis of glucose. When the cells are forced to synthesize glucose by this route, oxaloacetic acid in the cells becomes depleted. Because oxaloacetic acid is needed to remove acetyl groups from acetylcoenzyme A at the start of the citric acid cycle, the depletion of oxaloacetic acid results in an abnormal buildup in the concentration of acetylcoenzyme A. This buildup is intensified if fatty acid molecules are catabolized to produce energy for the glucose-starved cells, because the fatty acid cycle produces generous amounts of acetyl-coenzyme A.

When acetylcoenzyme A increases in concentration but cannot be removed effectively in the citric acid cycle, it undergoes other reactions. One of these reactions is the combination of acetyl groups to yield acetoacetyl-coenzyme A.

$$CH_3\overset{\displaystyle O}{\overset{\|}{C}}-SCoA + \overset{\displaystyle H}{\underset{}{\overset{|}{C}H_2}}\overset{\displaystyle O}{\overset{\|}{C}}-SCoA \xrightarrow{\text{in liver}} CH_3\overset{\displaystyle O}{\overset{\|}{C}}-CH_2\overset{\displaystyle O}{\overset{\|}{C}}-SCoA + HSCoA$$

acetylcoenzyme A   acetoacetylcoenzyme A

Cleavage of acetoacetylcoenzyme A results in three compounds that are collectively known as **ketone bodies**: acetone, acetoacetic acid, and β-hydroxybutyric acid, which does not actually contain a keto group. The general scheme for ketone-body formation is shown in Figure 24.3.

Figure 24.3.   *The formation of ketone bodies.*

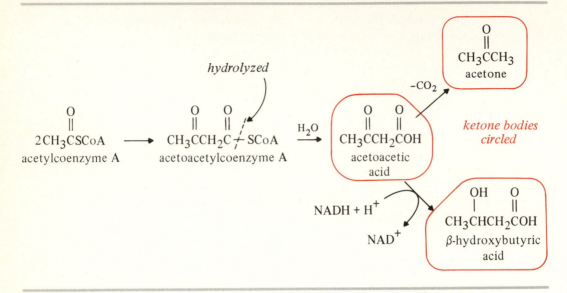

Problem 24.3.   Write the equation for the hydrolysis of acetoacetyl CoA to yield acetoacetic acid.

Ketosis

As the liver produces ketone bodies, they enter the bloodstream. A blood ketone-body level of 1 mg/100 mL is considered normal. **Ketosis** is the condition in which the liver produces abnormal quantities of ketone bodies. In ketosis, the blood level of ketone bodies becomes higher than normal—a condition called *ketonemia*, meaning ketone bodies in the blood. A diabetic may have a blood ketone-body level of 100 mg/100 mL. Other characteristics of ketosis are *ketonuria*, the presence of ketone bodies in the urine, and *ketone breath*, the sweetish smell of acetone on the breath.

In a case of ketosis, the overabundance of acetoacetic acid and β-hydroxybutyric acid in the bloodstream may also cause a person to suffer from *acidosis* (Section 10.9). Dehydration and excessive thirst may also occur, because the kidneys use more water to excrete the excess ketone bodies. In serious cases of ketosis and acidosis, coma and death can ensue. ■

■ 24.4 BIOSYNTHESIS OF FATTY ACIDS AND FATS (optional)

The reversal of the fatty acid oxidation cycle can be used for the biosynthesis of fatty acids, or **lipogenesis**, in the liver or adipose tissue. However, most fatty acids are synthesized by another route from acetylcoenzyme A. Because acetyl-coenzyme A is a product of other metabolic paths, fats can be synthesized from a variety of compounds. For example, about 30% of ingested carbohydrates are used to synthesize fat.

In the fatty acid synthesis cycle, acetylcoenzyme A is converted to malonylcoenzyme A, in a carboxylation reaction (addition of CO_2) that requires biotin (a vitamin), ATP, and CO_2.

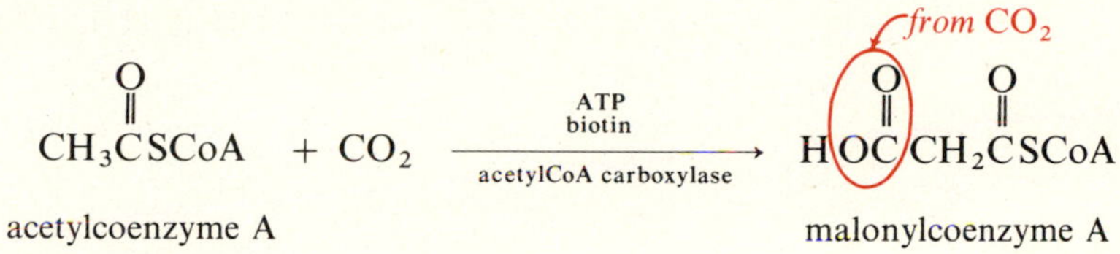

Figure 24.4.   *Biosynthesis of fatty acids. The reduction steps are accomplished by NADPH, a product of the pentose phosphate path of glucose metabolism (see Section 23.5.)*

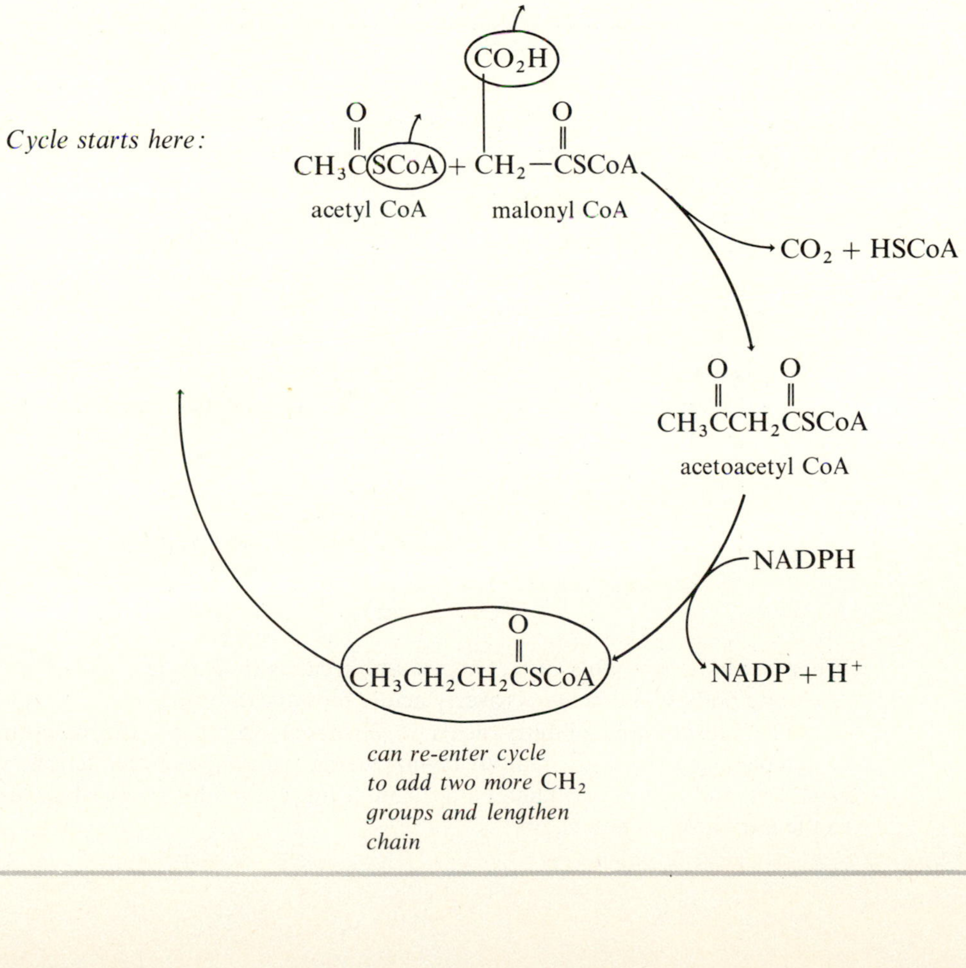

Malonylcoenzyme A reacts with acetylcoenzyme A to yield acetoacetyl-coenzyme A, which is reduced and sent through the cycle again to build up the fatty acid chain two carbons at a time. This reaction sequence is summarized in Figure 24.4. Finally, when the fatty acid chain is complete, the thioester is hydrolyzed:

$$CH_3(CH_2)_{14}\overset{\overset{\displaystyle O}{\|}}{C}-SCoA + H_2O \longrightarrow CH_3(CH_2)_{14}\overset{\overset{\displaystyle O}{\|}}{C}OH + HS-CoA$$

a thioester   a fatty acid
(can be used to
synthesize fats)

Problem 24.4.   Referring to Figure 24.4, write similar flow equations for the formation of the fatty acid caproic acid (*n*-hexanoic acid), $CH_3(CH_2)_4CO_2H$, from the thioester of butyric acid (*n*-butanoic acid), as it goes through the biosynthesis cycle again.   ■

SUMMARY

The digestion, or *lipolysis*, of triacylglycerols results principally in fatty acids and glycerol and is aided by *bile salts* (emulsifying agents) and *lipases* (lipolysis enzymes). Reformed triacylglycerols are transported as *lipoproteins* through the lymph system and bloodstream to the liver, where they are used to synthesize new triacylglycerols or other compounds, or are catabolized.

In the *fatty acid cycle*, fatty acids are broken down two carbons at a time. These two carbons are converted to acetyl groups in acetylcoenzyme A, which can enter the citric acid cycle. In their catabolism, fatty acids yield far more ATP units per mole than do carbohydrates.

Acetylcoenzyme A can also react to yield acetoacetylcoenzyme A, which can be converted to *ketone bodies*:

$$CH_3\overset{\overset{\displaystyle O}{\|}}{C}CH_2\overset{\overset{\displaystyle O}{\|}}{C}OH \qquad CH_3\overset{\overset{\displaystyle OH}{|}}{C}HCH_2\overset{\overset{\displaystyle O}{\|}}{C}OH \qquad CH_3\overset{\overset{\displaystyle O}{\|}}{C}CH_3$$

acetoacetic acid   β-hydroxybutyric   acetone
acid

When excessive amounts of fats are catabolized, as in diabetes, *ketosis* (excess of ketone bodies) and *acidosis* (overly acidic blood) can result.

The biosynthesis of fatty acids (*lipogenesis*) occurs by the reaction of acetylcoenzyme A with malonylcoenzyme A, subsequent reduction with NADPH, and hydrolysis. Each pass through this cycle adds two carbon atoms to the fatty-acid carbon chain.

KEY TERMS

lipolysis
lipase
monoacylglycerol

lipoprotein
fatty acid cycle
*ketone bodies

*ketosis
*lipogenesis

STUDY PROBLEMS

24.5. Write a general equation for the lipolysis of a fat to a monoacylglycerol.

24.6. (a) In what chemical form are fats absorbed through the intestinal wall?

(b) In what chemical form are they transported within the system in the lymph?

(c) In the blood?

24.7. Write complete equations, showing structures of glycerol and its products, for the following reactions in the catabolism of glycerol.

(a) glycerol + ATP $\longrightarrow$

         glycerol-1-phosphate + _____

(b) glycerol 1-phosphate + NAD$^+$ $\longrightarrow$

   _____ + _____ + _____

24.8. Complete the following equations, each representing one pass through the fatty acid oxidation cycle:

(a) $CH_3CH_2CH_2\overset{\displaystyle O}{\overset{\|}{C}}OH$ + HSCoA $\xrightarrow{\text{many steps}}$

(b) the acidic product from (a)

         + HSCoA $\xrightarrow{\text{many steps}}$

24.9. Complete the following scheme for the stepwise oxidation of a fatty acid:

$C_{17}H_{35}\overset{\displaystyle O}{\overset{\|}{C}}OH \xrightarrow{\text{HSCoA, ATP}}$

$C_{17}H_{35}\overset{\displaystyle O}{\overset{\|}{C}}-SCoA \xrightarrow{\text{FAD}}$

_____(a)_____ $\xrightarrow{H_2O}$ _____(b)_____ $\xrightarrow{NAD^+}$

_____(c)_____ $\xrightarrow{HSCoA}$ _____(d)_____

24.10. What product of the fatty acid oxidation cycle can enter the citric acid cycle to produce energy, CO_2, and H_2O?

***24.11.** Rewrite the following equations showing formulas (except for HSCoA) for reactions that occur in ketone-body formation. Circle the structures that are classed as ketone bodies.

(a) 2 $CH_3\overset{\displaystyle O}{\overset{\|}{C}}-SCoA \longrightarrow$
    acetyl CoA

               acetoacetyl CoA + HSCoA

(b) acetoacetyl CoA + $H_2O \longrightarrow$

               acetoacetic acid + HSCoA

(c) acetoacetic acid $\longrightarrow$ acetone + CO_2

(d) acetoacetic acid + NADH + H$^+$ $\longrightarrow$

             β-hydroxybutyric acid

***24.12.** (a) What chemical circumstance begins the series of reactions that leads to an excess of ketone bodies?

(b) List the causes, dietary and otherwise, for ketosis.

(c) Define ketonuria.

(d) Define ketone breath.

***24.13.** Complete the following equation for a reaction that occurs in lipogenesis:

$CH_3CH_2CH_2\overset{\displaystyle O}{\overset{\|}{C}}-SCoA$
     butyryl CoA

                  CO_2H
                  |
            + $CH_2-\overset{\displaystyle O}{\overset{\|}{C}}-SCoA \longrightarrow$
                malonyl CoA

***24.14.** It has been found that most naturally occurring fatty acids have an even number of carbon atoms. Using the diagram in Figure 24.4 (Section 24.4), show why this is true.

25 Metabolism of Proteins and Other Nitrogen Compounds

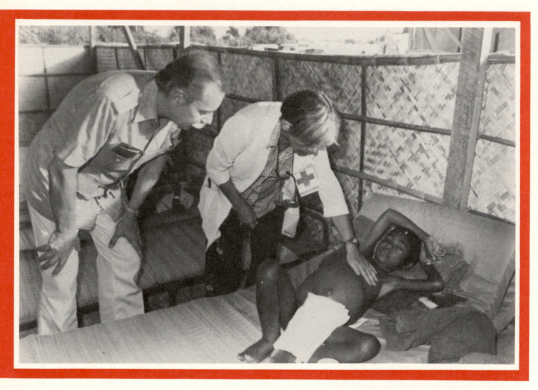

Medics examining a victim of kwashiorkor. (Courtesy of CARE.) This African child is suffering from kwashiorkor, a severe protein-deficiency disease of children. The child's swollen belly and puffy ankles are the result of a low blood protein level, which causes fluid to accumulate in the tissues. Other symptoms of kwashiorkor include wasting of the muscles and changes in the texture and color of the hair. If kwashiorkor progresses too far, death ensues.

Objectives   Define nitrogen balance, along with negative and positive nitrogen balances. □ Define glucogenic and ketogenic amino acids. □ Write an equation for a transamination reaction, and state the purpose of these reactions. □ Outline an oxidative deamination reaction. □ Write the net equation for the urea cycle, and explain its purpose. □ State the origin of uric acid in a system, and relate its presence to gout (optional). □ Describe the metabolic fate of hemoglobin (optional). □ Name two bile pigments, state their origin and fate, and relate these pigments to jaundice (optional).

We obtain dietary proteins from meat, eggs, milk, and plant material, such as soybeans and nuts. In the body, proteins are broken down into amino acids, which are then used to build the system's own proteins and to synthesize other amino acids, enzymes, some hormones, nucleic acids, and other nitrogeneous material. Amino acids can also be used to synthesize glucose or can be catabolized for energy.

In this chapter, we will consider a person's nitrogen balance, the digestion of proteins to yield amino acids, some of the biological reactions of these amino acids, and how excess nitrogen is converted to urea for elimination. We will also discuss briefly the formation of uric acid and the bile pigments, along with the role of their metabolic end products in gout and jaundice.

25.1   NITROGEN BALANCE

The body can store carbohydrates (as glycogen) or fats (as adipose tissue), but the body has no mechanism for storing excess protein, amino acids, and most other nitrogenous material. Therefore, a normal, healthy adult should ingest daily the same amount of nitrogen, principally from proteins, as is excreted. Such a person is said to have a *nitrogen balance*.

Nitrogen balance means that equal amounts of nitrogen (in compounds) are ingested and eliminated each day.

In cases of starvation, malnutrition, or certain diseases called wasting diseases, the body's muscles and other proteins are catabolized. A person with such a condition appears to be wasting away. This individual is said to have a **negative nitrogen balance**, or a net loss of nitrogen from the body—more nitrogen is eliminated each day than is taken in by diet. Postoperative patients commonly have a negative nitrogen balance; not enough nitrogen compounds are ingested or metabolized to maintain the body processes at an optimal level.

A healthy, growing child who takes in more nitrogen each day than is eliminated has a **positive nitrogen balance**. The net gain of nitrogen is used to synthesize new protein tissue, such as muscle and collagen. Human growth hormone and insulin are both instrumental in maintaining a positive nitrogen balance in a growing child. A person recovering from malnutrition or a wasting disease may also have a positive nitrogen balance while depleted protein is being replaced.

If a growing child does not obtain enough *complete* protein (protein containing all the essential amino acids), the child will develop a negative nitrogen balance. In extreme cases, this can lead to *kwashiorkor*, a protein-deficiency disease characterized by a swollen belly, patchy skin, discolored hair, loss of appetite, and diarrhea. The name kwashiorkor is from African words meaning first and second, which describe how a child develops the disease. Kwashiorkor often develops when a child (the first) is displaced from the mother's breast by a new sibling (the second) and is forced onto a starchy, protein-deficient diet.

Given a reasonable diet, the normal human body is capable of maintaining a proper nitrogen balance by changing the metabolic paths into which amino acids enter. For example, the body responds to a lack of a nonessential amino acid by synthesizing it from an amino acid present in excess. If excessive amounts of protein are ingested, the amino acids can be converted to other compounds and excess nitrogen can be excreted.

Problem 25.1. If an individual suddenly begins eating a higher-protein diet than usual, it takes a few days for the body to adapt to the elimination of the excess nitrogen. During this adaptation period, what type of nitrogen balance will the individual have?

25.2 DIGESTION AND FATE OF PROTEINS

Digestion of proteins begins in the stomach. The lining of the stomach is protected from self-digestion by a layer of mucopolysaccharides (polysaccharide-protein complexes). Enzymes called *proteases* catalyze the hydrolysis of large protein molecules into smaller peptides in the stomach, and then into individual amino acids in the small intestine. The resulting amino acids are absorbed through the intestinal wall.

The complete breakdown of proteins to amino acids is important. The body's immune system is based on recognition and elimination of foreign protein material (bacteria, for example). If dietary peptides were absorbed into the system from the intestine, the organism would produce antibodies to ward off this foreign protein. Food protein allergies, ranging from vomiting and rashes to anaphylactic shock, may be due to absorption of intact peptides.

Once absorbed, the amino acids become a part of the **amino acid pool**, which is not a physical storage site but simply a term used to describe the

Figure 25.1.   *Paths leading to and from the amino acid pool.*

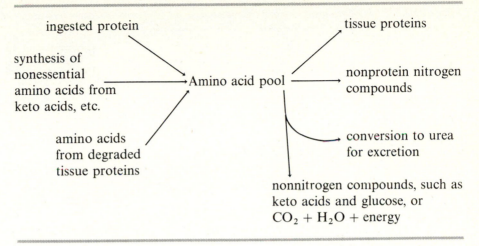

amino acids available in the system. The amino acid pool is maintained by the absorption of amino acids from digested proteins and by the natural break-down and turnover of the body's own proteins. Approximately half of the total body protein is broken down and replaced every 80 days. Individual proteins vary dramatically in their half-lives. For example, the half-life of insulin is 6–9 minutes, while that of muscle protein is about 180 days. The paths leading to and from the amino acid pool are summarized in Figure 25.1.

25.3 TRANSFORMATIONS OF AMINO ACIDS

Depending on the needs of the system, absorbed amino acids can be incorpo-rated into new proteins, converted into other amino acids in the liver, or oxidized to keto acids (with loss of NH_3). In addition, amino acids can be catabolized to products that enter some of the other metabolic sequences taking place in the body. The interrelationships of some of these metabolic paths are shown in Figure 25.2. We will not discuss the myriad of biological reactions of amino acids here but will briefly consider only a few of the most important ones.

Glucogenic and Ketogenic Amino Acids

The amino acids that can be converted to pyruvic acid and thus to glucose or glycogen in the liver are called **glucogenic amino acids**. Most amino acids are glucogenic; leucine is an exception.

Figure 25.2.   *Some relationships among the metabolism paths of proteins, carbohydrates, and fats.*

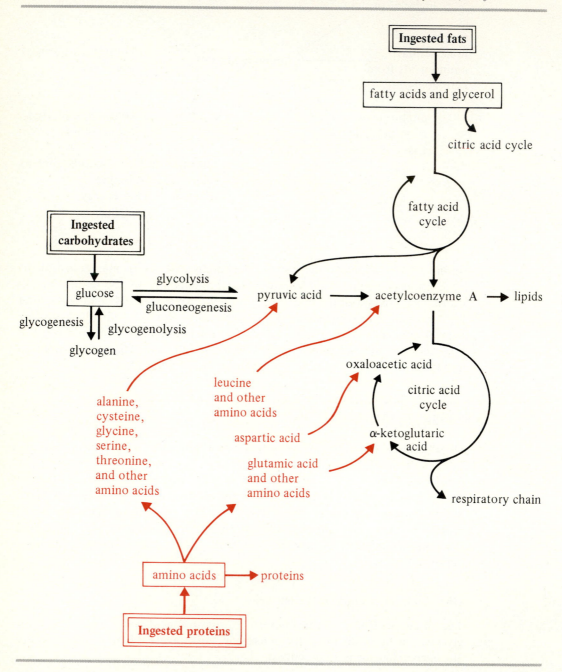

A glucogenic amino acid can be converted into glucose or glycogen:

$$\underset{\substack{\text{a glucogenic} \\ \text{amino acid}}}{\overset{\overset{\displaystyle NH_2}{|}}{RCHCO_2H}} \xrightarrow{\text{many steps}} \underset{\substack{\text{pyruvic acid} \\ \text{to liver}}}{\overset{\overset{\displaystyle O}{\|}}{CH_3CCO_2H}} \xrightarrow[\substack{\text{gluconeogenesis} \\ \text{in liver} \\ \text{many steps}}]{\text{oxaloacetic acid}} \text{glucose or glycogen}$$

Some amino acids—isoleucine, leucine, lysine, phenylalanine, and tyrosine—can be catabolized to acetylcoenzyme A or acetoacetylcoenzyme A, and thus can be converted to ketone bodies (Section 24.3). These amino acids are said to be **ketogenic amino acids**.

A ketogenic amino acid can be converted into acetylcoenzyme A and ketone bodies:

$$\underset{\substack{\text{a ketogenic} \\ \text{amino acid}}}{\overset{\overset{\displaystyle NH_2}{|}}{RCHCO_2H}} \xrightarrow[\substack{\text{many steps} \\ \text{in liver}}]{HSCoA} \underset{\text{acetyl CoA}}{\overset{\overset{\displaystyle O}{\|}}{CH_3C-SCoA}} \xrightarrow{\overset{\overset{\displaystyle O}{\|}}{CH_3C-SCoA}}$$

$$\underset{\text{acetoacetyl CoA}}{\overset{\overset{\displaystyle O \quad\; O}{\| \quad\; \|}}{CH_3CCH_2C-SCoA}} \xrightarrow{} \underset{\text{(acetone, etc.)}}{\text{ketone bodies}}$$

Transamination

Essential amino acids cannot be synthesized in the body; however, nonessential amino acids can be biosynthesized from other compounds. **Transamination**, the transfer of an amino group from an unneeded amino acid to a keto acid, is one way by which cells can synthesize a nonessential amino acid. Transamination reactions occur principally in the heart, brain, kidneys, and liver.

a transamination reaction:

An amino group has been transferred from one molecule to another.

$$\underset{\substack{\text{a keto acid} \quad\quad \text{an amino acid}}}{\overset{\overset{\displaystyle O}{\|}}{RCCO_2H} + \overset{\overset{\displaystyle NH_2}{|}}{R'CHCO_2H}} \xrightarrow{\text{transaminase}} \underset{\substack{\text{a new} \\ \text{amino acid}}}{\overset{\overset{\displaystyle NH_2}{|}}{RCHCO_2H}} + \underset{\substack{\text{a new} \\ \text{keto acid}}}{\overset{\overset{\displaystyle O}{\|}}{R'CCO_2H}}$$

Problem 25.2. Write the equation for the transamination reaction between glutamic acid (Table 20.1, Section 20.1) and a keto acid to yield isoleucine and a new keto acid.

Figure 25.3.   *Oxidative deamination of alanine.*

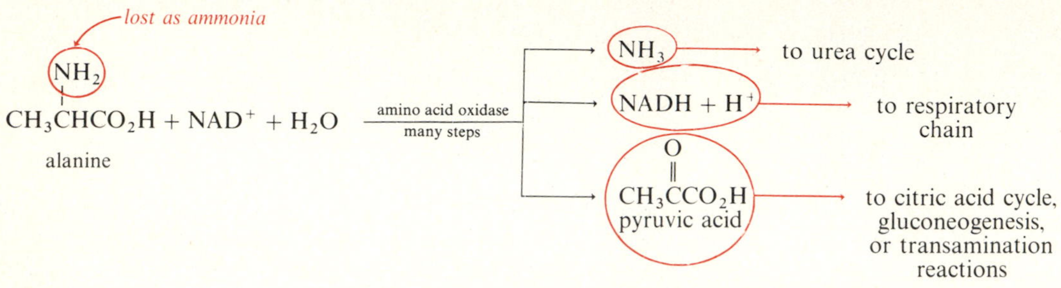

Oxidative Deamination

Amino acids can be oxidized directly to keto acids in the liver by the removal of the α-amino group as NH_3. This type of reaction, called **oxidative deamination**, is catalyzed by an enzyme called *amino acid oxidase*. A typical oxidative deamination sequence is shown in Figure 25.3.

Problem 25.3.   Glutamic acid (Table 20.1) can be converted into α-ketoglutaric acid by an oxidative deamination. Write the equation for this reaction.

$$HO_2CCH_2CH_2\overset{\overset{\displaystyle O}{\|}}{C}CO_2H$$

α-ketoglutaric acid

25.4   THE UREA CYCLE

Excess nitrogen must be excreted by a system. Fish and lower marine animals excrete unneeded nitrogen as ammonia. Birds, reptiles, and insects excrete nitrogen as uric acid. Higher animals, including human beings, excrete excess nitrogen principally in the form of urea. An adult male normally excretes about 30 grams of urea in a 24-hour period, plus small amounts of ammonia and uric acid.

ammonia   urea   uric acid

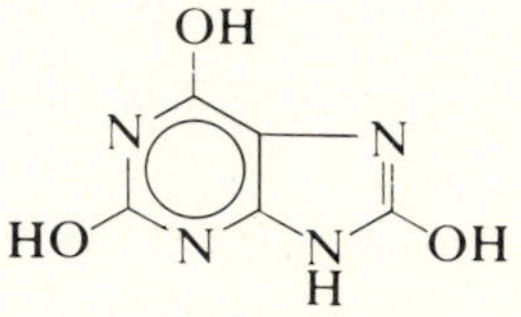

Figure 25.4.   *The urea cycle, also called the ornithine cycle.*

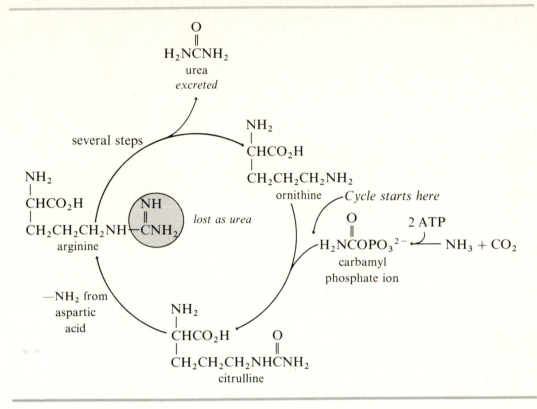

Amines and ammonia are converted to urea in the liver. The urea is then absorbed into the bloodstream, filtered out by the kidneys, and eliminated in the urine. The overall conversion of ammonia to urea may be represented simply, as shown in the following equation. However, as in most biological processes, the conversion actually occurs indirectly through a series of steps, as shown in Figure 25.4. This sequence of reactions is called the **urea cycle**, or **ornithine cycle**.

net equation for the urea cycle:

$$2\ NH_3 + CO_2 \xrightarrow[\text{in liver}]{\text{many steps}} H_2N-\overset{\overset{\displaystyle O}{\|}}{C}-NH_2 + H_2O$$

urea

Problem 25.4.   Referring to Figure 25.4, write the formulas for ornithine, the carbamyl phosphate ion, and citrulline. Circle the atoms in these compounds that will ultimately be eliminated as urea.

■ 25.5 METABOLISM OF PURINES AND PORPHYRINS (optional)

Purine Metabolism

Purine rings are found in nucleic acids. Organ meats and legumes are rich in purines. The product of purine metabolism is uric acid. In some people, the kidneys cannot remove the uric acid as quickly as it is formed, and sodium urate is deposited in the joints. This very painful condition is known as *gout*. Gout is controlled by decreasing the amount of purine-containing foods in the diet, by increasing the individual's fluid intake to help the kidneys remove the uric acid, and by administering drugs that reduce uric acid formation.

metabolism of purines:

purine skeleton,
as in nucleic acids
(*ring substituents not shown*)

uric acid
*The circled H's
are acidic.*

a sodium urate
implicated in gout

Porphyrin Metabolism

Hemoglobin, the oxygen carrier in erythrocytes, is composed of four globular proteins and *heme*, porphyrin rings bonded to Fe^{2+} (see Sections 16.7 and 20.4). The normal lifetime of erythrocytes is about 120 days, after which these blood cells undergo hemolysis, or rupturing, and are carried to the spleen. When an erythrocyte ruptures, it releases its hemoglobin.

The protein material from hemoglobin is metabolized in the same manner as is other protein. The metabolism of heme occurs principally in the endoplasmic reticula of liver cells. The released Fe^{2+} ion reacts with another protein to yield *ferritin*. In this manner, the iron is captured and stored to be recycled as needed.

The porphyrin ring system from heme is converted to pigments called **biliverdin** (*verd*, green) and **bilirubin** (*rubi*, red), as shown in Figure 25.5. These bile pigments are then converted to water-soluble carbohydrate complexes, transferred in bile to the intestines, and eliminated in the feces. The color of feces is due primarily to the presence of bilirubin. Some bilirubin is reabsorbed through the intestinal walls, then captured by the kidneys and eliminated in the urine. The yellowish color of urine is also due primarily to bilirubin.

Figure 25.5. *The porphyrin ring system is metabolized by opening of the large ring. The resulting chain of four nitrogen rings with their side chains (not shown here) forms a bile pigment.*

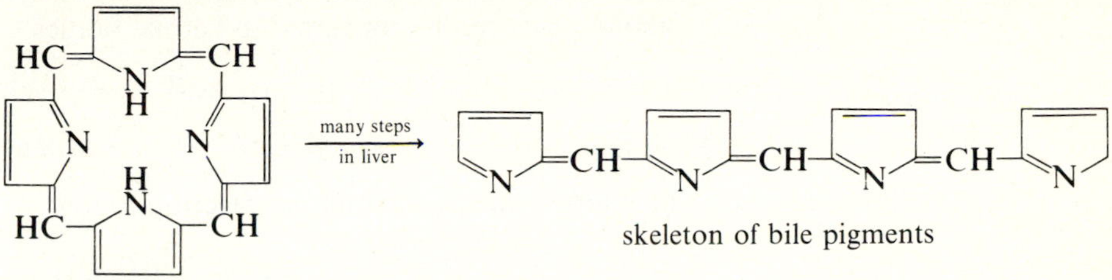

porphyrin skeleton,
as in heme

skeleton of bile pigments

hemoglobin
- 4 globins ⟶ amino acids
- Fe^{2+} ⟶ (protein) ferritin
- biliverdin ⟶ bilirubin

green   *red*

complexed with water-soluble carbohydrates and transferred to the intestines in bile

If insufficient bile is produced by the liver, bile pigments build up in the blood and give a yellowish cast to the eyes and skin. This condition, called *hepatic jaundice*, may arise from liver disease, such as hepatitis. An obstructed bile duct, which means that the bile pigments cannot be transported to the intestines, also causes jaundice. Another form of jaundice, *hemolytic jaundice*, arises from an overly rapid rate of erythrocyte death, in which case the liver cannot remove bile pigments from the blood as fast as they are formed.

Problem 25.5.   A patient with hepatitis or cirrhosis of the liver often has light, clay-colored stools. Explain why.   ■

SUMMARY

A person may exhibit a *nitrogen balance* (equal amounts of nitrogen ingested and eliminated), a *negative nitrogen balance* (net loss of nitrogen), or a *positive nitrogen balance* (net gain of nitrogen).

Proteins are hydrolyzed to amino acids in the digestive processes, and the amino acids are absorbed into the bloodstream. In the cells, the amino acids can be used to synthesize new protein or can be converted to other compounds. The free amino acids in the system collectively form the *amino acid pool*. Amino acids cannot be stored in the body but must be renewed relatively frequently.

Most amino acids can be converted to pyruvic acid, then to glucose and glycogen. Some amino acids can be converted to ketone bodies.

$$\text{glucogenic amino acid} \longrightarrow \text{pyruvic acid} \longrightarrow \text{glucose}$$

$$\text{ketogenic amino acid} \longrightarrow \text{acetyl CoA} \longrightarrow \text{ketone bodies}$$

Two paths by which amino acids can undergo reaction follow:

$$
\begin{array}{c}
\underset{\substack{\text{transamination}}}{\xrightarrow{\hspace{0.5cm} \overset{O}{\overset{\|}{R'CCO_2H}} \hspace{0.5cm}}} \quad \overset{O}{\overset{\|}{RCCO_2H}} + \overset{NH_2}{\underset{|}{R'CHCO_2H}}
\end{array}
$$

$$
\underset{\substack{\text{oxidative}\\ \text{deamination}}}{\xrightarrow{\hspace{1cm}}} \quad \overset{O}{\overset{\|}{RCCO_2H}} + NH_3
$$

$$\overset{NH_2}{\underset{|}{RCHCO_2H}}$$

Ammonia (NH_3) is, for the most part, converted in the *urea cycle* to *urea*, which is excreted in the urine. Purines are converted to *uric acid* for urinary excretion. The porphyrin ring system in heme is converted to bile pigments, such as *biliverdin* and *bilirubin*, which are excreted primarily in the feces. The released Fe^{2+} is stored in ferritin.

KEY TERMS

nitrogen balance	deamination	*ferritin
amino acid pool	urea cycle	*bile pigments
glucogenic amino acid	uric acid	*jaundice
ketogenic amino acid	*purine	
transamination	*porphyrin	

STUDY PROBLEMS

25.6. Several years ago, a rice-only diet for weight reduction was popular. Explain the consequences of such a diet in terms of nitrogen balance and essential amino acids.

25.7. List the paths to and from the amino acid pool in an animal system.

25.8 Why is it important that proteins be completely broken down to amino acids before they are absorbed through the intestinal wall?

25.9. (a) Define the terms glucogenic and ketogenic amino acids, using formulas in your answer.

(b) Suggest two reasons why a high-protein, low-carbohydrate diet can lead to acidosis and ketosis.

25.10. Complete the following equation for transamination:

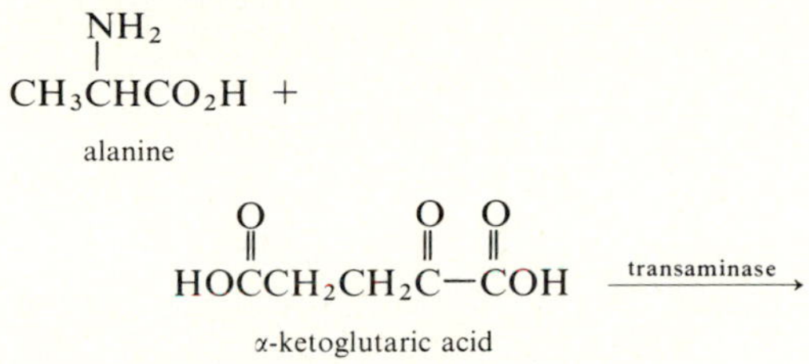

$$\underset{\text{alanine}}{CH_3\overset{\overset{\displaystyle NH_2}{|}}{C}HCO_2H} \; + $$

$$\underset{\alpha\text{-ketoglutaric acid}}{HOC\overset{\overset{\displaystyle O}{\|}}{}CH_2CH_2\overset{\overset{\displaystyle O}{\|}}{C}-\overset{\overset{\displaystyle O}{\|}}{C}OH} \;\;\xrightarrow{\text{transaminase}}$$

25.11. Complete the following equation for oxidative deamination:

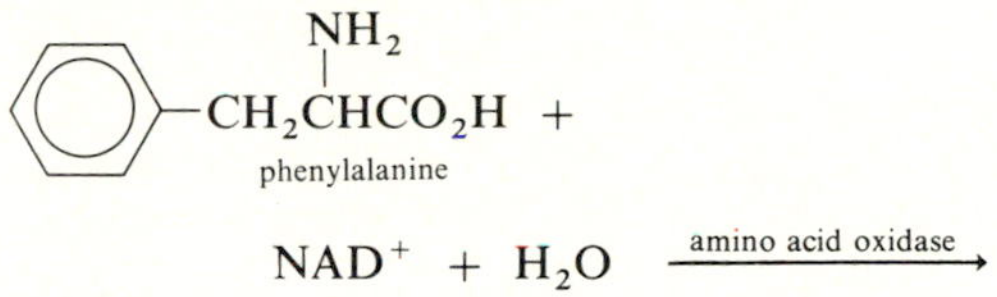

$$\underset{\text{phenylalanine}}{\bigcirc\!\!\!\!\!\bigcirc -CH_2\overset{\overset{\displaystyle NH_2}{|}}{C}HCO_2H} \; + $$

$$NAD^+ \; + \; H_2O \;\;\xrightarrow{\text{amino acid oxidase}}$$

25.12. (a) List the three principal compounds by which excess nitrogen is eliminated.

(b) Which is the principal nitrogenous waste product of humans?

25.13. (a) Write the net equation for the urea cycle.

(b) State the biochemical origins of the reactants in this reaction.

*25.14. (a) What type of compound is metabolized and excreted as uric acid in an animal system?

(b) What are some dietary sources of these compounds?

*25.15. When red blood cells die, their hemoglobin is metabolized. Describe the fate of (a) the four globins, (b) the organic portion of heme, and (c) the iron(II) ions.

*25.16 (a) Why do the skin and eyes develop a yellowish cast in cases of jaundice?

(b) List three causes of jaundice.

*25.17 Although Fe^{2+} ions are important in hemoglobin and in the cytochrome system, the dietary requirement of Fe^{2+} for an adult male is low unless he has suffered from internal or external bleeding. Explain why.

PART 5 Special Topics in Biochemistry

26 Chemistry of Heredity– Structures of the Nucleic Acids

Drs. J. D. Watson and F. H. C. Crick with a model of DNA. (From J. D. Watson, The Double Helix, *New York: Atheneum, 1968.) In 1953, the American biologist James D. Watson and the British physicist Francis H. C. Crick proposed a structure for DNA that explains the molecular basis of heredity. The Watson–Crick model is a double helix of two DNA molecules wound about each other and held together by hydrogen bonding between specific pairs of heterocyclic bases. In 1962, Watson and Crick were awarded the Nobel Prize for Medicine and Physiology jointly with Maurice Wilkins, who provided x-ray diffraction data for DNA.*

Define replication, transcription, and translation. □ Using word formulas, write the general structures of DNA and RNA. □ List the sugars and bases in DNA and in RNA. □ Write general structures for nucleotides and nucleosides. □ Describe the double helix of DNA, showing how specific base pairs are responsible for this structure. □ List the three types of RNA and their functions. □ List the two principal structural differences between DNA and RNA molecules. □ Describe bacterial and viral nucleic acids (optional).

It was recognized in the nineteenth century that the nucleus of a living, dividing cell contains stringy particles, which scientists called *chromosomes*. Human beings have 46 chromosomes in each cell. As early as 1884, scientists proposed that the chromosomes are the transmitters of hereditary traits. It was later proposed that each chromosome is composed of a huge number of genes, each of which directs the synthesis of a particular protein in our body.

$$\text{amino acids} \xrightarrow{\text{one gene}} \text{one type of protein}$$

Now we know that chromosomes and the chromatin from which chromosomes are formed are composed of proteins, along with polymers called **deoxyribonucleic acids**, or **DNA**. Depending on the species of plant or animal, DNA molecules have formula weights of from about 2 million up into the billions. These DNA molecules are responsible for the chemistry of heredity.

DNA molecules have a variety of unique properties:

1. When cells undergo normal cell division (*mitosis*), DNA duplicates itself in a process called **replication**. Thus, each cell nucleus in an organism contains identical DNA.

$$\text{DNA} \xrightarrow[\text{(mitosis)}]{\text{replication}} \text{identical DNA}$$

2. DNA molecules pass hereditary information to offspring in sexual reproduction by the combination of genes in a process called **recombination**. DNA formed by recombination is called *recombinant DNA*.

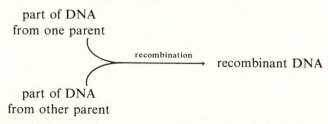

part of DNA
from one parent

recombination → recombinant DNA

part of DNA
from other parent

3. DNA molecules carry the code that controls the synthesis of each protein

molecule in an organism. This is accomplished by passing information from a portion the the DNA (a gene) to **ribonucleic acids (RNA),** in a process called **transcription**, which takes place in the cell nucleus. The RNA then leaves the nucleus to direct the synthesis of one particular protein, in a process called **translation**.

$$DNA \xrightarrow[\text{(in the nucleus)}]{\text{transcription}} RNA \xrightarrow[\substack{\text{(in the endoplasmic} \\ \text{reticulum)}}]{\text{translation}} \text{proteins from amino acids}$$

In this chapter, we will discuss the unique structures of DNA and RNA. In Chapter 27, we will discuss the dynamic processes involving these molecules: replication, recombination, protein synthesis, and related topics.

26.1   THE COMPONENTS OF NUCLEIC ACIDS

Nucleic acids are polymers. All nucleic acids contain repetitive units of three components: a sugar, a heterocyclic ring system called a "base," and a phosphate group. These three components are joined together in all nucleic acids in the following manner:

general structure of a nucleic acid:

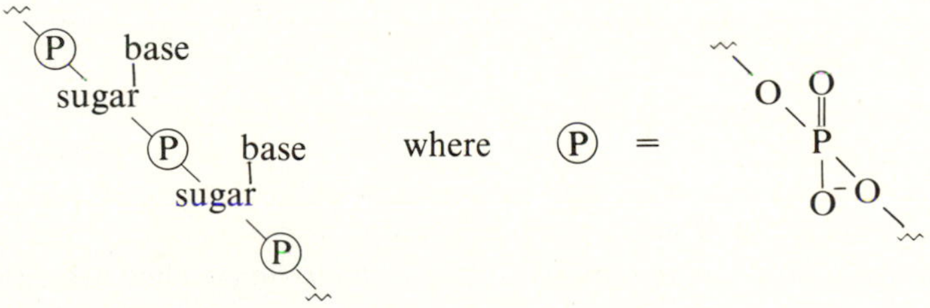

We will first discuss each of the three components, then consider how they are bonded together to form the nucleic acid polymers.

The Sugars

DNA and RNA contain only one type of monosaccharide each: DNA contains 2-deoxy-β-D-ribose, while RNA contains β-D-ribose. The names of the types of nucleic acids (**deoxyribo**nucleic acid and **ribo**nucleic acid) are derived from the names of these monosaccharides.

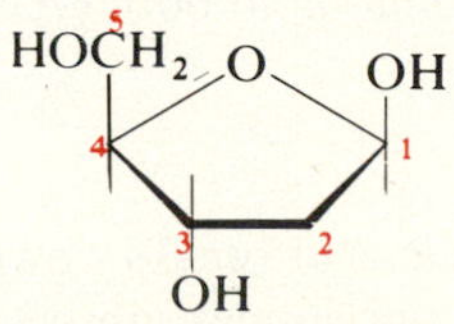

2-deoxy-β-D-ribose
in DNA

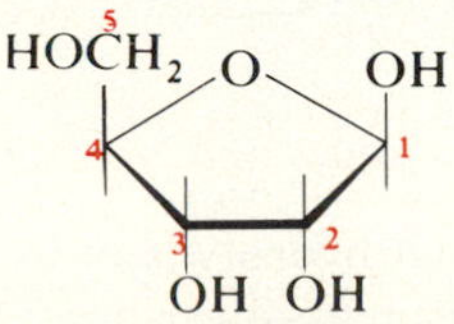

β-D-ribose
in RNA

Figure 26.1. *The principle bases in nucleic acids. Note that cytosine (**C**), adenine (**A**), and guanine (**G**) are found in both DNA and RNA. (The two-nitrogen hetero-cyclic ring in **U**, **T**, and **C** is called a pyrimidine ring; the heterocyclic ring system in **A** and **G** is called a purine ring system.)*

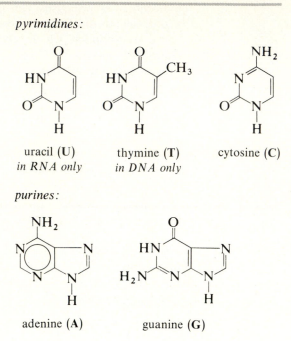

pyrimidines:

uracil (**U**)
in RNA only

thymine (**T**)
in DNA only

cytosine (**C**)

purines:

adenine (**A**)

guanine (**G**)

The Bases

The bases in nucleic acids are nitrogen heterocycles, cyclic compounds containing carbon and nitrogen in their rings. They are called bases because they react with acids to yield salts—for example, $R_3N: + HCl \rightarrow R_3NH^+ Cl^-$. Only four principal bases are found in DNA: *thymine, cytosine, adenine,* and *guanine.* Similarly, RNA contains *uracil, cytosine, adenine,* and *guanine.* The structures of these bases are shown in Figure 26.1. For convenience, these bases are often represented by their first letters; for example, **T** for thymine. Using these letters, we can list the bases in the nucleic acids as follows:

DNA contains **T, C, A,** and **G**.
RNA contains **U, C, A,** and **G**.

Note that **C, A,** and **G** are common to both types of nucleic acids.

Nucleosides and Nucleotides

Partial hydrolysis of a nucleic acid breaks down the polymer into **nucleotides** (phosphate-sugar-base). The nucleotides can be further hydrolyzed into **nucleosides** (sugar-base) and phosphate ions.

base
|
a nucleic acid $\xrightarrow{\text{hydrolysis}}$ (P)—sugar $\xrightarrow[\text{hydrolysis}]{\text{further}}$ sugar + phosphate ions

a nucleotide   *a nucleoside*

base
|

Nucleosides are composed of the sugar bonded by a β link to a nitrogen of the base. Nucleotides are simply the phosphates of the nucleosides.

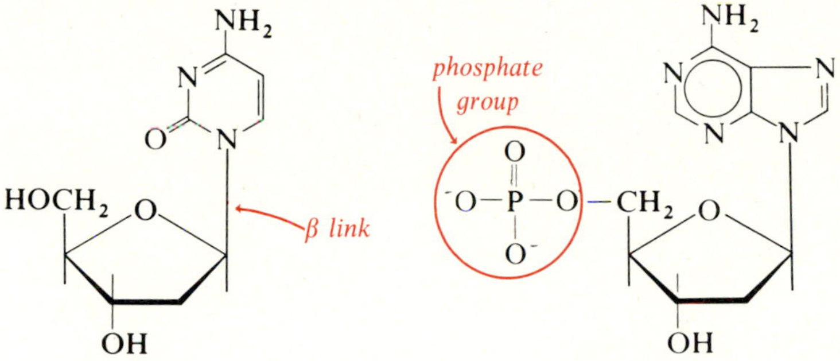

deoxycytidine
a nucleoside

deoxyadenosine phosphate
a nucleotide

Problem 26.1.   Adenosine triphosphate (ATP) is a nucleotide formed from the sugar β-D-ribose, adenine, and a triphosphate group. Without referring to any other section of this book, draw the structure of ATP.

26.2   THE DNA POLYMER

The primary structure of DNA consists of a series of nucleosides (sugar-base units) bonded together by phosphate groups. One polymeric molecule contains about 15 million of these nucleosides. A portion of a DNA chain is shown in Figure 26.2.

Like proteins, nucleic acids have secondary structures. In 1962, J. D. Watson, F. H. C. Crick, and M. Wilkins were awarded the Nobel Prize for their work in elucidating this secondary structure. *DNA is a double-stranded helix.* Two nucleic acid chains are wound about each other and held together by hydrogen bonds between pairs of bases, one from each chain, and by the attractions between the bases when they are stacked in the inside of the helix. Figure 26.3 shows two representations of the double-stranded helix. You can see that this helix resembles a flexible ladder with the hydrogen bonds between the two strands as the rungs.

The bases in one strand of DNA do not form random hydrogen bonds with bases in the other strand in the helix. *The hydrogen bonds are specific between pairs of appropriate bases.*

Figure 26.2.   *The primary structure of DNA: a backbone of sugar and phosphate units with a base bonded to each sugar unit.*

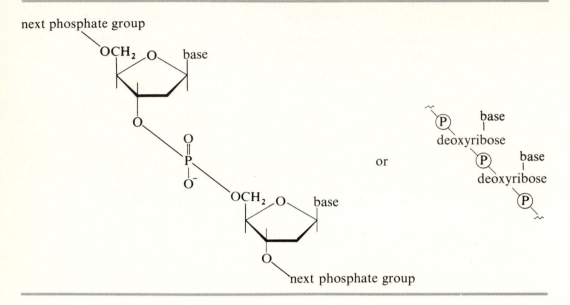

Guanine and *cytosine* form three hydrogen bonds between themselves, represented **—G≡≡≡C—**.

Adenine and *thymine* form two hydrogen bonds between themselves, represented **—A===T—**.

Figure 26.4 shows the hydrogen bonding between the paired bases guanine and cytosine (**—G≡≡≡C—**) to illustrate the unique compatibility of these two bases. Possible pairings of bases other than **G≡≡≡C** and **A===T** (guanine with adenine or guanine with thymine, for example) cannot form such favorable hydrogen bonds. Therefore, other possible pairings of bases are not found in the DNA helix.

Because adenine can hydrogen bond only with thymine, when an adenine base is present on one strand, a thymine must be opposite it in the other strand. The same holds true for guanine and cytosine. The sequence of bases in one strand dictates the sequence of bases in the other strand. The strands are completely complementary. *The order, or sequence, of bases in DNA forms the genetic code, and the complementary nature of the st unds is the chemical basis of heredity.*

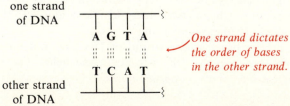

one strand of DNA

other strand of DNA

One strand dictates the order of bases in the other strand.

Figure 26.3. *DNA is a double-stranded helix held together by hydrogen bonding. (Photograph supplied through the courtesy of the Ealing Corporation, Natick, Mass.)*

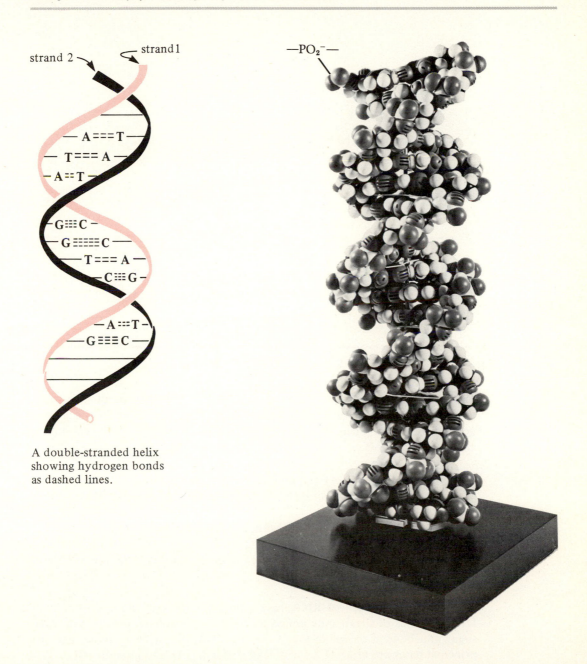

strand 2   strand 1

— A ≡≡≡ T —
— T ≡≡≡ A —
— A ═ T —

— G ≡≡≡ C —
— G ═══ C —
— T ≡≡≡ A —
— C ≡≡≡ G —

— A ═ T —
— G ≡≡≡ C —

A double-stranded helix
showing hydrogen bonds
as dashed lines.

—PO_2^-—

Figure 26.4. *The hydrogen bonding between guanine and cytosine, or* —G≡C—, *showing how the structures are ideally suited for three hydrogen bonds.*

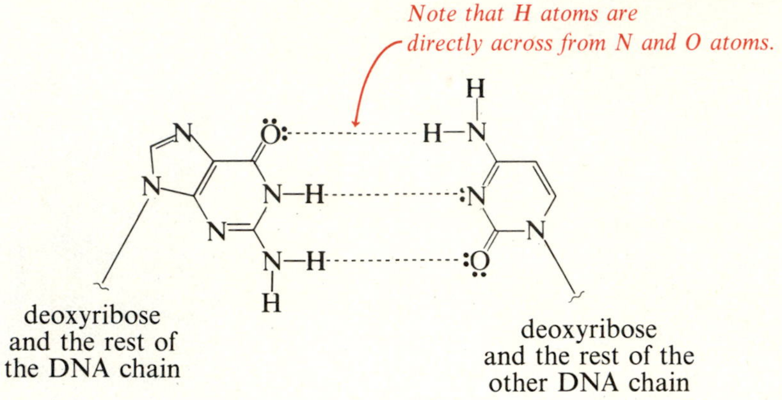

Note that H atoms are directly across from N and O atoms.

deoxyribose and the rest of the DNA chain

deoxyribose and the rest of the other DNA chain

Problem 26.2.   If one strand of DNA contains a partial base sequence of **T-G-A-C**, what is the partial base sequence in the complementary DNA strand?

26.3   TYPES AND STRUCTURE OF RNA

Ribonucleic acids (RNA) are the instruments by which specific protein molecules are synthesized, according to the instructions contained in the genetic code of the DNA. There are three principal types of RNA in a typical animal cell.

Messenger RNA (mRNA) leaves the nucleus and acts as the template, or pattern, for the incorporation of amino acids in the proper sequence in a growing protein chain.

Transfer RNA (tRNA) brings the proper amino acids to the mRNA as the protein is being synthesized.

Ribosomal RNA (rRNA) is a part of the *ribosomes*, granules that aid mRNA in protein synthesis.

These three types of RNA have primary structures similar to that of DNA: a series of sugar units joined together by phosphate groups, with each sugar unit bonded to one of four bases. In RNA, the sugar is ribose, and the principal bases are uracil (**U**), cytosine (**C**), adenine (**A**), and guanine (**G**).

$$\text{P} \quad \text{base}$$
$$| $$
$$\text{ribose}$$
$$\text{P} \quad \text{base}$$
$$|$$
$$\text{ribose}$$
$$\text{P}$$

Although the primary structure of RNA resembles that of DNA, there are two important differences in the overall structure of RNA molecules.

1. RNA chains are much shorter than DNA chains. tRNA contains 75–95 nucleotides, mRNA contains 75–300 nucleotides, and rRNA contains 100–3100 nucleotides; DNA contains millions of nucleotides.
2. RNA chains form single strands, not double helices. These single strands form their own secondary structures because of the hydrogen bonding within individual molecules. (For example, see Figure 27.4, Section 27.2.)

■ 26.4   NUCLEIC ACIDS IN BACTERIA AND VIRUSES (optional)

Bacterial DNA

Bacterial DNA does not always resemble the linear double-helical DNA of higher animals. For example, a double-stranded helix of bacterial DNA may have its ends joined together to form a circle, which then twists back upon itself to form a compact mass. Also, almost all the DNA of higher animals is found within the nucleus, but some bacteria contain small circular bits of DNA in the cytoplasm, outside the nucleus. These bits of DNA, which contain only a few genes, are called **plasmids**. Plasmids are relatively small molecules compared to the DNA in bacterial chromosomes.

Viral DNA and RNA

All organisms, plant and animal, are subject to viral infections. For humans, these infections range from exasperating conditions (such as colds, flu, and cold sores) to those that are fatal (such as rabies).

A **virus** is a small particle composed of either a DNA or an RNA molecule (not both) associated with polypeptides, some of which form a coat around the virus body. Viral DNA molecules contain relatively few genes (3–250) and occur in a variety of compact shapes arising from single-stranded or double-stranded nucleic acids, which may be linear or circular.

A virus cannot reproduce itself, nor can it synthesize its protein molecules. Therefore, a virus cannot truly be called a living thing. Some scientists say that

Figure 26.5.   *On the left is an electron micrograph of a colony of polio viruses, showing the regular geometric shape of these particles. (California University Virus Lab/Taurus Photos. Magnification 60,000.) On the right is an electron micrograph of* Herpes simplex *Type I viruses, showing a different type of viral shape. (© Martin M. Rotker 1982/Taurus Photos. Magnification 60,000.)*

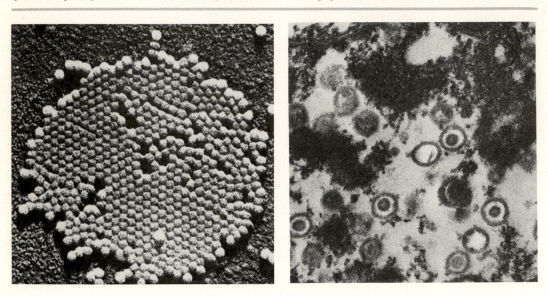

a virus is on the threshold of life. A virus reproduces by forcing another cell, called the *host cell*, to synthesize viral DNA, RNA, and proteins. In many cases, the virus forces the host cell to stop its normal activities and become a slave. This form of slavery usually results in the eventual death of the host cell.

The mechanisms of viral growth and reproduction are still not entirely understood. Some DNA viruses are capable of combining their DNA with the host DNA; the viral DNA may then remain in a *lysogenic state* (dormant) until activated by some outside source. *Herpes simplex I* (cold sores) and *Herpes simplex II* (genital herpes) are viruses of this type; both are generally dormant but can flare up into active viral infections every few weeks or months.

$$\text{viral DNA} + \text{host DNA} \xrightarrow{\text{enzymes}} \underbrace{\text{viral DNA}}_{\substack{\textit{can be active} \\ \textit{or inactive}}}\!\!-\text{host DNA}$$

Some RNA viruses that cause cancer in laboratory animals are capable of **reverse transcription**, in which viral RNA directs the host cell to synthesize viral DNA. Reverse transcription is abnormal to animal cells; animal DNA transcribes the genetic code to RNA, not the reverse. In the case of these cancer-causing viruses, the newly synthesized DNA may remain inactive for years or possibly for generations. ■

SUMMARY

Deoxyribonucleic acids (*DNA*) are found in the chromosomes in the cell nucleus and are the carriers of the genes. DNA can undergo *replication* and can also transmit genetic information to *ribonucleic acids* (*RNA*) in a process called *transcription*. RNA is needed for protein biosynthesis, or *translation* of the genetic code. DNA and RNA are polymers of repeating units of *nucleotides* (phosphate-sugar-base). The four bases in DNA are thymine (**T**), cytosine (**C**), adenine (**A**), and guanine (**G**). The four bases in RNA are uracil (**U**), **C**, **A**, and **G**.

DNA:

P — base (T, C, A, or G)
deoxyribose
P — base
deoxyribose
P

RNA:

P — base (U, C, A, or G)
ribose
P — base
ribose
P

A *nucleoside* (sugar-base) is a nucleotide minus its phosphate group.

DNA forms a double-stranded helix, with the two strands held together by —**G**≡≡≡**C**— or —**A**≡≡≡**T**— hydrogen bonds.

The three principal types of RNA are *messenger RNA* (*mRNA*), *transfer RNA* (*tRNA*), and *ribosomal RNA* (*rRNA*). These molecules are smaller than DNA and are not double stranded.

Bacterial double-stranded DNA may be linear or circular, and bacterial cytoplasm may contain *plasmids*. Viruses contain either DNA or RNA and cannot reproduce; instead, an infected host cell carries out the viral reproduction.

KEY TERMS

chromosome	transcription	tRNA
gene	translation	rRNA
deoxyribonucleic acid (DNA)	"base"	*plasmid
ribonucleic acid (RNA)	nucleotide	*virus
replication	nucleoside	
recombination	mRNA	

STUDY PROBLEMS

26.3. Define the following terms briefly in words:
(a) chromosome   (b) gene
(c) replication   (d) recombination
(e) transcription   (f) translation

26.4. What are the general functions in a cell of (a) DNA and (b) RNA?

26.5. Using a word formula, show the general structure of a nucleic acid.

26.6. Write a name and draw the formula for the monosaccharide in (a) DNA and (b) RNA.

26.7. Name the four bases in (a) DNA and (b) RNA.

26.8. Complete each of the following equations, insert dots for pertinent unshared valence electrons, and label the acid and the base.

(a) 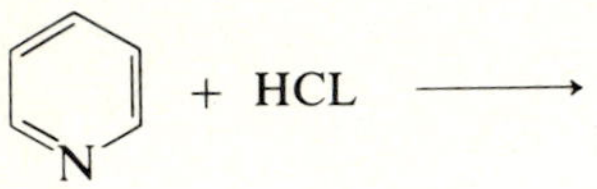 [pyridine structure] + HCL $\longrightarrow$

(b) 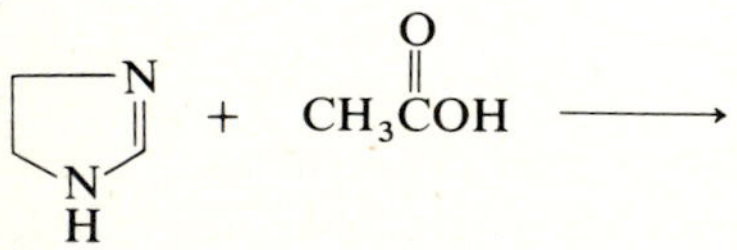 [imidazole structure] + $CH_3\overset{O}{\overset{\|}{C}}OH$ $\longrightarrow$

26.9. Using word formulas, show the structures of (a) a nucleotide and (b) a nucleoside.

26.10. Which of the following structures are nucleosides?

(a) 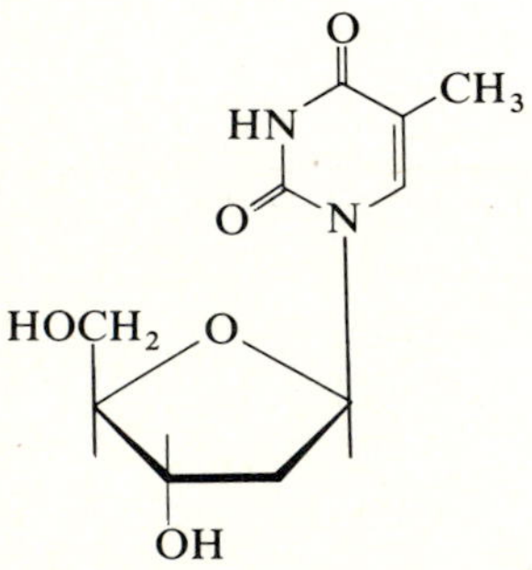 [structure with thymine base, HOCH$_2$ sugar, OH]

(b) 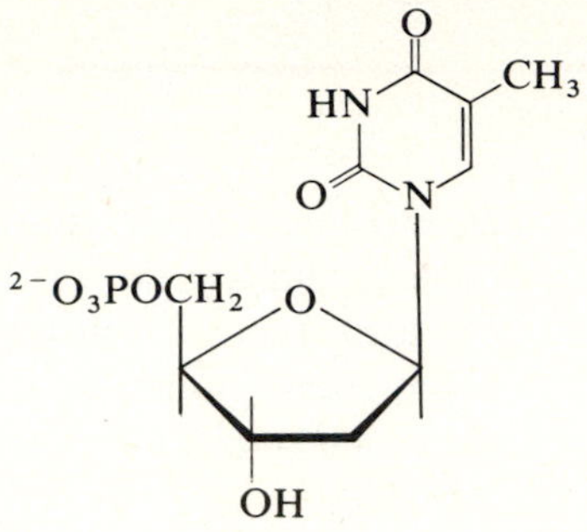 [structure with thymine base, $^{2-}O_3POCH_2$ sugar, OH]

(c) 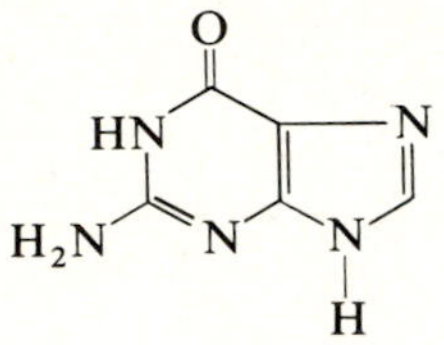 [guanine structure]

(d) 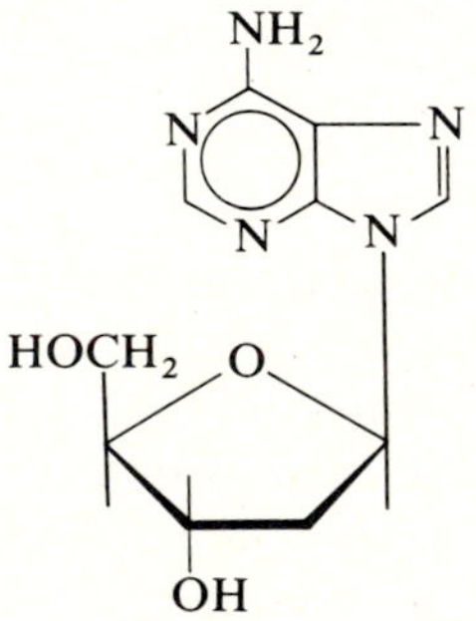 [adenine nucleoside structure, HOCH$_2$ sugar, OH]

26.11. Describe in words (a) the shape of DNA molecules and (b) what holds DNA in this shape.

26.12. Which of the following pairs of bases in DNA would you expect to be held together the most firmly by hydrogen bonding?
(a) guanine and cytosine
(b) guanine and thymine
(c) adenine and thymine

26.13. If a portion of one strand of DNA has the base sequence **C-G-A-T-A**, what is the base sequence of the complementary DNA strand?

26.14. Answer the following questions concerning the representation of DNA below.

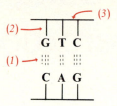

(a) What is meant by the dashed lines at (*1*)?
(b) What does line (*2*) represent?
(c) What does line (*3*) represent?

26.15. List the names and functions or occurrences of the three types of RNA.

26.16. Tell how the structure of an RNA molecule differs from that of a DNA molecule.

*26.17. What is a plasmid?

*26.18. List the components of two types of viruses.

*26.19. Which of the following sequences can arise from a viral infection?

(a) viral RNA $\xrightarrow{\text{host cell}}$ viral DNA

(b) viral DNA $\xrightarrow{\text{viral enzymes}}$ viral RNA

(c) viral DNA + host DNA $\xrightarrow[\text{recombination}]{\text{viral enzymes}}$

recombinant DNA

(d) viral protein $\xrightarrow{\text{host cell}}$ viral DNA

27 Chemistry of Heredity— The Dynamic Interactions of Nucleic Acids

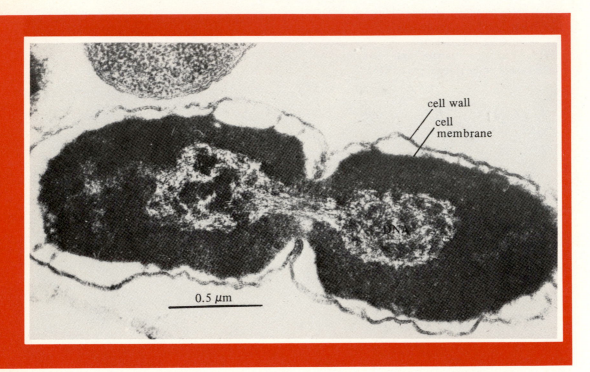

Division of a bacterial cell. [*Courtesy of M. E. Bayer and Cambridge University Press, from* J. General Microbiology, 53: 395 (1968).] *In normal cell division (mitosis), the DNA undergoes replication so that each daughter cell receives DNA identical to that in the mother cell. This photograph shows an E. coli cell undergoing mitosis. The light areas in the cytoplasm are chromosomal DNA molecules being distributed to the two daughter cells. Chapter 27 describes replication and other reactions of the nucleic acids.*

Objectives   Describe the replication of DNA, including the role of base pairing. ☐ Describe how DNA transcribes the genetic code to mRNA. ☐ Define codon and list the three types of commands that codons can give. ☐ Describe the process of translation, including the roles of ribosomes and tRNA. ☐ Define and describe the action of regulator genes and operator genes. ☐ Define mutation and mutagen (optional). ☐ List some hereditary diseases, their causes, and their effects (optional). ☐ Describe the process of genetic recombination in sexual reproduction and in the laboratory (optional). ☐ Define antigen and immunoglobulin (optional). ☐ Describe how genetic recombination allows each antibody-producing cell and its clones to produce a specific immunoglobulin (optional).

In Chapter 26, we discussed the structures of DNA and RNA. In this chapter, we will show how the double-stranded helix of DNA, with its pairing of bases, allows DNA to undergo replication and to transcribe the code to RNA for protein synthesis. In addition, we will discuss how protein biosynthesis is regulated; how antibiotics can interfere with bacterial reproduction; how errors in DNA lead to mutations and hereditary diseases; and how the body develops immunity to certain disease-causing organisms. You will find that all these topics are related to the base sequence in DNA and the pairing of the bases.

27.1 REPLICATION OF DNA

In ordinary cell division (*mitosis*), cells divide to form new cells with identical genes and, therefore, identical DNA. The hereditary information contained in the DNA polymer—the sequence of bases—can be transferred to new cells because DNA is capable of duplicating itself in replication. Replication is possible only because of the complementary nature of the bases and the resultant specific base pairings.

At the start of replication, the DNA helix begins to unwind. As it is unwinding, new nucleotides containing triphosphate groups line up along each single strand of the original DNA. *DNA polymerase* and other enzymes catalyze the successive polymerization of the nucleotides into a new DNA chain.

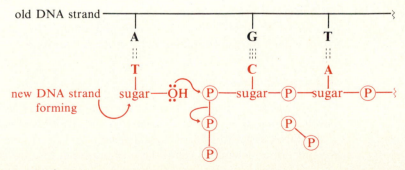

Figure 27.1. *DNA undergoes replication by unwinding and polymerizing complementary nucleotides, which form two new strands.*

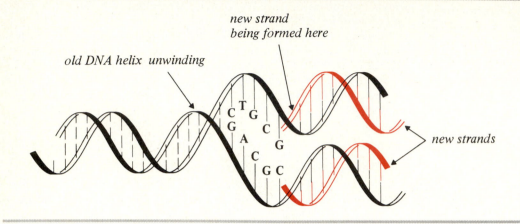

Because of the favorable hydrogen bonding between base pairs, the nucleotides in the newly forming DNA line up along the original strand in a complementary fashion—thymine to adenine and cytosine to guanine. The new DNA chain is the direct complement of the strand that synthesized it; it is identical to the second strand in the original DNA helix. Thus, as the helix unwinds, two identical DNA helices form. Each new helix contains an old strand and a newly synthesized strand (see Figure 27.1). When replication is complete, two identical double helices are present, where only one existed before. One DNA helix has produced its replica, which will carry the genetic code in a new cell.

27.2   THE BIOSYNTHESIS OF PROTEINS

A gene is a small part of a DNA chain that carries the instructions for the synthesis of one particular protein molecule. The instructions consist of the sequence of bases in the gene. DNA carries the genetic code, but it never leaves the nucleus of the cell to direct the synthesis of proteins. The instructions are transferred to mRNA (transcription), and finally the protein is synthesized from the information contained in the mRNA (translation).

Transcription of the Code

When the code is transcribed, *only a portion of the DNA helix unwinds, and one unwound strand is used as the template for the synthesis of mRNA.* That portion of the DNA constitutes a particular gene or series of genes. When the DNA unwinds, ribonucleotides (phosphate-ribose-base) line up along the unwound

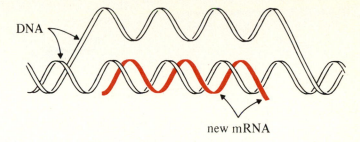

Figure 27.2. *Synthesis of mRNA from a portion of one DNA strand.*

DNA

new mRNA

strand and are polymerized to a molecule of mRNA, as shown in Figure 27.2. As nucleotides line up, guanosine becomes paired with cytosine and uracil with adenine, because of favorable hydrogen bonding. The result is an mRNA molecule containing a series of bases *complementary to the base sequence in the gene.*

transcription of the code:

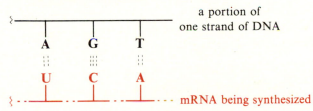

a portion of one strand of DNA

mRNA being synthesized

Problem 27.1. If a strand of DNA contains the base sequence **T-G-A**, what is the base sequence in the complementary mRNA molecule?

When the strand of mRNA is complete, it is released from the DNA and moves out of the nucleus into the endoplasmic reticulum of the cell. These channels are studded with rRNA-protein granules called **ribosomes**. A ribosome, usually represented as an acorn-like structure, is formed from two irregularly shaped segments that fit together with a crevice between them. When mRNA moves into the endoplasmic reticulum, many ribosomes become attached to its chain by means of the crevices. (See Figure 27.3.) These ribosomes aid in reading and transmitting the information contained in the mRNA and help hold the molecules in the proper positions as a protein is being synthesized.

Translation of the Code

Codons for the Amino Acids.   mRNA is the template for the growing protein chain. The sequence of bases in the mRNA determines the sequence in which amino acids will be incorporated into the protein molecule and thus deter-

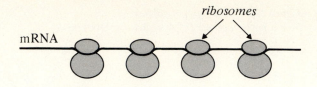

Figure 27.3.   *Ribosomes become attached to the mRNA molecule.*

mines the identity of the protein. The bases are read *three at a time*. A specific sequence of three bases in a row specifies one particular amino acid. Such a sequence of three bases is called a *codon*.

A **codon** is a series of three bases in a row in a mRNA molecule that signals (1) *initiation* of protein synthesis, (2) *elongation of the chain* by the incorporation of a particular amino acid into the growing protein molecule, or (3) *termination* of the synthesis of the protein molecule.

Table 27.1.   *Codon Assignments for the Amino Acids*

Codons[a]	Amino acid
UUU, UUC	phenylalanine (phe)
UUA, UUG	leucine (leu)
UCU, UCC, UCA, UCG	serine (ser)
UAU, UAC	tyrosine (tyr)
UGU, UGC	cysteine (cys)
UGG	tryptophan (try)
CUU, CUC, CUA, CUG	leucine (leu)
CCU, CCC, CCA, CCG	proline (pro)
CAU, CAC	histidine (his)
CAA, CAG	glutamine (gln)
CGU, CGC, CGA, CGG, AGA	arginine (arg)
AAA, AAG	lysine (lys)
AAU, AAC	asparagine (asn)
AUU, AUC, AUA	isoleucine (ile)
AUG	methionine (met) or N-formylmethionine (fmet)
ACU, ACC, ACA, ACG	threonine (thr)
AGA, AGG	arginine (arg)
GUU, GUC, GUA, GUG	valine (val)
GCU, GCC, GCA, GCG	alanine (ala)
GAU, GAC	aspartic acid (asp)
GAA, GAG	glutamic acid (glu)
GGU, GGC, GGA, GGG	glycine (gly)

[a] The codon **AUG** initiates synthesis of a peptide when preceded by an initiator region of mRNA. The codons **UAA, UAG,** and **UGA** are "nonsense" codons that signal termination of the synthesis instead of the incorporation of an amino acid.

The codons for the various amino acids have been determined experimentally. For example, a codon for the amino acid phenylalanine is uracil-uracil-uracil, or **UUU**. The three uracil bases in a row in mRNA signal that phenylalanine is to be placed in a protein at that spot. Table 27.1 lists the codon assignments, which are the same for all life forms that have been studied. Note that most of the amino acids have more than one codon. Also note that, in these cases, the first two bases in each codon are generally the same—for example, **UUU** and **UUC** for phenylalanine.

Incorporation of Amino Acids into Proteins. The proper amino acid is brought to the ribosome-mRNA site by tRNA (transfer RNA). (See Figure 27.4.) Part of the structure of a tRNA molecule is a recognition site that contains a sequence of three bases *complementary to one codon in mRNA*. This complementary sequence in tRNA is called the **anticodon**. For example:

codon bases (*in mRNA*):   **U   C   A**

anticodon complements (*in tRNA*):   **A   G   U**

The complementary relationships between the bases in DNA, mRNA, and tRNA are summarized in Table 27.2.

Table 27.2. Summary of Base Pairings Between Various Nucleic Acids

In DNA	In mRNA	In tRNA
adenine **A** ——— uracil **U**	——— adenine **A**	
cytosine **C** ——— guanine **G**	——— cytosine **C**	
guanine **G** ——— cytosine **C**	——— guanine **G**	
thymine **T** ——— adenine **A**	——— uracil **U**	

Example   Two codons for lysine are **AAA** and **AAG**. What are two anticodons for lysine?

Solution   The anticodons are the complements of the codons; therefore, the anticodons for lysine are **UUU** and **UUC**.

Problem 27.2.   Referring to Tables 27.1 and 27.2, determine the amino-acid sequence of a portion of a peptide formed from the following base sequence of DNA:

T A C C A T G G G C A T

Because of the complementary anticodon, the tRNA becomes bonded to the proper sequence (the codon) in mRNA. With the help of the ribosome, the

Figure 27.4.   *A tRNA molecule is shaped somewhat like a cross, held together by hydrogen bonds (the dashed lines) between complementary bases. The amino acid is esterified to the terminal ribose unit.*

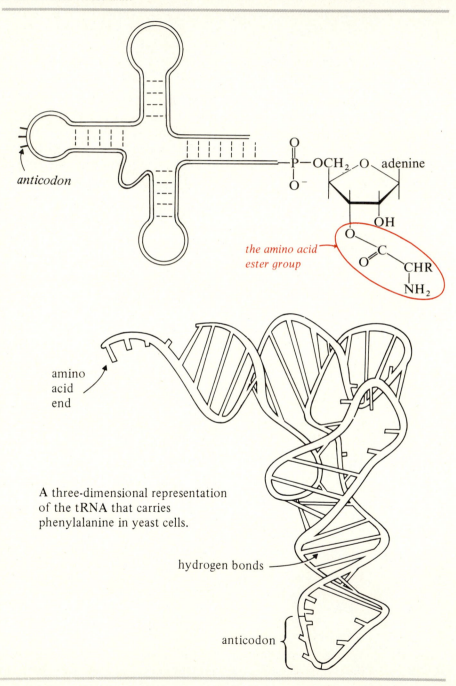

anticodon

the amino acid
ester group

amino
acid
end

A three-dimensional representation
of the tRNA that carries
phenylalanine in yeast cells.

hydrogen bonds

anticodon

Figure 27.5.   *The ribosome moves along the mRNA to the right, adding amino acids from tRNA molecules to the growing peptide chain.*

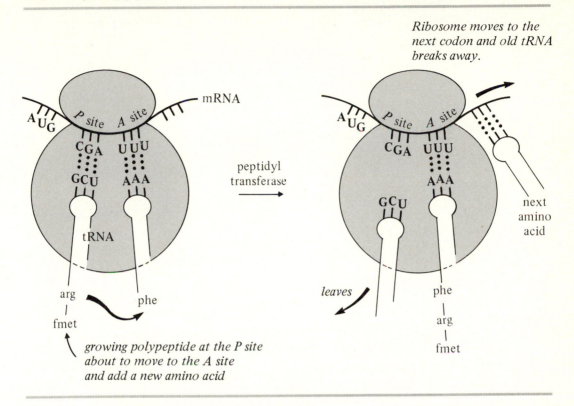

new amino acid is correctly positioned at a location called the **amino-acid binding site**, or the *A* **site**, on the ribosome. The ribosome also holds the growing end of the peptide at the **polypeptide binding site**, or *P* **site**. The end of the polypeptide chain becomes bonded to the new amino acid and thus is shifted to the *A* site in a reaction catalyzed by the enzyme *peptidyl transferase*. (See Figure 27.5.) The ribosome with the attached polypeptide moves on to the next codon, and the polypeptide site (old *A* site) becomes the new *P* site, ready for the next amino acid.

As one protein chain is being synthesized, other molecules of the same protein are synthesized in assembly-line fashion, with other ribosomes and growing proteins following the first. When a protein chain is complete, the ribosome drops off the end of the mRNA, and the completed protein molecule is released. (See Figure 27.6.)

Regulation of Protein Biosynthesis

It is evident that there must be a system for controlling the biosynthesis of proteins. Although all cell nuclei in an organism contain the same DNA and,

Figure 27.6.   *A series of ribosomes move along the mRNA chain. Amino acids are added as the ribosomes move along. When the ribosome reaches the end of the mRNA molecule, it drops off and releases the completed protein.*

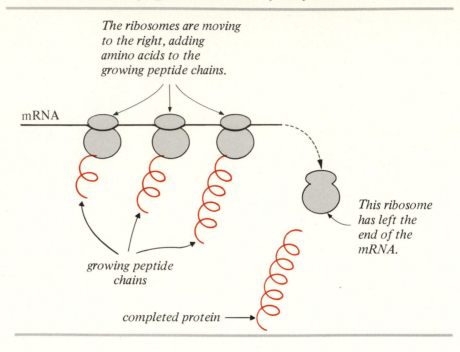

The ribosomes are moving to the right, adding amino acids to the growing peptide chains.

mRNA

This ribosome has left the end of the mRNA.

growing peptide chains

completed protein ⟶

thus, have the information needed to synthesize all proteins of the organism, the types of protein a cell synthesizes depend on the cell's function. In addition, the correct proteins must be synthesized only when the organism requires them.

The normal state of genes is that of *repression*, that is, "turned off." This is accomplished by the blocking and deactivation of a gene, called the **operator gene**, that initiates a particular mRNA synthesis. The blocking agent is a *polypeptide repressor molecule*, synthesized under the direction of a **regulator gene**.

repression:

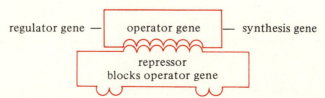

regulator gene — operator gene — synthesis gene

repressor blocks operator gene

When the signal comes—by means of a steroid hormone released by the endocrine system, for example—to activate that portion of DNA and to synthesize mRNA, an **inducer molecule** combines with the repressor molecule and changes its molecular shape. The altered repressor moves off the DNA, which can then synthesize the appropriate mRNA.

induction:

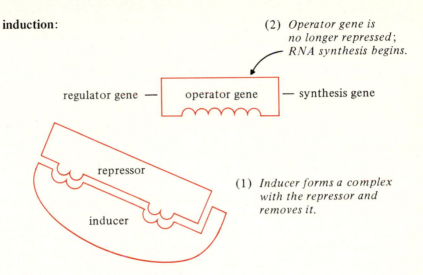

(2) *Operator gene is no longer repressed; RNA synthesis begins.*

regulator gene — operator gene — synthesis gene

repressor

inducer

(1) *Inducer forms a complex with the repressor and removes it.*

Antibiotics and Protein Biosynthesis.   Certain antibiotics function by inhibiting bacterial protein synthesis. Some of these antibiotics are listed in Table 27.3. Protein synthesis can also be inhibited in higher animals; compounds that have this biological activity are often poisons. For example, the poisonous mushroom *Amanita phalloides* contains α-amanitin, which blocks the synthesis of mRNA by binding with one of the RNA polymerase enzymes.

Table 27.3.   *Some Antibiotics That Interfere with Bacterial Protein Biosynthesis*

Antibiotic	Function
chloramphenicol	inhibits the enzyme *peptidyl transferase*
erythromycin	blocks the bacterial ribosome
puromycin	causes premature termination of the protein polymerization by mimicking and replacing a tRNA molecule
streptomycin	becomes associated with the bacterial ribosome, causes misreading of the codons, and, therefore, causes incorporation of the wrong amino acids
tetracyclines	block the *A* binding site, rendering it incapable of binding to tRNA

■ 27.3   MUTATIONS AND HEREDITARY DISEASES (optional)

The processes of the formation, replication, transcription, and translation of the genetic code of an organism are quite complex. Not surprisingly, errors do occur. New errors in the sequence of bases in DNA molecules are called **mutations** and can be passed on to succeeding generations. Spontaneous mutations are rather rare, partly because the DNA polymerase enzyme helps "proofread" the base sequence when DNA undergoes replication.

Most mutations arise from exposure to *radiation* (x, gamma, ultraviolet, etc.) or to *mutagenic compounds*.

A **mutagen** is a chemical compound that can cause changes in the base sequence of DNA.

Most mutagens are also carcinogens, or cancer-causing compounds.

Many mutations are minor and go unnoticed. Referring to the codons in Table 27.1, you will see that, in many cases, a change in the third base in a codon does not even signal a different amino acid.

Table 27.4.   *Some Hereditary Diseases and Their Causes*

Hereditary disease	Cause and effect	Comments
cystic fibrosis	deficiency of pancreatic digestive enzymes leading to poor digestion, especially of fats; other glandular problems leading to high salt levels in sweat, respiratory problems, etc.	often fatal because of respiratory problems
phenylketonuria (PKU)	deficiency of *phenylalanine hydroxylase;* inability to metabolize phenylalanine	see Section 15.5
Tay-Sachs disease	deficiency of *N-acetylhexosaminidase;* glycolipids are deposited in brain and eyes	see Section 19.3
sickle-cell anemia	defect in hemoglobin structure	see Section 20.4
galactosemia	deficiency of *UDP-galactose transferase;* inability to metabolize galactose	see Section 23.7
albinism	deficiency of *tyrosinase*, necessary for the formation of *melanin*, the pigment of skin, hair, and eyes	not usually fatal or incapacitating
hemophilia	deficiency of an antihemophilic globulin necessary for proper blood clotting	bleeding, especially internally, can cause death

the four codons for serine (ser):

U-C-U

U-C-A   *Regardless of the third base, serine will*
             be incorporated into the protein.
U-C-C

U-C-G

Other mutations can be lethal to a growing fetus. In between these two extremes are the genetic defects that lead to hereditary diseases.

Over 2000 hereditary diseases are known, but only a few are common. The most common hereditary disease is *cystic fibrosis*. About 5% of the population carries the hereditary factor for this disorder, and 0.05–0.1% of infants are actual victims. Sickle-cell anemia, phenylketonuria (PKU), and galactosemia are other examples of genetic disorders that have been discussed elsewhere in this book. These and some other hereditary diseases are listed in Table 27.4.

Many hereditary diseases and other fetal defects can be detected by *amniocentesis*, a procedure in which a sample of the amniotic fluid (the fluid that surrounds the fetus) is withdrawn, usually about three months after conception. The fetal cells found in the fluid are grown in a culture and then examined for chromosomal defects. ■

27.4   GENETIC RECOMBINATION (optional)

Sexual Reproduction

Replication of DNA results in identical DNA double-stranded helices. **Genetic recombination** is the formation of new and different DNA, called **recombinant DNA**, by the combination of preexisting separate DNA molecules into one molecule. This combination of DNA occurs naturally in sexual reproduction.

The first step in sexual reproduction is the formation of germ cells by *meiosis*, cell division in which the new cells (the germ cells) contain only half the usual number of chromosomes. These germ cells form the sperm cells of the male or ova of the female. Ordinary human cells contain 46 chromosomes; therefore, human sperm and ova contain only 23 chromosomes each.

In sexual reproduction, a sperm cell and an ovum unite to form a new cell, called a *zygote*, which has the normal number of chromosomes. When the zygote is formed, the chromosomes of the sperm and ovum undergo recombination to yield new DNA, which carries genes from both parents.

Artificial Recombination

Genetic engineering is the artificial recombination of DNA in the laboratory. This process usually involves the splicing of genes for a desired protein into

Figure 27.7. *How genetic engineering can be used to synthesize desired genes (foreign DNA), which then can be used to synthesize a desired protein.*

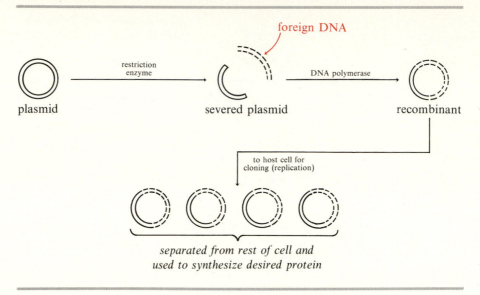

existing DNA molecules, such as those of bacterial plasmids (Section 26.4). The splicing is accomplished by cutting into the plasmid with *restriction enzymes* (bacterial enzymes that cleave nucleic acids), then inserting the foreign DNA and reclosing the plasmid, again enzymatically. The recombinant plasmid is returned to the host for cloning (replication). Then the genetic material can be isolated and used for protein synthesis. Figure 27.7 diagrams this procedure.

Until the 1980s, it was impossible to synthesize specific proteins on a commercial scale. Genetic engineering is allowing us to synthesize such medically important proteins as *insulin* (otherwise obtained from slaughterhouse animals), *human growth hormone* (HGH, obtained only from human cadavers), and *human interferon* (an antiviral agent normally produced only by human cells under the onslaught of viral attack). ■

■ 27.5 IMMUNOGLOBULINS (optional)

The human body protects itself against foreign invaders (bacteria, viruses, and so forth) in a number of ways. Collectively, the protective processes are called the *immune system*. One way the body responds to the invasion of **antigens**—large foreign molecules (not small ones, unless they are associated with a large molecule)—is to produce specific antibodies, called **immunoglobulins**

Figure 27.8. *Model of a typical immunoglobulin molecule.*

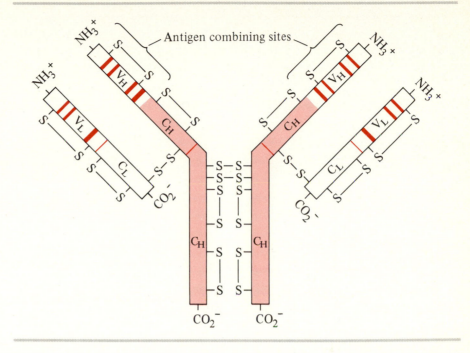

(Ig). Immunoglobulins deactivate specific foreign molecules, usually by precipitation, or *agglutination*. Two important types of immunoglobulin are *immunoglobulin M (IgM)*, produced shortly after antigen exposure, and *immunoglobulin G (IgG, gamma globulin)*, produced later for longer-term immunity.

All immunoglobulins are globular proteins with formula weights of about 160,000. Scientists have determined that an immunoglobulin molecule is shaped somewhat like a thick Y with a flexible hinge. Two identical antigen-binding sites are positioned at the ends of the arms of the Y, as shown in Figure 27.8.

The amino-acid sequences in the binding sites are variable. IgG molecules produced by different families of cells have different sequences of amino acids in their binding sites. Because the specificity of an immunoglobulin depends on this amino-acid sequence, one parent cell produces immunoglobulins for one antigen, while another parent cell produces immunoglobulins for a different antigen. The following discussion on the biosynthesis of immunoglobulins explains how similar cells can produce different amino-acid sequences at these recognition sites.

Figure 27.9. *Recombination of genetic information to code for the biosynthesis of a different recognition site.*

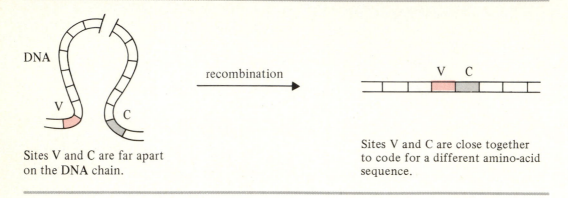

Sites V and C are far apart on the DNA chain.

Sites V and C are close together to code for a different amino-acid sequence.

Biosynthesis of Immunoglobulins

Immunoglobulins are synthesized by antibody-producing cells, primarily in the lymph fluids. It is hypothesized that the DNA in these cells undergoes genetic recombination early in the organism's life so that each antibody-producing cell has its own unique immunoglobulin code. Figure 27.9 shows how such a recombination could occur.

An antibody-producing cell is activated when a specific antigen becomes attached to receptors on the cell membrane. The activation initiates cell division and antibody production—each clone, or daughter cell, producing immunoglobulins identical to those of the parent cell. The formation of the cloned cells explains how a person can build up an immunity to a disease organism.

A newly active cell produces IgM for the first several days, then switches to the synthesis of IgG that has a binding site identical in structure to the binding site of the IgM. The concentration of IgG in the blood peaks at about three weeks—unless the organism is subjected to further antigen exposure, in which case the cells continue to produce IgG.

$$\text{inactive cell} \xrightarrow[\text{antigen}]{\text{specific}} \begin{array}{c}\text{cell division and} \\ \text{production of IgM} \\ \text{(for about 10 days)}\end{array} \longrightarrow \begin{array}{c}\text{production of IgG} \\ \text{with same binding site} \\ \text{(for 10 days–4 weeks)}\end{array}$$

Multiple myeloma is a malignant disorder of the antibody-producing cells in which an affected cell releases abnormally large amounts of identical but normal immunoglobulins. In the laboratory, myeloma cells can be fused to antibody-producing cells for the large-scale production of specific immunoglobulins. One use for these immunoglobulins is to assay for very small amounts of drugs, hormones, or enzymes in a patient's body fluids.

SUMMARY

In ordinary cell division, one DNA helix undergoes *replication* by an unwinding of the helix and the enzymatic polymerization of complementary deoxyribonucleotides, so that two helices are formed. In *transcription* of the code, a part of DNA (one gene) unwinds, and complementary ribonucleotides are polymerized to mRNA.

In *translation* of the code to synthesize a particular protein, the mRNA moves from the nucleus to the endoplasmic reticulum, where *ribosomes* become attached to its chain. Each set of three bases, or *codon*, in mRNA signals "start," "include a specific amino acid," or "stop." The proper amino acids are brought to the ribosomes by tRNA, which contain *anticodons* that direct the tRNA to the proper *A site.* The amino acid is transferred to a growing protein chain held at the *P site.* The ribosome moves on to the next codon, and the process is repeated. Some antibiotics function by interfering with a step in bacterial protein biosynthesis.

Protein biosynthesis is regulated by a repressor molecule (synthesized by a *regulator gene*) blocking the *operator gene.* An inducer molecule removes the repressor to allow the synthesis of mRNA.

Mutations, which are due to errors in the base sequence in DNA, can be caused by radiation or *mutagens. Hereditary diseases* are the result of harmful errors in the DNA base sequence.

Genetic recombination is the splicing of two different DNA molecules by sexual reproduction, by other natural processes, or by genetic engineering. *Immunoglobulins*, such as IgM and IgG, that are produced by different antibody-producing cells or their clones deactivate (agglutinate) specific *antigens.* The specificity may arise from an early recombination giving each antibody-producing cell its unique antigen-binding site.

KEY TERMS

replication	operator gene	*hereditary disease
transcription	regulator gene	*recombinant DNA
ribosome	repressor molecule	*genetic engineering
translation	inducer molecule	*antigen
codon	*mutation	*immunoglobulin
anticodon	*mutagen	*agglutination
binding sites		

STUDY PROBLEMS

27.3. A double helix of DNA can replicate itself exactly because

(a) hydrogen bonding occurs between specific base pairs and only those pairs.

(b) certain nucleotides are complementary to each other.

(c) either single strand of DNA can dictate the specific order of nucleotides incorporated into a new strand.

(More than one answer may be correct.)

27.4. Below we have represented a portion of a double strand of DNA. Which of the following statements are true?

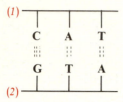

(a) When the strand unwinds for replication, hydrogen bonds are broken.

(b) In the replication, (1) will be used to form a strand identical to itself—that is, a new (1).

(c) In the replication, (2) will dictate that new nucleotides containing cytosine, adenine, and thymine will be lined up in that order.

(d) When the replication is complete, the original (1) and original (2) will be part of one helix and the new (1) and new (2) will be part of another identical helix.

27.5. (a) Describe in words what happens to DNA in normal cell division (mitosis).

(b) Compare this action of DNA with its action in translation of the genetic code.

27.6. If strand (1) of the DNA in Problem 27.4 were used to synthesize RNA, what would be the base sequence of this portion of the RNA chain?

27.7. Tell whether each of the following phrases refers to DNA, to mRNA, to tRNA, or to ribosomes:

(a) carries the codon that signals the incorporation of an amino acid into a protein molecule

(b) is a complement to a portion of a DNA strand

(c) is a complement to a portion of an mRNA strand

(d) remains in the cell nucleus

(e) forms an ester with a specific amino acid

(f) is the carrier of the genetic code from an old cell to a newly formed cell

27.8. (a) Define codons and describe their significance.

(b) Define anticodons and describe their function.

27.9. A DNA molecule contains the following sequence:

$$-\text{A T G G A G A A A}-$$

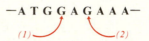

Referring to Table 27.1, predict the effect on the protein of each of the following changes:

(a) The **G** labeled (1) is left out of the DNA molecule.

(b) The **G** labeled (2) is changed to **A**.

(c) The **G** labeled (2) is changed to **C**.

27.10. Describe the functions of an operator gene and a regulator gene.

*27.11. (a) Define mutation.

(b) What are the causes of mutations?

(c) What are the possible results of mutations?

*27.12. (a) Define genetic recombination.

(b) Tell how genetic recombination is used in genetic engineering.

*27.13. Define the following terms:

(a) antigen   (b) immunoglobulin

(c) agglutination

*27.14. Which of the following statements are true?

(a) One parent antibody-producing cell synthesizes immunoglobulins with identical antigen-binding sites.

(b) The binding sites produced by one parent cell are different from those produced by other parent cells.

(c) The binding sites produced by daughter cells (clones) of a parent cell are different from those produced by the parent cell.

*27.15. List the similarities of and differences between IgG and IgM.

28 Chemistry of Blood and Urine

A blood chemistry instrument. (© Eric Kroll 1980/Taurus Photos.) Blood chemistry machines are used in clinical laboratories to determine the concentrations of various substances in blood samples. In a typical blood chemistry analysis, the computer output includes the blood levels of glucose, cholesterol, triacylglycerols, albumins, uric acid, carbon dioxide, certain enzymes, and many inorganic ions (such as calcium and sodium ions.) For comparison, the computer output also lists the normal concentration ranges of these substances.

Objectives   Diagram the parts of whole blood (plasma, serum, and clotting factors), and list the components of each. ☐ Describe the occurrence and functions of the different types of blood cells. ☐ List some causes of anemia. ☐ List the major blood types, and describe how blood type can be determined (optional). ☐ List the types and functions of plasma proteins. ☐ Diagram the steps in blood clotting, and describe the roles of anticoagulants. ☐ Write equations for oxygen transport and exchange; describe the role of pH at each step. ☐ Describe how carbon dioxide is transported and the reactions it undergoes. ☐ Describe how blood pressure and colloidal osmotic pressure aid in nutrient-waste exchange; list some causes of edema. ☐ List some normal and abnormal properties and components of urine, describing the causes of abnormalities.

Approximately 75% of a mammalian organism is composed of water. This water is found principally in the blood, the lymph, and the interstitial fluids, which are found between cells. These are all *circulating* fluids which remain in the body. Some body fluids are *noncirculating*; they do not remain in the system. Examples of noncirculating fluids are digestive juices and urine.

In this chapter, we will limit our discussion to the blood and the urine. We will discuss the composition and pH of the blood, blood types, how the pH affects the O_2-CO_2 exchange system, and how nutrients and waste products are exchanged at the capillaries. Then, we will consider how the kidneys function to transfer waste materials from the blood to the urine and how some diseases affect the composition of the urine.

28.1 PROPERTIES AND COMPOSITION OF BLOOD

The system of arteries, capillaries, and veins that carries blood throughout the body is called the *vascular system*. If the heart is included, the system is called the *cardiovascular system*. An adult human contains approximately 5.7 L (6 qt) of blood. The heart pumps this blood at the rate of over 16,000 L per day through about 60,000 miles of blood vessels. Blood serves a variety of functions in a mammal. It carries nutrients and oxygen to the cells and carries wastes and CO_2 away. It transports glucose, lipids, and amino acids. Blood also transports ions, hormones, and antibodies. In addition, blood helps regulate the organism's temperature and pH.

Whole blood is composed of *blood cells* suspended in a colloidal dispersion called *plasma*. The plasma contains the water-soluble components of blood (glucose, ions, etc.), colloidal globular proteins, and *clotting factors*. Blood without the blood cells and clotting factors is called *serum*. Serum can be observed as the straw-colored fluid oozing from a scab. Most clinical blood tests are performed on either blood plasma or serum.

Figure 28.1.   *The constituents of blood.*

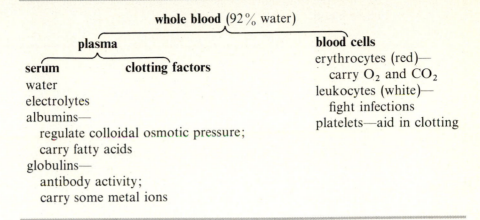

whole blood (92% water)

plasma

serum   **clotting factors**

water
electrolytes
albumins—
    regulate colloidal osmotic pressure;
    carry fatty acids
globulins—
    antibody activity;
    carry some metal ions

blood cells
erythrocytes (red)—
    carry O_2 and CO_2
leukocytes (white)—
    fight infections
platelets—aid in clotting

Table 28.1.   *Physical Constants and Composition of Whole Blood*

Physical constants
    specific gravity   1.054–1.060 (plasma alone, 1.024–1.028)
    pH   7.35–7.45

Composition
    blood cells:
        erythrocytes   4–5 million/mL
        platelets   5–10 thousand/mL
        leukocytes   250–400 thousand/mL

    proteins:
        albumins   4 g/100 mL
        globulins   2–4 g/100 mL
        fibrinogen   0.3 g/100 mL

    organic compounds:
        cholesterol   150–280 mg/100 mL
        glucose   60–100 mg/100 mL
        nitrogen compounds
          (urea, uric acid,
          bile pigments, etc.)   ~50 mg/100 mL

    electrolytes:
        sodium ion (Na^+)   136–145 mEq/L
        calcium ion (Ca^{2+})   4.5–5.5 mEq/L
        potassium ion (K^+)   2.5–5 mEq/L
        chloride ion (Cl^-)   100–106 mEq/L
        plus HCO_3^-, HPO_4^{2-},
          and others

Figure 28.1 outlines the general components of whole blood, while Table 28.1 lists the components in greater detail, along with the physical constants of whole blood.

28.2.   BLOOD CELLS

There are three types of blood cells. See Table 28.1 for the relative numbers of these cells in blood.

Red blood cells (erythrocytes) contain hemoglobin and are responsible for transporting O_2 and CO_2. Erythrocytes do not contain cell nuclei.

White blood cells (leukocytes) fight infectious bacteria. An elevated leukocyte count usually indicates a bacterial infection. Leukocytes do contain nuclei.

Platelets (thrombocytes) are necessary for the clotting of blood. Platelets do not contain nuclei.

In this section, we will discuss erythrocytes. Platelets will be discussed in Section 28.3.

Erythrocytes

Normal erythrocytes have about a four-month life span. They are synthesized in the bone marrow and destroyed in the liver and spleen at a balanced rate. The normal blood level of erythrocytes is 4–5 million per mL of blood. An *excess* of erythrocytes (up to 12 million/mL) is called **polycythemia**. This rare condition may be due to some impairment in oxygen transport or may occur in individuals living at high altitudes. A far more common condition is a *deficiency* of erythrocytes, called **anemia**.

Anemia can arise from loss of blood or a number of other causes, including the following:

Destruction of erythrocytes, which can be caused by carbon monoxide and some other toxins. Sickle-cell anemia (Section 20.4) is an example of a hereditary disease that results in erythrocyte destruction.

Decreased rate of erythrocyte synthesis, due to leukemia, multiple myeloma, Hodgkin's disease, radiation poisoning, or a deficiency in vitamin B_{12}, which is necessary for erythrocyte synthesis. *Pernicious anemia* arises from a deficiency in the gastric juices and the resultant inability of the intestines to absorb vitamin B_{12}.

■   Blood Types (optional)

It was observed early in the history of modern medicine that some recipients of whole-blood transfusions lived, while others died. The reason for some of

Photo 28.1.   *Donating blood.*
(Photo courtesy of American Red
Cross Blood Services—Northeast
Region, Needham, Massachusetts.)

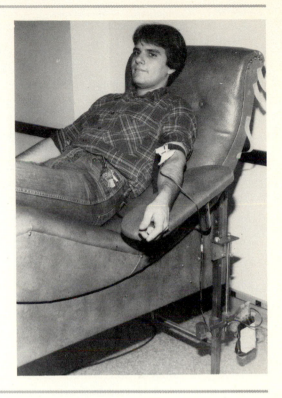

these deaths was finally traced to the incompatibility of a donor's blood and a recipient's blood because of a difference in **blood type**.

There are four major blood types: A, B, AB, and O. (There are some other rare blood types and other factors, such as the Rh factor, which we will not discuss here.) The serum of these four types of blood contains antibodies called *agglutinins* that recognize foreign erythrocytes as *antigens* (large foreign molecules that stimulate antibody production), bind to them, and cause them to precipitate. Type A blood, for example, has agglutinins that cause the erythrocytes of Type B or Type AB blood to precipitate.

A person with Type O blood is called a universal donor, because none of the other blood types contain agglutinins that cause it to precipitate. A person with Type O blood can safely *receive* only Type O blood, however, because this type of blood will cause Type A, B, or AB to precipitate. Type AB blood, by contrast, is universal-recipient blood, because it does not have agglutinins that precipitate A, B, or O blood. (See Figure 28.2.)

The antigen activity of erythrocytes arises from the different shapes of the polysaccharide coatings of these cells. The polysaccharide chain of a Type A erythrocyte has an end group of galactose with an *N*-acetyl attachment. Type B erythrocytes have plain galactose as the end group. Type AB blood contains

Figure 28.2. *Compatibility of blood types.*

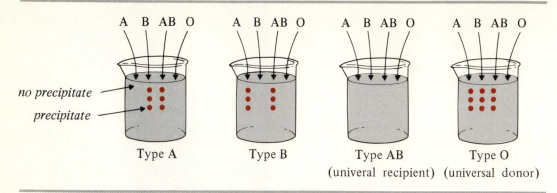

no precipitate

precipitate

Type A   Type B   Type AB   Type O
                              (univeral recipient) (universal donor)

both these end groups, and Type O erythrocytes are lacking the galactose end group entirely.

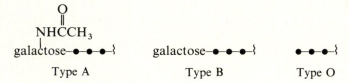

$$O$$
$$\parallel$$
$$NHCCH_3$$

galactose—●—●—●—{   galactose—●—●—●—{   ●—●—●—{

Type A   Type B   Type O

Problem 28.1. Blood types are determined by adding a sample of whole blood to standard serum samples of the various blood types and noting the presence or absence of a precipitate. When one particular sample is added to Type A, B, AB, and O serums, the serums of Types A and O show a precipitate, while those of Types AB and B do not show a precipitate. What is the blood type of this sample? (Refer to Figure 28.2.) ■

28.3 BLOOD PLASMA

Blood plasma is both a water solution and a colloidal dispersion containing a variety of organic compounds, including carbohydrates, enzymes, vitamins, and inorganic ions. Colloidal proteins account for about 8% by weight of plasma. These proteins are of three principal types:

albumins, which regulate the colloidal osmotic pressure of the blood (Section 28.5) and transport some metal ions and lipids.

globulins, some of which transport metal ions and others of which are antibodies (immunoglobulins).

clotting factors, necessary for the clotting of blood.

Figure 28.3. *The reaction scheme that leads to the formation of a blood clot.*

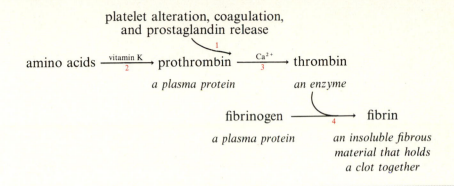

Clotting Factors and Anticoagulants

The formation of a blood clot involves a complex series of reactions. Blood clotting is initiated when the platelets are altered by an injury or an abnormal surface within a vein. The altered platelets clump together at the site and release prostaglandins, hormone-like substances that set into motion a series of reactions in which *prothrombin* is converted into *thrombin*. Thrombin is an enzyme that catalyzes the conversion of *fibrinogen* (a colloidal plasma protein) into an insoluble, bushy mass of threads called *fibrin*. The fibrin traps blood cells and solidifies into a clot, then finally into a scab covering an open wound. The steps in the formation of a clot are diagrammed in Figure 28.3.

Anticoagulants are agents that prevent blood clots. These compounds may be used in a clinical laboratory to prevent blood from clotting while tests, such as measurement of the sedimentation rate of red blood cells, are carried out. Anticoagulants may also be administered to a patient to dissolve clots within veins or to prevent their formation.

Anticoagulants act by interfering with the clotting mechanism. For example, *aspirin* acts as a mild anticoagulant by inhibiting the synthesis of the prostaglandins normally released by the platelets (step 1 in Figure 28.3). *Warfarin* is a rat poison that interferes with the action of vitamin K and thus with the synthesis of prothrombin (step 2). Rats that consume warfarin die from internal hemorrhages. Other antagonists of vitamin K are used medically as anticoagulants.

Salts that bind or precipitate calcium ions prevent the conversion of prothrombin into thrombin (step 3). Sodium citrate, a nontoxic compound that can be used intravenously, binds calcium ions into a soluble but nondissociated salt. Sodium oxalate can be used to precipitate calcium ions *in vitro* in blood testing, but it cannot be used internally because it is toxic.

$$Ca^{2+} + {}^{-}\overset{\overset{O}{\|}}{O}C - \overset{\overset{O}{\|}}{C}O^{-} \longrightarrow CaC_2O_4$$

oxalate ion   calcium oxalate
insoluble

Heparin is a polysaccharide that blocks the catalysis of thrombin (step 4) and thus interferes with clotting.

Problem 28.2.   Figure 23.3 (Section 23.4) shows the structure of citric acid. Write an equation to show how a citrate anion can bind Ca^{2+}.

28.4   OXYGEN–CARBON DIOXIDE EXCHANGE

Transport of Oxygen

The pH of normal blood is 7.35–7.45 (slightly alkaline). Recall from Chapter 10 that this pH is maintained primarily by a bicarbonate–carbonic acid buffer system controlled by respiration and the kidneys. You may want to review the discussions of blood buffers, acidosis, and alkalosis in Section 10.9.

blood   lungs

$$H^+ + HCO_3{}^- \rightleftharpoons H_2CO_3 \rightleftharpoons H_2O + CO_2$$

Kidneys can remove   exhaled
either H^+ or $HCO_3{}^-$.

The slight alkalinity of blood is necessary for hemoglobin (HHb) to pick up oxygen in the lungs and be converted to oxyhemoglobin. Recall from Chapter 6 that the partial pressure of O_2 is greater in the lungs than in venous blood. This difference in partial pressures also aids oxygen uptake by hemoglobin. Although one unit of oxyhemoglobin transports four O_2 molecules through the arteries to the capillaries, we usually represent oxyhemoglobin as $HbO_2{}^-$ as shown in the following equation.

in lungs:

High P_{O_2} *in lungs*   *Removal by* $HCO_3{}^-$ *in blood*
drives reaction to the right.   *also pulls reaction to the right.*

$$HHb + O_2 \longleftarrow HbO_2{}^- + H^+$$

hemoglobin   oxyhemoglobin

When oxygen-rich arterial blood reaches the capillaries, the oxygen must be released into the interstitial fluids before cells can absorb it. Cells consume O_2 and release CO_2 as a waste product; therefore, the environment surrounding a cell is low in O_2 but rich in CO_2. These relative concentrations help blood release O_2 to the interstitial fluids and take in CO_2. In addition, the fluids around the cells are more acidic than is arterial blood, because the released CO_2 reacts with water to form H_2CO_3. The increase in acidity shifts the $HHb \rightleftharpoons HbO_2^-$ equilibrium to the left, which also helps the release of O_2 from hemoglobin.

in tissues:

$$HHb \; + \; \boxed{O_2} \; \xleftarrow{\;\longrightarrow\;} \; HbO_2^- \; + \; \boxed{H^+}$$

to cells

Greater acidity drives reaction to the left.

The path of O_2 through the blood is summarized in Figure 28.4.

Transport of Carbon Dioxide

The fate of CO_2 given off by the cells is also summarized in Figure 28.4. When CO_2 is released into the blood, about 60% is taken up by hemoglobin to yield *carbaminohemoglobin*, $HHbCO_2$, which is returned to the lungs in the venous blood. In the lungs, CO_2 is released, and O_2 is taken up to begin the cycle anew.

in tissues:

$$HHb \; + \; CO_2 \; \rightleftharpoons \; HHbCO_2$$

from cells   carbaminohemoglobin
*in venous blood
and back to lungs*

The other 40% of the CO_2 released by the cells initially dissolves in the erythrocytes as HCO_3^-. However, these bicarbonate ions do not remain permanently in the erythrocytes. Instead, they are transported out of the erythrocytes, while chloride ions from the blood plasma are drawn in to replace them, in a process called the **chloride shift**.

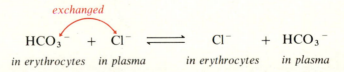

$$\underset{\text{in erythrocytes}}{HCO_3^-} \; + \; \underset{\text{in plasma}}{Cl^-} \; \rightleftharpoons \; \underset{\text{in erythrocytes}}{Cl^-} \; + \; \underset{\text{in plasma}}{HCO_3^-}$$

exchanged

Problem 28.3.   Without referring back to the text, describe how each of the following conditions would affect the $HHb + CO_2 \rightleftharpoons HHbCO_2$ equilibrium in the capillaries and/or in the lungs.
(a) A person breathes air that has a higher than normal carbon dioxide concentration.
(b) A person is sleeping, and the cells produce less carbon dioxide than when the person is working.
(c) A person is suffering from acidosis.

Figure 28.4.   *The chemical paths of oxygen and carbon dioxide in the arteries and veins. Both partial pressures and pH favor O_2 uptake and CO_2 release in the lungs and the reverse in the capillaries.*

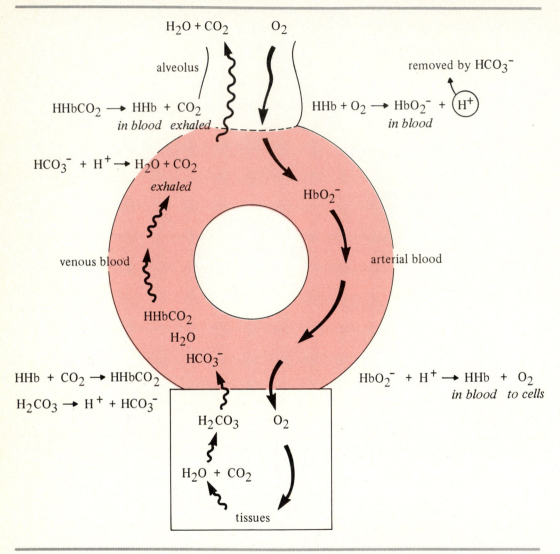

28.5   EXCHANGE OF NUTRIENTS AND WASTES

We have discussed the exchange of gases in the capillaries. Now let us consider the exchange of other substances. Blood plasma, but not interstitial fluids, contains colloidal proteins, principally in the form of *albumins*. Therefore, the blood plasma exhibits a *colloidal osmotic pressure* at the walls of the capil-

Figure 28.5.   *At the arterial end of a capillary, blood pressure forces fluid into the interstitial fluids. At the venous end, the colloidal osmotic pressure of the blood draws fluid into the capillary from the interstitial fluids.*

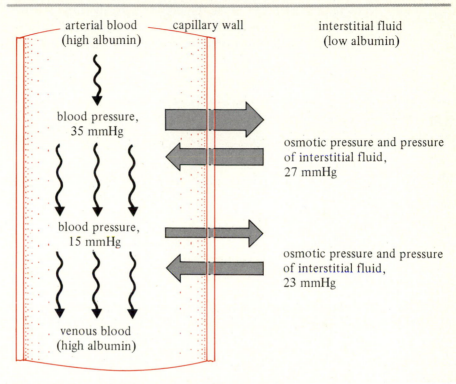

laries. Because of this colloidal osmotic pressure, there is a natural tendency for the interstitial fluids to flow into the capillaries by dialysis to dilute the more concentrated albumins in the blood. The actual direction of the flow, however, is also a function of blood pressure.

At the arterial ends of the capillaries, *the blood pressure is greater than the colloidal osmotic pressure.* The net result is that fluid is pushed out through the capillary walls into the interstitial fluids, where electrolytes and nutrients can be actively transported through cell membranes into the cells themselves.

At the venous end of the capillary, *the blood pressure is less than the colloidal osmotic pressure.* Thus, fluids containing excess electrolytes and small waste molecules are pushed into the capillaries. Figure 28.5 illustrates the opposing actions of blood pressure and colloidal osmotic pressure in the capillaries.

In a normal healthy person, the amount of fluid that leaves the capillaries and the amount that returns are equal. If the blood is low in albumin proteins, the process of dialysis is not as efficient; therefore, water collects in the tissues instead of being returned to the blood. The result is **edema**, a swelling of the feet, legs, hands, and possibly the rest of the body. Some causes of low levels of serum albumins and the resultant edema are *damaged kidneys*, which allow albumins to pass out of the blood and into the urine; *protein-deficient diet*, such as in kwashiorkor (Section 25.1); and *traumatic shock* (injury, burns, surgery, etc.), which increases capillary permeability and thus allows albumins to pass into the tissues.

28.6 PROPERTIES AND COMPOSITION OF URINE

Wastes are removed from the body by a variety of mechanisms. The lungs exhale CO_2 and H_2O. Sweat glands in the skin secrete water and salts. Bile pigments, cholesterol, salts, and undigestible material are excreted in the feces. The *renal system* (kidneys, bladder, urethra) removes water-soluble wastes in the urine. The kidneys also help control the electrolyte balance, the pH, and the water balance of the blood.

The blood passes through the kidneys at a rate such that it is filtered continuously. Almost all the filtered fluid returns to the blood, along with most of the dissolved electrolytes, amino acids, glucose, and so forth. About 1–2 L of fluid per day is passed as urine, ridding the body of urea (from metabolized protein), uric acid (from metabolized purines), excess salts, and other waste materials. If the kidneys fail to remove these wastes, the result is *uremic poisoning*, meaning urea poisoning of the blood. Uremic poisoning can be fatal; however, it can be treated by hemodialysis (Section 8.4).

Composition of Urine

Table 28.2 summarizes the physical constants and composition of normal urine. Normal urine is usually slightly acidic (pH 6); this acidity helps prevent

Table 28.2. *Physical Constants and Composition of Normal Urine*

Physical constants			
specific gravity		1.003–1.03	
pH		4.7–8.0	
volume		0.5–2.5 L/day	

Solutes (g/24 hours)			
urea	30	NH_4^+	0.5–0.8
creatinine	1–2	Cl^-	6–16
uric acid	0.7	SO_4^{2-}	2.5
Na^+	3–4	$H_2PO_4^-$	2.5
K^+	1.5	other compounds	2–3

dissolved salts from precipitating. However, the pH of normal urine can vary; for example, after a meal, urine is usually slightly alkaline.

The specific gravity of urine is used to estimate the quantity of dissolved solid material. Normal urine has a specific gravity within the 1.003–1.03 range. Drinking large amounts of liquid increases the volume of the urine, thus diluting it and causing its specific gravity to approach 1.000, the specific gravity of pure water. By contrast, high fever or other conditions of dehydration lead to more-concentrated urine and a greater than normal specific gravity.

Normal urine is clear but becomes cloudy on standing. If the turbidity is due to the normal precipitation of metal salts, it can be cleared up by acidification. If the cloudiness persists after acidification, then the presence of abnormal pus or protein is suspected.

Problem 28.4. What would be the probable effect of warming cloudy urine if the cloudiness is due to (a) salts or (b) albumin? Explain your answers.

Abnormal Constituents of Urine. Normal constituents of urine are considered abnormal when they are eliminated in excessive amounts. Table 28.3 lists some conditions that lead to abnormally high urinary levels of normal constituents. Also listed are some substances, such as pus or blood, that are considered abnormal in any measurable amount.

Except possibly after strenuous exercise, protein is rarely found in normal urine. The presence of protein, called *proteinuria* or *albuminuria*, can be indicative of kidney problems or heart disease, either of which can impair normal circulation to the kidney. Multiple myeloma, in which antibody-producing cells produce huge amounts of immunoglobulins (see Section 27.5), also results in proteinuria.

Volume of Urine

The volume of urine a normal person passes depends primarily on fluid intake and fluid loss by perspiration. The volume is regulated by a variety of hor-

Table 28.3. *Abnormal Constituents of Urine*

Substance	Cause or condition
High levels of:	
uric acid	leukemia, liver disease, gout
creatine	starvation, diabetes mellitus
bile	obstruction to flow of bile to intestines
glucose and ketone bodies	diabetes mellitus
Measurable levels of:	
blood	kidney or bladder infection or stones
pus (leukocytes)	kidney or bladder infection
protein	kidney or heart disease, multiple myeloma

mones, such as *vasopressin* and *aldosterone*. A high blood level of either of these hormones causes water retention and a decrease in urine production. On the other hand, a low blood level of either hormone, especially vasopressin, causes loss of water from the body by increased **diuresis**, or urine production. A person suffering from the rare disorder *diabetes insipidus* is deficient in vasopressin and may excrete up to 20 or 30 L of urine per day.

Some constituents of the diet can act as **diuretics**, or urine stimulants. Among these are caffeine and related compounds in coffee, cola, tea, or chocolate; ethanol in alcoholic beverages; and excessive dietary protein, catabolized to urea, a diuretic. Drugs such as hydrochlorthiazide may be prescribed as diuretics in cases of edema, hypertension (high blood pressure), or premenstrual tension due to water retention.

Abnormal Volume of Urine.   The total lack of urine, called *anuria*, occurs when some portion of the renal system is blocked because of infection, injury, malignancy, or other condition. A decreased flow of urine, or *oliguria*, may be the result of fever, kidney disease, or lack of fluid intake.

Polyuria, a larger than normal volume of urine, may occur in such conditions as diabetes or may result from high fluid intake or the use of diuretics. In the cases of high fluid intake or diabetes insipidus, the specific gravity of the urine is low because it is more dilute. In the case of diabetes mellitus, the specific gravity is high because of dissolved glucose.

SUMMARY

Blood and lymph are *circulating* fluids of the body, while digestive juices and urine are *noncirculating*. The system of heart and blood vessels is called the *cardiovascular system*. The components of blood are shown in Figure 28.1 and Table 28.1.

The three types of blood cells are *erythrocytes*, *leukocytes*, and *platelets*. An overabundance of erythrocytes is called *polycythemia*, while a deficiency is called *anemia*. The four principal blood types are A, B, AB (universal recipient), and O (universal donor).

Blood plasma contains water-soluble and colloidal substances. Plasma proteins include *albumins*, *globulins*, and *clotting factors*. When blood clots, *fibrinogen* is converted into *fibrin*. Anticoagulants interfere with the clotting mechanism.

The slightly alkaline pH of blood is controlled principally by the CO_2-HCO_3^- buffer system, which is regulated by the lungs and kidneys. Oxygen is transported in the blood in oxyhemoglobin (HbO_2^-). In the capillaries, the oxygen is released to the more acidic interstitial fluids.

$$HHb + O_2 \underset{\text{capillaries}}{\overset{\text{lungs}}{\rightleftharpoons}} HbO_2^- + H^+$$

Hemoglobin also returns most of the cells' CO_2 to the lungs.

$$HHb + CO_2 \xrightleftharpoons[\text{lungs}]{\text{capillaries}} HHbCO_2$$

Bicarbonate ions from dissolved CO_2 are transferred from erythrocytes to plasma in a process called the *chloride shift*.

The exchange of nutrients for wastes in the capillaries depends on the opposing forces of *blood pressure* and *colloidal osmotic pressure*. Edema can be caused by lack of plasma albumins.

The renal system rids the body of wastes in the urine. Abnormally high levels of some normal urinary substituents can indicate disease, as can the presence of blood, pus, or protein. Abnormalities in urinary volume are *anuria* (no urine), *oligouria* (little urine), and *polyuria* (large volume of urine). The volume is under the control of hormones, such as vasopressin and aldosterone, and may be affected by diet or diuretic drugs.

KEY TERMS

cardiovascular system	leukocyte	anticoagulant
blood cells	platelet	chloride shift
blood plasma	anemia	colloidal osmotic pressure
clotting factors	*blood type	renal system
serum	albumin	diuresis
erythrocyte	globulin	

STUDY PROBLEMS

28.5. Label each of the following body fluids as circulating or noncirculating:
(a) blood   (b) lymph   (c) urine

28.6. Define the following terms:
(a) blood serum   (b) blood plasma
(c) erythrocyte   (d) leukocyte
(e) platelets   (f) blood albumins
(g) blood globulins

28.7. (a) Define anemia.
(b) List three causes for this disorder.

*28.8. Referring to Figure 28.2, tell whether each of the following test samples is Type A, B, AB, or O blood:
(a) precipitates with Types B and O serum but not with A or AB serum
(b) does not precipitate with any of the serums

(c) precipitates with A, B, or O serum, but not with AB serum

28.9. List the three principal types of plasma proteins.

28.10. (a) What vitamin and what mineral (metal ion) are necessary to the blood-clotting process?
(b) Define fibrinogen, and describe its role in blood clotting.

28.11. Which of the following circumstances might interfere with normal blood clotting?
(a) low platelet level
(b) vitamin K deficiency
(c) calcium deficiency
(d) a dose of heparin

28.12. Write an equation that shows why

sodium oxalate can be used as an *in-vitro* anti-coagulant.

28.13. Write equations for (a) the carbonic acid–carbon dioxide equilibrium and (b) the carbonic acid–bicarbonate equilibrium. (c) Tell how the kidneys and lungs help maintain the pH of the blood at normal levels.

28.14. (a) Write the equation for $HHb\text{-}HbO_2^-$ equilibrium.

(b) Explain how a lack of oxygen in the atmosphere would affect this equilibrium.

(c) Explain how a lower than normal blood pH (acidosis) would affect this equilibrium.

(d) Give one reason why hemoglobin releases O_2 in the capillaries.

28.15. Write equations for the chemical reactions that carbon dioxide undergoes when it leaves the cells and enters the capillaries.

28.16. (a) At the arterial end of a capillary, which is greater—blood pressure or the colloidal osmotic pressure of the albumins?

(b) Which is greater at the venous end of a capillary?

(c) Describe the movement of interstitial fluids at each end of the capillaries.

28.17. What is the symptom of a low blood protein level? Explain why this symptom occurs.

28.18. List the functions of the kidneys.

28.19. List the effects on the urine of (a) drinking a large amount of fluid, (b) drinking very little fluid, (c) diabetes mellitus, (d) acidosis, and (e) eating a pound of liver.

28.20. Which of the following ions or compounds are normally found in the urine?

(a) NH_4^+ (b) urea   (c) uric acid
(d) Fe^{2+} (e) Cl^- (f) Na^+
(g) glucose   (h) proteins   (i) K^+
(j) phosphate ions

28.21. List four causes of proteinuria.

28.22. Describe the principal symptom of the lack of the hormone vasopressin in the bloodstream.

Appendix: A Review of Some Algebraic Manipulations

The ability to use simple algebra is essential to the study of chemistry. This review is intended for those students who may feel insecure about their mathematical background. We would also encourage these students to refer to a good beginning algebra text.

MULTIPLICATION

The process of multiplication may be represented in a number of ways.

$$5 \times 10 \qquad 5 \cdot 10 \qquad 5(10) \qquad (5)(10)$$

All mean 5 times 10.

If a number is used to multiply a series of terms, such as in the expression $10(a + b)$, then *each* term must be multiplied by the number.

$$10(a + b) \text{ means } 10a + 10b$$

FRACTIONS

A whole number is a number that does not contain a decimal or a fraction; examples of whole numbers are 1, 2, and 25. A number that contains a decimal or a fraction is not a whole number; 1.25 and $2\frac{1}{2}$ are examples.

A fraction is merely an expression of a division—that is, division of the numerator (the top) by the denominator (the bottom).

$$\frac{10}{5} \text{ means 10 divided by 5}$$

When another term is to be multiplied by a fraction, the term is multiplied by the numerator and divided by the denominator.

$$\frac{5}{9}(32) \text{ means } \frac{5 \times 32}{9}, \quad \text{or} \quad \frac{160}{9}$$

Be careful when carrying out arithmetic operations with fractions. Note that

$$\frac{x + y}{x} = \frac{x}{x} + \frac{y}{x}, \quad \text{or} \quad 1 + \frac{y}{x}$$

$$\frac{x + y}{x} \text{ does NOT equal } y.$$

The preceding equalities are more easily grasped with real numbers. For example,

$$\frac{6 + 5}{6} = \frac{11}{6} = 1.83 \text{ (not 5)}$$

A **ratio** is nothing more than a fraction. If we have 10 oranges in one basket and 5 in the other, the ratio of oranges in the two baskets is 10 to 5, or 10 : 5, or 10/5. The terms in a ratio, like those in a fraction, may be multiplied or divided by any number, so long as the numerator (top) and denominator (bottom) are multiplied by the same number.

$$\frac{10}{5} = \frac{10(100)}{5(100)} = \frac{2}{1} = \frac{8}{4}$$

When we speak of the ratio of atomic masses of hydrogen to oxygen in water, the ratio may be expressed as 1 : 8 or as 2 : 16.

$$\text{hydrogen : oxygen} = 1 : 8 = 2 : 16 \quad \text{or} \quad \frac{\text{hydrogen}}{\text{oxygen}} = \frac{1}{8} = \frac{2}{16}$$

DECIMALS AND SIGNIFICANT FIGURES

A decimal is simply a fraction of 10 or a fraction of a multiple of 10.

$$0.1 = \tfrac{1}{10} \qquad 0.01 = \tfrac{1}{100} \qquad 0.02 = \tfrac{2}{100} \qquad 5.3 = 5\tfrac{3}{10}$$

In solving problems involving either whole numbers or decimals, one should take note of the number of significant figures. This is particularly true when a calculator is used, because these calculators show more digits than are significant or meaningful. The definition of significant figures is best shown by examples.

0.02   *one significant figure*

0.023   *two significant figures*
          (Initial zeros are not significant.)

1.023   *four significant figures*

As may be seen in the examples, the placement of the decimal point and the number of *initial* zeros are immaterial in determining the number of significant figures. However, the intermediate or terminal zeros *are* counted as significant figures.

1.0300   *five significant figures*

1030   *three significant figures (but may be four)*

The answer to a mathematical problem should not contain more significant figures than are found in any term of the initial data.

$$6.19 \times 3.000 = 18.6 \text{ and not } 18.57$$

three   four   three   four

MULTIPLICATION AND DIVISION WITH DECIMALS

A calculator positions a decimal automatically. In hand calculations, however, the placement of a decimal in the answer must be determined.

When a number is to be multiplied by a decimal fraction, first carry out the multiplication and *then* place the decimal in the answer. To place the decimal, *add* the number of decimal places in each of the multiplied terms and use this sum as shown in the examples.

Example 0.05×25

Solution   1. Multiply.

$$5 \times 25 = 125$$

2. Sum the decimal places in the original terms.

0.05   25.

2 places + 0 places = 2 places

3. Insert the decimal in the answer.

1 25   or   1.25

2 places

To divide terms with decimals, first divide. Then *subtract* the number of decimal places in the denominator from the number of decimal places in the numerator.

Example $5.04/0.20$

Solution   1. Divide.

$$504 \div 2 = 252$$

2. Subtract the decimal places in the original terms.

$$5.04 \qquad 0.2$$

2 places − 1 place = 1 place

3. Insert decimal.

$$252 \qquad or \qquad 25.2$$

1 place

NUMBER-LINE SYSTEM FOR METRIC CONVERSIONS

One of the advantages of the metric system is that we can convert one unit to another without a lengthy calculation. Until you are familiar enough with metric units to make conversions in your head, you might find the number-line system helpful.

In this system, place the metric prefixes on a line with their appropriate numbers, as shown. To convert from one unit to another, simply move along the line, counting the number of spaces moved and noting the direction. If the move is to the right, move the decimal point to the right the appropriate number of spaces.

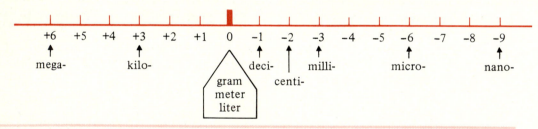

Example   Convert 250 milliliters to liters.

Solution   Start at milli-, and move *three spaces to the left* to liter (at zero). Thus, move the decimal three spaces to the left:

$$250 \text{ mL} = 0.250 \text{ liter}$$

3 2 1

Example   Convert 0.0053 centimeter to nanometers.

Solution   Start at centi- and count the spaces to nano-; move seven spaces to the right, adding zeros as necessary.

$$0.0053000 \text{ cm} = 53,000 \text{ nm}$$

Example   Convert 0.010 kilometer to centimeters.

Solution   From kilo- to centi- is a five-space move to the right.

$$0.01000 \text{ km} = 1000 \text{ cm}$$

PERCENTAGES

A percent is a fraction multiplied by 100, or parts per 100. To calculate what percent by weight of a mixture of nuts is peanuts, you would need the following information: (1) the weight of the peanuts in the mixture and (2) the total weight of all of the nuts together (peanuts included).

$$\%(w/w) \text{ peanuts} = \frac{\text{wt of peanuts}}{\text{total wt of nuts}} \times 100 = \frac{\text{wt of peanuts in g}}{100 \text{ g of mixture}}$$

Similarly, to calculate the approximate percent by weight of oxygen in water, divide the atomic mass of oxygen by the formula weight of water, and multiply by 100.

$$\%O = \frac{\text{atomic mass of O}}{\text{formula wt of H}_2\text{O}} \times 100$$

$$= \frac{16}{18} \times 100$$

$$= 89\%$$

ALGEBRAIC EQUATIONS

An algebraic equation lends itself to almost any mathematical manipulation. The equality between the two sides of the equation will still be present—*as long as the same operation is carried out on each side of the equation.*
    If $x = y$, then

$$23.7x = 23.7y$$

or
$$\frac{x}{8.2} = \frac{y}{8.2}$$

or
$$x + 693.1 = y + 693.1$$

or
$$(x + 5)^2 = (y + 5)^2$$

An equation may be rearranged by, for example, the addition or subtraction of a term from both sides of the equation:

original equation:
$$P_{\text{total}} = P_a + P_b$$

subtraction of P_b from both sides of the equation:
$$P_{\text{total}} - P_b = P_a + P_b - P_b$$

rearranged equation:
$$P_{\text{total}} - P_b = P_a$$

There is no need to go through the above steps to rearrange the original equation if you keep in mind that, when a term is shifted from one side of the equation to the other, the sign of the term is changed.

$$P_{\text{total}} = P_a + P_b$$

$$P_{\text{total}} - P_b = P_a \qquad \textcolor{red}{\textit{(Note change in sign of } P_b.)}$$

A term that is a multiplier or a divider may be shifted to the other side of the equation by multiplying or dividing both sides of the equation by that term:

original equation:

$$P_1 V_1 = P_2 V_2$$

division of both sides by V_1:

$$\frac{P_1 V_1}{V_1} = \frac{P_2 V_2}{V_1}$$

rearranged equation:

$$P_1 = \frac{P_2 V_2}{V_1} \qquad \textcolor{red}{\textit{Since } V_1/V_1 = 1, \textit{ the left side}}$$
$$\textcolor{red}{\textit{of the equation becomes } P_1.}$$

To shift a divider, multiply both sides of the equation by that term:

original equation:

$$\frac{V_1}{T_1} = \frac{V_2}{T_2}$$

multiplication of both sides by T_1:

$$\frac{V_1 T_1}{T_1} = \frac{V_2 T_1}{T_2}$$

rearranged equation:

$$V_1 = \frac{V_2 T_1}{T_2}$$

CANCELING

Identical terms or units in a fraction may be canceled:

$$\frac{(1.00 \ M)(100 \ \text{mL})}{1.5 \ M} \qquad \textcolor{red}{M \text{ appears in the numerator and}}$$
$$\textcolor{red}{\text{denominator and may be canceled.}}$$

Similarly, a multiplier or divider may be canceled if it appears on each side of an equation:

$$(1.50 \ M)(V) = (1.00 \ M)(100 \ \text{mL})$$

An added or subtracted term may also be canceled if the same term appears on each side of an equation. (This is true with some types of chemical equations as well as with algebraic equations.)

$$x + y - s = m + n - s$$

$$Na^+ + Cl^- + Ag^+ + NO_3^- \longrightarrow AgCl\downarrow + Na^+ + NO_3^-$$

EXPONENTS

An exponent of a base is the number of times that the base is to be multiplied by itself.

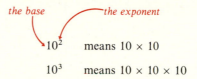

$$10^2 \quad \text{means } 10 \times 10$$

$$10^3 \quad \text{means } 10 \times 10 \times 10$$

Exponents to the base 10 are often used when extremely large or small numbers are encountered. Positive exponents are used to denote large numbers:

$$10^1 = 10 \text{ (1 with one zero)}$$
$$10^2 = 100 \text{ (1 with two zeros)}$$
$$10^3 = 1000 \text{ (1 with three zeros)}$$
$$5 \times 10^2 = 500 \text{ (5 with two zeros)}$$
$$5 \times 10^6 = 5{,}000{,}000 \text{ (5 with six zeros)}$$
$$6.023 \times 10^{23} = 602{,}300{,}000{,}000{,}000{,}000{,}000{,}000$$

Negative exponents are used to represent small numbers:

$$10^{-1} = 0.1 \text{ (decimal moved back one place from 1)}$$
$$10^{-2} = 0.01 \text{ (decimal moved back two places from 1)}$$
$$10^{-3} = 0.001 \text{ (decimal moved back three places from 1)}$$
$$5 \times 10^{-3} = 0.005 \text{ (decimal moved back three places from 5)}$$

For use in problem solving, exponents may be changed by proper manipulation of the multiplier.

$$1 \times 10^{-5} = 10 \times 10^{-6}$$
$$1 \times 10^3 = 0.1 \times 10^4$$
$$5.61 \times 10^6 = 56.1 \times 10^5$$

MULTIPLYING WITH EXPONENTS

When two numbers with exponents are to be multiplied, the exponents are *added*:

$$(10^5)(10^3) = 10^8 \longleftarrow 5 + 3$$
$$(10^{-3})(10^2) = 10^{-1} \longleftarrow -3 + 2$$
$$(2 \times 10^{-2})(5 \times 10^{-3}) = 10 \times 10^{-5} \longleftarrow -2 + (-3)$$

DIVIDING WITH EXPONENTS

When numbers with exponents are divided, the exponents are *subtracted* (denominator from numerator):

$$\frac{10^3}{10^2} = 10^1 \qquad 3 - 2$$

$$\frac{10^{-3}}{10^2} = 10^{-5} \qquad -3 - 2$$

$$\frac{10^6}{10^{-2}} = 10^8 \qquad 6 - (-2)$$

$$\frac{4 \times 10^6}{8 \times 10^{-2}} = \frac{4}{8} \times 10^8 = 0.5 \times 10^8$$

Answers to Problems

The answers to the chapter-end study problems are in the study guide.

Chapter 1

1.1. 5.0 gr × 0.0648 g/gr × 1.0 mg/0.001 g = 324 mg

1.2. 1.0 oz × 1.0 g/0.03527 oz = 28 g
2.0 cups × 1.0 qt/4 cups × 1.0 L/1.057 qt × 1000 mL/L = 473 mL

1.3. (a) 10°C   (b) 122°F

1.4. 40.5°C

Chapter 2

2.1. (a) mixture   (b) element   (c) compound   (d) compound
(e) mixture

2.2. (a) 250 g × 80° × 1.0 cal/g-deg = 20,000 cal
(b) 250 cal/deg   (c) 14.25 cal/deg

2.3. (a) 355 cal   (b) 2700 cal for H_2O
(c) Extra energy is required to overcome the strong attractions between water molecules.

Chapter 3

3.1. atomic no. = no. of p^+ = no. of e^- = 8
no. of n^0 = 16 − 8 = 8

3.2.

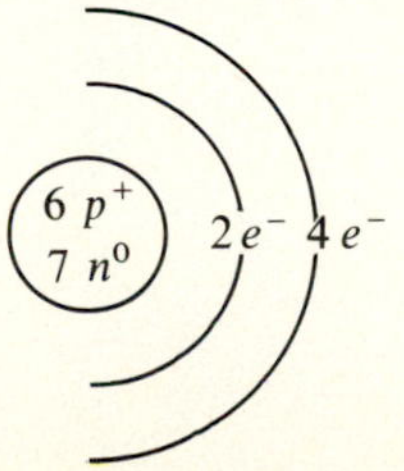

$6\ p^+$
$7\ n^0$ $2\,e^-\ 4\,e^-$

3.3. P: $2\,e^-$ $8\,e^-$ $5\,e^-$
S: $2\,e^-$ $8\,e^-$ $6\,e^-$

3.4.
	$1s$	$2s$	$2p$	$2p$	$2p$	$3s$	$3p$
Al:	↑↓	↑↓	↑↓	↑↓	↑↓	↑↓	↑

3.5. (a) 4   (b) 5   (c) 6

3.6. The outer electron shells of He and Ne are filled to capacity; therefore, these elements do not truly have any valence electrons. *Zero* means no valence electrons.

Chapter 4

4.1. (a) lose $1\,e^-$ (b) lose $2\,e^-$ (c) lose $3\,e^-$

4.2. $:\!\overset{..}{\underset{..}{O}}$ $+\ 2\,e^-\ \longrightarrow\ :\!\overset{..}{\underset{..}{O}}\!:^{2-}$

oxygen atom   oxide ion

4.3. (a) $2\,Na\ +\ Cl_2\ \longrightarrow\ 2\,Na^+\,Cl^-$

reducing   oxidizing   *oxidized*   *reduced*
agent   agent

(b) $SnCl_2\ +\ Cl_2\ \longrightarrow\ SnCl_4$
   (+2)   (0)   (+4)(−1)

reducing   oxidizing   *oxidized*   *Cl$_2$ reduced to 2 Cl$^-$*
agent   agent

4.4. (a) CaS   (b) Na_2O (c) Al_2O_3

4.5. (a) $Cu^+\,I^-$ (b) $Cu^{2+}\ +\ 2\,I^-$ (c) $2\,Cu^+\ +\ S^{2-}$
   (+1)   (+2)   (+1)

4.6. (a) potassium bromide   (b) silver oxide
(c) copper(I) iodide or cuprous iodide
(d) copper(II) iodide or cupric iodide

4.7. (a) $H\,(:\!\overset{H}{\underset{..}{N}}\!:)\,H$ (b) $H\,(:\!\overset{H}{\underset{H}{C}}\!:)\,H$

(c) $:\!\overset{..}{\underset{..}{Cl}}\,(:\!\overset{:\overset{..}{\underset{..}{Cl}}:}{\underset{..}{C}}\!:)\,H$
       $:\!\overset{..}{\underset{..}{Cl}}\!:$

4.8. (a) $(:\!N(::)N:)$ or $N\!\equiv\!N$

(b) $(:\!C(::)C:)$ or $\overset{H}{\underset{H}{}}C\!=\!C\overset{H}{\underset{H}{}}$

4.9. (a) $\overset{F^{\delta-}}{\underset{F^{\delta-}\ \ F^{\delta-}}{B^{\delta+}}}$ (b) $^{\delta-}O\!=\!\overset{\delta+}{C}\!=\!O^{\delta-}$ (c) $\overset{H^{\delta+}}{\underset{\delta+H\quad H^{\delta+}}{N^{\delta-}}}$

4.10. (a) nitrogen dioxide   (b) dinitrogen trioxide
(c) carbon disulfide

4.11. (a) SO_3 (b) P_2O_3 (c) PCl_5

4.12. (a) magnesium hydroxide   (b) ammonium carbonate
      (c) calcium bicarbonate   (d) potassium sulfate

4.13. (a) $MgCO_3$, magnesium carbonate
      (b) $Fe_2(SO_4)_3$, iron(III) sulfate or ferric sulfate
      (c) $(NH_4)_3PO_4$, ammonium phosphate

4.14. (a) $+2$ (b) $+3$ (c) $+5$ (d) $+7$ (e) $+6$

Chapter 5

5.1. (a) and (b) double displacement
     (c) single displacement   (d) combination

5.2. (a) endothermic   (b) more energy
     (c) No, energy would be absorbed

5.3. $H_2CO_3 + H_2O \rightleftharpoons HCO_3^- + H_3O^+$

5.4. The water becomes acidic through absorption of CO_2 from the air ($H_2O + CO_2 \rightarrow H_2CO_3$), followed by the reaction in Answer 5.3.

5.5. (a) Add NH_4^+: the equilibrium shifts to the left side of the equation (more NH_3 and H_2O, less OH^-).
     (b) Add NH_3: the equilibrium shifts to the right (more NH_4^+ and OH^-).

5.6. (a) $Mg(OH)_2 + 2\,HCl \longrightarrow MgCl_2 + 2\,H_2O$
     (b) $CaCO_3 + 2\,HCl \longrightarrow CaCl_2 + H_2O + CO_2$
     (c) $4\,Fe + 3\,O_2 \longrightarrow 2\,Fe_2O_3$

5.7. (a) 74.6 amu   (b) 94.3 amu   (c) 102.0 amu   (d) 28.0 amu

5.8. (a) $23.0/58.5 \times 100 = 39.3\%$ (b) 27.4%

5.9. (a) 44.0 g   (b) 70.0 g   (c) 23.0 g

5.10. $CH_3CH_2OH + 3\,O_2 \longrightarrow 2\,CO_2 + 3\,H_2O$

$$x = 0.5 \text{ mole ethanol} \times \left(\frac{3 \text{ moles water}}{1 \text{ mole ethanol}} \right) = 1.5 \text{ moles}$$

5.11. (a) 11.5 g   (b) 216 g   (c) 38.0 g

5.12. (a) 0.085 mole   (b) 2.78 moles   (c) 0.020 mole

5.13. $C_3H_8 + 5\,O_2 \longrightarrow 3\,CO_2 + 4\,H_2O$

$$\frac{100 \text{ g}}{44.1 \text{ g/mole}} = 2.27 \text{ moles } C_3H_8$$

$$g\ CO_2 = 2.27 \text{ moles } C_3H_8 \times \frac{3 \text{ moles } CO_2}{1 \text{ mole } C_3H_8} \times \frac{44.0 \text{ g}}{1 \text{ mole } CO_2}$$

$$= 300 \text{ g}$$

Chapter 6

6.1. Heating increases the kinetic energy of a gas so that the molecules or atoms hit the sides of the container more frequently and with greater impact. Thus, the pressure of the gas increases. If the pressure exceeds the strength of the weakest part of the can, the can explodes.

6.2. $P_2 = 2$ atm

6.3. 1.06 L

6.4. 0.78 atm or 593 mmHg

6.5. 180 mmHg

Chapter 7

7.1. (a), (c), and (d) are water soluble because each contains one or more groups that can form hydrogen bonds with water and none contains a large nonpolar group. (b) is water insoluble because it contains no polar group.

7.2. $CH_3CH_2\overset{..}{\underset{..}{O}}H\cdots\overset{..}{\underset{..}{O}}H_2$

7.3. $\dfrac{5.0\text{ g}}{1000\text{ mL}} \times 100\text{ mL} = 0.5\%\text{ (w/v)}$

7.4. 150 g

7.5. 3.0 g

7.6. 12% (v/v)

7.7. (a) 1.0 mg%   (b) 10 mg/L, or 10 ppm

7.8. (a) moles $= MV = (1.6\text{ moles/L})(0.100\text{ L}) = 0.16$
    (b) 0.16 mole $\times$ 40.0 g/mole = 6.4 g

7.9. 0.15 mole

7.10. (a) $(6.0\%)V_1 = (1.0\%)(200\text{ mL})$
          $V_1 = 33\text{ mL}$
      Dilute 33 mL of the first solution to 200 mL.
    (b) $(5.0\%)V_1 = (2.0\%)(2.5\text{ L})$
          $V_1 = 1.0\text{ L}$
      Dilute 1.0 L to 2.5 L.

7.11. 0.150 M

7.12. milliosmolarity $= 0.300 \times 1000 = 300$

7.13. (a) 0.5 M NaCl $\times$ 2 = 1.0 ⎱ same osmolarities; equal osmotic P
          1.0 M glucose $\times$ 1 = 1.0 ⎰
    (b) 0.20 M CaCl$_2$ $\times$ 3 = 0.6 ⎱ The osmotic pressure of the CaCl$_2$ solution is $\frac{6}{4}$ (or 1.5)
          0.20 M NaCl $\times$ 2 = 0.4 ⎰ times that of the NaCl solution

7.14. Sea water is "saltier," or *hypertonic*, with respect to body fluids. If sea water is ingested as a main source of water, water is drawn out of the tissues by *osmosis* and the person becomes dehydrated.

Chapter 8

8.1. Charged colloidal particles are attracted to the electrical plate of opposite charge and are neutralized there. Once the particles have given up their electrical charges, they can coalesce into larger particles and can precipitate.

8.2. (b), (c), and (d) all contain long, nonpolar chains and ionic ends, so they could act as emulsifying agents for oils in water.

Chapter 9

9.1. (a) $CaSO_4 \xrightarrow{\ H_2O\ } Ca^{2+} + SO_4{}^{2-}$
    (b) $HBr \xrightarrow{\ H_2O\ } H^+ + Br^-$
      or $HBr + H_2O \longrightarrow H_3O^+ + Br^-$

(c) $LiOH \xrightarrow{H_2O} Li^+ + {}^-OH$

9.2. (a) salt   (b) base   (c) acid

9.3. (a) $KOH + HNO_3 \longrightarrow KNO_3 + H_2O$

(b) $2 HCl + Ca(OH)_2 \longrightarrow CaCl_2 + 2 H_2O$

(c) $H_2SO_4 + 2 NaOH \longrightarrow Na_2SO_4 + 2 H_2O$

    or $H_2SO_4 + NaOH \longrightarrow NaHSO_4 + H_2O$

9.4. (a) contains Na^+, water soluble

(b) and (d) contain $SO_4{}^{2-}$, soluble

(c) contains Cl^-, soluble   (e) insoluble

9.5. (a) $H^+ + {}^-O_2CCH_3 \xrightarrow{H_2O} HO_2CCH_3$

(b) $Ag^+ + Cl^- \xrightarrow{H_2O} AgCl\downarrow$

(c) $2 H^+ + CO_3{}^{2-} \xrightarrow{H_2O} H_2CO_3 \longrightarrow H_2O + CO_2\uparrow$

*9.6. $Mg^{2+} + 2 CH_3(CH_2)_{16}CO_2{}^- \xrightarrow{H_2O} [CH_3(CH_2)_{16}CO_2]_2Mg\downarrow$

*9.7. (a) $Ca^{2+} + CO_3{}^{2-} \longrightarrow CaCO_3\downarrow$

(b) $Mg^{2+} + B_4O_7{}^{2-} \longrightarrow MgB_4O_7\downarrow$

Chapter 10

10.1. (a) $Ca: + 2 HNO_3 \longrightarrow Ca(NO_3)_2 + H_2\uparrow$

(b) $Mg: + 2 HCl \longrightarrow MgCl_2 + H_2\uparrow$

10.2. $CaCO_3 + H_2SO_4 \xrightarrow{H_2O} Ca^{2+} + SO_4{}^{2-} + H_2O + CO_2\uparrow$

10.3. $CH_3\overset{\overset{\displaystyle O}{\|}}{C}OH + NaHCO_3 \longrightarrow CH_3\overset{\overset{\displaystyle O}{\|}}{C}O^- + Na^+ + H_2O + CO_2\uparrow$

10.4. (a) $CH_3\overset{\overset{\displaystyle O}{\|}}{C}O^- + H_2O \rightleftharpoons CH_3\overset{\overset{\displaystyle O}{\|}}{C}OH + {}^-OH$

(b) $H_2PO_4{}^- + H_2O \rightleftharpoons HPO_4{}^{2-} + H_3O^+$

10.5. $CH_3\overset{\displaystyle\cdot\cdot}{N}H_2 + H_2O \rightleftharpoons CH_3\overset{\overset{\displaystyle H}{|}}{\underset{+}{N}}H_2 + {}^-OH$
   weak base

10.6. (a) $\underset{base}{Na_2O} + \underset{acid}{H_2O}$

(b) $\underset{base}{Na_2CO_3} + \underset{acid}{H_2CO_3}$

(c) $\underset{acid}{CH_3\overset{\overset{\displaystyle O}{\|}}{C}OH} + \underset{base}{NH_3}$

10.7. (a) $Mg(OH)_2 + 2 HCl \longrightarrow MgCl_2 + 2 H_2O$
   stronger base   *stronger acid*
   than $MgCl_2$ *than H_2O*

(b) $NH_3 + HNO_3 \longrightarrow NH_4{}^+ NO_3{}^-$
   stronger base   *stronger acid*
   than $NO_3{}^-$ *than $NH_4{}^+$*

10.8. $[H^+] = 1 \times 10^{-4.0} M = 0.00010 M$

10.9. (a) 3 (the $-CO_2H$ groups)   (b) $192.1/3 = 64.0$ g

10.10. 0.142 eq. NaCl $\times \dfrac{58.5 \text{ g NaCl}}{\text{eq. NaCl}} = 8.31$ g

10.11. (a) wt of 1.0 mEq $Ca^{2+} = \dfrac{40.1}{2} \times 0.001 = 0.020$ g

wt of 5.0 mEq $Ca^{2+} = 0.10$ g

(b) 0.096 g

*10.12. (a) $0.213\ N$ (b) $0.020\ N$

*10.13. (a) $3.0\ M$ (b) $0.17\ M$

*10.14. (a) $0.23\ N$ (b) 0.23 eq./L $\times 0.010$ L $= 0.0023$ eq.

10.15. $NaHCO_3 + Ca(OH)_2 \longrightarrow CaCO_3\downarrow + NaOH + H_2O$

or $2\ NaHCO_3 + Ca(OH)_2 \longrightarrow CaCO_3\downarrow + Na_2CO_3 + 2\ H_2O$

Chapter 11

11.1. $3\ \overset{0}{Cu} + 8\ H\overset{+5}{N}O_3 \longrightarrow 3\ \overset{+2}{Cu}(NO_3)_2 + 2\ \overset{+2}{N}O + 4\ H_2O$

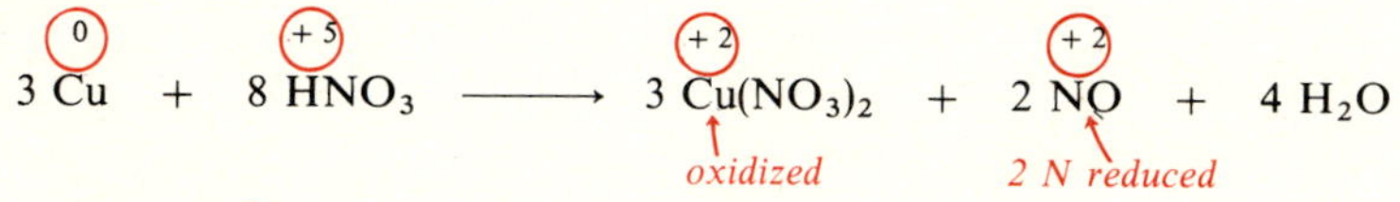

oxidized   2 N reduced

11.2. $2\ \overset{-4}{N}H_4\overset{+5}{N}\overset{-2}{O}_3 \longrightarrow 2\ \overset{0}{N_2} + \overset{0}{O_2} + 4\ H_2O$

2 N reduced   2 O oxidized

2 N oxidized

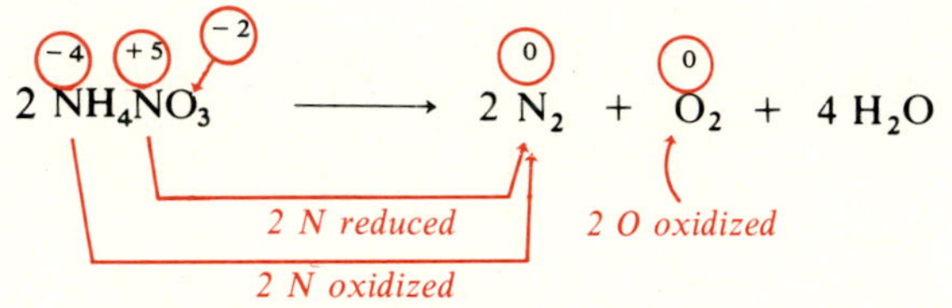

11.3. Cu is the reducing agent; HNO_3 is the oxidizing agent.

11.4. (a)

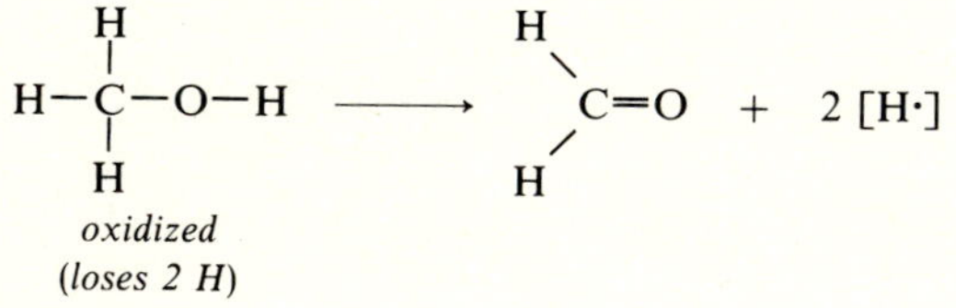

$$H-\underset{\underset{H}{|}}{\overset{\overset{H}{|}}{C}}-O-H \longrightarrow \overset{H}{\underset{H}{>}}C=O + 2\ [H\cdot]$$

oxidized
(loses 2 H)

*11.5. (a) 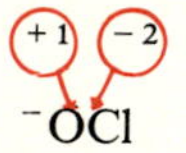 $\overset{+1\ -2}{^-OCl}$

(b) $OCl^- + Cl^- + 2\ H^+ \longrightarrow Cl_2 + H_2O$

Chapter 12

12.1. (a) $^{235}_{92}U$ (b) $^{37}_{17}Cl$

12.2. (a) $^{14}_{6}C$ (b) $^{11}_{6}C$

12.3. (a) $500\ \mu Ci$ (b) 36 mCi

12.4. (a) 0.150 rem   (b) 3000 mrems

12.5. 1.4 mrems

Chapter 13

13.1. (a) $ClCH_2CH_2Br$ (b) $ClCH=CHCl$

13.2. 120°

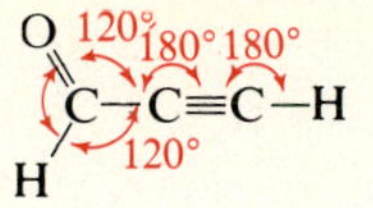

$$\overset{O}{\overset{\|}{C}}-\overset{}{C}\equiv\overset{}{C}-H$$

(with angle annotations: 120°, 180°, 180°, 120° around the structure, and H below the C)

13.3. (d) is the isomer (OH on end carbon)

13.4. (a) same compound (Cl on end carbon)

(b) isomers   (c) same compound (C=O in center position)

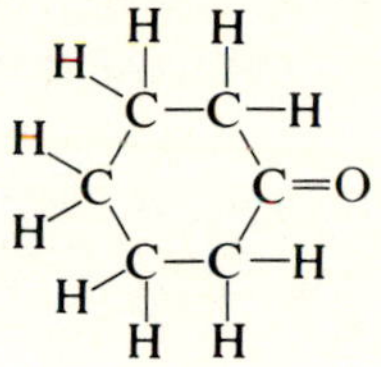

13.5. (a)

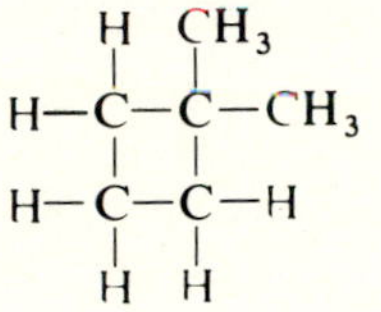

$$\begin{array}{c} H \quad CH_3 \\ H-C-C-CH_3 \\ H-C-C-H \\ H \quad H \end{array}$$

(b)

(ring structure with H, H atoms and C=O)

(c)

(ring structure with C=C double bond and H atoms)

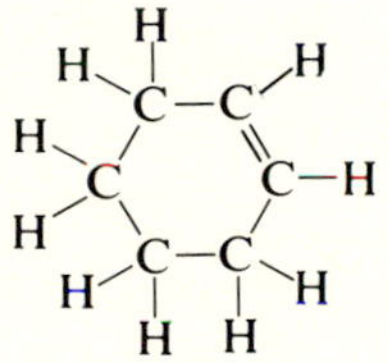

13.6. Only (c) represents structural isomers, Cl substituents in 1,3 and 1,4 positions as we move around the ring.

13.7. (a) RC≡CR   (b) $\overset{O}{\overset{\|}{R C R}}$

13.8. (a) $CH_3CH_2CH_2\overset{O}{\overset{\|}{C}OH}$ *carboxyl group*

(b) $CH_3\overset{NH_2}{\underset{|}{C}HCH_3}$ *amino group*

(c) H_2N —(ring)— CO_2H *amino, double bond, and carboxyl groups*

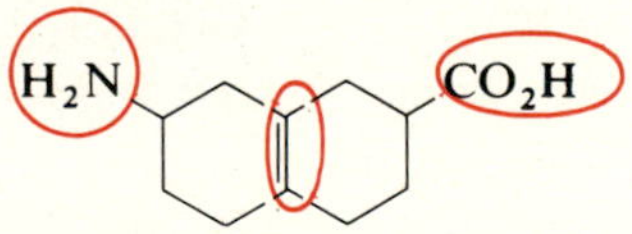

(d) (ring)—CH_2CH_2 OH *double bond and hydroxyl*

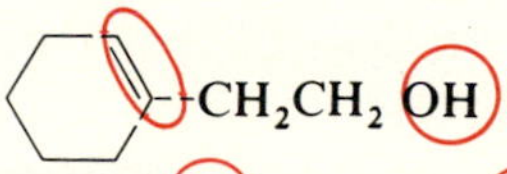

13.9. (a) CH_3CH_2 Cl (b) CH_3 CH_2OC H_2CH_3

(c) CH_3 $\overset{CH_3}{N}$ CH_3 (d) (ring with Br and =O)

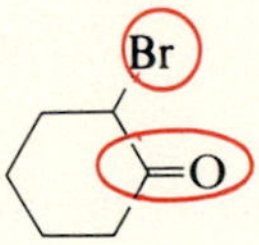

Chapter 14

14.1. (a) cyclopropane   (b) cyclobutane   (c) cyclopentane

14.2. (a) $CH_3CH_2CH_2\overset{CH_2CH_2CH_3}{\underset{|}{C}H}CH_2CH_2CH_3$

(b)
$$
\begin{array}{c}
\overset{\displaystyle CH(CH_3)_2}{|} \\
CH_3CH_2CH_2CHCH_2CH_2CH_2CH_3
\end{array}
$$

(c) 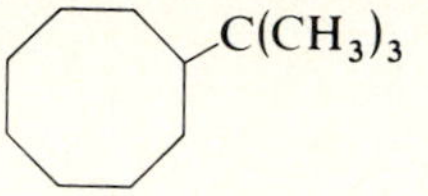
(octane ring with $C(CH_3)_3$ substituent)

14.3. (a) 2,2-dimethylbutane   (b) 1-ethyl-2-isopropylcyclohexane

14.4. (a) 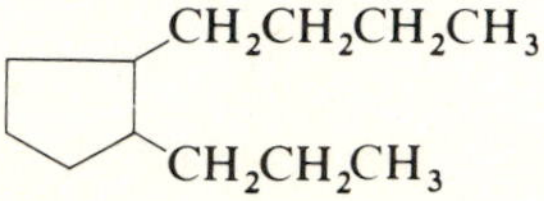
(cyclopentane ring with $CH_2CH_2CH_2CH_3$ and $CH_2CH_2CH_3$ substituents)

(b)
$$
\begin{array}{c}
\overset{\displaystyle CH_3}{|} \quad \overset{\displaystyle CH_3}{|} \\
CH_3C\!\!-\!\!-\!\!-\!\!CCH_2CH_2CH_3 \\
\underset{\displaystyle CH_3}{|} \quad \underset{\displaystyle CH_3}{|}
\end{array}
$$

14.5.
$$
\begin{array}{c}
\overset{\displaystyle H}{} \\
H\!:\!\overset{..}{\underset{..}{C}}\!:\!\overset{..}{\underset{..}{Cl}}\!: \quad + \quad H\!:\!\overset{..}{\underset{..}{Cl}}\!: \\
\underset{\displaystyle H}{}
\end{array}
$$

14.6. (a)
$$
\begin{array}{c}
\overset{\displaystyle CH_3}{|} \\
H_3C\!-\!C\!-\!CH_2Cl \quad + \quad HCl \\
\underset{\displaystyle CH_3}{|}
\end{array}
$$

1-chloro-2,2-dimethylpropane

(b) 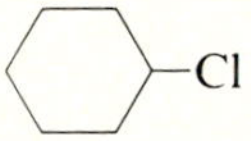
—Cl   +   HCl

chlorocyclohexane

In either case, replacement of any one H leads to the same product.

14.7. CH_3CH_2Cl + CH_3CHCl_2 + CH_3CCl_3
   chloroethane   1,1-dichloroethane   1,1,1-trichloroethane

   + $ClCH_2CH_2Cl$ + $ClCH_2CHCl_2$ + $ClCH_2CCl_3$
   1,2-dichloroethane   1,1,2-trichloroethane   1,1,1,2-tetrachloroethane

   + $Cl_2CHCHCl_2$ + Cl_2CHCCl_3 + Cl_3CCCl_3
   1,1,2,2-tetrachloroethane   pentachloroethane   hexachloroethane

14.8. (a) $CH_3CH_2CH_3 + 5\,O_2 \longrightarrow 3\,CO_2 + 4\,H_2O$ (b) CO_2, CO, H_2O, C

Chapter 15

15.1. (a) 3-chloro-1-butene   (b) cyclohexene

15.2. (a)
$$
\begin{array}{c}
\overset{\displaystyle Br}{|} \quad \overset{\displaystyle Br}{|} \\
CH\!\!=\!\!CCH_2CH_2CH_2CH_2CH_3
\end{array}
$$
   (b) 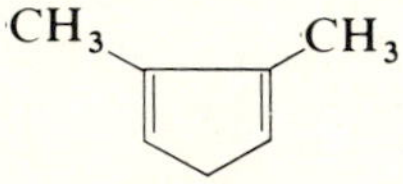
(cyclopentadiene ring with two CH_3 groups)

15.3. (a) *cis*-1,2-dichloroethene   (b) *trans*-1-chloro-1-butene
   (c) *cis*-1,3-dichlorocyclohexane

15.4. (a)

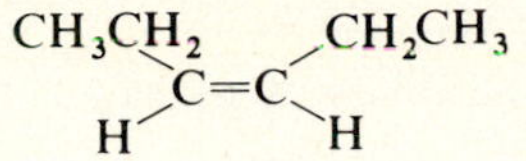

(b)

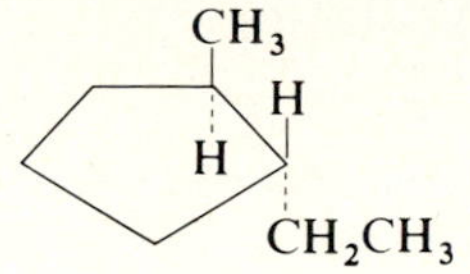

15.5. (a) $CH_3CH_2CH_3$ (b)  $CH_3\overset{\underset{|}{OH}}{CH}CH_2CH_3$ (c)

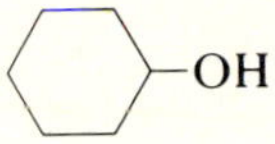

15.6. (a) 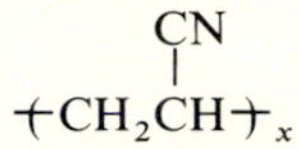 $+CH_2CH+_x$ (b) 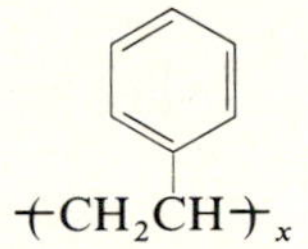 $+CH_2\overset{\underset{|}{CN}}{CH}+_x$

15.7. (a) *p*-chlorotoluene   (b) *m*-methylphenol or *m*-hydroxytoluene
(c) 2,4-dimethyl-2-phenylpentane

15.8. (a)

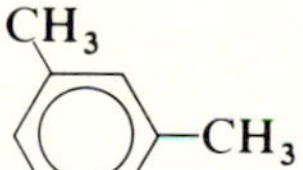

(b) 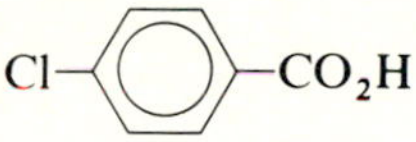 Cl—

—CO_2H

15.9.

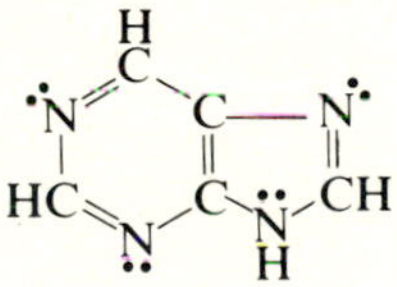

Chapter 16

16.1. (a) $CH_3\overset{\underset{|}{OH}}{CH}CH_3$ *hydroxyl (alcohol)*

(b) $CH_3OCH_2CH_2CH_2OH$ *ether*   *hydroxyl (alcohol)*

(c)

phenol

(d) $CH_2=CH—O—CH=CH_2$ *ether*   *carbon-carbon double bond (alkene)*

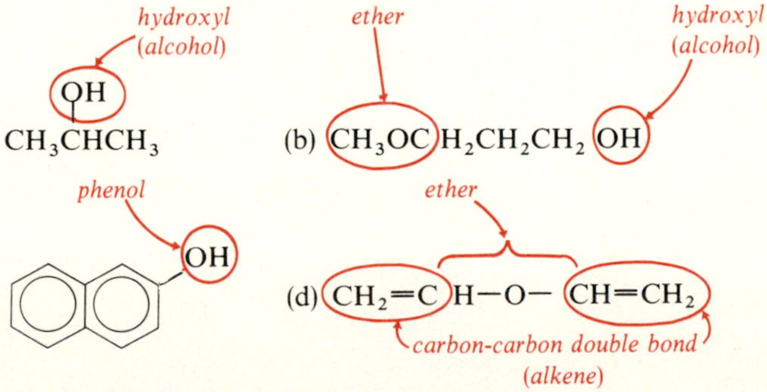

16.2. (a) 2-methyl-1-propanol or isobutyl alcohol, primary
(b) 2,3-dimethyl-3-pentanol, tertiary
(c) cyclohexanol, secondary
(d) 1-methyl-1-cyclohexanol, tertiary

16.3. Diethyl ether contains no partially positive H atoms, so there can be no hydrogen bonds between the ether molecules; therefore, diethyl ether is low boiling. Diethyl ether *does* contain an O atom with unshared electrons, so it can form hydrogen bonds with water; therefore, it is partially miscible with water.

$$R\ddot{O}:\text{---}H\ddot{O}:$$
$$\underset{R}{|} \quad \underset{H}{|}$$

16.4. (c) has a greater proportion of nonpolar CH bonds and would be the most soluble in a nonpolar solvent: like dissolves like.

16.5. ⬡—OH + H_2SO_4 $\xrightarrow{\text{heat}}$ ⬡ + $H_2SO_4 \cdot H_2O$

16.6. (a) $CH_3CH_2\overset{\overset{\displaystyle O}{\|}}{C}CH_3$ (b) ⬡=O   (c) $HO\overset{\overset{\displaystyle O}{\|}}{C}CH_2\overset{\overset{\displaystyle O}{\|}}{C}OH$

16.7. (a) diethylamine   (b) ethylmethylamine
16.8. (a) primary   (b) tertiary   (c) secondary

16.9. $CH_3\overset{\overset{\displaystyle O}{\|}}{C}OH + CH_3NH_2 \longrightarrow CH_3\overset{\overset{\displaystyle O}{\|}}{C}O^- + CH_3\overset{+}{N}H_3$

16.10. A glycine molecule contains an acidic $-CO_2H$ group and a basic $-NH_2$ group. Its molecules undergo an internal acid-base reaction to yield molecules that each contain a + and a − ionic charge. Thus, glycine molecules are very strongly attracted to one another, just as are the ions of a salt.

$$H_2\ddot{N}CH_2\overset{\overset{\displaystyle O}{\|}}{C}O\text{—}H \longrightarrow H_3\overset{+}{N}CH_2\overset{\overset{\displaystyle O}{\|}}{C}O^-$$

attracts $-CO_2^-$ *of other molecules*   *attracts* $-\overset{+}{N}H_3$ *of other molecules*

16.11. $(CH_3)_3\overset{+}{N}\text{—}H\,Cl^- + Na^+ + OH^- \longrightarrow (CH_3)_3N: + H_2O + Na^+ + Cl^-$
16.12. (a) $HOCH_2CH_2NH_2 + H_2O + Cl^-$
       (b) no reaction (no H on N to react with a base)

16.13. $Ag^+ + H_2\ddot{N}CH_2CH_2\ddot{N}H_2 \longrightarrow$ $Ag^+ \overset{\ddot{N}H_2}{\underset{\ddot{N}H_2}{\underset{\displaystyle |}{\overset{\displaystyle CH_2}{\underset{\displaystyle CH_2}{}}}}}$

Chapter 17

17.1. (a)

O ‖ COH *carboxyl*
⬡
CH ‖ O *aldehyde*

(b) $H_2N\overset{\overset{\displaystyle O}{\|}}{C}CH_2O\overset{\overset{\displaystyle O}{\|}}{C}CH_3$
amide   *ester*

17.2.

$$CH_3\overset{\overset{\displaystyle :\overset{..}{O}:}{\|}}{\underset{..}{C}}\overset{..}{\underset{..}{O}}H \cdots H\overset{..}{\underset{..}{O}}\overset{\overset{\displaystyle \overset{..}{O}}{\|}}{C}CH_3$$

$$CH_3\overset{\overset{\displaystyle \cdot\overset{..}{O}\cdot}{\|}}{C}\overset{..}{\underset{..}{O}}H$$
$$\vdots$$
$$H\overset{..}{\underset{..}{O}}\overset{\overset{\displaystyle \cdot\overset{..}{O}\cdot}{\|}}{C}CH_3$$

A pair of carboxylic acid molecules actually form a *pair* of hydrogen bonds:

$$CH_3C\overset{\overset{\displaystyle \overset{..}{O}:\cdots H\overset{..}{O}}{}}{\underset{\underset{\displaystyle \overset{..}{O}H\cdots:\overset{..}{O}}{}}{}}CCH_3$$

17.3. (a) hexanal   (b) phenylethanal, 2-phenylethanal, or phenylacetaldehyde
(c) 2-chloro-5-methyl-3-hexanone

17.4. (a) and (b) $CH_3CH_2\overset{\overset{\displaystyle OH}{|}}{CH_2}$ (c) $CH_3CH_2\overset{\overset{\displaystyle O}{\|}}{C}OH$ or $CH_3CH_2\overset{\overset{\displaystyle O}{\|}}{C}O^-$

17.5. (a) $CH_3CH_2\overset{\overset{\displaystyle OH}{|}}{CH}-OCH_2CH_3 \xrightarrow[-H_2O]{CH_3CH_2OH} CH_3CH_2\overset{\overset{\displaystyle OCH_2CH_3}{|}}{CH}-OCH_2CH_3$

                     a hemiacetal   *an acetal*

(b) $CH_3\overset{\overset{\displaystyle OH}{|}}{CH}-OCH_2CH_2OH \xrightarrow{-H_2O} CH_3\overset{\overset{\displaystyle O-CH_2}{\diagup}}{\underset{\underset{\displaystyle O-CH_2}{\diagdown}}{CH}}$

17.6. (a) chloroethanoic acid, 2-chloroethanoic acid, or chloroacetic acid
(b) 4-phenylbutanoic acid or 4-phenylbutyric acid

17.7. (a) $CH_3CH_2CH_2CH_2\overset{\overset{\displaystyle O}{\|}}{C}-OCH_2CH_2CH_2CH_3$

(b) $CH_3\overset{\overset{\displaystyle O}{\|}}{C}-O\overset{\overset{\displaystyle CH_3}{|}}{CH}CH_3$

17.8. (a) $CH_3CH_2\overset{\overset{\displaystyle O}{\|}}{C}OH + HOCH_2CH_2CH_3$

(b) cyclohexyl$-\overset{\overset{\displaystyle O}{\|}}{C}OH + HO-$phenyl

17.9. (a) $CH_3CH_2\overset{\overset{\displaystyle O}{\|}}{C}-O-$cyclohexyl   (b) $CH_3CH_2CH_2\overset{\overset{\displaystyle O}{\|}}{C}-OCH_2\overset{\overset{\displaystyle Cl}{|}}{CH}CH_3$

17.10. (a) 3-methylbutyl (or 3-methyl-*n*-butyl) acetate
(b) methyl *o*-hydroxybenzoate

17.11. (a)
$$\text{CH}_3\text{CH}_2\overset{\displaystyle O}{\overset{\|}{\text{C}}}\!-\!\text{OCH}_2\overset{\displaystyle \text{CH}_3}{\underset{}{\text{CHCH}_3}} + \text{Na}^+\ \text{OH}^- \longrightarrow$$

$$\text{CH}_3\text{CH}_2\overset{\displaystyle O}{\overset{\|}{\text{C}}}\text{O}^-\ \text{Na}^+ + \text{HOCH}_2\overset{\displaystyle \text{CH}_3}{\underset{}{\text{CHCH}_3}}$$

(b) $\text{CH}_3\text{CH}_2\text{OH} + \text{Na}^+\ {}^-\overset{\displaystyle O}{\overset{\|}{\text{O}\text{C}}}\!-\!\overset{\displaystyle O}{\overset{\|}{\text{C}}}\text{O}^-\ \text{Na}^+ + \text{HOCH}_2\text{CH}_3$

17.12. (a) $\text{CH}_3\text{CH}_2\text{CO}_2\text{H}$, propanoic (or propionic) acid
(b) $\text{HO}_2\text{C}\!-\!\text{CO}_2\text{H}$, oxalic acid

17.13. (a) $\text{CH}_3\text{CH}_2\overset{\displaystyle O}{\overset{\|}{\text{C}}}\text{NH}_2$ (b) $\text{CH}_3\text{CH}_2\overset{\displaystyle O}{\overset{\|}{\text{C}}}\!-\!\overset{\displaystyle \text{CH}_3}{\underset{}{\text{NCH}_3}}$

17.14. $\text{H}_3\overset{+}{\text{N}}\text{CH}_2\overset{\displaystyle O}{\overset{\|}{\text{C}}}\text{OH} + \text{H}_3\overset{+}{\text{N}}\underset{\text{CH}_3}{\text{CHC}}\overset{\displaystyle O}{\overset{\|}{}}\text{OH} + \text{H}_3\overset{+}{\text{N}}\underset{\text{CH}_2-}{\text{CHC}}\overset{\displaystyle O}{\overset{\|}{}}\text{OH}$

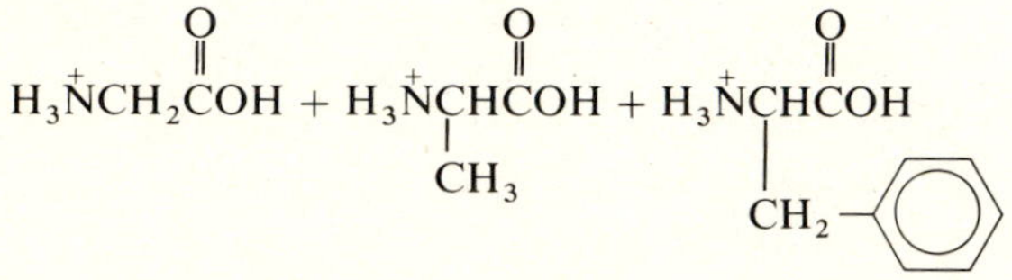

Chapter 18

18.1. (a) and (b) aldose and pentose (or aldopentose)
(c) ketose and triose (or ketotriose or triulose)
(d) aldose and triose (or aldotriose)

18.2. (a) D (b) L (c) L

18.3. (a) contains an aldehyde group, so would give a positive Tollens test.
(b) contains no group that can be oxidized by a very mild oxidizing agent, so would give a negative Tollens test.

18.4.

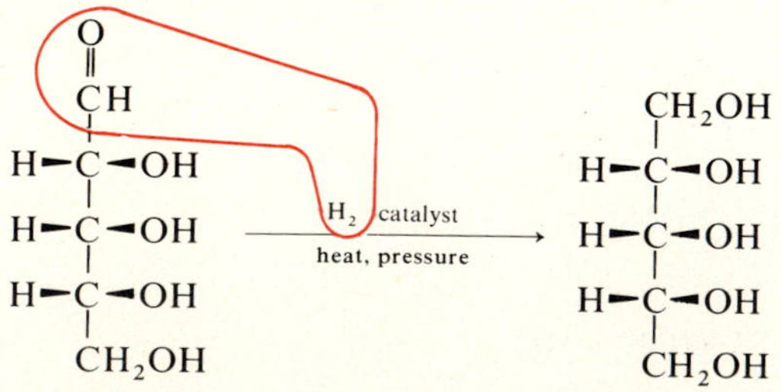

18.5.

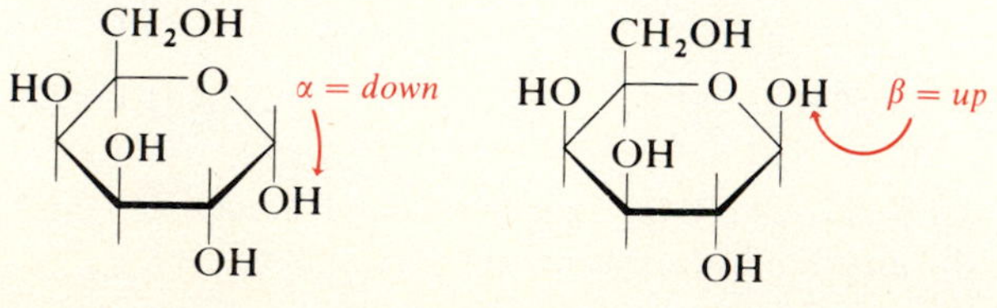

18.6.

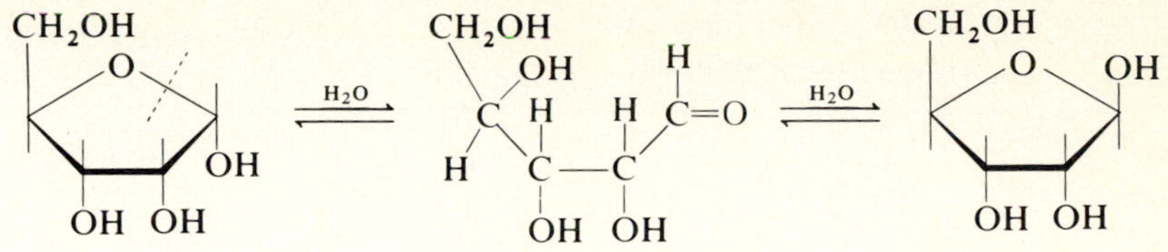

Start with either the α or β form. This is α-D-ribose.

open-chain aldehyde form

β-D-ribose

18.7.

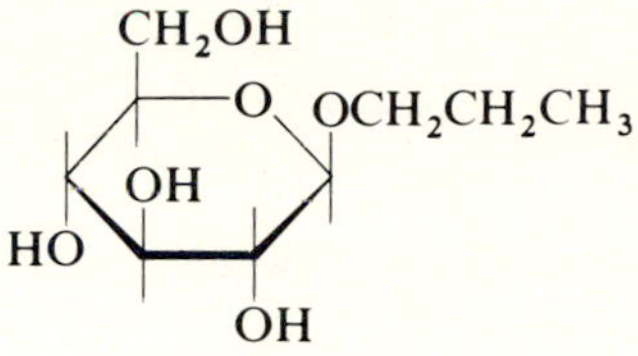

$+ H_2O \xrightarrow{H^+}$ D-glucose $+ CH_3CH_2CH_2OH$

18.8. (a)

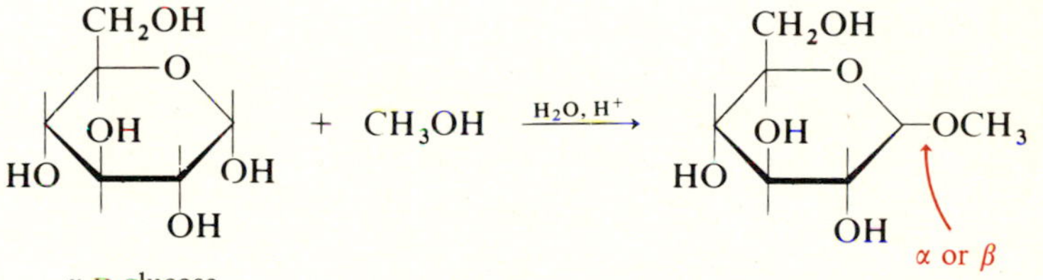

$+ CH_3OH \xrightarrow{H_2O, H^+}$

α-D-glucose

α or β

(b) In aqueous acid, α-D-glucose and β-D-glucose are in equilibrium. Therefore, *both* the α and β glycosides would be formed.

18.9. (a) maltose $\xrightleftharpoons{H_2O}$

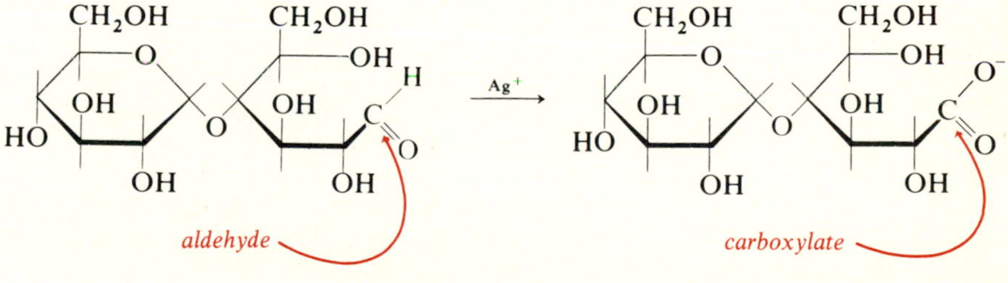

aldehyde $\xrightarrow{Ag^+}$ *carboxylate*

(b) Sucrose is a nonreducing sugar; therefore, no reaction occurs.

(c) lactose $\xrightleftharpoons{H_2O}$

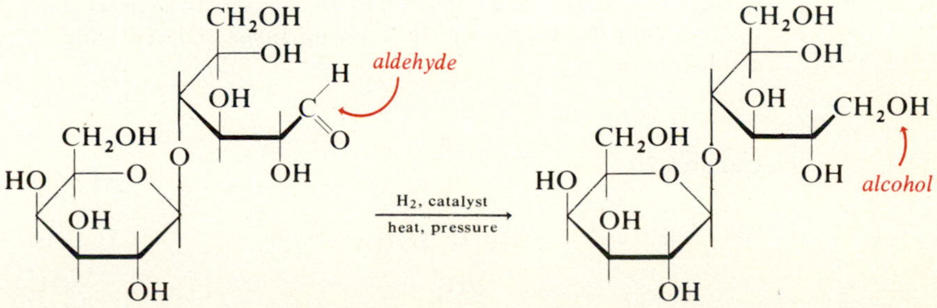

aldehyde $\xrightarrow[\text{heat, pressure}]{H_2, \text{catalyst}}$ *alcohol*

18.10. From Figure 18.7, we can see that the following compound (isomaltose) could result from hydrolysis:

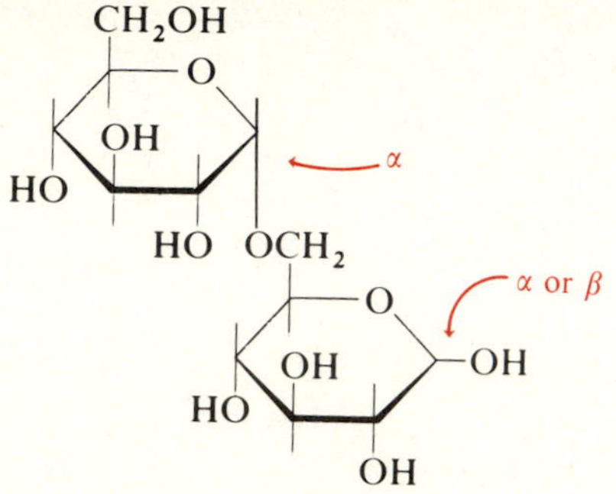
CH₂OH
O
OH
HO
α
HO OCH₂
α or β
O
OH
OH
HO
OH

18.11. glycogen + H_2O $\xrightarrow{H^+}$ maltose + isomaltose

Chapter 19

19.1.

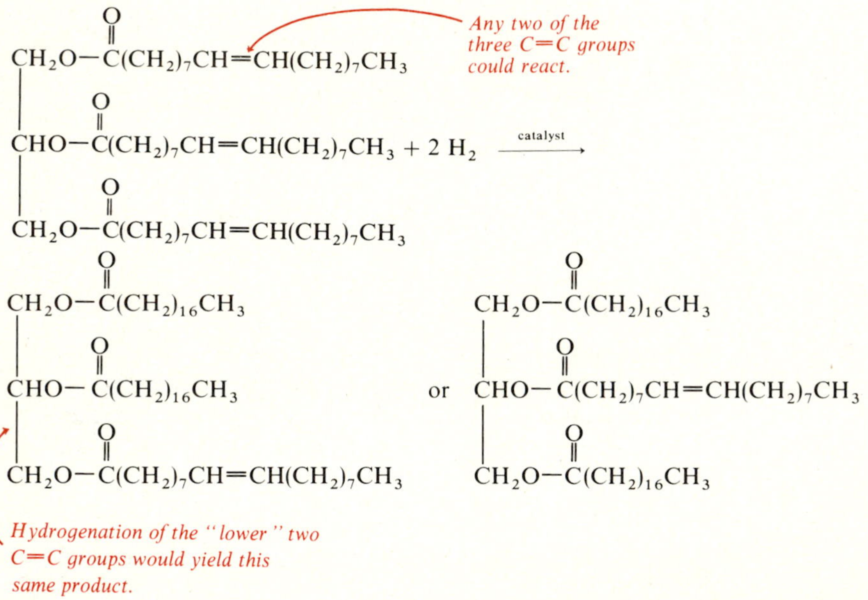
Any two of the three C=C groups could react.
catalyst
or
Hydrogenation of the "lower" two C=C groups would yield this same product.

***19.2.** Yes, a glycolipid could act as an emulsifying agent because it contains two long hydrocarbon chains plus a water-loving *sugar* end replacing the ionic end of other emulsifying agents.

Chapter 20

20.1. (a)

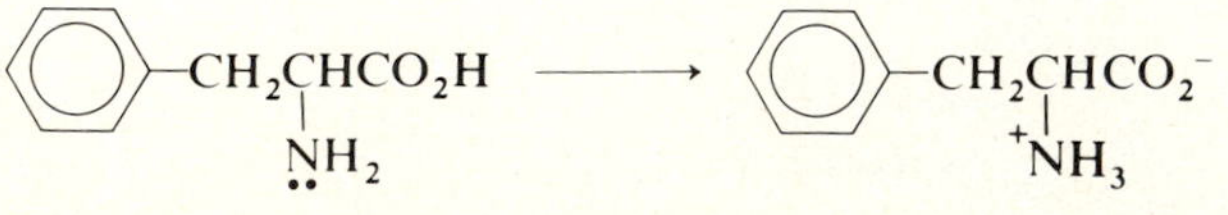

(b) 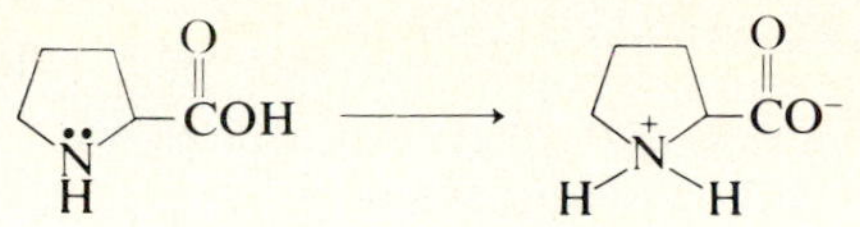 [structure: pyrrolidine ring with N-H and -COH, C=O] $\longrightarrow$ [structure: pyrrolidine ring with N^+ H H and $-CO^-$, C=O]

20.2. (a) CH_3CHCO_2H (or $CH_3CHCO_2^-$) $+ H^+ + Cl^- \longrightarrow CH_3CHCO_2H$

with NH_2 and $^+NH_3$ and $^+NH_3 Cl^-$ respectively below.

(b) $CH_3CHCO_2H + Na^+ + OH^- \longrightarrow CH_3CHCO_2^- \ Na^+ + H_2O$

with NH_2 below each.

(c) $H_2NCH_2\overset{O}{\underset{\|}{C}}-NHCHCOH + Na^+ + OH^- \longrightarrow$

with CH_3 below.

$H_2NCH_2\overset{O}{\underset{\|}{C}}-NHCHCO^- \ Na^+ + H_2O$

with CH_3 below.

20.3. $CH_3CH\overset{O}{\underset{\|}{C}}-NHCH_2\overset{O}{\underset{\|}{C}}OH \longrightarrow CH_3CH\overset{O}{\underset{\|}{C}}-NHCH_2\overset{O}{\underset{\|}{C}}O^-$

with NH_2 below left, $^+NH_3$ below right.

20.4. gly-ala-phe   ala-phe-gly
gly-phe-ala   phe-ala-gly
ala-gly-phe   phe-gly-ala

Chapter 21

21.1. (a) an oxidoreductase; more specifically, a dehydrogenase
(b) a transferase; more specifically, a transaminase

*21.2. (a) and (b) are both dicarboxylic acids and might inhibit the enzyme. Of the two, (b) is closer in structure to succinic acid, so could probably better block the active site.

Chapter 22

22.1. Because the reaction is the hydrolysis of the high-energy ATP, it is exothermic, just as is the hydrolysis of ATP to ADP.

22.2. $2 \ R\overset{OH}{\underset{|}{C}}HR + O_2 \longrightarrow 2 \ R\overset{O}{\underset{\|}{C}}R + 2 \ H_2O$

two

22.3. (a) gain of one electron: reduction
(b) neither; this reaction is a hydrolysis
(c) loss of two hydrogen atoms and two electrons: oxidation
(d) gain of one hydrogen atom and two electrons: reduction

Chapter 23

23.1. (a) *dehydration* (loss of H_2O) of an alcohol: the first reaction in step 7, in which glyceric acid 2-phosphate loses water

(b) *oxidation* of an aldehyde to a carboxylate: step 5, in which a phosphate of glyceraldehyde is converted to a diphosphate of glyceric acid

(c) This is simply an *isomerization*, in which a carbonyl group and a hydroxyl group are exchanged for each other. Step 2 (glucose 6-phosphate → fructose 6-phosphate) and the reaction prior to step 5 (dihydroxyacetone phosphate → glyceraldehyde 3-phosphate) are both this type of reaction.

23.2. (a) Steps 3 and 9 are the oxidation of hydroxyl groups to ketone groups. Step 5, in which CO_2 and an H atom are lost, is an oxidation of α-ketoglutaric acid. Step 7, in which two H atoms are lost from succinic acid, is the oxidation of that compound. (One easy way to determine oxidation reactions is to determine which reactions involve the corresponding reduction of NAD^+ or FAD.)

Step 4 is a decarboxylation (loss of CO_2), not usually considered an oxidation; however, this reaction does involve an internal oxidation-reduction. (The C in CO_2 has been oxidized; the C remaining behind gains H so is reduced.)

(b) Steps 3, 5, 7, and 9 are reductions of NAD^+ or FAD.

(c) Step 8 is a hydration (addition of water) of an alkene.

Chapter 24

24.1.
$$\begin{array}{c} CH_2O_2CR \\ | \\ CHO_2CR \\ | \\ CH_2O_2CR \end{array} + \ 3\ H_2O \xrightarrow{\text{lipases}} \begin{array}{c} CH_2OH \\ | \\ CHOH \\ | \\ CH_2OH \end{array} + \ 3\ RCO_2H$$

24.2. Step 1 is the loss of one ATP unit. In step 2, NADH enters the respiratory chain and produces three ATP units. Therefore, a net gain of two ATP units is realized.

*24.3.
$$CH_3\overset{O}{\overset{\|}{C}}CH_2\overset{O}{\overset{\|}{C}}-SCoA + H_2O \longrightarrow CH_3\overset{O}{\overset{\|}{C}}CH_2\overset{O}{\overset{\|}{C}}-OH + H-SCoA$$

Note that the thioester group is hydrolyzed; ketones do not undergo hydrolysis.

*24.4.
$$CH_3CH_2CH_2\overset{O}{\overset{\|}{C}}SCoA + \overset{CO_2H}{\underset{}{CH_2}}-\overset{O}{\overset{\|}{C}}SCoA \longrightarrow$$

$$CH_3CH_2CH_2\overset{O}{\overset{\|}{C}}CH_2\overset{O}{\overset{\|}{C}}SCoA + HSCoA + CO_2$$

NADPH

NADP + H^+

$$CH_3CH_2CH_2CH_2CH_2\overset{O}{\overset{\|}{C}}SCoA$$

Hydrolysis of this thioester yields $CH_3CH_2CH_2CH_2CH_2CO_2H$.

Chapter 25

25.1. The person will have a positive nitrogen balance until the metabolism adapts to ridding the body of excess nitrogen compounds.

25.2.

$$HO_2CCH_2CH_2\overset{\overset{\displaystyle NH_2}{|}}{C}HCO_2H + R\overset{\overset{\displaystyle O}{\|}}{C}CO_2H \xrightarrow{\text{transaminase}}$$

$$HO_2CCH_2CH_2\overset{\overset{\displaystyle O}{\|}}{C}CO_2H + R\overset{\overset{\displaystyle NH_2}{|}}{C}HCO_2H$$

25.3.

$$HO_2CCH_2CH_2\overset{\overset{\displaystyle NH_2}{|}}{C}HCO_2H + NAD^+ + H_2O \xrightarrow{\text{amino acid oxidase}}$$

$$HO_2CCH_2CH_2\overset{\overset{\displaystyle O}{\|}}{C}CO_2H + NH_3 + NADH + H^+$$

25.4.

$$\underset{\displaystyle CH_2CH_2CH_2NH_2}{\overset{\overset{\displaystyle NH_2}{|}}{C}HCO_2H} + H_2N\overset{\overset{\displaystyle O}{\|}}{C}OPO_3{}^{2-} \longrightarrow \underset{\displaystyle CH_2CH_2CH_2NH\overset{\overset{\displaystyle O}{\|}}{C}NH_2}{\overset{\overset{\displaystyle NH_2}{|}}{C}HCO_2H}$$

The other $-NH_2$ group in urea ($H_2N\overset{\overset{\displaystyle O}{\|}}{C}NH_2$) comes from aspartic acid; the carbonyl oxygen in urea comes from water in the final hydrolysis.

*25.5. Feces containing less bile pigments than normal are lighter-colored than normal.

Chapter 26

26.1. The structure is shown in Figure 22.4, Section 22.3. Note the similarity of this structure to deoxyadenosine phosphate, shown just before Problem 26.1.

26.2. A-C-T-G

Chapter 27

27.1. A-C-U (Note that uracil **U** in RNA pairs with adenine **A**.)

27.2. in DNA: **TAC-CAT-GGG-CAT**
in mRNA: **AUG-GUA-CCC-GUA**
peptide: (fmet, met, or " start ")-val-pro-val-

Chapter 28

*28.1. Only Types B and AB form a precipitate in both A and O serums; however, Type AB would also show a precipitate with B serum. Therefore, the sample is Type B blood.

28.2.

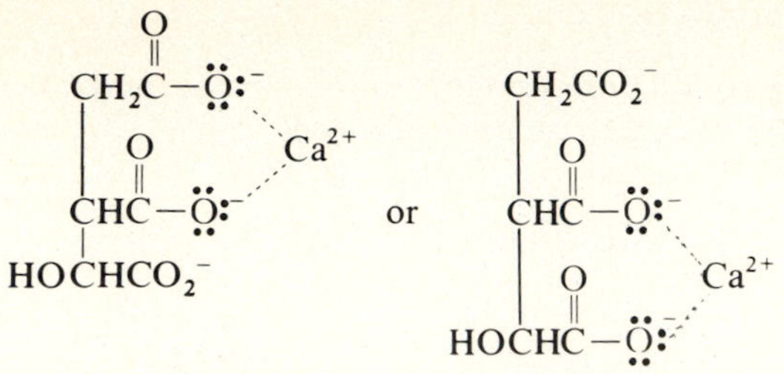

$$
\begin{array}{l}
\underset{\text{CH}_2\text{C}-\ddot{\text{O}}:^-}{\overset{\overset{\text{O}}{\|}}{}} \\
\quad\overset{\text{O}}{\underset{\|}{}} \qquad \text{Ca}^{2+} \\
\underset{\text{HOCHCO}_2^-}{\overset{}{\text{CHC}-\ddot{\text{O}}:}} \qquad \text{or}
\qquad
\begin{array}{l}
\text{CH}_2\text{CO}_2^- \\
\quad\overset{\text{O}}{\underset{\|}{}} \\
\text{CHC}-\ddot{\text{O}}: \\
\qquad\overset{\text{O}}{\underset{\|}{}} \qquad \text{Ca}^{2+} \\
\text{HOCHC}-\ddot{\text{O}}:^-
\end{array}
\end{array}
$$

28.3. (a) The greater partial pressure of CO_2 in the lungs means that more CO_2 would dissolve in the blood, shifting the equilibrium to the right.

$$\text{HHb} + CO_2 \;\rightleftharpoons\; \text{HHbCO}_2$$

In the lungs, CO_2 would not be as readily released. In the capillaries, less CO_2 could be taken from the cells.

(b) The cells give off less CO_2; therefore, the equilibrium in the capillaries is shifted to the left.

$$\text{HHb} + CO_2 \;\rightleftharpoons\; \text{HHbCO}_2$$

Because of the lower concentration of both HHbCO_2 and CO_2, the person would exhale less CO_2.

(c) An increase in acidity would not directly change this equilibrium, which does not contain H^+ (or H_3O^+) in its equation. However, the following equilibrium would be shifted to the left, increasing the amount of CO_2 in the blood as in part (a).

$$H_2O + CO_2 \;\rightleftharpoons\; H_2CO_3 \;\rightleftharpoons\; H^+ + HCO_3^-$$

28.4. (a) Salts become more soluble when the urine is warmed; the cloudiness decreases.

(b) Albumins undergo denaturation and precipitation when heated; the cloudiness increases.

Glossary

The number of the section where a word or term is discussed in detail is in parentheses. (Check the index for additional references.)

acetal: organic compound having the general structure $R_2C(OR)_2$ (17.3)

acetylcoenzyme A: an important biological compound that is involved in many anabolism and catabolism sequences (17.8)

acidosis: the condition in which the blood pH is lower than 7.35 (10.9)

active transport: movement of ions or molecules against a concentration gradient (22.2)

adipose tissue: an animal's fat deposits

aerobic: requiring oxygen (23.2)

albumins: globular proteins such as are found in blood and egg whites (20.4, 28.5)

alcohol: organic compound containing the —OH functional group (16.1)

aldehyde: organic compound containing the $-\overset{\text{O}}{\underset{\|}{\text{C}}}\text{H}$ functional group (17.2)

alkali: base; substance that can accept protons (Chapter 10)

alkalosis: the condition in which the blood pH is greater than 7.45 (10.9)

alkane: a hydrocarbon containing only single bonds (Chapter 14)

alkene: a hydrocarbon containing one carbon-carbon double bond and no other functional group (Chapter 15)

alveoli: air sacs in the lungs where O_2 enters the capillaries and CO_2 leaves

amide: organic compound with the general formula $R\overset{\text{O}}{\underset{\|}{\text{C}}}NR'_2$ (17.9)

amine: organic compound with the general formula R_3N, R_2NH, or RNH_2 (16.4)

amino acid: product of protein hydrolysis (20.1)

amphoterism: the ability of a molecule or ion to react with either acid or base (10.8)

anabolism: metabolism reactions in which small molecules are combined to yield larger ones (22.1)

anaerobic: not requiring oxygen (23.2)

analgesic: painkiller

anemia: deficiency of red blood cells (28.2)

anesthetic: an agent that numbs a part of the body (*local*) or that causes unconsciousness (*general*) (6.2)

anion: negatively charged ion (4.4)

antibiotic: a compound produced by one organism that is toxic to another organism (21.7)

antibody: a protein synthesized by certain cells that deactivates specific large foreign molecules (27.5)

anticoagulant: an agent that prevents blood clots (28.3)

antigen: a large foreign molecule that stimulates antibody production in an organism (27.5)

antimetabolite: a substance that inhibits a normal biochemical reaction; a specific enzyme inhibitor (21.7)

antiseptic: inhibiting the action of microorganisms; an agent that acts in this way

aqueous: referring to water

atomic mass: weighted average of the mass numbers of naturally occurring isotopes of an element (3.3)

atomic number: number of protons in the nucleus of an atom (3.3)

bile: a mixture of bile salts, bile pigments, and cholesterol secreted by the liver (19.5, 24.1, 25.5)

blood plasma: whole blood with the blood cells removed (28.1)

blood serum: whole blood with the blood cells and clotting factors removed (28.1)

buffer: a mixture that resists changes in pH when acid or base is added (10.8)

carbonyl group: the $C=O$ group (17.1)

carboxylate: salt of a carboxylic acid (17.5)

carboxyl group: the $-CO_2H$ functional group (17.4)

carboxylic acid: organic compound containing the $-CO_2H$ group (17.4)

carcinogen: a cancer-causing substance

cardiovascular system: system composed of the heart and blood vessels

catabolism: metabolism reactions in which larger molecules are degraded to smaller ones (22.1)

catalyst: a substance that speeds the reaction of other substances without itself being consumed (5.3)

cation: positively charged ion (4.4)

cellular respiration: oxidation of nutrients in cells (22.3)

cellular respiratory chain: the sequence of biochemical reactions in which oxygen is reduced to water (22.4)

chemotherapy: treatment of disease by the use of drugs or other chemicals

chromosome: DNA-protein complex in the nucleus of a dividing cell; contains the genes

citric acid cycle: biochemical reaction sequence that converts acetylcoenzyme A to coenzyme A, CO_2, and H_2O (23.4)

coenzyme: a nonprotein organic molecule that must be associated with an enzyme for the enzyme to be functional (21.3)

coenzyme A: an important coenzyme in metabolic sequences (17.8)

cofactor: a nonprotein part of an enzyme (21.3)

colloid: a homogeneous dispersion of one substance in another (Chapter 8)

covalent compound: a compound in which atoms are held together as molecules by covalent bonds (4.7)

decarboxylation: loss of CO_2 (17.5)

decomposition: breaking apart of molecules or nuclei

dehydration: loss of water (16.2)

denaturation: loss of higher protein structure because of heat, change in pH, etc. (20.5)

diabetes mellitus: a disorder of glucose metabolism, in which insulin is either deficient or not well utilized (23.6)

dialysis: the selective flow of small ions, small molecules, and water through a semipermeable dialyzing membrane (8.4)

diffusion: the movement of ions or molecules, especially in solution or in the gaseous state, so as to equalize concentrations (7.6)

digestion: hydrolysis of food molecules in the gastrointestinal tract (22.1)

distillation: a process in which a liquid is boiled, then its vapors are condensed and collected; used for purification or separation

electrolyte: ions or ionic compound (a partially ionized substance is a "weak" electrolyte) (9.1)

electronegativity: the magnitude of the attraction of an atom for its outer, valence electrons (4.10)

electron-transport chain: *See* cellular respiratory chain.

electrostatic charge: an electrical charge, such as an ionic charge, that remains at rest and does not flow as an electric current does

emulsifying agent: a substance that helps keep a mixture emulsified, or in a colloidal state (8.3)

emulsion: a colloidal dispersion of one liquid in another (8.1)

endoplasmic reticulum: cellular site of protein biosynthesis (22.2)

endothermic: refers to absorption of heat (5.2)

energy of activation: energy required to initiate a chemical reaction (5.2)

enzyme: a globular protein that catalyzes a biochemical reaction (Chapter 21)

equilibrium: a state of balance; a reversible reaction in which the rates of forward and reverse reaction are equal (5.6)

equivalent: the number of grams of a substance that will yield one mole of H^+ or OH^-, or one mole of ionic charge (10.6)

erythrocyte: red blood cell, which contains hemoglobin, the transporter of O_2 (28.2)

ester: organic compound with the general formula RCO_2R' (17.6)

ether: organic compound with the general formula ROR' (16.1)

exothermic: refers to liberation of heat (5.2)

fatty acid: a long-chain carboxylic acid (19.1, 24.1)

fatty acid cycle: the sequence of biochemical reactions in which fatty acids are converted to acetyl groups in acetylcoenzyme A (24.2)

formula weight: sum of the atomic masses of the atoms in the usual chemical formula for an element or compound (5.8)

functional group: a site of reactivity in an organic molecule—for example, $C=C$, $C\equiv C$, or an electronegative atom (13.7)

galactosemia: a disorder of galactose metabolism (23.7)

gastric: referring to the stomach

gene: carrier of hereditary traits; portion of DNA that directs the biosynthesis of a particular protein (Chapters 26 and 27)

gluconeogenesis: biosynthesis of glucose from small noncarbohydrate molecules (23.3)

glycolysis: conversion of glucose to pyruvic or lactic acid (23.2)

glycoside: a carbohydrate in an acetal form (18.6)

halogen: one of the elements of Group VII of the periodic table: F, Cl, Br, or I (3.6)

helix: a spiral, such as is formed by protein or nucleic acid molecules

hemiacetal: organic compound having the general structure R_2C-OR with OH on the central carbon (17.3)

hemoglobin: the conjugated globular protein in red blood cells that carries oxygen from the lungs to the tissues (20.4)

hemolysis: swelling and rupturing of erythrocytes by osmosis (7.6)

heterocycle: cyclic compound containing carbon and at least one other element in the ring (15.6)

hormone: chemical messenger secreted by endocrine glands and carried by the body fluids

hydration: addition of water (15.4)

hydrocarbon: a compound containing only carbon and hydrogen (Chapters 14 and 15)

hydrolysis: cleavage with water (17.7)

hydrophilic: attracted to and by water

hydrophobic: not attracted to or by water

hyperglycemia: high blood glucose level (23.6)

hypertonic: referring to a solution that has a greater osmolarity than does another solution (7.6)

hypoglycemia: low blood glucose level (23.6)

hypotonic: referring to a solution that has a lower osmolarity than does another solution (7.6)

intermolecular: refers to interactions between different molecules

interstitial fluid: fluid between the cells of an organism

intramolecular: refers to interactions within a molecule

in vitro: outside of an organism, such as in a test tube

in vivo: in a living organism

ion: electrically charged particle formed by an atom or by a group of covalently bonded atoms (4.1)

isomers: different compounds with the same molecular formula, but with different structures or spatial arrangements of atoms (13.3, 15.3, 18.2, 18.3)

isotonic: refers to a solution with an osmolarity equal to that of a second solution (7.6)

isotopes: forms of an element that have different mass numbers (3.3)

IUPAC: International Union of Pure and Applied Chemistry; often used to refer to the chemical nomenclature system developed by this group (14.1)

ketone: organic compound having the general structure $R_2C=O$ (17.2)

ketone body: compound produced from excess acetylcoenzyme A in the liver; often associated with diabetes (24.3)

kinetic energy: energy of motion (5.2)

kwashiorkor: a protein-deficiency disease of children (25.1)

leukocyte: white blood cell (28.2)

lipid: a water-insoluble biological compound that is soluble in nonpolar organic solvents (Chapter 19)

lipid bilayer: the double layer of phospholipid molecules that makes up the cell membrane (22.2)

lipogenesis: biosynthesis of lipids (24.4)

lipoprotein: lipid-protein complex (24.1)

lymph: fluid in the lymphatic system

lymphocyte: antibody-producing cell in the lymph

mass number: number of protons plus number of neutrons in nucleus (3.3)

metabolism: enzyme-catalyzed cellular reactions (22.1)

mitochondrion: cellular site of energy-producing oxidation-reduction reactions (22.2)

molarity: moles of solute per liter of solution (7.3)

mole: formula weight expressed in grams; 6×10^{23} formula units (5.8)

molecular compound: covalently bonded compound (4.7)

molecular formula: a formula that shows only the number and types of atoms in a molecule (4.11)

molecular weight: sum of the atomic masses of the atoms in a molecule (5.8)

mutagen: a substance capable of causing changes in the DNA of cells (mutations) (27.3)

neutralization: reaction of an acid and a base to yield a neutral solution or mixture (10.6)

nucleic acid: polymer that carries the genetic code and directs the biosynthesis of proteins (Chapters 26 and 27)

octet rule: the "rule of thumb" stating that atoms gain, lose, or share electrons so as to gain an outer shell of eight electrons (4.2)

orbital: region in space around a nucleus in which a particular electron or electron pair is likely to be found (3.5, 4.9)

ornithine cycle: *See* urea cycle.

osmosis: the flow of solvent through a semipermeable osmotic membrane from a dilute solution to a more concentrated solution (7.6)

oxidation: increase in oxidation number of an atom or loss of H (or gain of O) by an atom or molecule (4.3, Chapter 11)

oxidation number: the amount of positive or negative charge on a simple ion or a covalently bonded atom (4.3, 4.13)

oxidizing agent: a substance that causes an oxidation (4.3, 11.2)

partial pressure: the pressure of one gas in a mixture (6.3)

pathological: involving disease

peptide: "small protein" molecule (20.2)

phenol: organic compound containing —OH bonded to an aromatic ring (16.1)

plasma cell: blood cell (28.2)

polar: having a separation of + and − charge (4.10)

polymer: long-chain molecule containing repeating units (15.4)

porphyrin: a heterocyclic ring system found in heme, the cytochromes, and chlorophyll (16.7, 25.5)

precipitation: a solid material coming out of solution

prosthetic group: a nonprotein ion or molecule that is firmly held by a protein molecule and contributes to the protein's function

protein: types of polyamides that form muscle, skin, hair, enzymes, and so forth (Chapter 20)

purine: a heterocyclic ring found in nucleic acids (25.5, 26.2)

radiation: emission of energy or nuclear particles (Chapter 12)

reactant: a substance that undergoes a chemical reaction (Chapter 5)

reducing agent: a substance that causes a reduction (4.3, 11.2)

reduction: decrease in the oxidation number of an atom or gain of H (or loss of O) by an atom or molecule (4.3, Chapter 11)

renal: referring to the kidneys and urinary tract

respiration: the process of breathing; also, the oxidation of nutrient molecules within a cell

saponification: alkaline hydrolysis, generally referring to an ester (17.7)

saturated compound: a compound containing no double or triple bonds, or a fat containing no carbon-carbon double bonds (14.1, 19.1)

saturated solution: a solution in equilibrium with pure solute and containing as much solute as it can dissolve (7.1)

structural formula: a formula from which the order of the joining of atoms in a molecule can be deduced (4.11)

substrate: a substance that is acted upon

surfactant: wetting agent (7.2)

therapeutic: having healing power

thioester: organic compound with the general formula $R\overset{\displaystyle O}{\overset{\displaystyle \|}{C}}SR'$ (17.8)

thiol: organic compound containing the —SH functional group (16.3)

transamination: a reaction in which an amino group is transferred from one molecule to another (25.3)

triacylglycerol (triglyceride): an animal fat or vegetable oil (19.1, 24.1)

urea cycle: the sequence of biochemical reactions in which ammonia is converted to urea (25.4)

valence: the usual oxidation number of an element when it forms ions (4.3)

valence electron: electron in the outermost, unfilled shell of an atom (3.6)

vitamin: an organic compound that is essential in the diet, but only in very small amounts (21.3)

Index